Forestry Sciences

Volume 83

More information about this series at http://www.springer.com/series/5991

Rowland D. Burdon · William J. Libby
Alan G. Brown

Domestication of Radiata Pine

Springer

Rowland D. Burdon
Scion – New Zealand Forest
Research Institute Ltd
Rotorua

Alan G. Brown
Deakin, ACT
Australia

William J. Libby
Orinda, California
USA

ISSN 0924-5480 ISSN 1875-1334 (electronic)
Forestry Sciences
ISBN 978-3-319-65017-3 ISBN 978-3-319-65018-0 (eBook)
DOI 10.1007/978-3-319-65018-0

Library of Congress Control Number: 2017955956

Printed on acid-free paper

This Springer imprint is published by Springer Nature
The registered company is Springer International Publishing AG
The registered company address is: Gewerbestrasse 11, 6330 Cham, Switzerland

To the late Dr Max Jacobs ISO whose own pioneering work and leadership were so influential in the genesis of genetic improvement of radiata pine.

Preface

This is not just a story about radiata pine—it is also about people and institutions, aspects that have their own interest for many. More importantly, however, we consider the interactions between these aspects and the science and technology of domestication to be highly instructive. In the background, as a significant influence, have been shifts in prevailing economic ideology within the main countries involved.

In 1968, Professor Daniel Zohary of the Hebrew University, an eminent scholar of the domestication of wheat and of the role of plant domestication in the development of human societies, was visiting the Genetics Department in Berkeley. There, in working with one of us (WJL), he expressed envy of how those who studied forest trees in western North America were in a position to witness and track the process of domestication from its very beginning, instead of having to rely on fragmentary archaeological records.

In 1971, WJL spent a sabbatical year at the New Zealand Forest Research Institute (FRI) with Ib Thulin, RDB and others in FRI's Genetics and Tree Improvement (GTI) section. This greatly intensified WJL's interest in radiata pine, and he began to think about the radiata pine breeding being done in Australia and New Zealand in relation to its early domestication. During a mid-1971 tour of such work in Australia, guided by Ken Eldridge, he drafted his paper *Domestication Strategies for Forest Trees* (Libby 1973). Later that year, Thulin and WJL began to gather data for a paper on radiata pine, for the *Genetics of …* series prompted by the International Union of Forestry Research Organisations for many important forest tree species.

WJL returned to Berkeley but was focused on a revision of I.M. Lerner's book, *Heredity Evolution and Society*, based in part on a course that he was by then teaching. The work on the genetics of radiata pine progressed slowly, and stopped when Thulin died. But WJL continued to work on radiata pine as a research species, and this led to periodic returns to New Zealand.

During a 1991 visit to Rotorua and the FRI, the question of finishing *Genetics of Radiata Pine* was raised, and soon led to a year's return to Rotorua with the GTI section of FRI.

In the mid-1990s, several long careers (including our own) of geneticists, silviculturists, breeders, plant propagators and others who engaged in the active domestication of radiata pine were drawing to a close. So, while we all still could, this seemed a good time to jointly reflect on what we had done, and why.

This also seemed a time when the process of radiata pine domestication was passing through an important transition. The period from the mid-1950s to the mid-1980s was focused mostly on classical tree improvement as founded by the Scandinavian school, namely seed production by progeny-tested parents or families in open-pollinated seed orchards (Zobel and Talbert 1984; Libby 1973). As forest management increasingly responded to these better seedling trees, many programmes demanded even greater genetic control, and began substituting control-pollinated families or tested clones for orchard-run open-pollinated seeds. These in turn allowed revolutionary new ways of doing plantation forestry, including family forestry (Carson 1986) and full clonal forestry (Miller 1991; Ahuja and Libby 1993). These also required better and different propagation systems and, as greater genetic control of the propagules became available, these provide opportunities for different deployment options. Finally, such changes led to further modifications in breeding strategies, in breeding plans, and in management of the radiata pine resource.

The amount of information available on radiata pine's genetics had so increased between 1971 and 1992 that the "paper" had grown to be a book, and more. Among other things during 1992–1993, WJL was much influenced by the work of RDB on the performance of samples of radiata pine's native populations in New Zealand, and on the development of land races there from early introductions. Ken Eldridge had been doing similar work with the native populations in Australia, and had just completed a book on the domestication of eucalypts.

During 1992–1993, the demand for radiata pine wood on the Pacific Rim grew spectacularly, largely in response to the reduction in wood availability from western North America. This generated broad interest in the properties and husbandry of this promising new crop, and WJL attempted to develop the book to serve that broad interest group, as well as to serve a technical readership interested in the genetics and breeding of radiata pine. By 1994, when he again returned to New Zealand, it was clear that writing for this dual readership was difficult.

In a series of meetings at the New Zealand FRI in 1995, the idea of writing two books was proposed and then developed. This refocusing was greatly helped by the insight that not only has FRI greatly contributed to the early domestication of radiata pine through its work in breeding, silviculture and management of this species, but also that radiata pine has played a large, even dominant, role in the creation and development of FRI. With the 50th anniversary of FRI approaching in 1997, it seemed appropriate to focus a general-audience book on this important species. With the refocus in direction, it also seemed appropriate to include as authors two of the major participants in both understanding and creating radiata pine as a domesticated species, RDB from New Zealand and Ken Eldridge from Australia. Ken, however, had to withdraw from that role, and this led to the participation of AGB in the project.

The original intent, to produce a radiata pine addition to the *Genetics of* series, was effectively satisfied by the proceedings of a workshop held in New Zealand in late 1997 (Burdon and Moore 1997). This *Genetics of* series has included treatments of many important and interesting forest trees, each summarizing genetic information available for its species. While these and many other forest tree species were being passively modified in their native ranges by such human activities as selective harvesting (Burdon and Libby 2006), by 1997 it was clear that a much shorter list was being purposefully domesticated and extensively planted to serve human purposes and needs. Radiata pine is high on this short list. Thus, a different book focusing on the development, progress and extent of such purposeful domestication of radiata pine seemed appropriate.

After 1991, WJL's other commitments consumed almost all his time, so the completion of the remainder of the project—this book—has taken far longer than any envisaged, and fell to RDB. The delay in completion, however, has allowed us to cover further significant progress in domestication, to complement the earlier narrative.

The countries that are most prominent in our story are New Zealand, Australia and Chile.

We do not claim to have produced a complete or even a balanced account of the story. Two of us have been heavily involved in the New Zealand part of it, which has doubtless created a bias. In defence, we note that, except during the earlier years, New Zealand has been involved in more than its share of innovation. Reasons for that are various, and include the greater national focus on a single tree species, the pre-eminent role of a single, national forest service, and several biological factors.

In Australia, the plantations and associated research and development were widely dispersed institutionally, geographically and ecologically. Plantation development was of particular interest in South Australia from the latter part of the 1800s and milling of the logs began in 1902; problems of nutrition and site management arose early and have been addressed with significant success. An early start made on breeding has culminated in successful cooperative activity and in internationally adopted instruments (notably "SilviScan" for evaluating wood properties) and software.

In Chile, a great deal has happened, but relatively little is covered in formal publications, and still less in English. Moreover, Chile has tended to follow leads from New Zealand and Australia—often to great effect—although it is now involved in "cutting-edge" innovations.

Other countries, while they include significant growers, have not had the massive commitment to radiata pine that is conducive to sustained, ground-breaking research and innovation. Accordingly, our treatment of their work with this pine is often somewhat cursory and doubtless uneven. In general, our coverage of work done essentially in the footsteps of New Zealand work is deliberately limited.

A lot of work was done, especially early on, addressing the use and properties of radiata pine for pulp and paper products. Over the years, however, solid-wood uses have become increasingly dominant, especially for the forest grower, so our treatment is perhaps biased against pulping.

The central role of radiata pine in the development of industrial plantation forestry has provided the context for much research on forest operations, from land clearing, site preparation, through to wood harvesting. The same is true of huge advances in modeling and prediction of growth, wood outturns and product characteristics, which have been made available in decision-aid and planning software. Such software, which has become increasingly proprietary, is also becoming progressively more readily adapted for use with other plantation forest species and therefore less specific to radiata pine. Accordingly, we are not covering such developments in any detail.

With literature references we have, as far as we reasonably could, tried to use ones that are widely available, mainly journal articles or books or chapters therein. Also, we have often cited works that serve either as examples or as syntheses of earlier work, rather than attempting exhaustive literature reviews ourselves. We have also stuck, as far as was practicable, with references that are in English. There are, however, exceptions. In particular, we have relied significantly on bulletins published by the New Zealand FRI (now Scion), and leaflets and bulletins published by the Australian counterpart. The bulletins include formal conference proceedings. A preference for citing FRI/Scion publications fits with the general online, open-access availability, in contrast to the situation with publications from organisations that have suffered greater institutional changes. We have made minimal recourse to the "grey literature" comprising less formal conference or workshop proceedings. Moreover, some important material represents otherwise undocumented personal recollections.

The structuring into chapters on the basis of historical periods became less straightforward for the later periods. In defining the periods, we have given much weight to developments in genetic improvement. Even so, there is the difficulty that certain developments have occurred at different times among the main grower countries. Nevertheless, the defined periods are used as a basis for stocktaking of the progress to date of domestication, which is done at the end of each chapter from 2 to 7. Despite the importance of individuals and institutions in a history, our coverage of recent developments is more impersonal, in recognition of personal and commercial sensitivities.

This book was conceived for a varied readership. This included foresters in various countries, plant breeders within and beyond the area of tree breeding, members of the wider public with an interest in issues of sustainability, teachers in high schools and universities, and students in a range of disciplines. In countries extensively growing plantations or amenity plantings of radiata pine, it seems possible that curious citizens as well as visitors may like to learn why and how it came to be so frequently planted. To help cater for a broad mix of readers, we have signposted the book with a detailed breakdown into headings, supplemented by the use of boxes in order to outline concepts underlying more specialized technical topics. Nevertheless, it is planned to supplement this book with a shorter one, aimed more at a lay readership.

Readers may wish to dip into specific chapters rather than reading this book in its entirety. With this in mind, we have retained some overlaps between chapters, rather than relying simply on cross-referencing between sections.

In 2012 we learnt of another book that was nearing completion, "*Sustainable Management of* Pinus radiata *Plantations*" by Donald J. Mead (2013). Whereas our book is structured around historical periods, with strong emphasis on genetic improvement, Dr Mead's book is structured around technical themes, with main emphasis on silviculture. Despite overlaps, we see the two books as essentially complementary. We and Dr Mead have cross-checked factual statements, attempting to resolve material discrepancies. However, minor disparities, attributable for example to differences in criteria used or in baseline dates, will remain.

Regarding species names, we have mostly used common names rather than Latin (genus and species) names, but we have appended a list cross-referencing different names for individual species. Such cross-referencing, however, is particularly awkward for fungi, for which names are especially subject to revision and different names may be assigned to different phases in the life cycle of an organism, but the Internet is increasingly a source of updated information. Even with common names, incessant repetition of the full names can be very cumbersome in a work of this scale. Thus we have taken the liberty of generally referring to radiata pine as "radiata", in line with widespread conversational usage. We have done likewise for some other names in passages containing numerous repeated mentions.

We have included a Glossary for technical terms. Each Glossary term is boldfaced at first significant mention, with a superscripted "G".

Prior to publication, draft material was posted on the Internet and publicised widely among organisations and individuals, inviting corrections of factual errors and of significant omissions in respect of the scope declared in the Preface.

Photos, unless specifically acknowledged, are all drawn from the Scion (New Zealand Forest Research Institute Ltd) Image Library, and were largely taken by Harold Hiemke-Hemming, John Barran or David Blake.

References

Ahuja MR, Libby WJ (eds) (1993) Clonal Forestry. Vol. 1. Genetics and biotechnology (15 chapters, ix + 277 p); Vol. 2. Conservation and application (13 chapters, 240 p). Springer-Verlag, Heidelberg

Burdon RD Libby WJ (2006) Genetically modified forests: from Stone Age to modern biotechnology. Forest History Society, Durham, NC, USA, xii + 79 p

Burdon RD, Moore JM (eds) (1997). IUFRO '97 Genetics of radiata pine, Proceedings of NZ FRI-IUFRO conference 1–4 December and workshop 5 December, Rotorua, New Zealand. FRI Bulletin No. 203, 354 p

Carson MJ (1986) Control-pollinated seed orchards of best general combiners. p 144–149 *in* Williams GA, Wratt GS (eds), Agronomy Society of New Zealand Special Publication No. 5, Lincoln

Libby WJ (1973) Domestication strategies for forest trees. Can J For Res 3:265–276

Mead DJ (2013) Sustainable management of *Pinus radiata* plantations. FAO Forestry Paper No. 170. FAO, Rome, xv + 246 p

Miller JT (compil. & ed.) (1991) Proceedings, NZFRI/NZFP Forests Ltd Clonal Forestry Workshop, 1–2 May 1989, Rotorua, New Zealand. FRI Bulletin No. 160, 200 p

Zobel B, Talbert J (1984) Applied forest tree improvement. John Wiley & Sons, NY, 505 p

Acknowledgments

For starting the book WJL received financial support as a New Zealand-American and Australian-American Fulbright scholar, and later both financial and facilities support from the New Zealand Forest Research Institute (now Scion), and for finishing it RDB had office facilities at Scion.

A huge number of individuals have contributed in various ways to this book. Many are acknowledged by mention of their roles in the text, or by formal citations of their writings and other communications, or by photo credits.

In addition, we make special acknowledgment of the following: Bob Boardman for a variety of information, especially involving South Australia; Trevor Butcher for information on breeding activity in Western Australia; the late Ken Eldridge for a range of information on Australian tree breeding and radiata gene resources; Garth Nikles for information from Queensland; John N. Cameron and the late Mike Hall for information on genetic improvement and general forestry undertaken within the former APM Forests Pty Ltd; John Gleed of Tasman Forestry Ltd (later Fletcher Challenge Forests Ltd) for giving one of us (WJL) first-hand involvement in a programme of vegetative propagation and clonal forestry with radiata; the late Jack Henry, the late Peter Olsen and Wink Sutton for information on various aspects of the New Zealand forestry scene; Juan Carlos Carmona, Pablo Crespell, Veronica Emhart Schmidt, Claudio Balocchi and Keith Jayawickrama, for information on relevant developments in Chile; André van der Hoef for information on radiata breeding in South Africa; Cassie Carstens, Louis de Koek and Johan Vermaak for information on current planting of radiata in South Africa; John Mather for information on pitch canker in South Africa; Mario Michel for earlier information on the situation in Spain, and Paloma Moncaleán for information on recent developments there; and Don Mead, Tony Shelbourne and Mike Wilcox for checking and commenting on chapter drafts; Mike Carson for information and some suggestions; and John Dargavel for critical input on the structure of the book.

Others, who contributed information from early consultations, included the following: Frank McKinnell (Chaps 2 and 3); Bill Croft, Joe Engbeck, Deb Hillyard, Barbara Knowles, Sandy Lydon, Barbara McCrary, Bud McCrary, John McIntyre,

Connie Millar, Nikki Nedeff, Deborah Rogers, Winsome Shepherd and Vern Yadon (Chap. 2); John Bartle, Amy Brewer, Harry Bunn, John Harris, Eric Hopkins, Ian Johnson, Peter McKelvey, Charles Pawsey, Leon Pederick, Dennis Richardson and Syd Shea (Chap. 3); and Geoff Sweet (Chap. 4).

We also thank in particular Sarah Davies and Sierra De La Croix for help with the photo images, Greg Scott for help with draughting some figures, Chris Streatfield for posting draft material on the Internet, and John Moore for recent help with photography.

Among the many others who have contributed there will be those who have received less than due acknowledgment. To them we can only apologise.

Despite the input from others, responsibility for the content rests with us.

Contents

List of Abbreviations and Acronyms

ACT	Australian Capital Territory
AFLP	Amplified length polymorphism
APM	Australian Paper Manufacturers
BCFC	British Commonwealth Forestry Conference (1957)
BLUP	Best linear unbiased prediction/predictor
CF	Clonal forestry
CMPC	Companía Manufacturera de Papeles y Cartones (Chile)
CONAF	Corporación Nacional Forestal (Chilean forest service)
CORFO	Corporación de Fomento de la Producción (Chile)
CP	Control-pollinated
CRI	Crown Research Institute (New Zealand)
CSIR	Council for Scientific and Industrial Research (South Africa and formerly Australia)
CSIRO	Commonwealth Scientific and Industrial Research Organisation (Australia)
CV	Coefficient of variation
DNA	Deoxyribonucleic acid
DR	Dothistroma resistant breed (NZ)
DSIR	Department of Scientific and Industrial Research (NZ)
FAO	[United Nations] Food and Agriculture Organisation
FCC	Fletcher Challenge Corporation
FCNZ	Forestry Corporation of New Zealand
FRI	[New Zealand] Forest Research Institute
G x E	Genotype x environment interaction
GCA	General combining ability
GE	Genetic engineering
GF	Growth and form breed (NZ)
GTI	Genetics & Tree Improvement (New Zealand FRI)
IUFRO	International Union of Forest Research Organizations
LD	Linkage disequilbrium
LI	Long internode breed (NZ)

MAS	Marker-aided/assisted selection
MfA	Microfibril angle
MoE	Modulus of elasticity
NC	North Carolina, USA
NCSU	North Carolina State University
NGO	Non-governmental organization
NSW	New South Wales, Australia
NZ	New Zealand
NZFP	New Zealand Forest Products Ltd
NZFS	New Zealand Forest Service
OP	Open-pollinated
PCR	Polymerase chain reaction
PEG	Propagation and early growth (New Zealand FRI)
QTL	Quantitative trait locus
R&D	Research and development
RAPD	Random amplified polymorphic DNA sequence
RFLP	Restriction fragment length polymorphism
RNA	Ribonucleic acid
RNC	Red needle cast
[NZ]RPBC	[New Zealand] Radiata Pine Breeding Company Ltd
RWG 1	[Australian Forestry Council] Research Working Group 1 (genetics)
SA	South Australia
CA	Specific combining ability
SFA	Softwood forestry agreements
SSR	Simple sequence repeat
STBA	Southern Tree Breeding Association (Australia)
TFL	Tasman Forestry Ltd
TIMO	Timber investment management organization
UK	United Kingdom
UN	United Nations
UNDP	United Nations Development Programme
VM	Vegetative multiplication
WA	Western Australia

List of Figures

List of Tables

List of Boxes

Chapter 1
Introduction

Captain Jacques Cousteau's (1992) plenary address at the UN Conference on Environment and Development focused on human population and resource husbandry. He used as examples the past situation on Easter Island and the current situation in Haiti where, in both cases, the human population failed to renew its forest. Speaking of Easter Island, he noted that after eight centuries, this lush tropical island had become "… a barren, totally deforested piece of rock where a few hundred cannibals were hunting each other for survival." While the deforestation was not the only disaster to befall the island, it surely created a greater vulnerability to other factors. Easter Island's natural exuberance had expired largely under the load of too many consumers. All that remained were the proud statues and seemingly undecipherable tablets. We hope you agree with Captain Cousteau that this is not a good option.

Past and recent history has abundant examples of war, or threat of war, being used to acquire additional resources, but this also often proves to be a poor option. Indeed, the rises and declines of various past civilisations have been largely associated with destructive exploitation of forests (Diamond 2005). In recent times, many of the developed nations have increasingly used their political and economic power to buy somebody else's resources. This is better, particularly when there is fair trade between regions with natural shortages and regions with natural surpluses. But too often it has been a form of modern imperialism, with powerful nations acquiring the natural resources of less powerful nations. In such cases, it just spreads the drawdown of a resource in short supply and merely delays the ultimate depletion of that resource.

This book concerns another option, namely the creation of a wood resource that satisfies not only local needs for wood and wood products, but reduces the need to exploit forests in other regions. It is exemplified in the story of radiata pine (*Pinus radiata* D.Don), henceforth referred to as "radiata." The fact that the resources being conserved in other regions can be tropical rainforests makes the story even better.

Domestication is normally done in response to human needs (Zohary and Hopf 1988). For some of those institutions and individuals engaged in domestication, the profit motive in pursuit of greater values and/or the better control of costs in

R. Burdon et al., *Domestication of Radiata Pine*, Forestry Sciences,
DOI 10.1007/978-3-319-65018-0_1

producing a needed biological resource may be sufficient reason for doing it. A broader view would probably be that cost control and profits are means of encouraging domestication. Accordingly, we include some coverage of these aspects.

In this broader view, what often drives such things as domestication is perceived need, in this case mostly human needs for wood (Postel and Heise 1988). The ability of Earth's native forests to provide an adequate supply of wood as a renewable resource is under increasing constraints. These constraints result from various combinations of demands for harvested wood due to increasing human population; demands due to greater per-capita use associated with improved standards of living; conversion of native forests to agriculture, horticulture and other uses; inadequate regeneration of native forests following their naturally-caused or human-caused destruction; and where the reduction of wood harvest in productive native forests entails giving wildlife, watershed, aesthetic and conservation values precedence over wood harvest.

Wood is needed for various purposes. Historically, a dominant use was as fuel, with smelting of metals making particularly high demands on supplies. In the last two centuries or so, with dramatic rises in use of fossil fuels, one major pressure on forest resources was reduced. Today, however, the situation has changed again. The greatly increased human population and rising economic expectations are increasing the pressures for conversion of natural forest to agriculture, which entails release of carbon dioxide into the atmosphere and heavy losses of plant and animal species. The increased population and the global warming associated with the reliance of that population on fossil fuels create a two-fold need for use of wood. Producing wood sequesters atmospheric carbon and, in many cases, the use of wood is environmentally preferable to the use of substitutes such as aluminium, other metals, cement and brick, which entail fossil-fuels consumption, or of materials derived from fossil fuels (Koch 1992). This wood is increasingly coming from intensively managed plantations of fast-growing technically desirable species. It is thus both timely and appropriate that the domestication of radiata has been occurring.

Sustained or increased wood productivity of our intensively managed forests has several important benefits. Thereby, the standards of living in both developed and undeveloped regions can better be maintained or improved. To the degree that wood can be appropriately substituted for alternative materials, Earth's environment will benefit. And in that wood from intensively managed plantations can substitute for wood otherwise cut from native forests, those native forests can be dedicated to better serve alternative values, such as those associated with wildlife, watershed, aesthetics and conservation.

For example, it has been estimated that for every 420 hectares on which wood harvest is increased by one cubic metre per hectare per year, resulting thereby in wood not being harvested in native tropical rainforests, that stands on average to prevent or at least delay the extinction of an at-risk species due to rainforest harvesting (Libby 1994a, 1994b).

The domestication of radiata is leading to improvements in log and wood quality that make it appropriate for a wider range of uses, and/or more profitable wood-growing; to sustained or improved health of advanced breeds that make for safer and more stable plantation ecosystems; to increases in **harvest index**[G], which help allow greater percentages of plantation biomass to be usefully harvested; and to increases in volume growth per unit area. All of these factors also help greatly in reaping the above benefits of being able to use wood.

The scale of impact of radiata domestication is millions of hectares, and the leverage of that impact is many cubic metres of harvestable wood per hectare per year. This one programme can save thousands of Earth's species from extinction, while contributing as well to the quality of life of at least one of them, *Homo sapiens*.

1.1 Processes of Domestication

Domestication of plants and animals began in order to meet human needs, many of the needs relating to sheer survival of expanding populations, but some relating to lifestyle choices and even religious beliefs.

Early domestication of plants largely involved cereals, to provide more dependable food supplies for expanding populations than could be sustained in hunter-gatherer economies. Fruit and nut-producing trees and vines also have a long history of domestication, facilitated by learning that they could be grafted or rooted as cuttings to retain faithfully the characteristics of particular individuals. In some cases, domestication has entailed finding populations or varieties that lacked the toxins present in wild organisms, a classic example being the non-bitter almond that became widely cultivated. Domestication of animals provided dependable sources of meat, milk and fibre, and the animals could serve as beasts of burden to carry loads or pull vehicles, as well as becoming domestic pets.

Domestication of individual species has evidently occurred in interplays with domesticating other species. For example, domesticating grain species is believed to have led to domestication of the household cat to protect harvested grain from mice. Evidently for that role, the Libyan wild cat became tolerated and then domesticated, due also to some genetic mutation(s) that made those cats temperamentally tameable, and their descendants have spread worldwide (Driscoll et al. 2009). Another example is believed to be the domestication of a wolf species as dogs. While domesticated very early, during the human hunter-gatherer phase (Grimm 2015), dogs could both protect domesticated animals and serve as watchdogs for the more or less settled agriculture-based communities that developed from domestication of plants and/or animals (Box 1.1).

Box 1.1 Definition
Domestication is the use of purposeful selective breeding in conjunction with purposeful management, to obtain benefits that are complementary or even strongly synergistic. It entails increasing inputs in order to obtain greater returns.

There are the two main planks of domestication, namely intensified husbandry *per se* and genetic improvement, the latter being subject to the qualification that improvement achieved in one context need not necessarily be beneficial if the context changes. These two planks almost always operate interdependently, and are often highly synergistic. Genetic improvement can be specific to the environments created by intensive husbandry, and the success of such husbandry may depend heavily on genetically improved populations that usually become no longer fit to compete in the wild. A classic, and very recent, example of these effects is the case of the dwarf wheats of the Green Revolution (Evans 1998; Hedden 2003). Such varieties give greatly enhanced grain yields in response to generous fertiliser use, whereas in the traditional, long-strawed varieties such treatments lead to excessive straw production that tends to be followed by lodging (falling over) which vitiates potential responses in grain yield.

Historically, the genetic improvement of crops has often occurred by haphazard processes. The annual crop plants of modern agriculture have gone through hundreds and even thousands of generations of selection by humankind. Such selection, however, would often have been done subconsciously or else based on misconceptions that made the process much less efficient than is now possible with modern understanding of genetics, the science of heredity.

Forest trees have mostly been exploited as supplies of wood, and their domestication has come far more recently than that of the major crop plants, fruit and nut trees and traditional domestic animals. As with such plants (Evans 1993) and animal species (Diamond 2002), the list of forest-tree species that have been subject to intensive domestication is notably short. Because the domestication of forest trees is so recent (and incomplete), it is subject to far more documentation. Occurring as it has in the days of modern nation states, the roles of individuals, institutions, social and economic contexts and government policies in the domestication process are often both prominent and clearly traceable, with the lessons to be drawn. So too are the influences of advances in a range of technologies that were not traditionally associated with wood consumption. Indeed, a huge "game changer" was the discovery that wood could be pulped to provide the most convenient source of paper, which took effect in the 1atter half of the 19th century.

For a long time forest trees have been exploited for wood, as either fuel or timber, by cutting individual trees in natural forest. Domestication, where it has occurred, has often been spurred by exhaustion or impending shortages of natural wood supplies. In addition to meeting general wood supplies, domestication of forest trees has sometimes been directed at highly specific products, but such a pursuit can be outstripped by technological advances (as happened with growing naval oak timbers in Britain). More recently, the growing of wood crops has on occasion been subject to major private investment, sometimes in the nature of speculative investment bubbles. For forest trees, domestication belongs essentially in a context of plantation forestry. By contrast, natural forests that are perpetuated by natural regeneration would not rate as domesticated even if they are meticulously managed and may have evolved genetically in response to some management inputs.

Our unifying theme is *the beginnings of domestication of radiata, for which unusually good documentation exists*. Thus this book reviews human actions that altered the genetic structure of this important forest tree species, and it focuses on those few recent decades when the natural populations of radiata were both consciously and unconsciously sampled and modified to serve various human purposes. Additionally, this book illustrates how such domestication of forest trees was done from the latter half of the 20th century. However, it is a process that may continue for many generations, conceivably until civilization vastly changes or fails.

Requirements of a forest tree species for successful domestication are typically: ease of propagation and transplanting, modest requirements for site quality, reasonably broad climatic tolerances, rapid growth, and ability to produce wood of at least acceptable quality when grown rapidly on short rotations. These requirements are well met by radiata. To waive requirements for quite rapid growth rate and thriving on land of modest quality, a species' wood needs to be of very high value indeed.

We next consider the beginnings of the domestication of radiata as interactions of several processes and activities.

Population Genetics. Both human-caused genetic events and natural events with genetic consequences may be important in the domestication of a species or population. These include pre-human events affecting the macroevolution and paleohistory of radiata, more-recent human activities and other micro evolutionary forces within the native populations, and then the founding and development of plantations elsewhere, all of which can lead to the subsequent evolution of distinct "**land races**"[G]. Which native populations were sampled to found land races is an important variable, often crucial to how well early plantations succeeded in a country or region.

Management. The passive domestication of radiata began, as in most crops, when humans began to grow it in plantations with increasing knowledge and skill (Lewis and Ferguson 1993). Some genetic changes occurred as the forces of natural selection worked in new environments, as people planted radiata on soils and in climates that were different from those where the native population had evolved. The farmers and foresters who planted, cared for and harvested pines in those early days probably did not understand the science of genetics, which was not formally founded until 1900. Nevertheless, the removal or natural attrition of sick and malformed trees before they could effectively reproduce created local populations of trees increasingly better suited to grow in the new conditions. Thus, these so-called land races of radiata became not only more adapted to survive and grow in their new homes, but became more useful as well. Since the founding and subsequent development of land races of radiata have figured importantly in our domestication story, we cover it in more detail later in this chapter and in the following chapters.

Breeding. Once adequate knowledge was available about suitable climates, soils, nursery practices, establishment practices, and such silvicultural practices as thinning and pruning, then purposeful breeding of radiata could effectively begin. In interactive combination with cultural practices, breeding is the most powerful single force effecting domestication. Breeding strategies and plans have differed in response to the different goals of programmes in different regions and for different organizations. These differences also affected the traits that have been or will be selected to achieve those goals.

Propagation technology allows the new breeds, and **clones**^{G}, to be grown in large numbers, and employs both **sexual**^{G} and **vegetative**^{G} methods. As populations or species become increasingly domesticated, these methods of reproducing them are not only useful to achieve economies of scale, they may become increasingly necessary. In some cases, sites are utilised out of the ecological range of the species and naturally-seeded pines cannot successfully establish and compete with resident vegetation. In other cases, the trees' natural reproductive processes have become incidentally or even purposefully modified, and thus are less effective even within the ecological range of the domesticated population.

Deployment concerns do not become important until managers have well-known and diversified breeds, families, clones and defined **multiclonal varieties**^{G} available to them. This is a recent development, and we will consider the several important new deployment options made available by advances in breeding and propagation.

Domestication has tended to occur at two levels. The lower level, which might be called "passive domestication," entails proper choice of available genetic material (in this context, favoured provenances, land races and trees reproducing after thinning) and development of management practices. This level may even involve vegetative propagation, historically with easily-rooted cuttings or, for many fruit and nut species, often with grafts. Such propagation tended to focus on relatively few clones, preferred in part for their ease of propagation.

Even the lower level of domestication can certainly entail risks. Growing forests to meet current perceived needs can have its pitfalls, particularly if the needs are specialised and the delays in producing the end-product are long. In England, for example, large areas of oak were grown in the Crown Forests in the 17th and 18th centuries so as to produce specialised naval timber (ships' knees), but well over 100 years later, at the intended harvest dates, the anticipated wooden warships were totally obsolete. And a focus on very few preferred clones, as easily occurs with species that can be conveniently propagated as cuttings or grafts, has created severe risk exposure.

Embarking on intensive breeding represents a large upward shift in the level of domestication. An immediate ramification is adopting new propagation technology as tools for both doing the actual breeding and delivering genetically improved material on a commercial scale. Intensive breeding also favours far more precisely targeted deployment and husbandry of plants or animals resulting from breeding programmes. Pervading all phases of domesticating organisms is typically the theme of increasing inputs in both husbandry and genetic improvement in order to reap higher returns for the growers and society—or merely to maintain production in the face of new pests and diseases, nutrient or water shortages, or climatic change, thus meeting a need to "run fast in order to stand still". To this theme, radiata is no exception.

1.2 The Radiata Story

1.2.1 Historical Preview

Radiata has a peculiar and intensely localised type of natural habitat, native stands occurring in three disjunct locations on the Californian mainland coast and on two islands well to the south (Sect. 2.1). That habitat has meant a very limited natural range, which precluded any recognition as a significant timber species, but it allowed the evolution of a very high growth potential reflecting a very long and opportunistic growing season. However, there are large areas in other parts of the world, mainly in the Southern Hemisphere, where it does very well. This became clear upon it being imported, largely as a botanical curiosity, to far-flung lands. Introductions to Spain, South Africa, Australia and New Zealand occurred during 1840–1860, but only later in Chile (1880s) (Sects. 2.3, 3.3 and 5.1). Early domestication was favoured by ease of collection and storage of seed, and ease of growing in the nursery and good tolerance of transplanting, along with its growth rate. Its popularity escalated, initially for shelter plantings and for firewood, and then for timber production in the parts of New Zealand and Australia that encountered timber shortages by the late 19th century.

Despite establishment of some commercial plantations of radiata near the end of the 19th century and through the first decade of the 20th century, massive commitment to planting it began only after World War I, in the 1920s. Driving factors were variously: looming timber shortages as native timber supplies dwindled, especially in New Zealand; strong endorsement of radiata by some influential enthusiasts who had acquainted themselves with its capabilities; increasing appreciation of its ease of establishment and fast growth; availability of quality land where pastoral farming was beset by an undiagnosed stock health problem; the presence of degraded land in need of protection or rehabilitation; needs to combat unemployment; and a need for economic diversification beyond very narrow export bases in the two main radiata grower countries. Governments engaged heavily in this phase of the domestication process, which included developing its wood utilisation, by either direct investment or providing financial incentives. Even so, radiata was also the centrepiece of a private investment boom in afforestation, similarly beginning in the 1920s.

During the 1920s and 1930s there were some alarms and crises associated with greatly expanded growing of radiata, exposing (often belatedly) its climatic bounds, and revealing serious nutrient deficiencies that could, however, be corrected easily and to very good effect.

After World War II came challenges of processing and utilising the large resources of radiata wood becoming available. Massive industrial processing plants were built, albeit with confidence in their sustainability being locally complicated by some biotic alarms. In keeping with the scale on which radiata was being grown commercially, intensive genetic improvement was embarked upon from the early 1950s. Needs and scope for it were apparent, and were abundantly verified, but efficient capture of genetic improvement was on a learning curve.

Since the mid-20th century the domestication of radiata has been characterised by: extending its planting as a commercial crop to a vastly greater area than its natural extent (some 500 times); development of more sophisticated and cost-effective systems for establishment, tending and harvesting of crops, along with enhanced processing to be able to use the wood to better advantage; and programmes of increasingly intensive genetic improvement. Genetic improvement has variously achieved some of the benefits that might alternatively be achieved by improved growing practices, yet some of those practices have greatly facilitated it. Still, there have been challenges in achieving the desired integration of genetic improvement with forest-growing practice, some technical and some involving suspicious attitudes even among foresters. More specifically, the developments have entailed: greatly enhanced nursery and establishment technology; fine-tuning of mineral-nutrient status; improved collection and analysis of field data; consequent refinement of yield and outturn prediction in relation to site and tending regime for financial and operational planning; development of breeding strategies that would balance current needs and long-term sustainability; extending propagation technology well beyond natural modes of propagation; and using DNA technology for genetic improvement. Compared with intensive domestication of some other forest-tree species, growing radiata presented a more complex production system, on account of its greater range of end-products, thus entailing more complex domestication options. In these respects, work with domesticating radiata has included many of the cutting-edge developments in industrial plantation forestry, while affording important environmental benefits both on-site and even globally. We review not only these developments in several forest sciences, but also the institutional, social and political contexts in which they have occurred.

The pursuit of these benefits has often posed long-term challenges, because pursuing short-term financial gain can militate against resource sustainability and some general societal benefits. For the future, however, intensive domestication of radiata (and a quite short list of other forest trees) is seen not just as an economic and local environmental benefit. Within a wider environmental and humanitarian imperative, domestication programmes are often part of the task of meeting the material needs of the world's human population while preserving the natural biota and helping to keep the planet peacefully habitable.

1.2.2 Land Races

From the mid-20th century, with the serious domestication of radiata having begun, questions about the genetic base underpinning the domestication process have become interesting and topical. Among the most important of these are questions about the structure of genetic variation among and within the native populations, and about the origins of the putative land races of radiata. Development of land races has been part of the early process of domestication of this species, and they have formed a platform for its later, intensive domestication.

Understanding the origins of this increasingly important plantation crop has not only historical interest, but also both scientific and practical value. Such knowledge will allow tree breeders to better judge the extent to which they can continue to select within their local land race(s), and whether they may wish or need to bring genetic variability in from other land races and from remaining untapped variability of the native populations. In turn, conservationists can use the likely importance of keeping native genetic reserves to help develop an overall strategy for conserving native radiata populations.

Land races evolve as a result of human activity. These activities need not entail conscious attempts to create different populations. Various combinations of events and processes can change the genetic constitution of a population from that of its origin population(s) to a land race that is something different, unique and often relatively local. These processes include **founder events**^{G}; various natural selection pressures in the new environment; dysgenic and/or eugenic selection occurring naturally or through human intervention; and hybridizing/segregating mixes of trees from two or more genetically distinct populations that have been brought together where people have planted them in new places.

These various events and processes can and often do interact strongly with each other. For example, **eugenic**^{G} silvicultural practices, such as thinning out deformed or diseased trees well before they can reproduce, may be practiced on planted trees containing enhanced genetic variation due to recent hybridization, in an exotic environment in which the trees are encountering damaging physical and biotic agents that did not occur in the native environments where the origin populations evolved. When the human activity has been recognized as successfully tailoring the population to particular purposes, then the concept "land race" is transformed to "breed."

Three lines of evidence may be used to understand the genetic makeup of current land races and breeds. (1) Historical records concerning the number, origins and sizes of the founding population-samples. (2) Observations of current members of these land races, and of members of the native populations, most effectively made in appropriately designed **common-garden experiments**^{G}. (3) Biochemical and molecular genetic studies, comparing such things as the occurrence and frequencies of **alleles**^{G} (variants of **genes**^{G}) in the different land races and populations. We review the first of these lines in Chap. 2, and (2) and (3) in later chapters. For radiata, the available historical records, as good as some of them are, are nevertheless incomplete and equivocal. They need to be supplemented by common-garden and genetic-marker evidence to get a clearer picture of the native origins of present-day land races and breeds.

1.2.3 The Development of Plantation Management

For radiata, development of plantation management has been both straightforward and complicated. As we have already indicated, and recount later in more detail, the species produces seed that can be very easily collected and stored, it germinates and develops very well in the nursery, it transplants very well, its rapid growth reduces the need for weed control, it can grow consistently well across heterogeneous sites, and if neglected after establishment it can still produce a remarkably satisfactory crop because of its strong capacity for self-thinning.

On the other hand, the quest for tending regimes that optimize profitability has proved remarkably challenging. This optimisation problem remains ongoing, with issues that we address later.

1.2.4 The Step to Intensive Breeding

For most species important to human welfare, there is a time when it becomes ripe for breeding and, prior to that time, its domestication will be haphazard (or "passive"). The development of management knowledge, and the fiscal information stemming from that knowledge, are probably the main determinants of when a species' domestication shifts from passive to purposefully intensive. Moving on to an effective breeding programme requires not only the skills to select candidates for breeding effectively, to physically control and execute pedigreed matings, and to test the offspring; it also requires sustained funding and institutional support.

There are many examples in forestry where breeding programmes have been started without the requisite skills or support. Understaffed or underfunded breeding programmes have been conducted using parents of uncertain origin, and their test offspring were often planted on inappropriate sites, at too-close or too-wide spacing, and with erratic early survival and growth that more reflected the inadequate state of nursery and planting practices than it did the genetic qualities of the planting stock.

While the early radiata breeding programmes suffered from some of these problems, there were circumstances that made it more ripe for breeding than most other forest tree species in the world. Those special circumstances provide an important part of the explanation as to why radiata today is near or at the top of a very short list of tree species undergoing rapid and effective domestication.

1.2.5 The Political and Institutional Context for Technical Progress

Many economically important tree species are mostly grown and harvested within their native ranges, in some cases as plantations but often relying on natural regeneration. In strong contrast, radiata had been grown mostly as a plantation species out of its native range for nearly a century before breeding was begun. Thus, the knowledge of how to effectively raise and plant test offspring was already largely in place. In Australia and New Zealand, the two countries where its breeding was first undertaken, the radiata plantation estate was sufficiently large and valuable so that generous funding was provided and sustained—at least until recently. Perhaps importantly, those two countries were secure members of the British Commonwealth. Following World War II, at the time their tree-breeding programmes began, the British Empire was fragmenting. Postwar reconstruction in Australia put great emphasis on expanding the trained and educated workforce. Scientific, technical and professional people gravitated to New Zealand from the Empire and various other (often less stable) countries, and New Zealand also sent science graduates to obtain forestry training in Australia and Britain and occasionally other countries. The intellectual ferment among these scientists, technicians and foresters, with their wide and diverse experience, produced a sort of institutional hybrid vigour that was reflected in the radiata plantation forestry of that era. The development of radiata breeding programmes, especially in New Zealand, came under several international influences, most notably from Scandinavia and USA.

The breeding work began in the context of a broad portfolio of plantation-forestry and forest products research, which included developing greater expertise in forest inventory, nursery management, field establishment of plantations, soil properties and tree nutrition, general silviculture, forest diseases and pests, basic wood properties, sawing and drying technology, wood preservation, timber utilization, and wood pulping and papermaking. A focus on radiata plantations, which also drew on some early work in South Africa, established Australia and New Zealand at the forefront of both plantation forestry and the processing and utililisation of the resulting wood. Specific recognition came in the awards of two Marcus Wallenberg prizes to Australians, one for timber design standards and one for developing automated evaluation of detailed wood properties.

In South Africa important seminal contributions had earlier been made to plantation management systems, but the relatively minor importance of radiata there restricted South Africa's contribution to the strategies and practices of breeding this species.

In Chile, despite the great current importance of radiata there, timing and political factors led that country to being until comparatively recently more of a follower in our domestication story. However, Chile's need to diversify from a very narrow export base, the availability of much degraded land, and associated underemployment, became powerful drivers of using radiata. Governments of very different political hues have long continued to support its forestry sector.

1.2.6 The Current Radiata Resource

For every hectare of radiata now growing in the native forests of California and Mexico, there are now more than 500 ha of radiata trees planted and growing outside of its native range. It is planted in a number of countries, with major areas in four continents and New Zealand, and for a variety of human purposes (Table 1.1).

Among its main grower countries radiata is economically very important, and its wood is a major earner of foreign exchange for New Zealand and Chile. For none of the significant grower countries was it a free forest: instead, they created a

Table 1.1 Recent areas ('000 ha) of radiata plantation, and use categories, by country (Adapted from Mead 2013, table 1.2)

Country (year of data)	Area	Products/uses	
		Main	Secondary
New Zealand (2012)	1,543[a]	Export logs, sawtimber, pulp, plywood, board products	Energy, shelter, erosion control, posts & poles, horticultural bark
Chile (2011)	1,478	Pulp, sawtimber, board products, land rehabilitation, energy	Energy, edible fungi
Australia (2012)	772	Structural sawtimber, pulp, reconstituted wood, posts & poles	Energy, shelter
Spain (~2011)	~280	Sawlogs, agroforestry, farmland retirement	Energy, posts & poles
South Africa (2011)	40	Sawtimber, plywood, posts & poles	Energy
Ecuador (1990)	~20	Sawtimber, erosion control, agroforestry	Energy, edible fungi
Italy (2005)	6		
Argentina (2011)	5.5		
Others	~35		
Total	~4,180		

[a]Plus significant area equivalent of timber resource represented as shelterbelts, perhaps 10% or more of total

resource by *de novo* afforestation. Moreover, its acceptance as a timber tree was largely driven by individuals who enthusiastically championed it, despite no real historic repute and many negative perceptions. Its large-scale growing and processing was also led largely by men of unusual vision, who had the talent and drive to get these ventures carried out.

Apart from the purely economic benefits of radiata to its main grower-country economies have surely come wider benefits, social and environmental. Insofar as the availability of its wood may still help spare some of the tropical rainforest of Southeast Asia, the growers of radiata must also rate as ecological heroes.

References

Cousteau J (1992) Convince to conquer. Calypso Log. August 1992, p 3–6

Diamond J (2002) Evolution, consequences and future of plant and animal domestication. Nature 418:700–707

Diamond J (2005) Collapse: how societies choose to fail. Viking Penguin, New York, p 590

Driscoll CA, Clutton-Brock J, Kitchener AC et al (2009) The taming of the cat. Sci Am 300(6): 56–63

Evans, LT (1993) Crop evolution, adaptation and yield. Cambridge University Press, Cambridge, UK; New York, p 500.

Evans, LT (1998) Feeding the ten billion: plant and population growth. Cambridge University Press, Cambridge, UK; New York, p 247

Grimm D (2015) Dawn of the dog. Science 348(6236):274–279

Hedden P (2003) The genes of the green revolution. Trends Genet 19:5–9

Koch P (1992) Wood versus nonwood materials in U.S. residential construction: some energy-related global implications. For Prod J 42:31–42

Lewis NB, Ferguson IS (1993) Management of radiata pine. Inkata Press (Butterworth-Heinemann), Melbourne–Sydney, p 404

Libby WJ (1994a) Mitigating some consequences of giant sequoia management. pp 142–148 in Aune, PS (technical coordinator) Proceedings of the symposium on giant sequoias: their place in the ecosystem and society; June 23–25, 1992; Visalia, CA, USA. Gen. Tech. Rep. PSW-GTR-151. Pacific Southwest Research Station, USDA Forest Service, Albany, CA. p 170

Libby, WJ (1994b) The role of production forestry in preventing tropical species extinctions. P 5–9 In: Fins, L. (ed.) Nimbyism, genetic improvement and plantation forestry: growing wood for the 21st century. Proceedings Inland Empire Tree Improvement Cooperative. Moscow p 40

Mead DJ (2013) Sustainable management of *Pinus radiata* plantations. FAO forestry paper No. 170. FAO, Rome. p xv + 246

Postel S, Heise L (1988) Reforesting the earth. Worldwatch paper 83. Worldwatch Institute, Washington, p 66

Zohary D, Hopf M (1988) Domestication of plants in the old world. Clarendon Press, Oxford, p 249

Chapter 2
Early History: 7,000,000 Years Ago to 1901 C.E.

When domesticating a plant or animal species, it is useful to know something about the history of that species, and particularly something about the evolutionary forces that may have modified its genes and how alternative versions of its genes (called **alleles**^{G}) are organised within and among populations. Furthermore, it is useful to know how much of its natural genetic variation is likely to have been included in the populations being domesticated, and which native populations have not yet contributed potentially useful genetic variants to the breeding lines in use. Besides being useful to breeders, such knowledge is often interesting and satisfying in its own right, and we hope you will find that to be the case with radiata pine.

2.1 Five Small Native Forests[1]

The present and prehistorically recent native forests of radiata pine (henceforth called "radiata") occur in only five general locations, three coastal in California and two on Mexican islands (Fig. 2.1). These five locations differ substantially from each other with respect to soil, elevation and winter rainfall (Table 2.1). They share a special variant of a regional Mediterranean-type climate, namely a cold ocean current to windward that produces fogs among the trees on most days during the almost rainless summer. The reduced evaporation due to fogs and the supplementary soil moisture from fog drip allows the pines to grow even though the amount of rainfall in these populations is relatively low. Their temperatures are usually moderated throughout the year by onshore winds, and generally stay above 0°C and below 30°C, although occasional days with lows down to −8°C and highs up to 40°C have been recorded, with little or no damage to the native

[1]This section represents a minor revision, with some updating, of Libby (1997).

R. Burdon et al., *Domestication of Radiata Pine*, Forestry Sciences,
DOI 10.1007/978-3-319-65018-0_2

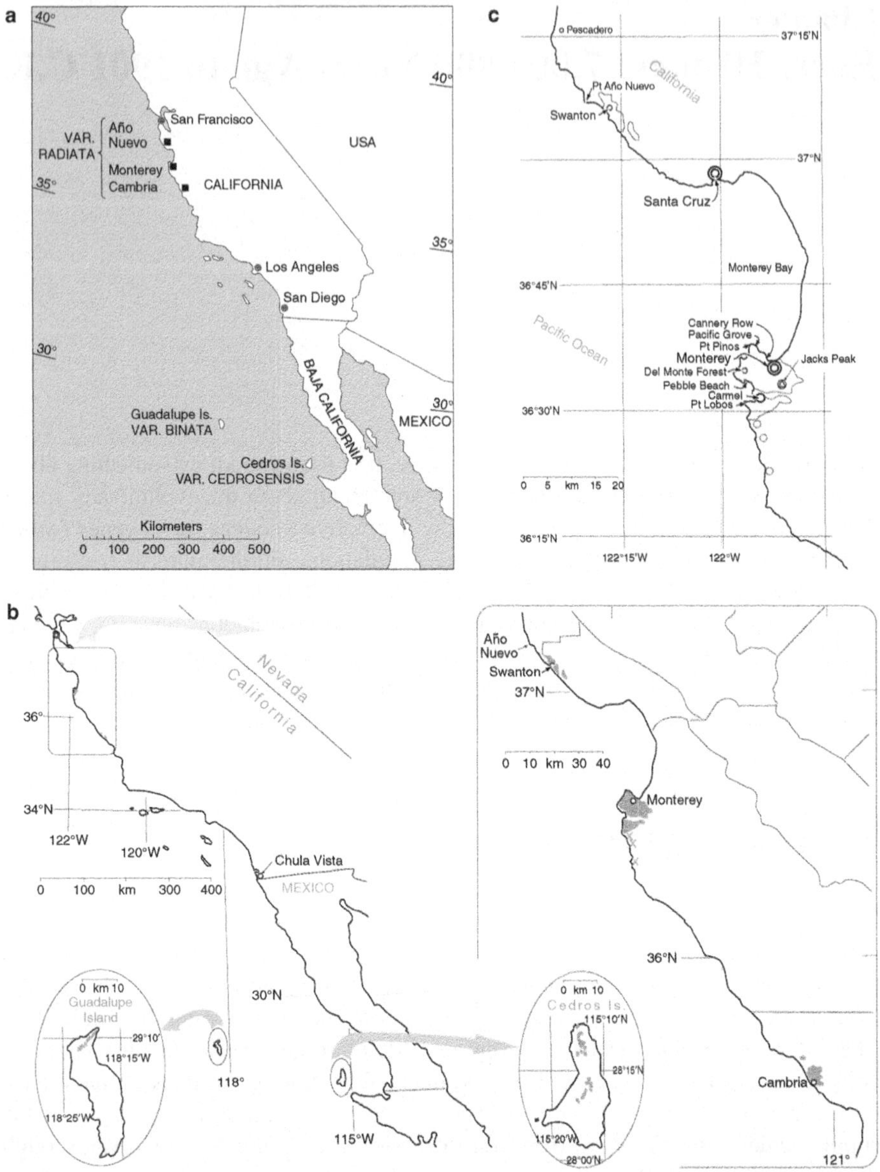

Fig. 2.1 Maps showing locations of (a) native stands of radiata (after Burdon, 2004, permission Elsevier), (b) over total range (after Critchfield and Little 1966) and (c) detail in Monterey locality (After Forde 1964a)

trees. Effective rainfall is typically limited to late autumn to early spring, with a peak in midwinter (Scott 1960). In some years, one or both island populations receive occasional but substantial rainfall from "runaway" summer tropical storms (Oberbauer 2006) (Box 2.1).

Table 2.1 Particulars of native populations of radiata (After Forde 1966; Burdon 2001; Rogers 2002; Oberbauer 2005, 2006)

Population	Lat. (°N)	Rainfall (mm/yr)	Altitude (m asl)	Exposure	Soil - parent material and depth and texture	Extent (ha or trees)	
						Current	Historic
Año Nuevo	37	675–900?	10–330	Varied	Mainly mildly calcareous argillites, depth variable	450	450
Monterey	36½	400–700?	10–580	Varied	Very varied: granites, sediments, and marine terraces of varied weathering status.	3000	7,400
Cambria	35½	450–575?	10–200	Mainly mild	Sandstone, sandy loam, variable depth and drainage.	800	1,400
Guadalupe	29	330–510?	330–1,155	Severe	Basaltic, rocky loam	<200 trees	640
Cedros	28	150?	275–640	Locally severe	Mainly ancient sediments or metamorphics, soils skeletal	130	130

1. Precipitation figures shown include allowances for altitudinal gradients (cf Forde 1966), but effect on hydric regimes of additional rainfall at higher altitudes may be offset on mainland by less influence of coastal fog. 2. All mainland populations have small areas of coastal sand. 3. Guadalupe trees do not include very recent regeneration. Question marks denote considerable uncertainty.

Box 2.1 Natural Range
Radiata occurs naturally in five discrete populations, three on the Californian mainland and two on Mexican islands. While geographically separated, the populations share a peculiar, localised variant of a Mediterranean climate caused by a cold ocean current. Differences exist among the populations, but the magnitude of the differences between pairs of the populations varies widely. Historically, domesticated radiata evidently came from only the two northernmost populations.

All five populations have some remarkably sharp boundaries, with few or no large outlying native trees beyond the outer edge of the forest. The native populations generally occur as mosaics of even-aged stands, which form part of a complex fire ecology with fires of widely varying severity (Stephens et al. 2014). Radiata forests regenerate vigorously after severe fire (Fig. 2.2) or other massive

Fig. 2.2 Dense radiata regeneration after a 1987 fire on Monterey Peninsula (Photo WJL March 1990)

disturbance, and this pine is thus a good coloniser of cleared sites both within its native forests (Forde 1966) and as an exotic plantation species (Bannister 1965).

The mainland populations have been subject to considerable human disturbance (Libby 1997), while the Guadalupe population has been catastrophically affected by introduced goats (Ledig et al. 1998; Oberbauer 2006), leaving the Cedros population as the only one nearly free of human influences.

2.1.1 Mainland Populations

In all three of these native mainland populations, radiata pine now grows from just above sea level to upper elevations of about 200–300 m (500+ m at Carmel Highlands in the southern Monterey population), stopping at an inland boundary where coastal fogs are dissipated in the hot summer sun. All natural stands are within 8 km from the sea. Prior to burning or clearing by humans, each of these pine forests was nearly continuous on the north- and sea-facing slopes, and windward from its interior edge towards the coast. At the lower elevations, the pines grow on the dunes or bluffs very close to the sea. There, buffeted by wind and burned by salty moisture, the mature trees grow low to the ground or pressed against the face of the bluff, yet these stressed, sea-edge pines are typically laden with cones. The trees reach their best development near the bottoms of sheltered valleys, with such sites in all three mainland populations containing large, tall and (often) straight pines. Radiata grows on a variety of soils, but relatively shallow soils of sandstone, shale or other sedimentary origins are the most common (Lindsay 1932; Scott 1960) (Fig. 2.3).

Fig. 2.3 View from roadside of native radiata stand on a comparatively infertile site on Huckleberry Hill, Monterey Peninsula (Photo RDB 1972)

2.1.1.1 Extent

The radiata population on and near the Monterey Peninsula, near 36½°N latitude, is by far the largest of the five, with undeveloped forest land currently totaling about 3800 ha (9400 acres). Various categories of developed forest land (urban, suburban, golf courses, city parks, etc.) occupy another 4100 ha. These are generally characterised by interrupted and depleted native stands mixed with planted trees of both native and exotic species. Shortly before humans arrived on the scene, the Monterey population occupied an uncertain but probably slightly larger area than the current natural extent of 7900 ha (Jones and Stokes Associates Inc. 1994). This population is now substantially urbanised, with Monterey, Carmel, Pacific Grove and several extended communities being largely or wholly contained within the population limits.

The Cambria population, about 130 km (80 miles) south-east of Monterey, is the second-most extensive of the native populations. It currently occupies about 900 ha, reduced by various human activities from about 1400 ha. A mature nearly-square 260-ha stand about 5 km north of the main population, although genetically consistent with the main Cambria population, may have been planted.

Fig. 2.4 Native stand of radiata at Cambria, showing sapling regeneration in foreground (Photo RDB 1972)

As at Monterey the Cambria population is now substantially urbanised, with the town of Cambria occupying the central part of the main population (Coffman 1995). In addition, current and recent ranching has reduced numbers of trees per hectare north and east of the urban limits within its recent natural range (Fig. 2.4).

The northernmost of the five populations, about 80 km north of Monterey, has its northern edge opposite Point Año Nuevo (now an island) and is thus identified by that name. Its area of about 450 ha (calculated by M.H. Bannister, NZ Forest Research Institute, from distributions mapped by Forde 1964a, 1966) is essentially the same today as it was prior to the arrival of humans, but details of its distribution have been moderately changed by human activities.

2.1.1.2 Associate Tree Species and Soils

In the Monterey radiata pine population, Douglas-fir and redwood also occur in mixture with radiata on a few sites with deep moist soils, mostly near the southern end of this population but rare elsewhere. Granites underlie about 13% of the population (Dunning 1916; Jones and Stokes Associates Inc. 1994). Monterey cypress (*Cupressus macrocarpa*) (also called macrocarpa [cypress]) occurs as two very small sea-edge native populations, both flanked by and grading into radiata forest. Coast live oaks are increasingly common on the more inland sites, and recently appear to be replacing radiata where fire and other disturbances have long been excluded. Near the centre of the Monterey Peninsula, pockets of bishop pine (*Pinus muricata*) (also called muricata pine) and another cypress (*Cupressus goveniana*) (Gowen cypress) occupy strongly-weathered marine terraces (Rogers 2002), where the soil has developed an impervious layer and has become more acidic and infertile. (The Gowen cypress also occurs on a similar site further south within the Monterey radiata population.) The two pines grow intermixed at the edges of this muricata pine population, and groups of radiata tower above the muricata pine within it along stream edges and in pockets where the soil is better. Natural hybrids of muricata and radiata pines and their hybrid derivatives are very rare, if indeed they exist at all (Griffin 1970 p. 90).

In the area occupied by the Cambria population radiata is the only **conifer**[G]. In all three mainland populations, coast live oak is the most common tree growing in association with radiata; several other **hardwoods**[G] are less-frequent associates. Except for a small area of coastal sand, the Cambria population occupies a single sedimentary formation, which produces a sandy loam, although rock outcrops remain.

In the Año Nuevo population, radiata grows in nearly pure stands on some slopes, and intermixed with Douglas-fir and coast redwood on the better sites, those with deeper, moister soils. Soils are derived from marine sediments and are sometimes appreciably calcareous (Rogers 2002). The Año Nuevo radiata also grows with knobcone pine (*Pinus attenuata*) in a narrow, discontinuous transition zone on ridgetops along the inland boundary of the radiata distribution. Natural hybrids between these two species (named *Pinus* × *attenuradiata*), and various second- and later-generation derivatives of such hybrids, occur at low frequency in and near this transition zone.

2.1.2 Island Populations

The two island populations, like the Cambria pines, have no other associated conifers and, like all three mainland populations, have distributions on the islands closely correlated with the local occurrence of persistent summer fog (Fig. 2.5).

Fig. 2.5 Aerial view of the main northern radiata subpopulation on Cedros Island. The pines occur on the windward slopes of the ridges, with hard desert below them on the lee slopes. The bare ground in the canyon between the pines is the result of earlier copper and gold mining (Photo WJL 1964)

The southernmost population (28°N latitude) occurs on Cedros Island, which in only very recent geologic time separated from mainland Baja California, and was only recently (Axelrod 1980; Bannister; McDonald 1983) accepted as belonging within radiata rather than bishop pine (*P. muricata*). Thus, the native plants and animals on this island are largely the same species as those on similar mainland sites. Although other tree species are native on Cedros Island, the pines grow in essentially pure stands with few or no other associated trees. They occur on about 130 ha, in two similar-sized subpopulations. One subpopulation is on the high central ridges of the island and the other, 15 km distant, is on the upper ridges and windward cliffs of the island's northern point. The often thin soils in these subpopulations are mostly of sedimentary origin, but some in the northern subpopulation are derived from granites or mineralised metamorphic rocks.

Unlike the mainland populations, which mostly grow at relatively low elevations, the pines on Cedros Island occur only on the ridgetops and upper windward slopes (Fig. 2.5), at elevations between about 275 and 640 m (Libby et al. 1968). The lee-side transitions are spectacular, changing in a distance of 2–3 m from moist pine forest with ferns in the understory to hard desert with widely-spaced plants such as prickly pear and barrel cactus. The abundance or absence of fog drip is responsible for these transitions. As wind-blown fog crests the ridge through the pine forest, moisture condenses on the pine needles and drips to the ground. Tree-rings observed on old cut stumps in the northern Cedros Island population have an unusual pattern (WJL personal observation 1968). The earliest (central) rings are very narrow and then, over a 10–20-year period, the rings become progressively wider. Apparently the young pines condensed and dripped more moisture each year as their growing crowns became taller and broader and thus intercepted larger quantities of fog. In some years, no rain has been recorded in the distant fishing village on Cedros Island, and in such years perhaps all life-sustaining moisture had to be wrung from the passing fogs. Still, some of these pines have reached reasonable sizes, with measurements of 32 m height for one and 77 cm diameter for another (Libby et al. 1968).

Of the five natural populations, that on Guadalupe Island is now by far the smallest in both area occupied and number of living trees, because of its relict status resulting from the introduction of goats. It exists over a linear distance of about 8 km on the single ridgetop and windward cliffs of the northern quarter of the island (Fig. 2.6), being remnants of a largely pure stand estimated to have covered 650 ha (Oberbauer 2005). As on Cedros Island, most of the pines grow at a substantial elevation above sea level, between about 485 and 880 m. A small sub-population is much higher, near the main peak between 1120 and 1155 m elevation, 1.5 km distant from the rest and separated by a barren rocky slope. At present, the pines throughout the Guadalupe population are widely scattered. In 1964, there were only 383 pines over the entire 8-km length of the population, and as of 2002 there were only about 200 alive (Rogers 2002; Rogers et al. 2005). Regeneration of the pines effectively ceased well over 100 years ago as a result of the introduction of goats. However, following eradication of goats in

Fig. 2.6 The lower (northern) part of the Guadalupe Island population of radiata, showing the pines on the upper windward slope and ridgetop (Photo WJL 1964)

2004, very promising regeneration of pines has occurred (Vargas-Hernández et al. 2013).

Guadalupe is an oceanic island, and has been far distant from the mainland ever since it emerged about seven million years ago. This island is the partial rim of a volcano that rises about 5000 m from the ocean floor (Moran, 1996). Relatively few plant and animal species have colonised the island. The pine probably arrived as seeds in still-closed cones that were attached to floating logs, between one and four million years ago. The only two tree species currently growing with the pines are island oak and a palm. The palm and the latitude (29°) suggest a warm site. However, because of the cold ocean current near the windward side of the island, many days are foggy and chilly in and near the pines; ice and snow have frequently been observed there (Moran 1996). The soils, mostly of basaltic origin, are very thin. But the pines' roots penetrate the fragmented volcanic rock and some trees have attained large sizes. The tallest pine measured on the island in 1965 was 33 m and the largest diameter, attained by a different tree, was 211 cm at breast height. This latter is the largest diameter recorded in any of the five native populations of radiata (Libby et al. 1968). Unlike those mainland radiata pines that are exposed to onshore winds, which lean and twist or even grow prostrate, the Guadalupe Island pines are unusually straight and often surprisingly tall despite all growing in extremely exposed windy conditions.

2.1.3 Paleohistory and Taxonomy

While paleohistory and taxonomy are not the same thing, systematic taxonomy attempts to reflect evolutionary lineages by its constructed hierarchical relationships among species and larger groupings; thus the former supports the latter even though taxonomic groups are based on much more than fossils. Various systematic-taxonomy treatments have been published for the pine (Price et al. 1998). Each differently used and interpreted the various shared and divergent traits likely to have resulted from shared ancestry followed by migration into different environments and then independent adaptive evolution. This investigation is still very much ongoing, with new techniques (especially DNA analysis) and information resulting in further advances in our understanding of the evolutionary history and the modern relationships of the pines.

2.1.3.1 Fossils and Evolution

The lineage of the over 110 tree species now classed in the genus *Pinus* (Richardson 1998) came into existence on Earth at least 130 million years ago, perhaps as long as 225 million years ago, during or before the Cretaceous Period of the Mesozoic Era (Millar 1993). Pines apparently differentiated from earlier conifers in mid-latitudes of "Laurasia", the northern supercontinent that later separated into Eurasia and North America (Mirov 1967; Mirov and Hasbrouk 1976; Millar and Kinloch 1991). In North America, they first spread east and west in a broad temperate zone, then episodically north to arctic regions and south to present-day Mexico and Central America and to islands in the Caribbean Sea (Millar 1999). They also spread throughout Eurasia, barely touching the Southern Hemisphere in Sumatra, but nowhere did they make it further south (Mirov 1967). Prior to their recent introductions by humans, pines did not grow on the many Southern Hemisphere sites for which they are wonderfully adapted, simply because they had not previously managed to get there.

As with most species, exactly when, where and how radiata became a new species are not known. The immediate ancestors and closest relatives of radiata appear to have evolved between about 30 and 15 million years ago, in the uplands of present-day Mexico (Axelrod 1967a; Millar 1993, 1999). During this time, the pines that became the muricata, knobcone and radiata species (Millar 1986) migrated west from higher ground to near the Pacific Ocean. The present-day Sea of Cortez, which might have blocked such a migration, did not then exist (on many maps, this is called the Gulf of California). These ancestral pines, along with some cypresses, then migrated north-west and met Douglas-fir and redwood migrating south from their more-northern origins. Fossils of evolving radiata have been found in the coastal area of present-day California in various strata deposited between seven million and about ten thousand years ago (Jones and Stokes 1994), and biochemical evidence suggests that radiata or its immediate ancestors had evolved as a separate lineage somewhat before seven million years ago (Millar 1999). Thus, the radiata species appears to be about 30 times as old as the human species (cf Lerner and Libby 1976).

The coast of modern-day California has been geologically active during the past 10 million years, with the coastal strip of land bumping and grinding north-westward along the gigantic San Andreas Fault. Furthermore, during this period, the level of the ocean has fluctuated greatly as global ice ages came and went. Any species that lived near that coast had to be fairly mobile, migrating inland as the ocean rose and flooded the lower-lying land, and then back as the ocean level fell and the coastline moved out. Many of these species were also island-hoppers as islands were cut off from the mainland, or rose from the sea, or were inundated (Valentine and Lipps 1967; Weaver and Doerner 1967). Although these habitable sites were coming and going with geologic swiftness, their climate was equable, particularly compared to the much greater climate changes occurring during the last ten million years in more inland areas. Nevertheless the general California climate was changing to become seasonally hotter and drier, and radiata was adapting to these changes (Axelrod 1981; Millar 1999).

Fossils of radiata have been found from near Tomales Bay (north of San Francisco) to Chula Vista (near the Mexican border), and in many locations in between that are not currently occupied by radiata (Jones and Stokes 1994; Millar 1999). At least two scenarios are possible. In one, the radiata forest was much larger in the past than it is today, and today's three mainland populations are the greatly reduced remnants of that larger forest (Axelrod 1967a, b; Griffin 1972; Jones and Stokes 1994). In a second scenario, relatively small discrete populations have been more or less typical throughout most of the past million or more years of radiata's history. The modern populations of radiata (and probably others now **extirpated**^{G}) have migrated back and forth as more-or-less discrete populations, within the coastal region between and perhaps beyond the present locations of Tomales Bay and Chula Vista, up and down the coastal slopes, and perhaps on and off islands, as climates and sea-levels changed (Millar 1999). It is not currently possible to determine whether either or some combination of these two scenarios is correct, but the genetic architecture of modern radiata (Burdon 1992) seems to favour the second.

The pine fossils from these California mainland sites are mostly cones, twigs and needles, with some chunks of wood. In the strata including fossils from radiata forests that existed a million or more years ago, most of the cones are small and symmetrical and the needles are mostly in bundles of two. The radiata pines that today exist in the Monterey and Año Nuevo populations bear cones that are generally much larger than those fossil cones, and cones in the warmer and drier Cambria population are even larger. It has been suggested that the larger seeds contained in the large cones store more nutrients; these give germinating seedlings a better start in the harsh Cambria environment (Axelrod 1980). Yet the cones and seeds on the even harsher and drier Cedros Island are, on average, the smallest in the species.

With few exceptions, cones from the present-day mainland trees are strongly asymmetrical (Fig. 2.7) and **reflexed**^{G}, with large woody scales on the side of the cone away from the branch. Mature needles of modern mainland radiata are usually in bundles of three, with bundles of four or five being common on juvenile trees (as many as nine needles in a bundle have been observed). By contrast, trees on or from the two modern island populations of radiata have small symmetrical

Fig. 2.7 Cones of modern radiata, all from representative trees grown in the same common-garden plantation in California. Opened and partially-opened cones are in the upper row, and closed cones in the lower. There are four cones from each of the five populations, left to right, Año Nuevo, Cambria, Cedros Island, Guadalupe Island and Monterey. Note the average differences in cone size, cone shape, and thickness of the cone scales among the populations, and the variation in these traits among cones within each population. Each cone is from a different tree (1996 photo Iris C. Libby 1996)

cones and mature needles in bundles of two, with three-needle bundles being common on juveniles. A few trees from the island populations have cones that are larger, and/or have asymmetrical scales swollen on the outside, like those of the mainland populations. But the average cone sizes, cone shapes and needle numbers are clearly and distinctly different between the mainland and island populations, based both on trees growing in the native populations and on samples from them grown together in plantations (Moran 1996).

For several traits, evolution seems to have proceeded more slowly on Baja California's offshore islands than it did on the mainland, a sharp contrast to the situation Darwin and others found on the Galápagos Islands. In the Galápagos, evolutionary changes have occurred with dramatic rapidity; plant and animal species found on different islands in the Galápagos Archipelago differ greatly from each other, and from their more slowly-evolving mainland relatives. For many animal and plant species that currently exist on both California's mainland and the offshore islands, the island populations seem more like their fossilised mainland ancestors than like their current living mainland relatives (Mason 1932; several chapters and published discussion in Philbrick 1967 particularly the contributions of Raven). The pattern of fossil and modern radiata is consistent with this generalisation.

The greater frequency and severity of fire during the transitions between moist cool glacial periods and hot dry interglacial periods is currently considered to have been a major force affecting the evolution of pines, the pines evidently having been most abundant during these transition periods (Millar 1999). The evolution of larger asymmetrical cones with particularly large woody scales facing out seems to have been an effective fire adaptation, better allowing some of the seeds beneath the thick insulating scales to survive a hot fire. The trait of keeping their cones closed until they are opened by heat is not only a good adaptation to fire, but it has incidentally provided protection to radiata seeds as the cones floated to and from islands. Large fires are less frequent on the offshore islands than along the nearby mainland, in part because the islands are smaller targets for lightning starts, and in part because ground-strike lightning is less commonly generated over the cool offshore currents than in the updraft conditions of coastal mountains. There have, however, been forest fires in both the Cedros and Guadalupe pine stands during the past century (Libby et al. 1968). Whether these were human-caused or were ignited by lightning strikes from occasional summer tropical storms is not clear.

Larger, thickened scales may give some of the seeds protection against squirrels as well (Linhart 1978). Since squirrels and other cone-eating mammals are less common or absent on the islands, this may have also contributed to the differences in cone size and morphology between the island and mainland populations. It is also possible, even likely, that the reduction in summer rainfall that occurred during the last million years of the evolution of these pines selected for larger seeds in larger, more asymmetric cones, particularly in the Cambria population (Axelrod 1980; Millar 1999).

Why mainland radiata evolved an additional needle per bundle is anybody's guess. The current mainland and island populations differ in many other traits that cannot presently be easily studied in fossils, and the alternative forms of some of these traits in the island populations are likely to be useful as radiata is being domesticated.

2.1.3.2 The Naming of Radiata

The first attempt by Western science to acquire the pine from Monterey began in the 1780s. A gardener named Collignon (spelled Colladen or Colligon in some accounts) accompanied the French "Lapérouse" expedition of reconnaissance, exploration and biological collection. In September 1786, he observed and collected many plant species from the Monterey Peninsula. He later sent a package, including a mixture of pine seeds, cones and needles thought to be from Monterey, to Paris' Museum of Natural History (McKelvey 1994). The European naming of radiata did not begin until 1812, when the French taxonomist G. Loiseleur des Longchamp named that mixture of pine specimens *Pinus californiana*. Later, it became apparent that this sample contained parts of several different species of pine, some of which were from radiata. Because of this confusion, this name was not retained (Millar 1986). Another historical version has Loiseleur describing *P. californiana* from a tree in the Jardin des Plantes in Paris, grown from one of the seeds in a second package sent by Collignon directly to the Jardin. Because it had subsequently died and was thus unavailable for confirmation, the name assigned by Loiseleur was passed over (from Veitch's 1881 Manual, 1900 edn; Boardman 1996). Indeed, this botanical name has now been formally rejected (Anon. 2013a).

Thomas Coulter, an Irish botanist and adventurer, collected cones close to the beach at Monterey in 1829, and sent them to England. There is no record of any seeds being included in the cones or in the collection. In 1835, the species was scientifically described from the cones sent by Coulter and officially named by David Don, an English taxonomist. Don called it *Pinus radiata*, for the esoteric feature of radiating lines on the exposed faces of the cone scales.

As is so often the case with both scientific and common names, this species bore multiple names in the hundred or so years following its recognition by Europeans (Gordon 1875; Elwes and Henry 1910; Dallimore and Jackson 1954). For a time, *P. insignis* was its dominant scientific name. This name, which means "distinguished" or "remarkable," was given to it by David Douglas in 1831, to emphasise the rapid growth that makes the species remarkable (Coffman 1995). Douglas had made a thorough collection and set of observations of the pine at Monterey but, unfortunately, his collection and description of the trees lay in a herbarium box with his assigned name until it was accepted and published in 1838 by Loudon, another taxonomist (Millar 1986; Boardman,1996). According to the rules of taxonomy, the name *P. radiata* was validly published first and thus, after a bit of academic bickering, it was accepted over Douglas' and Loudon's better-described *Pinus insignis*.

Since 1835, various collections of herbarium material have been made from radiata pine in California, and various taxonomists have identified populations of other pine species as belonging in the radiata species, or have placed some of the radiata populations in other species. That now seems sorted out, and today it is clear that the five populations at Año Nuevo, Monterey, Cambria, Guadalupe Island and Cedros Island are all radiata, and that these five are the only extant native populations of *P. radiata*.

There has been some continued taxonomic quibbling about subspecific names (Millar 1986). The varietal names *P. radiata* var. *cedrosensis* and *P. radiata* var. *binata* are generally accepted for the pines from Cedros and Guadalupe Islands, respectively. The *binata* varietal name recognises the fact that the needles are in bundles (**fascicles**$^{\mathbf{G}}$) of pairs, which however is typical of the pines of both islands. The name *P. radiata* var. *macrocarpa* has sometimes been used for the pines of the Cambria population, in recognition of the much larger cones in that population. However, the three mainland populations are generally placed together, in *P. radiata* var. *radiata*. Nevertheless, as we will show in later chapters, radiata pines from the Año Nuevo and Monterey populations also differ appreciably in several traits, in some of which trees from Monterey are more like trees from Cambria than like trees from Año Nuevo. Rather than get tangled up in further taxonomic detail, it seems sensible to just remember that each of the five populations is non-trivially different from all of the others, and that they are all radiata.

While the name *P. radiata* has endured, several features of the species have been highly conducive to taxonomic confusion. In the genus *Pinus* the difference between needles being generally grouped in threes (on the mainland) and twos (on the islands) usually corresponds to substantial species differences. And the large tree-to-tree differences in size and shape of cones, which are among the favourite criteria of herbarium taxonomists, have also proved confusing. Indeed, even within the mainland populations confusion persisted among some taxonomists for nearly 100 years as to whether one or two species were involved (Forde 1964b). Despite the wide tree-to-tree variation, recent population-architecture studies show that unusually high percentages of the genetic variation occurs among populations compared with other conifer species, Thus among-population differences, apart from being confusing for taxonomists, are of special significance as sources of genetic variation.

This species was and is known by many common (non-scientific) names. The Native Americans and (perhaps) the Native Mexicans had their local names for these useful trees, as do Spanish-speaking, French-speaking, English-speaking and other-speaking peoples who now grow them. American-speaking people in California most often call them "Monterey pine," although "Cambrian pine" is sometimes used for those in the Cambria Forest. American speakers often drop the tilde from Año when discussing or writing about the Año Nuevo native population, which may bring a gasp or laugh from Spanish speakers. For the domesticating populations, "radiata pine" has become generally accepted, but is conversationally abbreviated to radiata.

2.1.4 Native American Impacts

While Native Mexicans no doubt visited and even lived on some parts of Cedros Island, particularly near the Cerros Mountain springs and along the coast, the pines are remote from these locations. Some reliable springs were close to the pines on Guadalupe Island, lying below the main population and fed by fog-drip from them. But the pines and springs on Guadalupe are a long hard climb from safe landing beaches and it is 250 km from the nearest point on the mainland. Thus, it seems likely that Native Mexicans rarely visited the pines on these two islands, and had little impact on them when they did.

Native Americans have seasonally or permanently lived in all three mainland populations for millennia before European activity there. Some areas had been occupied for at least 3000 years, others for perhaps as long as 30,000 years (Mann 2011). The Spaniards called all Native Americans living along California's central coast the Costeños, or "coast people" (later Costanoans or Costenones). However, the so-called Costeños were not a cohesive nation such as the Apache or Navaho. About 40 different groups wandered from shore to marsh to oak grove in this coastal region, hunting and gathering deer, birds, salmon, acorns, grass seeds and shellfish. They spoke 12–14 distinctive languages, and communication among the different language groups was limited. Different peoples inhabited each of these radiata populations, the Ohlone at Año Nuevo (B. McCrary pers. comm. to WJL 1996), the Rumsen at Monterey (Croft pers. comm. to WJL 1996; Engbeck pers. comm. to WJL 1996; Lydon 1996) and the Chumash at Cambria (Coffman 1995).

It seems likely that their cultures, and thus their interactions with the pines, differed not only among these tribes but also, over time, within them. We know little about them today, as their cultures and population numbers quickly collapsed following European contact (Howard 1972). The policy of the Spanish missions was to convert these pagans to Christianity, and part of this policy was to stamp out so-called pagan practices and all memory and record of them. While this policy was highly successful in obliterating their cultures, a combination of cruelty and disease succeeded in converting most of them from the living to the dead, rather than from pagan to living Christian (LeBoeuf and Kaza 1981).

These Native Americans no doubt used dry pine branches and cones for fuel, and they cut or basal-burned live pines for light construction or even for their great longhouses. However, their main impact on the evolution of these pines was probably through area-burning fire, in some cases used intentionally and skillfully, and in others as unintentional wildfires that escaped a campfire, religious ritual or small purposeful burn. In the hunting-and-gathering cultures of the coastal Native Americans, the skillful and purposeful burning was most frequently done by the women of the tribe. This practice preceded the arrival of Europeans, but did not cease upon their arrival. For example, a 1783 letter from the Governor of Alta California to the mission fathers at Carmel and Santa Cruz requested them to warn the Christian Indians, and particularly the old women, to curb their destructive practice of burning in the forest (Gordon 1977; Hillyard 1996).

The largest population of the Costeños, about 4000 at the arrival of the Spanish in 1769, was on the Monterey Peninsula. There, and nearby at Point Lobos, the Rumsen people set fire to the pines and burned them down to get at the seeds in their cones. In general, the pine seeds were obtained by roasting the closed cones, after which they were eaten whole or crushed into meal (LeBoeuf and Kaza 1981). Preferential felling of those trees with the larger cones and seeds might well have selected against the evolution of even larger cones at Monterey and Año Nuevo, compared to the situation at Cambria. They probably chose trees with delayed cone-opening (serotiny) as well. Their frequent area-burning fires would have intensified selection for serotinous cones. Such frequent fires would have been of lower intensities than previous natural fires, which were less frequent but often stand-replacing. These lower-intensity fires might have been particularly effective in selecting those trees with thicker bark (at least on the lower bole), a trait that is most developed in the Monterey population (Burdon et al. 1992a, b; Stephens and Libby, 2006). So, as in the story of The Three Bears, the fires needed to be neither too hot nor too cool, but somewhere near just right. Whether the coastal Native Americans figured this out, or whether fires of about the right frequency and intensity for the pines just happened, would be interesting to know.

The Ohlones living among and near the Año Nuevo pines were among the largest populations of the Costeños, perhaps second only to the population at Monterey. They enriched berry crops by burning the local blackberry sites within the pine population, and they also burned the coastal bench areas to facilitate deer hunting and favour the grasses used for basketry and gathered for their edible seeds. Escaped fires apparently burned well beyond these intended sites, and some of these frequently-burned areas were noted as being nearly free of pine in the diary kept in 1769 by Father Crespi, a member of the first party of European visitors to the area. He also noted that the smaller houses in the village at Whitehorse Creek were built of split pine wood (LeBoeuf and Kaza 1981), although the major structures there were probably built of redwood and/or Douglas-fir.

Whether and how much the Chumash used fire as a management tool at Cambria is uncertain, but some combination of natural fires and perhaps human-set fires is likely to have been a particularly important selective force in evolving the large asymmetrical cones that characterise the Cambria population. While the traits of the mature bark are not well known from fossil evidence, modern mainland pines differ from the island pines in that the mainland trees consistently have much thicker mature bark, at least on the lower bole, a likely adaptation to the more frequent low-intensity ground fires in all three mainland populations. Such fires would select for thicker bark, while infrequent but intense stand-replacing fires would little affect the evolution of bark thickness.

Although large, intense fires can and do occur in the native populations, the foggy climate there seems likely to make such fires unusual. Natural fire-return frequencies on the Monterey Peninsula have been estimated to have been in the range of 70–150 years (Cylinder 1996), while fires set by Native Americans were apparently much more frequent (Greenlee and Langenheim 1990). Adaptations such as the asymmetrical heavy-scaled cones and thicker bark may well have largely occurred as a result of the more frequent use of fire by Native Americans in these forests. A period of 3000 or so years seems ample to have **natural selection**^{G} favour fire-adapted trees. However, preferential felling of trees with larger cones for eating the heavier seeds would complicate the selection pressures.

2.1.5 Spanish, Mexican and Californian Impacts

While there are tales and some evidence of earlier visits to the Monterey region by ships from China (Howard 1972), Spaniards were the first Europeans to see and then extensively use the native radiata forests. And, of course, the Native Americans remained as an important presence after the arrival of the Spanish, and their influence on the forest continued both independently and as labourers on the ranches and settlements developed by the colonists.

The pines at Monterey and Año Nuevo had been sighted and recorded by Juan Cabrillo on 16 November 1542. Next, 60 years later, Sebastian Vizcaino and his crew spent two weeks ashore, studying the area near the harbour at Monterey (Encyclopedia 1979; LeBoeuf and Kaza 1981; Clark 1991), and making the journal notation: "… great pine trees, smooth and straight, suitable for the masts and yards of ships" (Larkey 1972). Father Antonio, chaplain aboard Vizcaino's ship, noted a headland as they sailed north on 3 January 1603 and commemorated the sighting by naming it "Punta de Año Nuevo," the Point of the New Year (since then, it has become Año Nuevo Island). However, a permanent onshore European presence didn't begin among the pines for another 160 years, when a Spanish base was established at Monterey in 1763 to counter the British victory over the French in eastern Canada (Weber 1992). Six years later, while trekking north-west from San Diego on behalf of Spain in 1769, Gaspar de Portolá noted and recorded the pine in the vicinity of Cambria. His expedition proceeded to and through the pines at Monterey, and later stopped in the middle of those at Punta de Año Nuevo. There they feasted on the abundant blackberries, which, according to Father Crespi, a member of Portolá's exploration party, had the joint effect of giving many of the men diarrhoea and curing the scurvy beginning to affect them.

As he passed near present-day Cambria during his 1793 coastal voyage, George Vancouver noted that "very large" trees "with spreading branches" were "distributed in detached clumps," indicating a fragmented pine forest well before substantial European use began in those stands. Very little of the Cambria pine population has escaped intensified human influence during the past 150 years. By the mid-1800s, the entire Cambria area was contained in a single Mexican "rancho." As deduced from an 1850 map prepared by Julian Estrada of his Rancho Santa Rosa, some of the pine forests then shown as continuous have later been substantially reduced or further fragmented. Sawmilling and lumber-exporting occurred some or all of the time in the latter half of the 1800s and continued until 1971, when the last sawmill in the Cambria forest ceased operation. However, the activities of cattle ranchers and dairy farmers, who cleared and burned the forest to increase pasturage, seem to have had a larger impact on the area occupied by the Cambria forest than did 100-plus years of locally patchy logging (Coffman 1995).

The naming of Monterey perhaps gives some insights as to how it was seen and then used (Clark 1991). When Juan Cabrillo observed the bay in 1542, he named it "Bahia de los Pinos," the Bay of Pines. He described this bay as extending from Point Pinos (the north-western point of the modern Monterey Peninsula) to what later became known as Año Nuevo Point, thus including part of the Monterey and all of the Año Nuevo populations of radiata in his Bay of Pines. Sixty years later, Sebastian Vizcaino renamed the bay "Bahia de Monterrey," in honour of Gaspar de Zuñiga y Azevedo, then Viceroy of the Americas' New Spain. Azevedo, also the 5th Condo de Monterrey, had been born in Monterrey, Spain, a city at the foot of a forested mountain named Monte Rey, or Mountain of the King. The extra "r" was later added to be sure people pronouncing it gave emphasis to "the KING." (The New Americans later de-emphasised the king.) But there may be more to this than the title and birthplace of Vizcaino's commander. Vizcaino may have thought it particularly appropriate to commemorate his viceroy at this place because the wooded hill on the peninsula resembled the wooded mountain near Azevedo's home. Later, in 1850, Mariano Guadalupe Vallejo gave some weight to the idea that the presence and nature of the pine played a part in the naming, noting that another meaning of "monterey" is "king of the forest," and possibly also "King's wood." In Vallejo's opinion, the name was not only honouring Vizcaino, but "honoring also the neighboring forest of massive pines and other trees."

After arrival of the Spaniards, human use of the Monterey radiata forest became the most intense of that at any of the five native populations. The easy access to good timber influenced the siting of the Spanish base there in 1763. Father Junipero Serra, head of the California missions, arrived and began settlement on 3 June 1770, establishing the Monterey and Carmel missions in 1770 and 1771. Spain designated the town of Monterey as capital of both Baja (Lower) and Alta (Upper) California in 1776. Following Mexico's 1822 secession from Spain, it remained capital of both regions under Mexican rule until 1846, when a naval expedition captured the town for the United States. This occupation lasted for two years, until all of Alta California became part of the United States in 1848. The oldest known radiata on the Monterey Peninsula in 1994 was 192 years old (Nedeff 1994), indicating that most or all extant pines there have begun their lives since the Spanish colonization.

The good harbour and excellent offshore fisheries had quickly resulted in the town of Monterey becoming a major fishing port and seafood-processing city, well described in John Steinbeck's famous 1945 novel *Cannery Row*. These activities resulted in the radiata forest in the vicinity of the town of Monterey being logged more frequently than any other forest in the western United States. Those radiata stands entered by selective loggers, in the Spanish era and later, typically suffered severe **highgrading**^{G} (called "creaming" in New Zealand), as the straightest and strongest trees were selectively cut for shipmasts and fort palisades, and for building construction ranging from mission buildings to canning factories (Curator, Carmel Mission pers. comm. to WJL 1962).

Nearing the end of the Spanish and Mexican colonization, 300 years after its sighting by Cabrillo, R.H. Dana described the Monterey area in his 1840 *Two Years Before the Mast* thus: "… the shores are extremely well wooded (the pines abounding upon them) …" (Clark 1991). An 1827 watercolour picture, however, indicates few pines in a landscape on the Monterey Peninsula where now there is heavy forest (Hillyard 1996). During Mexican rule, much land on the peninsula had been cleared for pasture (Fig. 2.8), as hides and tallow became important trade items from the 14 ranchos located in the Monterey region. Perhaps Dana and the watercolour artist were viewing different parts of the Peninsula—or perhaps Dana unknowingly described the aggressive retaking of abandoned grazing land by young pines.

In the 1860s, most of the pines on 2800 ha in the north-western part of the Monterey Peninsula were cut and milled at the Sawmill Gulch sawmill (Larkey 1972; Jones and Stokes 1994), perhaps providing easy picking of cones for seed export during that period. Beach whaling, beginning in the early 1860s, was sufficient to make the beaches near the present Pebble Beach Golf Course "white with bones" (Larkey 1972), and cones may have also been easily gathered there from the pines cut and burned to render the whale oil.

Later, fuel was needed for trains, and to provide heat for the homes in town, for the military presidio and for the canneries (Fig. 2.9); pine cordwood was the

Fig. 2.8 Present-day "New Monterey," including the future site of the Monterey Aquarium in the right foreground. In this 1880 photo, a few pines remain in the developing town, and near the ridgeline, with pasture in between. (Photo C.W.J. Johnson, permission Pat Hathaway Photos, Monterey)

Fig. 2.9 Stacked radiata cordwood, probably used to fuel trains and/or the canneries. (1901 photo C.A. Culp, permission Pat Hathaway Photos, Monterey)

main fuel on the Peninsula until about 1910 (Dunning 1916). Entire forest stands were clearfelled during these decades to create additional pasture and for fuel. On the main part of the Monterey Peninsula, cutting of the pine was unregulated and often heavy prior to 1896 (Dunning 1916). The railroad had arrived in Monterey

Fig. 2.10 An 1880 photo of just beyond the Monterey train depot. Seed collectors during the 1850s, 1860s and 1870s might well have stayed in Monterey, and collected cones from the scattered trees in the pastures above town. (Photo C.E. Watkins, ca. 1880, permission Hathaway Galleries, Monterey)

Fig. 2.11 An 1885 view across the town of Monterey from the Presidio, showing thoroughly cleared land, grazing cattle, and scattered pines in and near the town. (Photo C.W.J. Johnson, permission Pat Hathaway Photos, Monterey)

in 1874, and was later extended elsewhere on the peninsula (Fig. 2.10). Thus, rather than the earlier **dysgenic**[G] selection of the mission and fort period, chance and location often decided which few trees would remain following such forest clearing (Fig. 2.11). Some of these surviving trees were later to become the parents of extensive new pine forests that followed the major fires that occurred on the Monterey Peninsula between 1900 and 1907, and that invaded the former grazing land as it was converted to urban and suburban homesites. The Del Monte Forest, occupying much of the peninsula, was about 80% forested by 1915, about a quarter of that forest being about 15 years old owing to the 1900 fire. However, over a third of that forested area was composed of uniform stands of 40–60-year-old pine, these mostly having regenerated naturally after the extensive logging in the 1860s and 1870s (Dunning 1916).

Not all the native radiata stands in the Monterey population experienced these fates. For example, the forest on the Old Capitol Site remained a forest when plans to build the capitol of the brand new State of California in Monterey fell through, and the seat of state government moved to San Jose, then to Benicia, and finally to Sacramento. This forest was burned in one or more of the 1900–1907 conflagrations (Jones and Stokes 1994), but seeds from the burned cones had established a new native forest of radiata by 1910. Similarly, the radiata pines in Pescadero Canyon, near Carmel, grow on the poor soils of a series of elevated marine terraces. It is unlikely that builders in search of good timber would have chosen many or even any of the relatively small pines growing on those soils.

In 1915, Samuel F.B. Morse arrived to administer the development of much of the Monterey Peninsula. He had the vision to keep the forest intact during that development. (However, following World War II, from 1946 through the 1950s, sawmills in Pacific Grove and Carmel cut local radiata for home construction—Jones and Stokes 1994). By 1970, some 2000 homes were nestled within the forest matrix, along with three golf courses and such amenities as the Pebble Beach Lodge and equestrian facilities. In 1971, the S.F.B. Morse Botanical Reserve was dedicated in the upper part of Pescadero Canyon to preserve the enduring qualities of the forest so valued by Morse (Larkey 1972).

It is now unclear which stands within the native populations are still a good sample of the radiata forest as it existed before the arrival of humans. The genetic constitutions of most stands were probably changed substantially by the Spanish-era highgrading, or by new stands on cleared land being founded by only a very few parents. As one example, the present-day Point Lobos State Park has an essentially continuous stand of radiata, and one might think it has been preserved as a good sample of the native forest. However, not only did the Rumsens burn the pines to get at their cones on Point Lobos, and later engage in ranching there under the control of the Carmel Mission, but gold prospecting, granite quarrying, shore whaling, abalone canning, dairying, farming and other disruptive human activities, followed by several decades of complete fire suppression, occurred within the boundaries of this presently-protected "native stand" (Drury and Neasham 1954).

The whalers, about 20 Portuguese, rendered the whale oil on shore between 1861 and 1884, an important period for the export of radiata seed. While not specifically mentioned, it seems likely that pines would have been cut and used as fuel to boil the whale oil, as well as to provide wood for other purposes in the whalers' small village. The radiata seed sent overseas during this period might have been easily collected there, as well as near the Pebble Beach shore whalers, in Sawmill Gulch, and near the town of Monterey. On Point Lobos, the whalers' pigs, goats and cows browsed through the forest, and land was cleared for gardens. Monterey cypress (macrocarpa) trees were planted near the whalers' cottages in 1875 (Drury and Neasham 1954), raising the possibility that pines may also have been planted (perhaps dug locally, perhaps not). Coal was later briefly mined in the area (1888–1896), but after the whaling and seed-export period.

In the 1890s, Japanese built a small town and cannery on the Carmel Bay side of the point to support abalone diving and canning, an enterprise that lasted until 1928. Much of the rest of the point was devoted to dairying and some farming between 1897 and 1930. Beginning in 1916, several movies that required a number of small forest fires to be set were made on Point Lobos.

In 1933 Point Lobos was made a State Park, and many of the buildings were demolished to return it to a more natural state—but during World War II in the early 1940s the U.S. military set up coast defences there (Lydon 1996). Following these various disturbances, this tough radiata species retook cleared parts of the point (Fig. 2.12), and it now appears to be a natural forest.

Another present-day forested park, at Jack's Peak, was in historical times largely open ground. In 1994, an elderly Monterey resident recalled planting (pers. comm. to WJL) radiata seedlings there as a boy, for a Boy Scout project. There is also a substantial plantation of now-large radiata and macrocarpa cypress trees

Fig. 2.12 Aerial view across Point Lobos and Carmel Bay with the Monterey Peninsula in the background. This 1949 view shows Point Lobos after the military returned it to full park status, at a time when the pine forest is retaking the point. One may see that the main Monterey Peninsula is nearly fully reforested. (Photographer unknown, permission Pat Hathaway Photos, Monterey)

Fig. 2.13 A 1905 street scene in Carmel (corner of Eldorado and Acacia), showing radiata trees planted in 1903. (Photo R.A. Cohen, permission Pat Hathaway Photos, Monterey)

near the inland edge of the population high above Point Lobos, with a mystery as to who planted it and why. Individual homeowners and municipalities have, since about the turn of the century, replanted radiata near homes on cleared properties and along streets (Fig. 2.13). The California highway department has certainly planted non-native radiata along highways within the native stands since the mid-1900s. Many of the seeds for the mid-20th-century and later urban and highway plantings came from New Zealand plantations.

Although we have learned of these examples of documented and likely radiata planting within the Monterey population, such activities were unusual before about 1900. Furthermore, despite such post-1900 plantings, almost all the present-day mature forest in the Monterey population was established by natural regeneration, particularly following the 1901–1907 fires, with the trees that were left during earlier logging and land-clearing serving as the parents. Thus, the radiata pines in the present forest at Monterey, while in some cases are the offspring of trees in essentially undisturbed stands, are more likely descended from the survivors of earlier dysgenic logging, or from trees left on the edges of cleared pastures or of clearfell logging. Important for our story of early domestication, most of the seeds that were collected from the Monterey population prior to about 1880, which gave rise to the land races in Australia, New Zealand and elsewhere, came from trees that had regenerated naturally in the native populations prior to these dysgenic influences.

The story at Año Nuevo is quite different (LeBoeuf and Kaza 1981, B. McCrary pers. comm. to WJL 1996). The major genetically-important differences among the three mainland radiata populations were probably influenced by the abundant redwood and Douglas-fir near Año Nuevo, the scarcity of these two species near Monterey, and their absence near Cambria. Near Año Nuevo, from the earliest Spanish mission-builders on, those seeking timber for missions, forts, general building construction and many other uses, would have chosen the straight durable redwoods, or the strong timbers of Douglas-fir, over the generally crooked, decay-susceptible, and only moderately strong radiata. Thus, while highgrading and its attendant dysgenic effects tended to dominate several cycles of radiata logging at Cambria and Monterey, where radiata was effectively the only available species, highgrading by Spanish, Mexicans and Californians probably affected the genetic constitution of the Año Nuevo radiata population little, if at all. Earlier construction of longhouses and other structures by the Ohlone peoples may have creamed the best radiata trees, but this is speculation.

The Spanish mission, built at nearby Santa Cruz during 1791–1794, soon began large-scale cattle raising, and by 1814 there were 2900 head of cattle in and near the pines. Russians were also active there hunting sea otters, and the Portuguese engaged in shore whaling from this region between 1862 and 1895. After Alta California became American, Isaac Graham (a cousin of Daniel Boone and himself a famous frontiersman) bought Rancho Punta del Año Nuevo, and in 1851 established a sawmill. There was extensive logging of redwood and Douglas-fir nearby, and milling of these species within the Año Nuevo population by the New Americans from about 1860 on. William Waddell milled and exported lumber from a harbour and dock near the middle of the radiata population from 1864 through the 1870s, but little of it was radiata. (Waddell himself was killed on the property by a grizzly bear in 1875.) In 1908, a railroad was built to and beyond the small town of Swanton, within the Año Nuevo population, with steep grades that needed Shay engines to bring more-distant redwood and Douglas-fir down from the mountains of the Coast Range. Logging continues within this population to the present, as does occasional milling of the native Año Nuevo radiata.

In 1861 the Steele brothers, a leading dairy family, leased and then bought 2800 ha in and near the pines. Using mostly Native American labourers, they produced cheese and butter. In the early 1860s, during the American Civil War, they made a gigantic wheel of cheese 6 m in circumference, 46 cm thick and weighing 1750 kilograms. This was then successfully transported from Año Nuevo to San Francisco by road, where it was carved and auctioned, with proceeds going to the forerunner of the Red Cross. Thus there must have been a respectable road in the 1860s, and seeds could have gone by the same route from the pines at Año Nuevo to scientists and seed dealers in San Francisco.

While some radiata trees of questionable origin have been planted along the highway, the current radiata component of the forest at Año Nuevo seems likely to be descended with little modification from pre-European native forest. Although the New Americans probably changed the genetic structure of the radiata forest very little at Año Nuevo, they did change its shape. Some areas of forest were converted to pasture or cropland. Other areas returned to forest as the New Americans ceased the burning practices of the Native Americans and instituted instead fire protection and fire suppression. For example, the barren area noted by Father Crespi now supports a substantial stand of naturally regenerated radiata. More recently, the percentage of radiata has increased in some of the mixed stands, with young pines encroaching on areas previously occupied by the preferentially logged Douglas-firs and redwoods in those stands (Forde 1962, 1966; Mirov 1967). A recent population study of turpentine composition (Burdon et al. 1997) suggested that rather than being relatively stable, this population had been reduced to a limited "core" area from which recolonisation has occurred.

On Cedros Island (Libby et al. 1968; plus interviews by those authors with local Mexican officials), there has been some early-1900s exploration and mining for copper and gold in the mineralised rocks in the canyon below the northern subpopulation. Some nearby pines were cut for mine timbers and probably for fuel. But relatively few trees were taken, and the Spanish and Mexican owners of Cedros Island seem to have influenced these pines only a little.

In sharp contrast, the recent influence of humans on the Guadalupe Island pines, while indirect, has been devastating. The island was sighted and recorded during Vizcaino's 1602 voyage, visited by galleons returning from the Philippines en route to Acapulco on the Mexican mainland, and charted and surveyed in 1837 by a French expedition (Moran 1996). The pines were not "officially discovered" until 1875 (Elwes and Henry 1910, Millar 1986). There was an old ranch, or perhaps several ranches, on Guadalupe Island near the springs, and the pines were probably used by the ranchers. The tenure and extent of the ranches are not clear, and only remnant adobe ruins remain (Moran 1996).

The most important human-mediated event occurred in the early 1800s, when whalers released goats on the island. Their purpose was to have a supply of fresh milk and meat available on subsequent visits, with little likelihood of crew jumping ship to such an inhospitable place. By 1880, these goats had multiplied to exceed the carrying capacity of the island, numbering in the hundreds of thousands and concentrating near the springs below the pines. Mice arrived in 1875, and these ate and continue to eat the seeds of the pines (Moran 1996). In 1893, there was still evidence of dense pine forest that had clothed most of the upper north-western slopes, and of one or more large forest fires (Moran 1996). Except for two saplings growing on an outcrop inaccessible to the goats, for over 100 years no new pine seedling was able to grow large enough to be safe from the goats. Meanwhile, the large pines died of various natural causes. The dense forest of radiata that was observed a century or so ago on Guadalupe Island was reduced to scattered survivors. The springs that might have supplied visiting Native Mexicans with fresh water in earlier times became unreliable or completely dry; the fog drip that charged their aquifers waned as the large pines died and the survivors' seedling offspring fell victim to the starving goats (Libby et al. 1968; Moran 1996).

There is recent good news for the local flora, including the pines. Successful efforts by the Mexican Government to eliminate the goats have resulted in many young pines regenerating among the old survivors (Fig. 2.14).

Fig. 2.14 Young seedlings thriving on Guadalupe Island in 2008 after elimination of goats in 2004 (Courtesy Deborah L. Rogers)

2.2 Early Institutional and Political Factors in Plantation Forestry

In the 1800s, the earliest collectors of most plant species were professional botanists. After studying what they needed from the collection at hand, they often exchanged seeds and plants with fellow botanists. The more promising species among their collections were usually sent to interested fanciers and progressive landowners, or to commercial nurseries. Later, for some of those species, commercial seed merchants became involved in larger-scale seed collections. Records were generally better kept and maintained by the botanists for their earliest collections than for subsequent exchanges and for commercial collections. In general, this pattern was true for radiata.

While a few organizations planned and established plantations of radiata before 1900, this activity generally began in earnest in the first half of the 20th century. In the latter half of the 19th century, the important institutions affecting the later development of plantation forestry with many species were those that acquired the seed and planted the first generation or so of the tree species grown in each region. These early trees not only drew attention to radiata as a candidate for the 20th-century plantations through their relative performances, but also often supplied locally-grown seed for later plantations. In these latter cases, and particularly for radiata, they became the founders of each region's land race(s). Thus knowing their origins is important for understanding the subsequent domestication of radiata.

In Australia, New Zealand and later Chile, "old boy" networks among botanic gardens, and then seed dealers, nurseries and landowner enthusiasts, served to move and establish these early radiata collections across large regions in each country. In particular, landowner enthusiasts responding to the then-current craze for new conifers (Shepherd 1990) drove the early introductions. Typically, it began at specimen-tree scale, in parks and gardens, and as trial windbreaks. This was followed in a few years by some landowners establishing plantations including this remarkable pine. Governments then became formally involved, often in response to perceived future timber shortages and the demonstrated success of the pioneer plantations.

2.2.1 Australia

The colonial history of the continent (1788–1900) is important background to the region's forest policies and administration; until the end of this period Australia was a set of essentially autonomous colonies. Of these colonies, South Australia was the first to have a forest service of its own—one of the first public forestry departments in the British Empire. It was formally established in 1883, and promptly engaged actively in plantation forestry. Victoria, however, did have as State Government Botanist Dr Ferdinand von Mueller (Home et al. 1998), based at the Melbourne Botanic Garden, who enthusiastically promoted radiata (http://adb.anu.edu.au/biography/mueller-sir-ferdinand-jakob-heinrich-von-4266) and who also had influential connections. He owned property in the Mt Lofty Ranges of South Australia, where F.E.H.W. Kirchauff, a Member of the House of Assembly of that state, was his neighbour.

We will now address the story in South Australia, leaving development of institutions in the other states until the next chapter.

Even before human impacts, less than 1% of South Australia supported closed-canopy high forest. Aboriginal practices then affected these and other native Australian forests in important ways, particularly through the extinction of some of the large animal species (Flannery 1994); but the early activities of European colonists depleted the already very restricted South Australian forests at an alarming rate. Just 34 years after the founding of the colony, attention of some far-sighted people was focused on the lack of natural forest resources. In response, an initiative to establish a forest service came from the Surveyor-General, G.W. Goyder, and in 1870 this proposal was carried successfully to the South Australian Parliament by Kirchauff (Lewis 1975).

Among other things, Kirchauff's legislative action creating the forestry department clearly included the establishment of plantations as a supplement to natural regeneration in designated forest reserves. An *Act to Encourage the Planting of Forest Trees* quickly followed, in 1873. In 1875, Parliament created the Forestry Board, with Goyder as Chair, and put 79,000 ha in its care. South Australia at this point was the first of the colonies in the South Pacific to establish an organization charged with the development of timber plantations. The Woods and Forests

Department replaced the Forestry Board in 1882, becoming the first state organization on Earth to designate plantations of exotic species as the primary source of forest products.

It was soon decided to grow quality timber on the forest reserves, with preference to be given to good species with rapid growth. By 1881, enough had been known to focus on "Mediterranean-climate" pines (maritime pine (*Pinus pinaster*), Corsican pine (*Pinus nigra*), Canary Island pine (*Pinus canariensis*) and Aleppo pine (*Pinus halepensis*)) as being well adapted to quickly produce good-quality timber on many of South Australia's forest reserves (Lewis 1975; Boardman 1988, 1996).

2.2.2 New Zealand

The islands that are now New Zealand were settled by Maori, evidently beginning around 800 years ago. During the following several centuries, the Maori population increased greatly, and severe impacts on the native fauna and forest came within 200 years (Halkett 1991). This began in earnest with Maori using fire for hunting moas, large flightless birds, including the largest bird surviving at the time. All nine species of moa were evidently hunted to extinction, which is thought to have been virtually completed for the giant species during the 14th century. Also, many other native birds were either hunted to extinction, or their habitats were so modified as to help cause their extinction (Flannery 1994). After the moa-hunting phase, some forest destruction may have continued partly because of accumulation of fuels from what the moas might have consumed if still present. The most severe impacts on the vegetation and birds probably came within 200 years of initial settlement, but there would have been continued forest clearance by Maori, largely to promote growth of bracken fern, of which the roots were a food source, and for ease of travel. Thus between 1200 and 1800, the native forest that had covered about 75% of the land area was reduced by a third (Sutton, 1991). However, the introduction of the true potato in 1772 or 1773 by Europeans led to a new round of forest destruction (Cameron 1964). Whereas the staple cultivated vegetable, the kumara, was suited only to special sites, the potato, being far more tolerant of cool sites, was grown in shifting cultivation in ash beds created by felling and burning of forests. Indeed, it is thought that the rate of such destruction may have matched that wrought subsequently during European settlement for farming.

Beginning in the 1790s, a few Europeans lived with the Maori or in small enclaves. The first were mostly spar-cutters and other timber cutters (Simpson 1973), plus a few deserting sailors, and then came increasing numbers of sealers, whalers, missionaries, traders and early settlers. In 1840, some of the Maori chiefs and the British Crown signed the Treaty of Waitangi. Both the treaty and the events that precipitated it catalysed large-scale immigration, largely from Great Britain. The 1840s and subsequent years featured settlers attempting to get clear title to real estate, and then building homes, clearing land, establishing crops and herds, and dealing with the immediate problems of survival (Sinclair 1959; Belich 1994).

In several North Island regions, mutual distrust between colonists and Maori was building. The 1858 census marks the first time colonists outnumbered Maori. The Maori Wars of the 1860s particularly disrupted development in central North Island, but the South Island was essentially free of such preoccupation. Even so, there were numerous landowners in both islands who for various reasons were actively planting exotic tree species. Among those species radiata featured from after about 1860, albeit almost incidentally at first.

The New Zealand political character was evolving rapidly during this period, which included increased and often organised introduction and generally informal trials of many exotic forest-tree species. Although forestry and politics are frequently indirectly linked, one noteworthy happenstance closely linked the two in 1876. A forests bill had been introduced in 1873 by Julius Vogel, the aggressive and expansionary national treasurer of that era. Vogel had hoped that good husbandry of the forests, already important national assets, would strengthen the country's development (Roche 1990). While the bill was passed in 1874 it was greatly weakened by provincialist politicians. This was doubtless among the factors that led Vogel to manoeuvre the abolition of the provincial governments in 1876, leaving a single national government. This contrasted strongly with the political development of Australia, among other things resulting in the early centralization of forestry research in New Zealand and its fragmentation in Australia. By the end of the 19th century, New Zealand was perhaps the most radical nation in the world, described as "a laboratory in which political and social experiments are every day made for the information and instruction of the older countries of the world" (Sinclair 1959). This general willingness to innovate had effects on forestry as well as on many other New Zealand institutions.

To satisfy the colonial settlers' demands for cropland, pasture and wood, as was the pattern elsewhere, the most accessible forests were cleared. Although there was then substantial demand for timber for construction and export, the pace of forest conversion was such that only a small percentage of New Zealand's forest was harvested for timber during the 19th and early 20th centuries. A much larger percentage was simply felled and burnt to make way for pasture. In such circumstances, little institutional attention was directed to reforestation with native species, much less to establishing plantations of exotic species. However, a provincial bill in support of tree planting in Canterbury had its purposes extended to the entire country by the 1871 *Forest Trees Planting Encouragement Act*. The central government established a small forestry department with the successful passage of the *1874 Forest Act*, but with the relative abundance of apparently inexhaustible native forest, the Forestry Department was not supported well and was disbanded in 1888 (Roche 1990, p. 98).

In 1892, the Minister of Lands was also made Commissioner of Forests, and four years later a State Forest Branch was set up in the Lands Department. Its primary duties were modest afforestation and operation of nurseries. Its nurseries were both for its own plantings and to supply the public as a way to encourage private tree-growing (Allsop 1973). In the 1890s, the government took its next tentative steps in evaluating exotic forest species, by planting research plots in the

Otago and Canterbury regions of South Island, and then beginning the Whakarewarewa Forest near Rotorua on North Island. These trials focused on proven timber species, such as Corsican pine and larch from Europe, and Douglas-fir and ponderosa pine from North America. At Whakarewarewa, radiata pine was included as a shelter tree, protecting the valuable young seedlings of the more favoured species in the trials (Hegan 1993).

From very early days of intensive European settlement private commercial plant nurseries were established readily, doubtless in response to lively demand.

2.2.2.1 Canterbury

Unlike in the rest of 19th-century New Zealand, where there was an abundance of timber in the native forests, there were problems of timber supply in the eastern part of South Island. Also, various of the British landholders (freeholders or leaseholders) in this region had arrived with sufficient capital, and could afford the start-up expenses of planting substantial numbers of trees. Moreover, in that windswept region, establishing trees to shelter homesteads was often a pressing need. Furthermore, there was often enthusiasm for collecting recently discovered tree species, and conifer species were favoured because their seeds were usually relatively easy to import in a viable state.

In a response to the local timber shortage, and encouraged by the 1871 Canterbury *Forest Trees* bill, many plantations of species better-known than radiata were established in this region in the 1870s. By the 1890s, however, the first sawmilling of surprisingly large young radiata pines that had been used in shelterbelts or included in mixed plantings took place and in response, some commercial radiata plantations were being established in various localities (Burdon and Miller 1992).

2.2.2.2 The Wellington Botanic Garden

Sir James Hector arranged for most of the currently-known well-documented importations of radiata seeds into New Zealand during the late 1860s and the 1870s (Burdon and Miller 1992). While serving as Director of both the Wellington Botanic Garden and of the Geological Survey, he used these joint offices to effect an evenhanded distribution of these early radiata acquisitions throughout New Zealand (Shepherd and Cook 1988; Shepherd 1990). This was professionally brilliant, and surely served to distribute similar proportions of the two native provenances to many locations. Although they had been preceded by some significant introductions, Hector's distributions are likely to have been much the most important. His country-wide distributions of seed and/or planting stock accounts well for what appear to be similar proportions of Año Nuevo and Monterey ancestry among the regional land races within New Zealand (Burdon 1992; Burdon et al. 1997).

Several groves and individual trees from the 1870s acquisitions of radiata still exist in the Wellington Botanic Garden. These are surely among the best, and are probably **the** best, of documented samples of the founding generation of radiata in New Zealand. How appropriate that they are on the site from which Hector made those early distributions! Unfortunately, the records do not clearly tie particular plantings to particular native origins. However, with molecular fingerprinting techniques now available, they may soon be used to tie these century-old veterans to the precise native stands where they were conceived as seeds and in whose environments they evolved.

2.2.3 South Africa

South Africa had very small areas of original native forests, and much of those forests had already been destroyed by the 1870s (King 1938). Despite the limited history of forestry, or because of it, a Superintendent of Plantations was appointed at Cape Town, in 1875, in recognition of the need for plantations. The Cape *Forest Act* was passed in 1888. Wattles (*Acacia* spp.) were also planted for fuel and huts from 1898. The Woods and Forest Department of Cape Colony made vain efforts at natural regeneration of commercial native forests (King 1938). Otherwise, they were evidently monitoring plantation areas more than being actively involved in species testing and afforestation.

Meanwhile, early plantations of exotic tree species were being established, mostly by private industry, whose plantings commenced in 1870—wattle for tan bark, and wattle and eucalypts for mining timbers (Sherry 1971).

No date of radiata's first importation has been traced, and thus when, how and from where it arrived remains unknown. It seems that radiata arrived there before it got to Australia and New Zealand, since radiata trees were producing seed there by 1857 (Poynton 1960). While small plantings and individual trees of radiata did well in much of the Cape Colony no major planting of it appears to have been done in the 19th century. The first record of establishing an actual plantation of radiata was in 1885 at Tokai Forest Reserve, Cape Town (Donald 1993). Little other concrete information seems available for this period.

2.2.4 Chile

One might think, with the substantial ship-traffic stopping in Chile on the way to or from California, and with the trading directly between Chile and California in the mid-1800s, that radiata would have been repeatedly introduced to Chile during or even before this period. In puzzling about this, J.S. Krebs (1973) invoked a quote from James Michener's 1968 *Iberia*: "It was unusual to find a Spaniard who spoke with such love of rivers and meadows and mountains, for Spain more than any other European nation has abused its land." The Spanish emphasised livestock raising, and other aspects of agriculture, but colonial Spain had very little if any record of afforestation. However, there was more to it than that. Unlike South Australia and the Canterbury region of New Zealand, there was abundant high-quality timber available in Chile. Furthermore, pre-Columbian native building construction (Mann 2011) was only slightly modified by the Spanish tradition of building with mortar and stone. Thus, in spite of the relatively much greater danger of collapse and injury associated with mortar and stone buildings during earthquakes, the demand for wood in building construction was not great.

The dominant culture in Chile in the mid-1800s was Spanish. The Spanish did bring their native maritime pine (*P. pinaster*) early, and it was widely planted. Furthermore, it frequently reproduced naturally in the vicinity of the planted trees, and did this so often that it was soon considered by many to be native to Chile. Even if a few radiata pines did make it to Chile during this period, it seems their fate was to be largely unappreciated or ignored (Krebs 1973).

Although the Spanish had been in Chile for many years, the area south of the River Bío Bío long remained the frontier, under the control of the native Araucanians. This also was the approximate northern edge of continuous forest, unattractive and even repellent to Spanish more comfortable in the Mediterranean-climate woodlands and chaparral further north (Krebs 1973). In 1881, the holdout Mapuche tribe finally surrendered at Nacimiento and the conquest of the Araucanian frontier was complete (Clapp 1995). Meanwhile, beginning in 1856, a substantial German settlement had been established farther south, in Valdivia, with its people and influence soon expanding elsewhere. Shortly thereafter, experimentation in Chile with exotic trees began in earnest (Junge 1953; Clapp 1995), as did a greater use of wood in building construction. While preoccupation with the native Araucanians perhaps delayed attention to reforestation and afforestation opportunities in Spanish-culture Chile, German-culture Chileans were at the forefront of these new ventures.

2.3 Early Plantings

People introduce non-native plants and animals for a variety of reasons. Among these reasons are one-upmanship among fanciers, curiosity, aesthetic perceptions, need, hope of economic gain, nostalgia and professional activity. Whatever the motives for original introductions, the follow-up introductions and planting generally depended on how the species performed, in terms of their growth, site tolerances and uses. With respect to motivations to introduce radiata in the 19th century, one-upmanship and curiosity seem to have dominated initially, to be slowly superseded by its ability to fill needs and eventually by hope of economic gain. As radiata became massively planted as an exotic its aesthetic appeal to people generally disappeared, despite the magnificence of some of the earliest plantings. While reports of the giant conifers of western North America often fired the imagination, those were just a few of the many conifers that had recently been discovered there. Elsewhere in the world, notably in northern India, other impressive conifers were being discovered too. It became fashionable for well-to-do British landowners to collect newly-discovered exotic conifers. And, while many were probably competitors, these collectors eagerly swapped seed and other propagules in order to broaden their collections (Box 2.2).

Box 2.2 Evolution of Planting Status
Radiata was initially planted as a botanical curiosity during a period when Europeans were discovering many "new" conifers. Many botanic gardens, and landowners eagerly collected such new discoveries, for specimen-tree plantings. In Australia and New Zealand, in particular, the success of specimen-tree plantings led to more extensive plantings for shelter and fuel, and eventually for producing timber. In Chile, this process was foreshortened, following a later initial introduction.

This fashion spread to the British colonies of the Southern Hemisphere. It doubtless ensured that many obscure and even unlikely species, such as radiata, were tried as exotics there. Curiosity seems a likely motivator, in both Britain and her colonies, but mostly for a relatively few well-informed fanciers. It must have powerfully complemented the role of the botanic gardens that were generally established promptly in those colonies. These factors doubtless ensured that many obscure and unlikely species, such as radiata, were tried as exotics alongside the wider-spread and more valued New World conifers such as redwood and Douglas-fir. Radiata's natural range was so small that most 19th-century colonial gardeners and farmers would not have even heard of this obscure and often unprepossessing pine.

European exploration and colonization was occurring along the Pacific coast of North America in the 17th and 18th centuries. The early trade routes from Europe to the New World and beyond generally had ports of call in Spain, and ships usually called in Chile and/or South Africa. It is curious that radiata was not effectively introduced into regions where it has now become an important plantation tree, particularly in Spain and Chile, until the middle of the 19th century or even later. It seems logical that new botanical findings from California would have been tried in those similar climates long before they were tried in Australia and New Zealand. Yet, surprisingly, they were not, at least not at an intensity that stimulated further plantings or the keeping of long-term records of the trials.

Most of the new British colonies had an abundance of native forests. Cutting these to clear land for homes and farms, incidentally using some of the felled trees to satisfy immediate needs for wood, was more a focus of thought and action than planting new species for future harvests. South Australia and New Zealand's Canterbury were the two colonial settlements that were short of timber almost from their founding, and thus forest-tree plantations were begun early there.

Hope of economic gain motivated the establishment of few if any forest plantations in the Southern Hemisphere countries in the 18th and 19th centuries, although such activity was then long-established in China, Japan and parts of Europe.

Nostalgia motivated many Southern Hemisphere colonists to bring or send for familiar plants and animals, mostly from their previous European homelands. Yet superimposed upon the nostalgic motivation was the strain of curiosity and collectorship.

Gold miners, heading to the newer mid-19th-century finds in Australia and New Zealand, were the majority of the earliest immigrants from California, and it has been speculated that some of them might have brought tree seeds with them. But miners in those gold-rush days were not typically motivated by nostalgic commitment to something as long-term as a tree nor by botanical collectorship, and the records do not point to their having played any part in the global spread of radiata.

So radiata was only incidentally included in the various activities that resulted in many species being introduced and extensively tried in many countries around the world. It was largely professional activity, in the form of botanic-garden and arboretum collections, followed by the networking connections to tree-planting enthusiasts, which gave radiata its first chances to excel.

For reasons we hope will become clear below, these early introductions have taken on enormous importance in the development of radiata as a domesticated crop.

During the 1850–1880 period, when several of the land-race-forming seed collections were sent to New Zealand from California, all three mainland native populations of radiata were known. More important, all were accessible, and various enterprises in and near them were engaged in active commerce and trade. Not only was Monterey a thriving and developing community, but a stage route went north to San Francisco through the heart of the Año Nuevo population, and Rancho Santa Rosa had long been engaged in trade from Cambria. During this period, trade was active between California and much of the rest of the world (stops by passing ships at Guadalupe and Cedros Islands were recorded, but infrequent, and the pine stands were inaccessible). Thus, radiata seeds could have been shipped directly to the various countries beginning to plant radiata, or through the intermediaries of seed dealers and botanic gardens in third countries. The accounts indicate some but surely not all of what in fact occurred.

There is some reason to wonder if, prior to 1857, goldminers and/or ship ballast might have passively brought seeds of radiata to Australia, New Zealand and elsewhere, most likely contained in closed cones and then released when the cones dried out and were exposed to heat (Harris 1991). Such passive importation, or even a purposeful introduction by a miner or other early traveler, has not been documented, nor would it have been likely to be reliably documented. The first documented trees and seeds did not come directly from California, but came from England and Scotland, in 1857 (Fielding 1957a, 1957b; Shepherd and Cook 1988; Shepherd 1990). At least some of them probably came from stored quantities of David Douglas' 1831–1832 seed collections in or near Monterey, or from cones produced by trees raised in England and Scotland from Douglas' seeds (Shepherd 1990).

Despite many records of importations of radiata seeds to various countries, very little is explicitly recorded concerning the native-population origins of land-race stocks. Laboratory results and circumstantial considerations point to land races having originated entirely from the two northernmost populations, Año Nuevo and Monterey, the contribution of the former being very disproportionately high in relation to the population sizes (Burdon et al. 1997 and references therein; Aragonés et al. 1997). Californian ancestry of domesticated populations, and perhaps even of individual trees, should become identifiable in greater detail using using recently-developed DNA markers.

2.3.1 European Seed Collections and Plantings

Douglas was not the first to send radiata seeds to Europe. During 14–24 September 1786, personnel from the two ships of the French "Lapérouse expedition" not only fulfilled orders to evaluate the Spanish military garrison and to make botanical and zoological collections at Monterey, but they also cut and loaded all the wood the ships could hold, which probably included some radiata. Three months later, on 27 January 1787, two packages containing the Monterey collections were dispatched by Collignon, the expedition's botanist, to Paris from Macao. A year after that, in February 1788, the two ships called at Botany Bay in Australia, but there is no record of radiata seeds being left there. Soon afterwards, both ships sank in the Solomon Islands, and all on board were lost (McKelvey 1994).

The radiata seeds from Monterey in the package sent by Collignon to the Paris Museum were mixed up with collections from other species. One plant, recorded as from this Paris Museum collection and identified as *Pinus californiana*, was still living outdoors in the Jardin des Plantes in 1812. It seems likely that this tree from the museum collection was some pine other than radiata, perhaps Italian stone pine (Elwes and Henry 1910). However, Collignon's second package was sent directly to the Jardin des Plantes, and seeds from that package were planted there in 1787. Collignon's notes, apparently included with those seeds, indicate that they were collected "from the shores of Monterey Bay." Although their existence was perhaps unknown to the taxonomist Loiseleur, the 12 trees that grew from them were radiata. They, or perhaps their offspring, survived in the Jardin des Plantes until 1830, when they were killed by a severe spring freeze. In Loudon's 1838 publication taxonomically describing radiata and naming it *P. insignis*, he noted that another tree of *P. insignis* from that Collignon collection had been growing in Dublin, Ireland, and it was also killed in severe spring weather in 1830. However, seeds or possibly seedlings from the trees growing in Paris were planted in the Royal Horticultural Society's Garden in London about 1829 (Boardman 1996), thus continuing Collignon's collection line at least there and perhaps elsewhere as well.

At least a few of the radiata pines in and around London survived killing frosts in 1838 and 1839, and similar frosts in England killed or severely damaged even large radiata pines in 1867 and again in 1908 (Elwes and Henry 1910). The frosts were and are even more severe in central France. All of the radiata pines in or around Paris were killed by such frosts in the 1830s, and radiata pines planted near Paris in the mid-1970s suffered the same fate. The frequency of such radiata plantings is not clearly recorded for those generally mild regions of Europe, which, however, are subject to periodic cold-air masses from the interior of the continent. It is possible that some later radiata pines, descended through two or more generations from the Collignon collection in the Jardin des Plantes, Dublin and perhaps elsewhere, were available to produce seeds and offspring for some of the exports from various places in Europe during the period 1850–1900.

While botanical collections had been made during 1825 through 1828 in the Monterey region, it is not clear that any radiata seedlings from them were planted in Europe (Millar 1986). From some of the radiata seeds collected in 1831 and 1832 by David Douglas in or near Monterey, trees were planted in 1833 in and around London. Most of these young pines had died by 1839, but some at the Horticultural Society of London's garden in Chiswick and elsewhere survived those first winters (Boardman 1996). In 1838, Richard Forest's Nursery in Kensington, England, had *P. insignis* for sale at prices from 21 to 100 shillings each (Shepherd 1990), and young radiata pines were being sold in France and Germany as well (Bannister 1973).

Douglas, being a Scot, also sent seeds to the Edinburgh Botanic Garden. Most of the radiata from that shipment that were subsequently planted in northern England and interior Scotland died in the severe winter of 1860–1861, and later plantings in this region similarly failed. In the context of European experience with conifers, those planted near the western coast of Scotland, in Ireland and in Wales grew exceptionally well (Elwes and Henry 1910).

In 1839, K.T. Hartweg also sent a collection of herbarium specimens and seeds from Mexican conifers to the Horticultural Society of London (Boardman 1996). Hartweg's collection reportedly included *P. insignis* from "southern California" (possibly Cambria, which like Monterey was then still part of Mexico) and also the two-needled *Pinus binata* variety from Guadalupe Island. However, the lack of literature references to Hartweg in relation to that variety suggests that the reported collection from Guadalupe Island was of no ongoing significance.

It is possible that seedlings from the Hartweg collection, as well as those from the Douglas collection, were later planted in the Horticultural Society's "pine-tums" at Woburn, Hackney, Syon House, White Knights, Pain's Hill and especially at Dropmore in Buckinghamshire. These pineta helped spark a world-wide interest in the horticultural cultivation of pines and other conifers, and spread their use in shelterbelts and hedgerows (Boardman 1996).

The Dropmore radiata, planted in 1839, had reached 27 m height in 1891 after only 52 years, impressive by European standards. Interestingly, some of the radiata planted at Dropmore in 1839[2] were rooted cuttings, reported in 1882 to be somewhat smaller than the seedlings but otherwise growing as well as the seedlings (Elwes and Henry 1910; Bannister, 1973). By 1910, radiata had become a proven and valued tree in shelter and ornamental plantings in south-western

[2]It appears that nurserymen of the time would often endeavour to extend scarce supplies of novelty seed by multiplying seedlings as cuttings.

England, Wales and Ireland. Its high growth rate was attracting speculation as to its possible rough-timber uses, such as for pit props in mines, or even for finer uses from older closely-grown trees (Elwes and Henry 1910).

Two decades after the Dropmore plantings, seeds and small trees began to be sent to the Southern Hemisphere colonies in large numbers. In his 1966 manuscript, Joseph Ewan (1973) set the scene, as follows: "It is not easy to realize the zeal of the Englishman for his garden and stove a century ago. While orchids, aroids and other choice exotics were the prizes for the well-to-do, the middle class were gaining leisure and some affluence with the Victorian graces, cultivating their perennial borders, mulching the favorite conifer or tending the tender subjects in his glasshouse … William Lobb, then, was in the parade of collectors for great horticultural establishments led by the heroic David Douglas who was followed by Thomas Coulter and Karl Theodore Hartweg."

Lobb collected as an agent of Veitch who began his nursery operation in Exeter and purchased the Royal Exotic Nursery in Chelsea, London, in 1853. Lobb went first to South America, and later arrived in California in mid-1849, at the height of the gold rush. The following spring, Lobb and T.L. Andrews went to Monterey and the nearby Santa Lucia Mountains. According to Ewan (1973), "Veitch had instructed Lobb to collect seeds of several of the conifers discovered by Douglas and Coulter which were not yet introduced into English gardens." Certainly the high point of the year 1850 was his successful shipment of Santa Lucia fir to Veitch and its introduction into horticulture. It seems likely that Lobb would have included radiata seeds in that shipment.

In 1851 Lobb was mostly in the upper Sacramento River valley, where among other things he discovered incense-cedar. In 1852 he returned to northern California from the Columbia River through Oregon, collecting conifer seeds along his route. He was back in Monterey in 1853, where he learned of the discovery of giant sequoia, journeyed to the discovery site at North Grove Calaveras, and returned to England with the first seeds, seedlings and specimens of the most massive tree seen by humans before or since—his most famous collection. But did he also bring radiata seeds from Monterey?

Lobb returned to California in early spring of 1854, and from then through 1856 little is known of his whereabouts. He reportedly visited Cedros Island during this period, but the records are not clear on that possibility. He may have been ill with a tropical disease contracted during his earlier collecting period in South America. In 1857, he left the employ of Veitch and collected for Clapton Nursery in 1858. By 1862 he was no longer active, and he died in San Francisco in 1864.

In preparing his 1973 monograph on Lobb, Ewan was frustrated in obtaining correspondence between Lobb and Veitch that documented all of the seeds that were collected and shipped. Ewan noted the secretiveness enforced by Veitch on Lobb with respect to details of the species Lobb collected and shipped to England, and the evident destruction of Lobb's letters and the firm's records before it went out of business. That secretiveness has impeded scientists 140 years later from learning interesting and possibly important details of the work accomplished. Radiata (or its synonyms) is nowhere mentioned among those records of Lobb's collections from Monterey that still existed in 1965 (Ewan 1973), although the 1910 book *The Trees of Great Britain & Ireland* (Elwes and Henry 1910) implies that radiata seeds were in Lobb's 1850 and 1851 consignments to Veitch. As noted below, Veitch subsequently made at least two shipments of radiata seeds to help found the plantations in New Zealand, and possibly those in Australia as well. As it is highly likely that those were collected by Lobb, it would be useful to know when and, more important, where Lobb made those collections. It is also interesting that Lobb's fame was based on his acquisition of Santa Lucia fir and giant sequoia seeds, and that records of his radiata collections were not even maintained. Yet, a century later, the impact of his radiata collections on Earth's forest resources dwarfs those of his other collections.

Near the close of the century, early documented shipments of radiata seeds to Chile took a different but perhaps genetically similar route. Those shipments were also accomplished through connections with Europe rather than with North America. Arturo Junge, who owned an ornamental nursery near Concepción, had received his training at the horticultural school in Erfurt, Germany. In 1886 he ordered seed samples of 50 conifer species for Chile from three European suppliers: Schmidt in Erfurt, Von Spreckelsen in Hamburg and Vilmorin in Paris (Junge 1953). Those seeds arrived in late 1886 or early 1887, to be sown in the winter of 1887. Seeds supplied by Vilmorin as Douglas-fir (Contesse 1987) produced seedlings that were clearly another species, but the seedlings performed outstandingly, and were identified by a Dr Neger as radiata.

In summary, it seems likely that most of the mid-to-late-19th-century radiata seeds in and disseminated from Europe came from Lobb's collections for Veitch, or perhaps from some other unknown collector working for Veitch. Some of the seeds, however, came directly from the Douglas collections, and/or secondarily from trees in England and Scotland grown from the Douglas and/or Hartweg collections, and/or even tracing back to Collignon's 1787 collection. Importantly, it appears that most or all of them trace their likely origins to somewhere in the Monterey population. Additional trees, of the *binata* variety of radiata, are recorded as growing at Kew in England in 1910, originating from collections sent from Guadalupe Island in 1875, 1892 and 1896 (Elwes and Henry 1910), well after the main period of seed export to the Southern Hemisphere.

2.3.2 Australia

The date for Australia's first importation of radiata is unclear. Several early importations, of seed or seedlings, were of unspecified pines (Shepherd 1990) which may or may not have included radiata. Shepherd (1960) found, in the archives of the Sydney Botanic Garden, a record "1 insignis plant from Kew, 1 insignis plant from Veitch in 1854." A more specific record involves the arrival, dated as 13 December 1857, on a ship from England (Plymouth and London); one potted "*P. insignis*" was received by the Sydney Botanic Garden, and other potted trees and a consignment of seeds of *P. insignis* went on to the Hobart Royal Botanic

Gardens in Tasmania and the Melbourne Botanic Gardens in Victoria. Von Mueller listed radiata as growing at the latter in 1858 (Fielding 1957a, 1957b).

It is very likely, however, that several other importations of radiata had arrived in Australia before 1857. There are unsubstantiated reports that a radiata was growing on the grounds of Entally House, near Launceston, Tasmania, dating back to the 1840s. The Sydney Botanic Garden received a mixed lot of California pine seeds in 1851. Other shipments of radiata seeds seem to have arrived during the early- and mid-1850s without currently-known documentation. For example, the Hobart Botanic Gardens recorded *P. insignis* as present there after 1855, with seeds from cones produced by these trees being sent to the Melbourne Botanic Gardens in 1865. *P. insignis* was listed for sale in 1857 by Adcock's Nursery in Geelong, and by five others by 1865 (Brookes and Barley 1992), and as under cultivation in the Gardens in 1858 (but not in 1857) (Mueller 1858a, b). Moreover, a radiata plant, reputedly three years old, is recorded as having been imported from Sydney to New Zealand in 1859 (Shepherd 1990), which would presumably put Australian possession of radiata seeds back to 1856 or before.

In 1861, von Mueller distributed 400 seedlings of "*P. insignis*" from the Melbourne Botanic Gardens; in 1862 he recorded receipt of a shipment of unspecified California pines from C. Walker, a San Francisco nurserymen; and in 1866 he distributed 4500 "*P. insignis*" seedlings (S. Maroske, Melbourne Botanic Garden, pers. comm. to K.G. Eldridge). In these early days, radiata was being planted for shelter, firewood and as amenity trees, particularly in Victoria and Tasmania. By 1866, potted radiata pines donated by von Mueller had been planted in an avenue in South Australia's Adelaide Botanic Gardens, and they were reported as thriving and over 16 m tall by 1879. A large purchase of radiata seeds from "Europe" was recommended (and presumably made) in 1869. An 1891 report by von Mueller, who had become an enthusiastic promoter of this species, noted radiata as having been "… extensively distributed through the Colony of Victoria and also some other parts of Australia since 1859 …" (Mueller 1895).

For a while, there was speculation that Australia's several land races of radiata all traced back to the few trees received by the Melbourne, Hobart and Sydney (New South Wales) Botanic Gardens in 1857. But the Hobart Botanic Garden records, the nursery advertisements, and the extensive distributions beginning in 1859 by von Mueller and others, make it unlikely that Australia's land races originated from only those few trees.

In addition to active private nurseries (Brookes and Barley 1992), four state nurseries, producing stock of numerous forest-tree species, had been established in Victoria by 1901, the first at Mt Macedon in 1872. Several species were established in an adjacent 50-acre plantation commencing in 1873; radiata was added there in 1880 and soon after at two other sites. Unfortunately the Macedon nursery and associated plantations were burnt in 1883 (Algar 1988; Moulds 1991).

Early plantings of radiata in New South Wales were made in substantially different climates (Grant 1989): a 0.8-ha private plantation near Albury in 1878, 83 trees among maritime pines (*P. pinaster*) in a small pilot planting by the state in 1885 near Cootamundra, and further very small plantings at Gosford in 1888 and

1894. In Western Australia the first recorded trial plantings of pines had been made on coastal dunes near Bunbury in 1897; they failed (R. Underwood pers. comm. 2012). In following years plantings of radiata had mixed success, continuing to do poorly on the poor sandy soils where it was mostly planted. In both of these states significant follow-up planting of radiata occurred only slowly. We have traced no such early records of radiata plantations in Tasmania.

Among the Australian states, South Australia is the only one where there are sufficient records for this period to warrant more detailed treatment, which now follows. Because of their small area of diminishing native forests, South Australian foresters had to learn to manage forest plantations early. The first period in the development of plantation forestry in South Australia focused on the most pressing needs: to conduct species trials, and to mill locally-grown wood of radiata and other candidate species to demonstrate the feasibility of the proposed plantation enterprise (Boardman 1988).

Credit for recommending the extensive testing of radiata in the early 1870s perhaps belongs to Edwin Smith. Smith was a Scottish-trained nurseryman who saw the potential in this remarkable pine. He was also well-placed to be effective, being both brother-in-law and advisor to Goyder, the Surveyor-General (Carron 1985).

A curious note concerning Australia's 19th-century radiata trees takes place in South Australia about 1878. Forest Conservator J. Ednie Brown began a set of large-scale species-introduction trials. While seed acquisitions of most of the included species are well documented, radiata by any name is not mentioned in those records. Bob Boardman (1996), a student of the origins of South Australia's land races, has suggested that the radiata seeds did not need to be purchased. He reasoned that shortly before the species-introduction trials were established at Bundaleer and Mt Gambier, the young trees in the avenue of radiata in the Adelaide Botanic Garden were probably bearing many cones. Unlike seeds of most of the other species in the trial, seeds of radiata did not need to be purchased abroad; they could just be picked from the trees in Adelaide. Anyway, the first recorded forest plantation in South Australia was at Bundaleer in 1876 (Lewis 1975), perhaps in advance of any systematic species-introduction trial, and radiata was presumably included.

There is a record of a purchase of "*Pinus radiata* true to name" by South Australia in 1881, from Lawson's Seed Merchants in Edinburgh, but where the supplier obtained the seed from is conjectural. By good luck, adjacent compartments of an 1882 plantation on the Wirrabara Forest Reserve were planted with seedlings of "regular *Pinus insignis*," probably from seed collected from the Adelaide trees or from other trees tracing to von Mueller, and seedlings of "*Pinus radiata* true to name" from the 1881 importation of seed from Edinburgh. That plantation was harvested in 1920, the first major sale of plantation timber in South Australia (Boardman 1996). As such, it attracted much attention. By more good luck, the Conservator at the time was Walter Gill, also a professional photographer. His several photographs, included in the 1920–1921 Annual Report of the South Australian Woods and Forests Department, included views of the adjacent compartments that had been called "*Pinus insignis*" and "*Pinus radiata*" when planted. They showed trees that had no noticeable differences in average appearance, a conclusion reinforced by the recollections of Robert Stolz, a cadet forester in 1920 at Wirrabara and familiar with that early plantation. Furthermore, the pictured trees from these two seed sources of radiata were fully consistent with trees later grown in the region from known samples of the Monterey population, and did not closely resemble trees later grown there from the other native populations (Boardman 1996).

This history, and the performance of the trees in the 1882 Wirrabara plantation, has led Boardman (1996) to infer that all of the radiata planted in South Australian plantations in the late 19th and first half of the 20th centuries traced their lineages exclusively to the Monterey population, and conceivably to the original collection made by David Douglas. This possibility cries to be checked by molecular techniques that are still becoming fully available. If so, this perhaps narrow genetic base was propagated and continued through several generations of South Australian plantations. It seems likely that South Australia did not receive the major seed importations that other Australian states did during the first half of the 20th century.

2.3.3 New Zealand

Perhaps the most remarkable things about radiata's introduction into New Zealand are how soon that introduction occurred after colonization began in earnest (Sinclair 1959), and then how quickly this little-known tree was put to widespread use.

There have been some unsubstantiated reports of early importations of radiata to New Zealand. While Shepherd (1990) postulated that Andrew Sinclair may have been involved in introducing radiata to New Zealand as early as 1844, we have found confirmation to be notably absent. Shepherd also mentioned old radiata pines in Isel Park, Nelson, with a date of about 1850; However, personal inspection (RDB) of the map of the old trees and the recorded planting dates (Douglas 1975) gave no compelling reason to believe that any radiata were planted there before the early 1860s. Shipments documented only as including seeds of various conifers were received in New Zealand during 1854–1857 from Veitch and Sons, Veitch's seed company. Those shipments may well have included some radiata pine seeds. A plausible case of introduction around 1856–1858 involves trees at Pah Farm, Auckland, involving a landowner with links to San Francisco (John P. Adam pers. comm. 2012). It is based on photographs of radiata trees some years after planting, on the appearance in the photographs of those trees beside a house, and on circumstantial evidence as to the date of the photographs.

The first clearly recorded introduction of living radiata was in 1859, to J.B.A. Acland for the Mt Peel sheep station in South Canterbury. It was a three-year-old potted plant from Shepherd & Co.'s Darling Nursery in New South Wales (Shepherd 1990). Also it is specifically documented that Veitch shipped radiata seeds to New Zealand in 1859 and again in 1863 (*op. cit.*). In all, about 20 importations of radiata seeds into New Zealand were recorded during the period 1859–1885 (Shepherd and Cook 1988; Shepherd 1990). Nurserymen's newspaper advertisements started featuring radiata from 1862 (D.J. Mead, pers. comm. 2013).

The earliest recorded importations were small, on a specimen-tree scale, but the success of radiata as a young tree quickly led to larger importations beginning in the mid-to-late 1960s.

Unfortunately, none of the records available from those shipments indicate where the seeds were collected (Shepherd and Cook 1988; Shepherd 1990). Most shipments originated from and were labeled merely “San Francisco” or “Santa Rosa” or “Europe.” Furthermore, these records do not indicate how many or which sub-populations were sampled (particularly interesting would be samples of native stands occupying ecologically different sites within the same population), nor how many trees provided seeds for each collection. Principles of evolution and population biology were not an established part of botanical thinking or commerce in that period. While several of these records are good as to the quantities of seeds received, questions of place of origin and numbers of seed parents apparently were not considered.

The “San Francisco” collections do have some further documentation (Shepherd and Cook 1988; Shepherd 1990). As demand for radiata seed built, an 1868 consignment was from Professor Albert Kellogg through the California Academy of Sciences in San Francisco. Subsequent larger consignments from “San Francisco” were from Miller & Sievers, a commercial seed, nursery and florist business in San Francisco. As mentioned in Sect. 2.1.3.2, the earlier botanists in North America and Europe seem to have received most or even all of their radiata collections from Monterey. By the 1860s, however, there was much commerce between San Francisco and the dairy farms and sawmill in the nearest radiata population, at Año Nuevo. We (WJL and RDB) have personally noted that closed, seed-containing cones have appeared more abundant at Año Nuevo than at Monterey, and that may have been the case in the 1860s and 1870s as well. Thus our recent observations on cone abundance and these historical records make it possible, even likely, that the radiata seeds sent to New Zealand came first from a few small collections in the Monterey population, and later from larger collections that might have originated in any of the mainland populations, but seem likely to have been collected at Año Nuevo. (Subsequent research, see Burdon 1992; Burdon et al. 1997, makes this very plausible.)

The available historical records document a minimum estimate of how many introductions were made to New Zealand, and provide data on how large some of them were. Kellogg’s records, which might have been the best for the consignments from California, were evidently destroyed in the fire that followed the 1906 San Francisco earthquake. The documented shipments indicate a range from as few as two seed parents to about 200,000 seeds, probably from several to many trees; if all of those imported seeds were sown and germinated at typical rates, they could have produced about 500,000 seedlings.

After 1882, most of the radiata planted in New Zealand almost certainly came from seeds collected from cone-bearing trees in New Zealand. By then abundant closed cones containing viable seeds would have been available locally. Locally collected seeds would then have come disproportionately from the earlier of the then-extant introductions that, being older, would have been producing heavier crops of cones. A shift in seed sources was indicated by the subsequent absence of radiata in tree-seed acquisition records and in the trend to lower prices in nursery lists. The price of radiata seedlings fell sharply, presumably as inexpensive local seeds replaced expensive imported seeds. Thus, the New Zealand land races were most likely founded by the population samples imported prior to 1882, and were then shaped largely by subsequent natural (and semi-natural) selection in New Zealand.

There appear to have been several networks involved in the early introductions of radiata. A loose network evidently existed very early on between New Zealand landowners and nurserymen in Sydney, the latter having evidently obtained radiata seed and at least one small potted tree through Veitch in England. Another loose network may have existed among general merchants, between Auckland and San Franciso, with some migration each way; for them, movements of plant material would have been an incidental activity. Later on, there was the documented, country-wide network run by Hector from the Wellington Botanic Garden. Other networks, however, could well have existed.

2.3.3.1 Canterbury

While over 50% of New Zealand was forested at the time of European colonization, the windswept plains of Canterbury, in eastern central South Island, were essentially treeless. "The dream of Canterbury was the dream of a tree, and thousands upon thousands were planted across the Plains to provide shelter for agriculture and settlement, and to create the now-green city of Christchurch" (Hegan 1993).

New Zealand's first clearly recorded radiata, planted in 1859 at the Mt Peel sheep station in South Canterbury (Sect. 2.3.3) is still thriving (Mead 2013, fig. 1.2). Measured at slightly over 50 m tall (Burstall and Sale 1984), its recent breast-height diameter measurement (R. Woollons pers. comm. to RDB 2011) was just over 3 m (based on 9.43 m circumference), but this diameter is inflated by some fluting from both ground level and large low limbs. In the same district, on the property of a Mr Grey on the present outskirts of the township of Geraldine, two specimen trees of radiata were planted in 1860 (Burstall and Sale 1984) among European hardwoods. One of these trees, the "Grey pine" (Fig. 2.15) is still alive and is also enormous. Nearly 50 m tall, with a very cylindrical bole, it has a breast-height diameter of 2.67 m; unfortunately, in 2006 it suffered significant snow damage, but it is still showing good vitality.

Notable early planting of radiata in South Canterbury took place at Albury Park, initially in 1868, and on a larger scale a few years later with a stand that was still in good condition at around age 105 years (Fig. 2.16).

A key player in the early radiata acquisition and subsequent acceptance drama was T.W. Adams, who emigrated from England in 1862 at age 19. After three years working for an earlier settler, he purchased a farm at Greendale, west of Christchurch. He started planting trees in 1866. His trials were systematic, with good records of plantings and their growth, and he wrote about them in a series of articles appearing mostly in Canterbury publications between 1886 and 1913. By 1910, his plantations covered over 60 ha and contained 800 species of timber

Fig. 2.15 The "Grey pine," Geraldine, New Zealand, during the 1960s

trees, fruit trees and shrubs in trials and in his arboretum. He had received seeds from botanists and collectors all over the world. As they grew, his trees are likely to have been sources of both information and seeds for the plantations then being established in Canterbury by the Selwyn Plantation Board. His passion for forestry was legendary, but there are reports that his five sons would sometimes have preferred to play sport on Saturdays rather than plant trees (McKelvey 1991).

Radiata may not have technically qualified as a suitable tree in the terms of the 1871 Forest Trees Planting Encouragement Act. Rather, radiata was often planted in the 1870s by "independent farmers who had dared to plant trees which were then officially considered fit only for firewood" (Simpson 1973). There was doubtless a rapid increase in plantings of radiata to provide windbreaks and other shelter on the windswept Canterbury Plains. Prime requirements for a shelter tree were rapid growth and bulk; radiata quickly proved to provide both of these, and within 10 years of its introduction it had become the species of choice for shelter and then firewood in Canterbury (Hegan 1993). Advertisements from a Canterbury nursery listed 500,000 radiata seedlings for sale in 1873, and 100,000 in 1876 (Shepherd 1990). Unfortunately, no records of the origins of these seedlings have

Fig. 2.16 Early exotic radiata planting (in background), Albury Park, South Canterbury, New Zealand, being felled at around age 105 and milled into lumber (Photo RDB 1993)

been found, nor have we been able to confirm the actual numbers. A "plantation" area (in unstated species) of 3284 ha was reported in Canterbury as of 1881 (Burdon and Miller 1992). By the 1890s, the Selwyn Plantation Board was systematically establishing plantings on the Canterbury Plains.

2.3.3.2 Auckland

While it now appears that the first plantings of radiata in New Zealand may have been in the Auckland region, there are better-documented plantings that were made there in the following few years. During the late 1850s and early 1860s Sir George Grey regularly imported plants and seeds (including pines) to Auckland from Sir William Hooker, Director of London's Kew Gardens. Indeed, very old radiata pines, which appear to have originated from the Monterey population, are still present on the property once owned by Grey on Kawau Island. In 1862, David Hay advertised metre-tall *P. insignis* plants from his Auckland nursery (Shepherd 1990). These were probably mostly sold and distributed as individual specimen trees.

One of the earliest documented plantings in New Zealand took place in 1866 in the Auckland area of North Island. Mr W. McLaughlin, a retired merchant from Peru whose hobby was tree planting, imported radiata seeds from "Santa Rosa," California, in 1865. He raised seedlings that year on his Puhinui Estate near the Auckland suburb of Papatoetoe and planted them out as a shelter belt the following year. He later cut out an orchard and planted radiata in plantation form as an investment. This plantation was regarded as having outstanding growth and form. But when it was eventually felled in 1959 (New Zealand Forest Service 1959), no seeds had reportedly been collected from it for use elsewhere (Healy 1982). (It is intriguing that Santa Rosa, California, is where a famed fruit-tree breeder, Luther Burbank, did his major work—but he did not begin this work until 1875—Encyclopedia 1979).

The record of McLaughlin's seed import suggests that Rancho Santa Rosa in the Cambria population was the source of this earliest and highly successful for-investment plantation in New Zealand. However, Martin Bannister and Margot Forde, both familiar with the native radiata populations, visited the plantation before it was felled, and Forde was firmly of the opinion that it was Monterey-origin stock (M.H. Bannister pers. comm. to RDB 1965). Otherwise, the native-population origin of this early Auckland plantation remains an intriguing mystery. The same applies to how the seed came to be imported, although it is suspected that a network of general merchants between Auckland and San Francisco would have been involved.

Another documented 1866 planting in Auckland, involving six trees of radiata among 18 other exotic conifer species, revealed its much superior growth rate, with trees up to 18 m tall by 1879 (Gillies 1879).

Later on, the multiple, widespread distributions of seed and/or nursery seedlings by Hector from the Wellington Botanic Garden contributed to plantings of radiata in this region, presumably along with various unrecorded private importations.

2.3.3.3 Elsewhere in New Zealand

In the 1860s, most radiata planted in parts of New Zealand other than Canterbury were for shelterbelt and aesthetic purposes (Burdon and Miller 1992). The distributions from the Wellington Botanic Garden (Sect. 2.2.2.2), plus various private acquisitions, were no doubt used in a great variety of ways. Of particular note, plantations of radiata and maritime pines were established in 1873 near Matamata, south of Auckland in the eastern Waikato Region, presumably from private seed importations. When compared with the performance of maritime pine there, the greatly superior growth and straightness of the radiata contributed importantly to the decision to extensively plant radiata pine in the 1920s (Page 1923).

2.3.4 South Africa

Three exotic pine species were effectively introduced into South Africa well before radiata—Scots pine sometime before 1714, and maritime pine and Italian stone pine (*Pinus pinea*) in the late 1700s, with commercial pine plantations started in 1825 (Poynton 1977). Viable seeds were recorded from radiata pines in the Caledon District in 1857, indicating an early and unrecorded importation as seeds or young plants, probably before 1850 (Poynton 1960, 1977).

Radiata and several other pine species were recorded as growing at the Cape Peninsula by 1865, long before colonists discovered gold in South Africa. It is possible that some South Africans were also seed or potted-tree customers of Veitch & Sons, or received additional seed of radiata from other sources, at about the same times that the first radiata pines were arriving in Australia and New Zealand. However, its larger-scale importation, or perhaps the local collection of seed from the few radiata planted earlier, began in the late 1870s or early 1880s. The first mention of its use in plantations was in 1883 (Scott 1960), 250 radiata seedlings (possibly inbred) derived from the seeds of a single Cape Town Botanic Gardens tree of unknown origin being recorded in the Tokai Plantation nursery (Poynton 1977). However, Poynton also noted that very little commercial planting of radiata was done before 1910.

2.3.5 Chile

The first recorded introduction of radiata to Chile had occurred inadvertently, in seed ordered by Arturo Junge from Vilmorin in Paris, when a seedlot labelled Douglas-fir proved to be radiata (Sect. 2.3.1). Its very rapid growth and its identification as radiata presumably led to an order specifically for radiata. Of recent German extraction, Junge kept meticulous records. He paid 9 marks for 227 grams of radiata seed, received as *P. insignis* from Erfurt, Germany, in 1890 (Krebs 1973). That was the first recorded case of intentional importation of radiata pine into Chile. Compared with the dominant Spanish-colonial view, north-western Europeans of the mid-to-late 1800s had an almost reverential attitude toward nature, trees and forests; Junge seems to have shared this attitude. His journals include details on hundreds of tree species that he and his nursery staff received and tried in the vicinity of Concepción. Of these, only radiata, macrocarpa cypress, bluegum (*Eucalyptus globulus*) and the Lombardy poplar clone showed sufficient promise for widespread planting. Radiata's attractiveness as an ornamental was quickly apparent, and Junge's nursery donated some to the city of Concepción for planting, including on the site of the University of Concepción (Krebs 1973). At least some of these donated trees were planted on the Cerro Caracol within the city and were still standing in 1978 (RDB personal observation), and may well be the oldest radiata in Chile. During the same visit to Chile RDB observed some very large, old radiata pines in Parque Saval, Valdivia, where the original German settlement occurred.

Radiata's performance in species trials around Concepción led naturalist Federico Albert to both recommend and provide seedlings of radiata to landowners in coastal areas in the mining region. Since the local coastal forests that had supplied timbers for the mines were nearing exhaustion, both pines and eucalypts were planted to replenish the supply (Clapp 1995). Sr Cousiño, one of the mine owners, also had radiata planted in his botanical park ("Parque Cousiño"), at Lota, slightly to the south of Concepción. The park had been developed during 1862–1874. In the early 1970s, the tallest radiata pines in the park were 40 m high, and of a diameter too large for available increment corers. The caretaker claimed they were 130 years old, thus planted in the 1840s, but there were no corroborating records. A smaller radiata at the park edge was cored and had 65 rings, so its age was not noteworthy and it may have been younger than the largest individuals. It is possible that all of these radiata pines were added to the park plantings well after its original establishment, and thus are probably from the Junge importations (Krebs 1973).

In addition to the radiata pines on Cerro Caracol and near Concepción, plantations near Lota were personally noted by RDB in 1978 to have relatively small cones strongly suggestive of the Monterey native population (Fig. 2.7). Those recent Lota plantations were presumably descended from plantings made there during 1907–1912 (Contesse 1987). Seedlots collected further to the south, in the nearby Arauco peninsula, during the 1970s, contained smaller seeds than seedlots from other regions (CONAF Centro de Semillas records), also suggesting a Monterey origin. This in turn suggests that stands in that part of Chile had originated from one or more early importations by Junge.

Junge had received his training at the horticulture school in Erfurt, from where he ordered his radiata seeds. The supplier there, Schmidt, apparently got seeds from England. It is not clear whether, by 1890, these English-source seeds were still coming from the Douglas collection and their offspring, or from some other lineage growing in England, or perhaps they had been acquired more recently from somewhere in California. Common-garden trials and DNA studies may help sort this out.

2.3.6 Spain and Portugal

While Spain was slow to begin to recognise and use radiata from its Californian colony, at least a few were planted in early decades after the colony was lost. Even earlier, there is a record of an importation through France in 1840 for planting in an unidentified Spanish botanical garden (Mead 2013). A famous specimen tree was planted in 1860 in a park at Lequeitio, 40 km west of San Sebastian; it measured over a metre in diameter in 1928 (Scott 1960). This was reportedly the parent and then ancestor of a high proportion of later plantings in Spain (L.D. Gea pers. comm. 2011). Similarly, a radiata planted tree sown in 1883 at Camancha, on the Portuguese island of Madeira, was 30 m tall at age 24, and others there were growing "with extraordinary rapidity" by 1910 (Elwes and Henry 1910).

Recent DNA research (Aragonés et al. 1997) indicates that the radiata land race in the Basque Country of Spain originated mainly from the Año Nuevo population.

2.3.7 *The British Isles*

Radiata pines from the Douglas and possibly Collignon collections of Monterey-population origins were being grown with marginal success in coastal England and Scotland during the latter half of the 19th century. Conceivably, some trees of Cambria origin (the Hartweg collection) and some from among three or more Guadalupe Island collections may have made their way from sexually competent parents at Kew and various pineta into general use by the turn of the century.

Radiata pines, probably of mostly Monterey origins, were performing much better in Wales, coastal Scotland and particularly Ireland during this period, in some areas "thriving amazingly" (Elwes and Henry 1910). All plantings during this period, however, seem to have been only on an amenity scale.

2.3.8 *Other Countries*

In most other regions, such as those bordering all coasts of the Mediterranean Sea, in East Africa, and in South America, extensive trials of this species had not yet begun by 1900. Interestingly, two very large radiata pines, clearly of Guadalupe Island origin, were identified growing in La Jolla, near San Diego, California (WJL), but no record of their planting date or seed collection could be found.

2.4 Summary of Domestication Progress

The activities of Native Americans doubtless had some influence on the genetic makeup of mainland populations, especially Monterey. There, the practices of frequent burning and harvesting of seed for food would have entailed selection through survival of future parent trees with thicker bark and smaller cones and seeds, but cannot be counted as actual domestication. The Spanish colonists then had modest influences, with some selective cutting of the straighter trees around Monterey which would have been dysgenic, and some cutting of mine timbers on Cedros Island. Then US settlers basically "mined" native stands at Monterey, for some sawn timber but mainly as fuel for whale oil extraction and railroad engines, which would have tended to be dysgenic where trees were selectively logged for timber.

First steps towards domestication arguably occurred with English nurserymen in the 1830s propagating the species as rooted cuttings, presumably to extend the small available amounts of seed as a novelty species. Introduction of radiata elsewhere began mainly during 1840–1860, initially as an ornamental or a botanical curiosity, followed by larger plantings for producing shelter and firewood. Its introduction to Chile, however, was not until 1884, and was apparently accidental.

While some substantial exotic plantings of radiata were soon established, and it was locally an important source of native sawn timber at Monterey and Cambria, only very small amounts were sawn into lumber elsewhere before 1900, in areas within Australia and New Zealand that had very limited supplies of native timber. Indeed, no major commitment to its development as a timber crop was made before 1920. Long before then, however, local exotic stands had become the basis for self-sufficiency to meet national seed requirements, at least within New Zealand.

Impending exhaustion of virgin timber supplies in several countries was giving impetus to create large resources of forest plantations. In this connection, the work of enthusiastic proponents had helped set the stage for a dramatic upsurge in planting radiata which came after World War I in New Zealand. It had become a favourite quick-growing tree for shelter and firewood there and elsewhere in the Southern Hemisphere, and in near-coastal California for parks and new residential areas.

References

Algar WH (1988) Forestry and forest products. Australian Academy of Technological Sciences and Engineering, Parkville, p 60. Reprinted from (pp. 193–252) Technology in Australia 1788–1988 for Australian Forest Development Institute on the occasion of The International Forestry Conference for the Australian Bicentenary 1988

Allsop F (1973) The first fifty years of New Zealand's forest service. Government Printer, Wellington, p 123

Anon (2013a) To reject *Pinus californiana* Loisel. (Pinaceae). Taxon 62(6):1321

Anon (2013b) Updated Chile forest incentive scheme winners and losers. South Hemis For Ind J 18(3):4

Aragonés A, Barrena I, Espinel S et al (1997) Origins of basque populations of radiata pine inferred from RAPD data. Ann For Sci 54:697–703

Axelrod DI (1967a) Evolution of the Californian closed-cone pine forest. Philbrick, pp 93–149

Axelrod DI (1967b) Geologic history of the Californian insular flora. Philbrick, pp 267–315

Axelrod DI (1980) History of the maritime closed-cone pines, Alta and Baja California. University of California Publications in Geological Sciences Vol 120: p 143

Axelrod DI (1981) Holocene climatic changes in relation to vegetation disjunction and speciation. Am Nat 117:847–870

Bannister MH (1965) Variation in the breeding system of *Pinus radiata*. In: Baker HG, Stebbins GL (eds) The genetics of colonising species. Academic Press, 353–374

Bannister MH (1973) The origins of radiata pine in cultivation. What's new in forest research 2. New Zealand Forest Research Institute, Rotorua, p 4

Bannister MH, McDonald IRC (1983) Turpentine composition of the pines of Guadalupe and Cedros Islands, Baja California. N Z J Bot 21:373–377

Belich J (1994) Making peoples: A history of the New Zealanders, from polynesian settlement to the end of the nineteenth century. University of Hawaii Press, Honolulu.

Boardman R (1988) Living on the edge—the development of silviculture in South Australian pine plantations. Aust For 51:135–156

Boardman R (1996) 1 October letter to WJL

Brookes M, Barley R (1992) Plants listed in nursery catalogues in Victoria 1855–1889. Ornamental Plant Collections Association Inc., South Yarra, p 316

Burdon RD (1992) Genetic survey of *Pinus radiata*. 9. General discussion and implications for genetic management. N Z J For Sci 22:274–298

Burdon RD (2004) Genetics of *Pinus radiata*. In: Burley J, Evans J, Youngquist JA (eds) Encyclopedia of Forest Sciences 3. Elsevier Academic Press, Oxford, pp 1507–1516

Burdon RD (compil.), Miller JT (ed) (1992) introduced forest trees in New Zealand: recognition, role and seed source. 12. Radiata pine (*Pinus radiata* D.Don). FRI Bulletin 124. New Zealand Forest Research Institute. Rotorua, p 59

Burdon RD, Bannister MH, Madgwick HAI et al (1992a) Genetic survey of *Pinus radiata.* 1: Introduction, description of experiment, and basic methodology. N Z J For Sci 22:119–137

Burdon RD, Bannister MH, Low CB (1992b) Genetic survey of *Pinus radiata.* 2: population comparisons for growth rate, disease resistance and morphology. N Z J For Sci 22:138–159

Burdon RD, Broekhuizen P, Zabkiewicz JA (1997) Comparison of native-population and New Zealand land-race samples of *Pinus radiata* using cortical oleoresin monoterpenes, pp 50–56 in Burdon RD and Moore JM (eds) IUFRO '97 Genetics of Radiata Pine. Proc NZFRI-IUFRO Conference 1–4 December and Workshop 5 December, Rotorua, New Zealand. FRI Bulletin 203.

Burstall SW, Sale EV (1984) Great trees of New Zealand. AH & AW Reed. p x + 283

Cameron RJ (1964) Destruction of the indigenous forests for maori agriculture during the nineteenth century. *N Z J For* 9:98–109

Carron LT (1985) A history of forestry in Australia. Australian National University Press, Canberra, p x + 355

Clapp RA (1995) The unnatural history of the Monterey pine. Geogr Rev. 85(1):1–19

Clark DJ (1991) Monterey county place names. Kestrel Press, Carmel Valley, p 737

Coffman T (1995) The Cambria forest. Coastal Heritage Press, Cambria, p 77

Contesse (González) D (1987) Apuntes y consideraciones para la historia del pino radiata en Chile. Boletín de la Academía Chileno de la Historia No. 97a [1986]: 351–373

Critchfield WB, Little EL (1966) Geographic distribution of the pines of the world. U.S. Department of Agriculture, Forest Service, Miscellaneous Publication 991, Washington

Cylinder P (1996) 11 October 1996 report at Carmel Pitch Canker Symposium, and personal communication to WJL

Dallimore W, Jackson AB (1954) A handbook of coniferae including Ginkgoaceae. Third edn, with supplement. Edward Arnold & Co, London, p xvi + 686

Donald DGM (1993) Radiata pine in southern Africa. Lewis and Ferguson, pp 353–364

Douglas B (1975) Isel park—a woodland garden. Royal New Zealand Institute of Horticulture, National Diploma of Horticulture thesis. p 101+

Drury A, Neasham VA (1954) History at Point Lobos. In: Drury A, Engbeck JH (eds) Point Lobos: interpretation of a primitive landscape. University Extension. University of California, Berkeley, pp 67–73. 80

Dunning D (1916) A working plan for the Del Monte Forest of the Pacific Improvement Company. MS thesis in Forestry, University of California, Berkeley. p 78

Elwes HJ, Henry A (1910) The trees of Great Britain & Ireland. *Pinus radiata.* Vol V. Privately published in Edinburgh, pp 1079–1084

Encyclopedia 1978–79 The new illustrated Columbia encyclopedia. Vol 1–24. Columbia University Press, NY

Ewan J (1973) William Lobb, plant hunter for Veitch and messenger of the big tree. University of California Publications in Botany 67. University of California Press, Berkeley, p 36

Fielding JM (1957a) The breeding of Monterey pine in the Australian Capital Territory. Paper to 7th British Commonwealth Forestry Conference, Australia and New Zealand, p 19

Fielding JM (1957b) The introduction of Monterey pine to Australia. Aus For 21:15–16

Flannery TF (1994) The future eaters. Reed Books, Port Melbourne, p 423

Forde MB (1962) Variation in the natural populations of Monterey pine in California. PhD dissertation. University of California, Davis, p 291

Forde MB (1964a) Variation in native populations of *Pinus radiata.* 1. Sampling methods and branch characters. N Z J Bot 2:213–236

Forde MB (1964b) Variation in native populations of *Pinus radiata* 4. Discussion. N Z J Bot 2:486–501

Forde MB (1966) *Pinus radiata* in California. N Z J For 11:20–40

Gillies TB (1879) Notes on the growth of certain trees on scoria soil near Mount Eden, Auckland. Trans R Soc N Z 12:357–358

Gordon BL (1977) Monterey bay area: natural history and cultural imprints, Second edn. Boxwood Press, Pacific Grove, p 300

Gordon G (1875) The pinetum. A synopsis of all the coniferous plants at present known, 2nd edn. Simpkin Marshall & Co, London, p xxii + 353

Grant TC (1989) History of forestry in New South Wales 1788 to 1988. The Author, Sydney, p 320

Greenlee JM, Langenheim JH (1990) Historic fire regimes and their relationship to vegetation patterns in the Monterey Bay area of California. Am Midl Nat 124:239–253

Griffin J (1970–1990) (Personal communications to WJL, Dr James Grifffin, University of California Hastings Natural History Reserve)

Griffin J (1972) What's so special about Huckleberry Hill on the Monterey Peninsula? In: Howitt BF (ed.) Forest heritage: a natural history of the Del Monte Forest. Del Monte Forest Foundation, Pebble Beach, pp 3–7. p 56

Halkett JC (1991) The native forests of New Zealand. GP Publications, Wellington, p 149

Harris JM (1991) Radiata pine in New Zealand. Chapter 1. In: Kininmonth JA, Whitehouse LJ (eds) Properties and uses of New Zealand radiata pine. NZ Forest Research Institute, Rotorua, pp 1-1–1-7

Healy B (1982) A hundred million trees. The story of N.Z. Forest Products Ltd. Hodder and Stoughton, Auckland, p 222

Hegan C (1993) Radiata, the prince of pines. N Z Geogr 20(Oct–Dec):88–114

Hillyard D (1996) 11 Oct. 96 talk at the Carmel Pitch Canker Symposium, and personal communication to WJL

Home RW, Lucas AM, Maroske S et al (1998) Regardfully yours: selected correspondence of Ferdinand von Mueller. Volume 1, 1840–1859. Peter Lang, Bern, p 842 plus two subsequent volumes

Howard D (1972) Indian shell-midden sites. In: Howitt BF (ed.) Forest heritage: a natural history of the Del Monte Forest. Del Monte Forest Foundation, Pebble Beach, pp 39–43. p 56

Jones & Stokes Associates Inc. (1994) Final. Monterey pine forest ecological assessment: historical distribution, ecology, and current status of Monterey pine. September 12. (JSA 94-083) Sacramento Ca, USA. Prepared for The Nature Conservancy, Sacramento Ca, and the California Department of Fish and Game, Monterey, Ca

Junge A (1953) Historia del 'pino insigne' en Chile, y el premio 'Arturo Junge Sahr'. Chile Maderero 3(10):4

King NL (1938) Historical sketch of the development of forestry in South Africa. *J S Afr For Assoc* 1:4–16

Krebs JS (1973) Monterey pine: an introduced species in Chile. PhD dissertation. Department of Geography, University of Colorado. Boulder, p 386

Larkey FB (1972) Footnotes to the history of the Del Monte Forest. In: Howitt BF (ed.) Forest heritage: a natural history of the Del Monte Forest. Del Monte Forest Foundation, Pebble Beach, pp 43–48. p 56

LeBoeuf BJ, Kaza S (1981) The natural history of Año Nuevo. Boxwood Press, Pacific Grove, p 425

Ledig FT, Vargas-Hernádez JJ, Johnson KH (1998) The conservation of forest genetic resources: case histories from Canada, Mexico, and the United States. J For 96:32–41

Lerner IM, Libby WJ (1976) Heredity, evolution and society, 2nd edn. W.H. Freeman & Co, San Francisco, p 431

Lewis NB (1975) A hundred years of state forestry—South Australia 1875–1975. Woods and Forests Department, Adelaide, (A Centenary Publication) Bulletin No. 22. p 122

Libby WJ (1997) Native origins of domesticated radiata pine. Burdon & Moore, pp 9–24

Libby WJ, Bannister MH, Linhart YB (1968) The pines of Cedros and Guadalupe Islands. J For 66:846–853

Lindsay AD (1932) Monterey pine (*Pinus radiata* Don) in its native habitat No. 10. Bulletin, Commonwealth Forestry Bureau, Canberra, p 57

Linhart YB (1978) Maintenance of variation in cone morphology in California closed-cone pines: the role of fire, squirrels and seed output. Southwest Nat 23:29–40

Lydon S (1996) March e-mail to WJL. Professor of History, Cabrillo College

Mann CC (2011) 1491: New revelations of the Americas before Columbus, 2nd edn. Vintage Books, New York, p 553

Mason HL (1932) A phylogenetic series of the California coastal closed-cone pines suggested by the fossil record. Madroño 2:49–55

McKelvey P (1991) Thomas William Adams 1842–1919. Early farm forester. N Z J For 36 (2):23–25

McKelvey SD (1994) Early botanical explorations 1790–1850. Fremontia 22(3):14–19

Mead DJ, (2013) Sustainable management of *Pinus radiata* plantations. FAO Forestry Paper No. 170. FAO, Rome. p xv + 246

Millar CI (1986) The Californian closed cone pines (Subsection *Oocarpae* Little and Critchfield): a taxonomic history and review. Taxon 35(4):657–670

Millar CI (1993) Impact of the Eocene on the evolution of *Pinus* L. Ann Mo Bot Gard 80:471–498

Millar CI (1999) Evolution and biogeography of *Pinus radiata*, with a proposed revision of its quaternary history. N Z J For Sci 29:335–365

Millar CI, Kinloch BB (1991) Taxonomy, phylogeny, and coevolution of pines and their stem rusts, pp. 1–38. In: Hiratsuka Y, Samoil JK, Blenis PV, Crane PE, Laishley BL (eds) Rusts of Pines. Proceedings of the 3rd IUFRO Rusts of Pines Working Party Conference. 18–22 September 1989, Banff, Alberta. Foresty Canada Information Report NOR-X-317. Edmonton, Alberta

Mirov NT (1967) The genus *Pinus*. Ronald Press, NY, p 602

Mirov NT, Hasbrouck J (1976) The story of pines. Indiana University Press, Bloomington, p 148

Moran R (1996) The flora of Guadalupe Island, Mexico. California Academy of Sciences, San Francisco, p 190

Moulds FR (1991) The dynamic forest. Lynedoch Publications, Richmond Victoria, Australia, p 232

(von) Mueller F (1858a) On a general introduction of useful plants into Victoria. Trans Philos Inst Vic 2:93–109

(von) Mueller F (1858b) Annual report of the government botanist and director of the Melbourne Botanic Garden. Government Printer, Melbourne

(von) Mueller F (1895) Select extra-tropical plants. Government Printer, Melbourne, p 390

Nedeff N (1994) Address to the 22 April 1994 Monterey CA symposium—Monterey Pine, Our Vanishing Ecosystem

New Zealand Forest Service (1959) Old exotic stand, Auckland. Newsletter 9(2), Feb. 1959, Supplement, p 2

Oberbauer T (2005) A comparison of estimated historic and current vegetation community structure on Guadalupe Island, Mexico. In: Quatrini L, Steele C, Glassow M (eds) Proceedings of the Sixth California Islands Symposium, Ventura, California, December 13, 2003. National Park Service Technical Publication CHIS-05-01. Institute for Wildlife Studies, Arcata, pp 143–153

Oberbauer TA (2006) La vegetación de isla Guadalupe: entonces y ahora. In: Peters E, Santos Del Prado K (eds) Restauración y Conservación de la Isla Guadalupe. Instituto Nacional de Ecología, Mexico, pp 39–53

Page PM (1923) Pine plantations, Matamata estate, Waikato, New Zealand. In: Goudie HA (eds) The remarkable Pine: *Pinus radiata* in New Zealand. State Forest Service Circular No. 3. Government Printing Office, Wellington, pp 11–14. p 14

Philbrick RN (ed.) (1967) Proceedings of the symposium on the biology of the California islands. Santa Barbara Botanic Garden, Santa Barbara, p 363

Poynton RJ (1960) Notes on exotic forest trees in South Africa. (2nd edn, revised). Department of Forestry, Government Printer, Pretoria, p 186

Poynton RJ (1977) Tree planting in South Africa. Volume 1. The pines. Department of Forestry, Republic of South Africa, p 576

Price RA, Liston A, Strauss SH (1998) Phylogeny and systematics. In: Richardson DM (ed) Ecology and biogeography of *Pinus*. Cambridge University Press, Cambridge, p 49–68

Richardson DM (ed.) (1998) Ecology and biogeography of *Pinus*. Cambridge University Press, Cambridge, p xvii + 527

Roche MM (1990) History of New Zealand forestry. New Zealand Forestry Corporation in association with GP Books, Wellington, p 466

Rogers DL (2002) *In situ* conservation of monterey pine (*Pinus radiata* D.Don); information and recommendations. Genetic resources conservation program report No. 26, Division of Agriculture and Natural Resources, University of California. p xi + 80

Rogers DL, Vargas Hernández JJ, Matheson AC et al (2005) Conserving the pines of Guadalupe and Cedros Islands, Mexico: An international collaboration. In: Aldemaro R, West SF (eds) Environmental issues in Latin America and the Caribbean. Springer, Dordrecht, p 31–54

Scott CW (1960) *Pinus radiata*. FAO Forestry and Forest Products Studies No. 14. FAO, Rome, p 328

Shepherd RW (1990) Early importations of *Pinus radiata* to New Zealand and distribution in Canterbury to 1885: implications for the genetic makeup of *Pinus radiata* stocks [2 parts]. Horticulture in New Zealand 1(1): 33–38; 1(2):28–35

Shepherd W, Cook W (1988) The Botanic Garden, Wellington. Millwood Press, Wellington, p 396

Sherry SP (1971) Black wattle. University of Natal Press, Pietermaritzburg, Republic of South Africa, p 451

Simpson TE (1973) Kauri to radiata: origin and expansion of the timber industry of New Zealand. Hodder & Stoughton, Auckland, p 442

Sinclair K (1959) A history of New Zealand. Penguin Books, Harmondsworth, p 320

Stephens SL, Libby WJ (2006) Anthropogenic fire and bark thickness in coastal and island pine populations from Alta and Baja California. J Biogeogr 33:648–652

Stephens SL, Piirto DD, Caramagno DF (2014) Fire regimes and resultant forest structure in the native Año Nuevo Monterey pine (*Pinus radiata*) forest, California. Am Midl Nat 152:25–36

Sutton WRJ (1991) New Zealand radiata pine plantations: an example of sustainability. Plenary paper, Society of American Foresters National Convention, San Francisco, p 5 (metric version)

Valentine JW, Lipps JH (1967) Late Cenozoic history of the southern California islands. Philbrick, p 21–35

Vargas-Hernández J, Rogers DL, Hipkins V (2013) Restoration of *Pinus radiata* on Mexico's Guadalupe Island. In: Bozzano M, Jalonen R, Thomas E, Boshier D, Gallo L, Bordács S, Smith P, Loo J (eds) Genetic considerations in ecosystem restoration using native species. A thematic study of the state of the world's forest genetic resources. FAO, Rome, p 236–239

Weaver DW, Doerner DP (1967) Western Anacapia: a summary of the Cenozoic history of the northern Channel Islands. Philbrick, p 13–20

Weber DJ (1992) The Spanish frontier in North America. (see p. 237 for Monterey base). Yale University Press, New Haven, p 579

Chapter 3
The Early Plantation Period: 1902–1951

The first half of the 20th century is when most large-scale radiata plantation programs were initiated, mostly on sites where soil and climate seemed appropriate. Most of these large early plantations were in New Zealand, Australia and, somewhat later, in Chile. There, near Rotorua in New Zealand, Concepción in Chile and Mt Gambier in South Australia, radiata was found to thrive near south latitudes 37° and 38°, similar to those of the northern mainland native populations. However, this close matching of latitude proved not to be essential; plantations also succeeded over substantial areas at higher latitudes (in the mid-40s in Europe and New Zealand) and at lower latitudes (to around 32–33° in Australia and South Africa). In addition, radiata has been found to perform acceptably at high altitudes (generally 2500–3500 m) in some localities close to the Equator.

Radiata entered the 20th century gaining increased interest and attention in a few places, such as the Canterbury region of New Zealand, and near Concepción in Chile. It had already generated much enthusiasm in South Australia. Elsewhere, however, skepticism and even scorn for this reputedly rough, weedy tree were the common attitudes of people considering species for use in timber-producing plantations. Approaching the end of the 19th century, radiata, having been mostly appreciated as an ornamental species, did become accepted as one that could provide rapid shelter from wind or sun, with incidental use as a tolerably good firewood.

3.1 Why Large Plantation Programmes?

As travel became safer and more frequent, and as distant lands and the things that lived in them became better known, great enthusiasm for keeping and growing things exotic spread from northern Europe to other parts of the world, and particularly to the British colonies of the Southern Hemisphere. During the last half of

R. Burdon et al., *Domestication of Radiata Pine*, Forestry Sciences,
DOI 10.1007/978-3-319-65018-0_3

the 19th century, many species of both plants and animals were moved between continents and tried by individuals, by acclimatisation societies and by governments. While a few sizeable plantations of timber-growing trees were established in the Southern Hemisphere, for example in South Australia (Lewis 1975), South Africa (Poynton 1977), or Canterbury in New Zealand (Roche 1990), most plantings of exotic trees during this period were in parks, in formal and informal garden settings, in small woodlots, or in windbreaks.

During this period of enthusiasm for exotic species, many tree species were typically considered for introduction and trials in a region. Climate and soils provided a first approximation when deciding whether to include radiata at a site. It turned out that these were not stringently limiting, nor did a lack of close similarity to the peculiar coastal California conditions preclude success at a new site (Scott 1960; Burdon 2001a). Experience from trial and error was the most reliable guide, and by 1901 landscape and garden plantings in many locations were available as such guides.

Southern Hemisphere regions, except for South Australia, did not begin serious consideration of plantations for the primary purpose of producing timber until the end of the 19th century or later. By then, Europe and eastern Asia already had several centuries of experience in plantation establishment and management. The European plantations in particular then provided both the models and some of the trained foresters for large-scale plantations of conifers in Southern Hemisphere countries.

Several populations of conifers that have evolved near the west coast of North America (particularly in British Columbia, Washington, Oregon and California) are among the fastest-growing and achieve the largest sizes of any conifers on Earth, both in their native forests and when planted in similar conditions elsewhere.

With few exceptions, native Southern Hemisphere conifers grow much more slowly than do some of the conifers native to the Northern Hemisphere. Among Australian conifers, hoop pine (actually, an *Araucaria* species, not *Pinus*) produces valuable wood at reasonable growth rates in plantations, and the two main Queensland kauri (*Agathis*) species attain large sizes. But on both individual-tree and per-hectare bases, all grow much less wood per year, and/or grow it less consistently, than the best western American conifers. In sharp contrast, Australian species of eucalypts are among the fastest-growing temperate and subtropical angiosperms on Earth. Both at home and abroad, these grow far taller and attain larger diameters than any other broadleaved trees, including any from the tropics (Eldridge et al. 1993). Why western North America and Australia have produced substantial numbers of Earth's fastest-growing and largest tree species is still a matter of speculation and debate, but it is undoubtedly so.

In retrospect, five factors were important in setting the stage for large Southern Hemisphere forest plantations in the 20th century.

- Scarcity of native forests in a region. This was the case in the Canterbury region of New Zealand, in much of South Africa, in South Australia and in a few other localities.

- The unsustainable clearing and exploitation of native forests. While continuing and even accelerating after 1901, these two processes were by 1901 well advanced in Australia, Chile, New Zealand and parts of southern and eastern Africa, but were becoming increasingly recognised as serious problems.
- Problems of or with the native forests. The native forests in many regions, once cut, did not regenerate satisfactorily, and/or they did not then grow at rates capable of meeting projected local needs for wood. In most regions where such problems occurred, it took time and experience to comprehend them.
- Outstanding performance of some exotic species. Small early plantings of several species of Northern Hemisphere conifers and Australian hardwoods grew much faster than did the native species. The greatly superior growth capabilities of species from these two regions were not readily predictable, and it took many observations over several decades to first identify them and then to gain confidence in them.
- Suitability for appropriate sites. With further experience, it became clear that some of these exotic species grew well on classes of land that occur over large areas, were relatively inexpensive and were often unsuitable for agriculture.

Not all of these five factors occurred or were important in every region, and those that did influence decisions to create plantations often became evident at different times. Where and when enough of them occurred in various temperate and subtropical regions of the Southern Hemisphere, large-scale forest plantations with exotic species were a common result. Radiata proved to be a particularly good fit in the last two factors.

3.1.1 Australia

From the time of first European settlement in 1788, colonists in most regions of Australia found extensive native forests, and soon began using them. Many of these forests were dominated by trees in the hardwood genus *Eucalyptus*, of which there are hundreds of different species and which occupy a great diversity of sites (Eldridge et al. 1993). These trees provided a wide range of useful material, but also posed such problems as high growth stresses leading to distortion of timber freshly sawn from logs, collapse on drying, and high density leading to difficult working with hand tools. Also, they very often showed site sensitivity whereby they would not grow consistently well over large areas. Additional important and useful species of other hardwood genera were also present, as were several valuable softwood (conifer) species, particularly in Queensland, New South Wales and Tasmania. The most accessible stands of valuable conifers were decimated for sawlogs and through forest clearing for agriculture in the 20th century, and a general shortage of softwood timber stands was increasingly anticipated (Hutchins 1916). Conifer wood is preferred for construction and several other purposes, mostly because it is generally easier to saw, dries with much less degrade, and is often stronger per unit weight than wood from angiosperms. It was envisaged that plantations would be one answer to this shortage. Gaining momentum in 1919, this recognition led eventually to substantial programs to plant conifers, mostly exotic, in all of Australia's six states and the Australian Capital Territory (Hall 1951; Carron 1985). Many of these programs provided economic relief to the unemployed during the depression years in the late 1920s and 1930s, and sometimes the growing trees improved catchment cover, reducing erosion and improving water quality. The focus in those decades was often more on gaining the experience to achieve successful establishment of plantations than on creating a new high-quality wood resource. A feature of the plantation experience in most states until about 1930 was that many attempts were made to grow plantations on lands regarded as unsuited to agriculture, and most of these sites proved to be far from ideal for forest plantations also. Some such sites were eventually found to have particular nutrient deficiencies that could be effectively corrected, or to be better suited to species other than radiata.

By 1939, native and exotic conifers provided wood for about half of the annual consumption of timber in Australia, yet only some 5% of the standing volume in the exploitable native forests was in conifers. Most native conifer species were by then known to be relatively slow-growing, hoop pine being the best-known exception (Hall 1951). While recognised much earlier, the trend toward increasing imbalance between the demand for coniferous wood and its local supply, with a consequent need for increased imports, was inescapable. The logical response was to increase plantation forestry on larger areas, with attention to appropriate conifer species, and to efficient plantation establishment and silviculture.

By 1950, the Australian softwood plantation estate of about 120,000 ha was 87% state owned, with only 13% in private hands (Scott 1960). At this time the area of reserved native forests, which were off-limits for harvesting timber, was about 11 million hectares.

3.1.2 New Zealand

In the several centuries before the arrival of Europeans, activities of Maori colonists and their numerous descendants had greatly reduced the forest cover of New Zealand (Sect. 2.2.2) from about 75% to 53% of the land area (Sutton 1991). European colonists and later immigrants often treated the remaining native forests as impediments to agriculture, in particular pastures for their imported sheep and cattle, and large additional areas of native forest were cleared (Sutton 1984).

During the Maori and early European colonial periods, the fate of most native trees on land being cleared was to be burned on site. But some wood of native trees was used locally, and it was increasingly milled and sold. By the late 1800s and in the early 1900s, not only did native forests provide most of the wood and sawn timber needed for various uses in New Zealand, they also supplied sawn timber, masts and spars for export and had thus become an important part of New Zealand's economy. A consequence was overexploitation of kauri and other native species.

As the availability of kauri wood fell behind rising demand, its price climbed sharply during 1899–1922, with harvest peaking in 1905 and exports beginning to fall in 1907. As prices of native timber rose, timber imports began flooding in, and they were increasingly competitive in the domestic market. These changes drew the official attention of a Royal Commission (1909) on Timber and Timber Building Industries. This 1909 commission considered, among other things, the effects of logging practices such as selective cutting and clearfelling, and different milling techniques and standards. The commission found and made clear that changes in logging and/or milling practices could not return New Zealand to its former position as a net exporter of wood, and that even the domestic needs for wood could not be satisfied by logging native species. Pre-1907 volumes of sawn timber from New Zealand's native forests have never since been available to either the New Zealand domestic or export market (Simpson 1973).

Although not fully realised in 1909, the combination of climates and soils in New Zealand made it an excellent location for effectively growing several exotic tree species as crops. The New Zealand Government considered the 1909 commission findings and then appointed its first and only Royal Commission on Forestry, in 1913 (Royal Commission 1913; Sutton 1984; Roche 1990). Besides evaluating the indigenous forests and forwarding several important recommendations concerning them, that commission made a strong recommendation to greatly increase exotic plantations (Royal Commission 1913). Furthermore, the 1913 commission, perhaps surprisingly, suggested the previously underrated radiata as one of the promising exotic species for their recommended plantations. This recommendation was accepted and implemented, resulting in the expansion of the plantation estate from less than 8000 ha of mostly government plantations in 1912 to over 300,000 ha of both public and private lands planted with exotic species by 1935, albeit with a substantial contribution from private investment (Healy 1982).

Prior to the 1909 and 1913 Royal Commissions, the Forests Branch of the Lands Department, after some initial trial plantings of exotic species, had begun afforestation with exotic species around 1900, in conjunction with the Prisons Department. Many of their early plantings were made to serve particular end-uses, for example hickory for tool handles; catalpa for fence posts and railroad crossties ("sleepers"); and European larch, Corsican pine and Douglas-fir for high-quality sawn timber. This approach was soon abandoned, in part because of some inaccurate matching of species to site, and also because it was recognised that specialty uses often change faster than trees can be grown to serve them. Sawn products, broadly specified, increasingly seemed a more robust goal. This recognised that pulpwood and other lower-value wood resources could be obtained as by-products from sawmilling (slabs, edgings, sawdust), and from upper logs in good trees and cull logs from malformed trees. This evolving policy was affirmed by the New Zealand Forest Service in 1934 (Sutton 1991).

After 1919 planting of exotic forest accelerated rapidly in New Zealand, in the "first planting boom" (Fig. 3.1). In particular, two very large plantation programmes were undertaken then in central North Island. One was by the New Zealand Government, establishing the Kaingaroa Forest south-east to south of

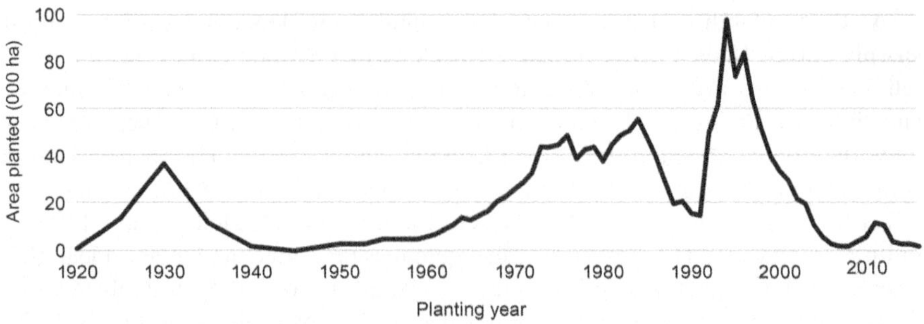

Fig. 3.1 Areas of new planting per annum of commercial forest plantations in New Zealand from 1920 onwards. Up to around 1965 the proportion of radiata was *ca.* 75%, thereafter *ca.* 90% (From NEFD 2016, permission Ministry of Primary Industries)

Rotorua. The other, generally west and north-west of the Kaingaroa Forest, was by private-subscription organizations that coalesced around the largest one, which was called Perpetual Forests (Healy 1982), and which then went on to absorb some other, smaller organisations. This concentration of plantations in central North Island had both causes and effects. An important cause was the availability of a substantial area of land in that region that was relatively inexpensive, allowed easy establishment of radiata and several other species, and supported their subsequent good growth. An important effect was a later concentration of wood-processing facilities that allowed more efficient utilization of the wood resource (Sutton 1995).

One of the causes of the pumice lands in central North Island being inexpensive was "bush sickness," in which livestock that grazed on this apparently-suitable land did not thrive; rather, they often sickened and died. Bush sickness was later found to be due to cobalt deficiency in the soil and thus in the plants being grazed; given this understanding, it could then be effectively remedied by small amounts of cobalt applied to the pastures. By the time this was learned, however, over 200,000 ha of these pumice lands were already planted or committed to forest plantations. A period followed when many influential people viewed the afforestation as a mistake, but the apparently successful plantation program was too gigantic to gracefully back away from. The scale of this enterprise gave plantations a fair trial, and from the late 1960s opinion strongly shifted to the view that such plantations were and are a good use of not only those lands but of other soils as well.

Smaller concentrated plantation forests were developed by the state in about a dozen locations throughout New Zealand, each generally combining opportunity with local need. Two such, presented as examples, were at Woodhill and Golden Downs. In the 1920s, at Golden Downs in the hinterland of the town and port of Nelson in northern South Island, returned servicemen from World War I were settled on the land, but farming generally failed. The land was then acquired by the state and over 11,000 ha effectively planted to several exotic timber species including radiata. About 30 km north-west of Auckland, on the "wastelands" at Woodhill (and some other places), shifting coastal sand dunes were encroaching on neighbouring farms. The sand was stabilised with preplantings of marram grass and yellow lupin beginning in 1932, and then successfully planted to radiata (Restall 1964; Whitehead 1964) in an operation that extended over many years. This dune-stabilisation project also provided a sawn-timber supply, and the developing plantations eventually sheltered recreation sites close to the beach for the Auckland-area population. By 1951 sand-dune plantings of radiata had reached 3800 ha (Whitehead 1964), increasing to over 36,000 ha by 1984 (McKelvey 1999) including large areas of stable dune.

This era completely changed the nature of wood production in New Zealand. The 1913 Royal Commission and a 1925 Forest Service analysis both predicted future wood needs for New Zealand with remarkable accuracy, but underestimated both the supply of wood available from native forests and the rates of growth of the plantations. As a result, New Zealand became the first nation to create a forest plantation resource *before* it was needed (Sutton 1984, 1995). As the exotic plantations established during 1925–1935 reached harvest age, the annual cut of radiata increased rapidly. By 1951 radiata contributed almost 40% of the estimated outturn of sawtimber, the balance being almost entirely native species (NZ Forest Service 1952). Later, in 1960, the exotic cut from that first planting boom had begun to exceed the harvest from native forest, more than doubling the latter in 1961 (Sutton 1995) as native-forest harvests decreased and exotic harvest increased. Within State Forests that ratio rose to seven times as much by 1968 (Anon1968).

3.1.3 Elsewhere

South Africa's native timber resources were limited from before European settlement, and the native species were not readily amenable to plantation forestry. Indeed, by 1900 the need for a large, renewable timber resource was becoming pressing. Trialling of numerous exotic timber species (Poynton 1960) was a natural consequence, and radiata did very well on suitable sites.

In Chile, the picture was regionally mixed. There had been abundant native timber, especially in the south, but this was dwindling fast in the northern parts of the naturally forested region of the country. By 1900, there was a strong demand for mining timbers in the coal-mining areas located immediately to the south of Concepción, which prompted the establishment of plantations of both radiata (Contesse 1987) and blue gum (*Eucalyptus globulus*). By 1930, however, regional depletion of native timber had increased, while very large areas of land in the Coast Range had become severely degraded by unsustainable farming. There were also large areas of sandy to scoriaceous river basin that had very little agricultural potential. The scene was therefore set for the government to embark on encouraging large-scale afforestation.

3.2 The Evolution of Institutions that Supported Plantation Development

The early afforestation programmes with radiata in Southern Hemisphere countries were driven mainly by a combination of historical or impending sawn-timber shortages and by the relative performance of the candidate species, augmented in some cases by entrepreneurial quests for handsome profits. The institutions involved, however, were varied, which largely reflected the different structures of countries' governments and ownership and availability of land in the various countries. Governments typically took the early lead in afforestation, although the choice of species was sometimes strongly influenced by the enthusiasm of individuals for radiata or for other species. Central, state/province or local governments were variously involved, sometimes independently among different agencies at particular levels of the same governments. Supplementing or even preceding government programmes, small-scale private plantings began quite early, especially in New Zealand, but large-scale entrepreneurial plantings by private companies generally began later.

3.2.1 Australia

Consequences of independent plantation programs being conceived and executed in the different Australian states include different emphases, initiation times, program intensities and responses to local needs and conditions. This situation had both positive and counterproductive effects.

The federation to form the Commonwealth of Australia occurred in 1901. While tariffs, customs, defence, etc. were assigned to the Federal Government, the different states retained many of the colonies' former rights, including management of "unoccupied" lands and reserved forests. While in many respects (including aspects of forestry) Australia has become increasingly cohesive since 1901, the independence of the state forest services from the central government and in some respects from each other has been resolutely defended (Carron 1985, 1993).

The Federal Government also established a Forest Products Laboratory in 1919, which became part of CSIR ([Australian] Council for Scientific and Industrial Research) as Division of Forest Products 1928–1971), when organisational arrangements then changed; the Division contributed to a variety of developments that assisted in the marketing and use of radiata. A fundamental step was the publication of a *Handbook of Structural Timber Design* (Langlands and Thomas 1939) that facilitated development and industry acceptance of stress and proof grading of structural timber some two decades later, and of national standards (Algar 1988).

The Federal Government's role in forestry substantially expanded in the 1920s, culminating in July 1930 with the Act to establish the (Australian) Commonwealth Forestry Bureau. The Third British Empire Forestry Conference, held in Australia and New Zealand in 1928, had endorsed the proposed legislation. Silvicultural research was identified as a field needing attention, but progress was greatly slowed by the Depression. Importantly for our story, though, A.D. Lindsay was sent to California to study tree species of known or potential commercial interest, and delivered a detailed report on radiata in its native habitat (Lindsay 1932). In 1933 the first research officer in the Bureau, M.R. Jacobs, was appointed, in Canberra. In 1937, Federal and South Australian authorities agreed to establish a cooperative research station at Mt Burr, SA, and the following year J.M. Fielding was appointed as research officer there. In 1946, after World War II, Fielding was posted in Canberra, and C.K. Pawsey was appointed to replace Fielding at Mt Burr.

Experience gained in Australia with exotic forest trees, mainly up till 1950, was described at the Seventh British Commonwealth Forestry Conference in 1957, for which all of the state and territory forest services prepared papers; information from these has been used below and was summarised by Streets (1962).

Almost all the significant planting during this period was undertaken by public-sector organisations. One private-sector venture, South Australian Perpetual Forests, was incorporated in 1926 and had planted 2500 ha of radiata in the south-east of South Australian within 10 years; other smaller private-sector groups followed in the same region (Carron 1985).

3.2.1.1 South Australia

In 1873 the South Australian State Government had passed the *Forest Trees Act No. 26*, which by means of a bonus encouraged landowners to plant trees; in 1881 this inducement was replaced by the free distribution of seedlings, a scheme that lasted until 1924. The South Australian forest service is considered to have had its genesis in the *Forest Board Act No. 8* in 1875; by 1881 annual planting, including of radiata, had been proceeding at three regional centres; in 1883, the South Australian Woods and Forests Department was designated. The second Conservator in the Department, Walter Gill, had been appointed in 1890; he ably served in that role for 33 years. The well-illustrated annual reports of the Department in that era, in which the role of radiata in South Australia was consolidated, are testimony to his capacity. As trees in the early plantings approached merchantable size, emphasis on and facilities for utilisation of the wood they were growing increased. In this, the Department played a very active role. Commencing in 1903, it built a series of progressively more sophisticated sawmills that, from 1925, included seasoning kilns. Greater volumes of small-diameter logs became available as routine commercial thinning was instituted by Geoffrey Rodger, appointed Conservator in 1935; utilisation capacity for such material was further increased in 1941 when a small private-sector pulp mill was built at Millicent, 50 km from Mt Gambier (Lewis 1975; Carron 1985).

The Department appointed its first "Investigation and Research Officer," M.A. Rankin, in 1932. Much of the early effort following that appointment was in support of and then response to a Parliamentary Committee of Enquiry (which in 1936 reported as a State Royal Commission on Afforestation). This Commission's recommendations endorsed the directions in which plantation development was proceeding in South Australia. Another notable development was setting up, in collaboration with the Commonwealth Forestry Bureau in Canberra, the cooperative Mt Burr Forest Experimental Station near Mt Gambier in 1937, which gave a base for research in the south-east of the State (Boardman 1988), with much pioneering research on radiata.

3.2.1.2 Victoria

Victoria's Ferdinand von Mueller was prominent in the development of the forestry profession in southern Australia (South Australia, Victoria, New South Wales, Western Australia and Tasmania) over almost four decades of the late 19th century (Gillbank 1993). He was concerned not only with the conservation, management and effective use of the major natural forests of Victoria, but "since Victoria forests were deficient in softwoods … to introduce a wide range of pines" (Mueller 1858). He vigorously contributed to numerous reports and enquires. In 1888 G.S. Perrin was appointed Conservator, but he had few resources at his disposal. In 1885 Inspector-General Ribbontrop of the Indian Forest Service critically reviewed forestry in the state and endorsed softwood afforestation, but not radiata because "its wood is of low character" (Carron 1985). Plantations of radiata had been established at Mt Macedon in 1880, and more at Creswick and You Yangs in the late 1880s (RBG 1995). Victoria's Government finally passed a *Forests Act* in 1907, forming a state Forests Department and supporting more plantation establishment. That 1907 Act also led to the opening of the Forestry School at Creswick in 1910 (Moulds 1991; Dargavel 1995; Collett 2010).

In 1950, just as the period for this chapter drew to a close, Australian Paper Mills Ltd—which was also operating a pulp and paper mill at Maryvale—commenced planting radiata pine. In 1951, APM Forests Pty Ltd, a subsidiary company, was formed and the scale of its operations, including the planting of radiata, quickly increased (Chandler 1957; Sinclair 1991).

3.2.1.3 New South Wales

In New South Wales (NSW), as in Victoria, much of the modest forestry effort of the 19th century had been devoted to minimising harmful exploitation and alienation of the native forests in an atmosphere in which schemes for human settlements abounded. A weak 1909 *Forest Act* was supplanted in 1916 by the *Forestry Act*, which established the State's Forestry Commission. On appointment as Commissioner in 1926, Norman Jolly scaled up NSW's 1914–1924 softwood afforestation program, choosing to concentrate on radiata in the Southern Tablelands of the state (Carron 1985) where G.J. Rodger was able to plant about 700 ha annually up to 1935. Planting was then much curtailed while results were reviewed, before expanding immediately after the 1939–1945 war on proven sites.

3.2.1.4 Other Australian States: Queensland, Western Australia and Tasmania

In Queensland a small Forests Branch had operated from 1900 within its Lands Department until 1906, when the Queensland forest service achieved independent status. Stemming the tide of alienation and harmful exploitation of forests was its major early preoccupation in the following decades; but preliminary work on the cultivation of hoop pine commenced in 1917. In 1924 the service was reorganised along lines that remained for many decades. In pursuit of a source of softwood sufficient for Queensland's domestic needs, a plantation program establishing 2000 ha per year was planned. Hoop pine was preferred where appropriate, but slash and loblolly pines were planted on the infertile coastal plain soils that occur for hundreds of kilometres north from the NSW border (Carron 1985). First planted in 1918, by 1932, 90 ha of radiata had been established, mainly in highlands west of Brisbane. However, for most of the available land conveniently located for markets, slash pine (*Pinus elliottii*) from south-eastern United States was far better adapted.

In Western Australia (WA), in 1916, 37 years after Victoria's von Mueller had urged the establishment of a WA forestry administration, David E. Hutchins, a prominent forester with experience in Africa, expressed a similar sentiment in a critical report to the WA Minister for Mines (cf Hutchins 1916). Hutchins was then made responsible for the resulting Forestry Department. C.E. Lane Poole, who had been Inspector of Forests in Sierra Leone, was appointed as Inspector-General of Forests in 1916, and drafted the Forests Act which was passed in 1919. This Act has been cited as a model of Australian forest legislation. Among other things, it included provision to establish plantations to meet the anticipated softwood requirement of WA, which then did proceed on a modest scale mainly with maritime pine (Anon 1969; Carron 1985; Dargavel 2008). In the seven years to prior 1916, 240 ha of radiata had been planted, much on unsuitable sites; thus little more was planted until after 1950.

In Tasmania, G.S. Perrin had been appointed Conservator of Forests in 1886, but it was not until 1920 that a state forest service was formed. Tasmania's initial *Forestry Act of 1920*, substantially amended in 1946, led to conifer planting on a small scale commencing in 1921 (RBG 1995), but from 1922 focused planting on "wasteland" in the west and north-west. By 1932 it was clear that those plantations would not be successful. The purchase of better land for additional radiata plantations, former farms on the north coast, commenced in 1937 and continued until 1948. After a pause planting resumed, on this more-suitable land until 1960, by when about 6000 ha had been established. The first commercial thinning commenced in those better plantations in 1950 (Carron 1985; Elliott et al. 2008).

3.2.2 *New Zealand*

By 1890, a realization had begun that native conifer forests could not continue to supply New Zealand's needs for wood, a concern reinforced by a government-sponsored

Timber Conference in 1896. This led to a revival of the Forests Branch within the Lands Department (Sutton 1995). Systematic planting by the state began on a modest scale about 1900, at seven locations, two in central North Island near Rotorua, and four scattered around the South Island.

The importance of having these early plantations available for inspection by the 1913 Royal Commission (Royal Commission 1913; Roche 1990) was enormous, as this allowed it to consider the plantation option with first-hand site-specific knowledge. During February through May 1913, the six commissioners conducted hearings in many parts of the country, observed native forests and exotic plantations over an itinerary covering about 11,000 km, listened to testimony from many interested and knowledgeable people, discussed and drafted their findings, and delivered their catalytic report 108 days after their appointment. It was an intense and magnificent effort, and its effect was to change the face and future of New Zealand. Their report had several important foci: conservation of substantial areas of natural forest; management of destructive animals such as deer; management of native forests for their recognised multiple resources; and, emphatically, the usefulness of exotic species in substituting for native species whose wood resource was coming into short supply. For this latter role, various eucalypts and pines were identified as among the most promising candidates, which included some of the better-known European and North American species. Thus, while radiata was endorsed, it was only one among several species considered and recommended for plantation use.

3.2.2.1 The State Forest Service

World War I soon delayed effective action on the 1913 recommendations. After the war, the State Forest Service was established in 1919, and it advertised worldwide for a director (Allsop 1973; Roche 1990).

Small but persistent groups advocating forest **conservation**^{G} and even **preservation**^{G} had gained only limited support in the late 19th century. However, their views were heard by the 1913 Commission and incorporated in some of its recommendations concerning indigenous forests. These views later resonated with the Commissioner of State Forests, Sir F.H.D. Bell, appointed in 1918, and with the new Director of the Forest Service appointed in 1919 (Allsop 1973; Roche 1990). Indeed, one of the first actions of the Forest Service was to conduct a national survey of the remaining indigenous forest resources.

L. McIntosh Ellis, a 32-year-old Canadian trained in both North American and European forestry traditions, was chosen as Director of the new Forest Service, and Arnold Hansson, a 31-year-old Norwegian forest consultant, joined as Chief Inspector.

Ellis had been strongly influenced by the North American conservation movement, and brought vision, realism and enthusiasm to the position (Allsop 1973; Sutton 1987). Initially, he focused on management of the native forests. However, Ellis soon realised that the native species could not meet future needs; the first forest experiment station was then established in Westland, South Island, to investigate the restocking of logged indigenous forests with exotic species (Allsop 1973).

By 1925, Ellis had focused on exotic plantations as the simultaneous solutions to the problems of New Zealand's wood-supply and of indigenous-forest conservation. He calculated the plantation area needed to meet the nation's future demand for sawn timber while incrementally reducing and then eliminating the need to unsustainably log the remaining native forests. He had far outdone his North American mentors, whose successors 50 years later were still fighting over issues of preservation versus logging in North American forests. Ellis set a goal of 125,000 ha to be planted by the state by 1935, and recommended additional planting by local authorities and the private sector (Sutton 1995). The pace of planting accelerated dramatically, with the state and the private sector each planting about 6000 ha per year during the late 1920s; this area rose to a combined total of 36,000 ha planted during 1932, the peak year of this first planting boom (Sutton 1986).

Ellis had recognised that the best way to protect the native forest from destructive or unsustainable logging is to provide a better and cheaper alternative, and his first goal of 125,000 ha of new plantations was meant to accomplish that. But his vision went beyond that substantial goal. In 1927, he proposed a 25-year campaign of planting idle, waste and deteriorated lands to reach a goal of two million planted ha, to make New Zealand a major wood supplier on the world market. However, both biological and financial problems were dampening enthusiasm for tree planting, with few tangible benefits yet at hand to provide a counter force. Ellis' larger vision was not supported, and he left New Zealand in 1928 to become a forestry consultant in Australia (Allsop 1973; Sutton 1987).

He was ably replaced by E. Phillips Turner, who stoutly and successfully defended the afforestation campaign. In this defence, he cited Ellis' long-term economic and conservation goals, as well as the newer and more urgent rationale of forest planting being an appropriate way to provide employment during the immediate crisis of the depression.

Turner retired in 1931 and A.D. McGavock became the Forest Service's third Director, at a time when an Economy Commission was recommending its abolition. McGavock successfully defended it, and among his most powerful arguments was the need to provide an exotic-species wood supply to meet the shortfall expected from the indigenous forests. It is sometimes thought that the conservation and preservation of extensive native forest in New Zealand was an unforeseen and serendipitous effect of New Zealand's love affair with radiata. But McGavock, like Ellis, clearly saw the exotic plantations as an effective and necessary tool to achieve such conservation of native forests. In his 1934 policy review, he stated that conservation of the native forests, particularly of those protecting water and soil, was of greater national importance than the timber that might be extracted from them, which could be supplied by the exotic plantations (Allsop 1973).

McGavock, however, had misgivings about the long-term biological stability of exotic species in widespread plantations, viewing the exotic plantations as a stopgap until enough could be learned about native-forest species and ecosystems to once again rely on them. This view, shared by others then and now, probably contributed to the end of the first planting boom after Ellis' target for a stopgap timber supply was achieved (Allsop 1973).

The target of 125,000 ha of state plantations was achieved in 1931, well ahead of schedule, the planting program having been greatly assisted by the government using forest employment to provide jobs during the developing economic depression (Sutton 1995). Unfortunately, the rush to establish plantations, and the use of relatively untrained workers as a form of constructive relief work during the Depression, resulted in several problems. Planting on unsuitable sites was common during those early years, as were poor planting techniques, inadequate follow-up tending, and a longer list of both biological and management problems (Scott 1960; Allsop 1973). Errors of planting the wrong trees for a site—inappropriate species and inappropriate provenances of species—and often faulty planting technique were judged to have been as costly as destructive fires. The Forest Service had responded to some of these observed and perceived problems by restricting radiata to 30% of its plantings after 1930 (Allsop 1973); which in hindsight proved to be a loss of nerve that created more problems than it avoided (Burdon 2001b).

The planting boom effectively halted by 1935, partly through targets having been exceeded, and partly through financial stringency—despite the earlier role of providing relief work.

3.2.2.2 Other Public Agencies

The Lands Department and then the Forest Service were not the only public agencies engaged in afforestation, although they were by far the main ones. County Councils, notably Ashburton in Canterbury, often established their own plantations, sometimes on land acquired for other purposes. Among city corporations Dunedin was prominent in establishing its own plantations. In Canterbury, the Selwyn Plantation Board, which had been created in the 1890s, represented joint interests of several local bodies. The Railways Department established plantations of their own on Railway Reserves. And from 1932 onwards, the Lands and Survey Department had become engaged in using trees to stabilise and reclaim moving sand dunes (Restall 1964). In all those cases radiata was prominent among the species used.

The year 1919, as well as seeing the birth of the State Forest Service, saw the death of T.W. Adams, and his will contained an important bequest: £2000 and 40 ha to be used in the establishment of a School of Forestry at Canterbury University College. A Diploma of Forestry was established in 1921; the University of New Zealand recognised the School of Forestry at Canterbury in 1924 and established annual funding for it. The 40 ha contained Adams' arboretum and many research plantings. A survey in 1981 identified 700 trees belonging to 196 species and varieties still on this site (McKelvey 1991). However, government funding for the forestry school ceased in 1933, and it closed the following year (McKelvey 1991). The same fate befell a similar forestry school at Auckland, which had been set up despite there being barely room for one such school in New Zealand.

3.2.2.3 Private Enterprise

Large-scale entry of the private sector was noteworthy, and at that time unusual. It began in 1917, when Landon Smith and Douglas Wylie stopped in Rotorua to visit H.A. Goudie. Smith and Wylie were investors then interested in acquiring land for farming, and were returning to Auckland after evaluating some land, characterised by pumice soils, that they had just deemed unsuited to their purposes because livestock mysteriously failed to thrive there. Goudie, then Conservator of Forests in Rotorua, had 20 years' experience of experimental and operational forest plantings around Rotorua. He enthusiastically recommended that they consider growing radiata instead of more traditional farm crops. The fortuitous timing of their meeting, and Goudie's persuasive zeal, started a project that led from Perpetual Forests, through New Zealand Forest Products Ltd (Healy 1982) and later to Carter Holt Harvey Ltd, now controlled by the Rank Group. Of huge importance to the success of their early forest plantings was Owen Jones, an Oxford-educated forester who was Forest Administrator for Perpetual Forests.

After their catalytic meeting with Goudie, Smith and Wylie then spent six months in Australia, looking at well-established plantations in South Australia, Victoria and New South Wales, and then near Christchurch in the South Island, New Zealand. Most importantly, they met R. Dalrymple-Hay, Commissioner of Forests in South Australia, who gave them a tour and information about South Australia's radiata forests. Dalrymple-Hay and other Australians thus provided further encouragement for using radiata as a basis of a commercial venture. Smith and Wylie then bought 1275 ha of land near Putaruru in central North Island from the Selwyn Timber Company, one of several in a group then known as New Zealand Redwood Forests Ltd. They began selling publicly-offered stock in March 1923 as Afforestation Ltd. That went so well that New Zealand Perpetual Forests was incorporated in May 1923, after only two months! Their brochures indicated that stock shares would not only create personal income for the investors, but would create a "national benefit to New Zealand." Few of the small investors ever derived much personal income from this venture (Healy 1982), partly because of less than laudable roles of individuals who we leave unnamed, but the second clause has proven prophetic.

In reading the history of radiata plantation forestry in New Zealand, one gets the admiring impression that Perpetual Forests, for example, had achieved remarkable organization in the original afforestation, and that the maintenance of the growing trees had been well looked after. B. Healy's 1982 book *A Hundred Million Trees* concentrated not on the dedication and positive forestry contributions of Owen Jones and other foresters and woodsworkers who had accomplished all this, but mostly applauds the businessmen, lawyers and politicians who made it all financially, legally and politically possible. However, it also documents how some of these same businessmen, lawyers and politicians, through a combination of power-hungry arrogance, chaotic greed and mutual suspicion, made a series of disastrous moves. These manoeuvrings almost wrecked the private radiata plantations as a viable commercial venture, and small investors suffered the financial consequences.

Perpetual Forests and the companies that it absorbed were not the only success story. For instance, Matea Forests Ltd achieved remarkable success in establishing radiata on high, cold sites behind some remoter parts Kaingaroa Forest.

3.2.2.4 Research and Development

A.R. Entrican was another key figure during this period. He was an engineer, appointed to the newly established Forest Products Branch of the Forest Service in 1921. His whole career was in the New Zealand Forest Service, serving as its Chief from 1939 to 1961. While Ellis had created the plantation resource, Entrican embarked on figuring out how to have it efficiently used. Also, he later emphasized and supported training and education, from woodsworkers through rangers to foresters, and succeeded in professionalising plantation forestry (Sutton 1987). His vision and drive laid the foundation for industrial forestry in New Zealand, and assisted the development of radiata as the main crop in that enterprise (Sutton 1984).

In the late 1930s, under Entrican's leadership, the state became actively involved in finding at least some solutions to problems then causing a loss of confidence in the radiata plantations, and then in gaining better public acceptance of those plantations (Orr 2017). Sawmills at that time used techniques developed for the large-diameter medium-density old-growth logs still being cut from the native forests. Those techniques were poorly suited to the smaller-diameter lower-density logs from the plantations, and thus they produced products of uneven and often poor quality (Allsop 1973). A state-owned experimental and demonstration sawmill was established at Waipa, Rotorua, in central North Island near some of the early radiata plantations. By 1939, it was converting radiata logs into increasingly better, and better understood, sawn products. Those advances in knowledge and techniques required research not only into techniques of sawing, but also into appropriate grading, seasoning and preserving radiata's sawn products (Sutton 1984). The grading rules, developed by J.S. Reid of the Forest Service, were a major innovation in that they accommodated the distinctive but variable patterns of timber defects common in plantation-grown radiata. The Forest Service policy of making its sawing and marketing techniques freely available contributed greatly to the acceptance of plantation-grown wood (Allsop 1973). The success of this state sawmill (Fig. 3.2) in accomplishing these goals created a climate supportive of further research, in retrospect one of the keys to the developments that followed.

Early forestry and forest-products research efforts addressed a broad range of issues and problems, but were episodic and uncoordinated. By the mid- to late 1930s most such efforts were abandoned because of the financial stringencies caused by the Depression, followed by the manpower demands of the 1939–1945 World War II. After the war, availability of trained foresters and scientists from Europe and British colonies to New Zealand created the opportunity for setting up an organised research establishment, as Ellis had suggested in the 1920s. A woodwasp epidemic that had been raging in the radiata plantations in the 1940s gave

Fig. 3.2 Aerial photo of Waipa Sawmill near Rotorua, New Zealand, built by the State Forest Service to pioneer large-scale processing of exotic species. Opened in 1939, it became the largest sawmill in the Southern Hemisphere. Note the very large yard for drying sawtimber in the open

this some urgency. The forest-products research group agreed to move from Wellington to Rotorua where, in 1947, a landmark event was the founding of the New Zealand Forest Research Institute as part of the New Zealand Forest Service. There, the study of wood and its uses was physically and administratively linked to all facets of research on the growing of forests and those, in turn, to research on a broad range of forest-related topics (Allsop 1973).

3.2.3 Chile

In Chile, much early emphasis was placed on raising cattle and growing wheat. Soil and/or climate proved to be unfavourable or marginal for these enterprises in substantial areas, with erosion and other soil deterioration becoming increasingly serious by the decade. While these problems were developing, ornamental and other plantings of eucalypts, maritime pine and radiata were growing surprisingly well in some of the problem regions, attracting both local and official attention.

The owners of many of the degraded farms and ranches were forced to sell their land, often to financially-solvent businesses and others who then began planting new forests. During the early 20th century, accumulating experience was leading to an understanding of what the land could and could not sustain. Limited areas (about 400 ha) of radiata plantation were established by 1907 south of Concepción near Lota, in a coal-mining area, but the species was not a success as a mining timber (Contesse 1987).

Chilean culture itself was evolving. The earlier colonial Spanish culture had been modified by the influence of northern Europeans and others, and by several generations of living in Chile. The destruction of their excellent native forest with both match and saw had been as careless of the future as had been the case in New Zealand. Just as in New Zealand, this was followed by recognition of both the need for a reliable supply of wood and an increasingly obvious need to better protect soil on erosive and degraded sites. Thus, many Chileans of Spanish as well as northern European descent became deeply involved in forestry. These forestry activities are properly viewed as something fairly recent and distinctively Chilean, rather than in the context of earlier colonial Spanish or other immigrant cultures (Krebs 1973). With respect to the development of plantation forestry, however, this developing Chilean culture had important influences from the German forestry tradition.

Large-scale planting of forest trees (Contesse 1987) began almost 15 years later in Chile than in New Zealand. By 1932, modest areas had been planted to radiata, but they then summed to less total area than New Zealand planted in that single year. However, in 1931, Chile passed a *Forest Law* that subsidised plantation forestry. Since then, successive governments have repeatedly identified the forest sector as a major national project. The area of radiata plantation began to massively increase, the first planting boom peaking in the 1940s. Even more than in New Zealand, the profit motive was a prime mover, such that by 1960, 85% of the radiata plantation estate in Chile was privately owned (Scott 1960). Chilean Governments encouraged private planting by exempting plantations from taxation for 30 years, by distributing seedlings, and by other supportive measures (Scott 1960).

Thus, while Chile came out of the plantation starting blocks over a decade after New Zealand, this gap was greatly narrowed as New Zealand paused for an economic depression and then a war.

3.2.4 South Africa

As in Australia, a federal system of government was adopted, in this case with the four self-governing colonies becoming provinces within the Union of South Africa in 1910. Following union, a national Chief Conservator was appointed, although the provinces continued to have their own forest services under various names and operating unevenly with respect to forest management. In 1913 a Research Section was founded, in which work was later done by I.J. Craib that profoundly influenced tending regimes for radiata. A Forest School had been established in 1906 in the Cape Colony but operated only for a few years. However, professional forestry training started at Stellenbosch University in 1932 (King 1938; Poynton 1979).

The Woods and Forests Department of Cape Province actively engaged in afforestation, establishing most of the radiata plantations (Donald 1993). Other provincial governments engaged in planting radiata, but with little lasting success. However, where radiata remained healthy, in the winter- and evenly-distributed rainfall areas of Cape Province, plantings increased considerably after World War II.

3.2.5 *Spain*

Spain had a very late start in conifer plantation forestry, both compared to northern Europe and to several Southern Hemisphere countries. A major afforestation program of about a million hectares was begun in 1940, utilising both native and exotic species (FAO Secretariat 1958; Scott 1960). Planting of radiata pine was mainly in what is now the Basque Autonomous Region, near the northern coast from the western Pyrenees, but with significant plantings further to the west in the adjoining province of Galicia. Choice of radiata was governed by its need for adequate rainfall and mild summer temperatures, but also by the frost-tenderness of the more-commonly planted bluegum. In the east of that area there were major losses of radiata to frost in 1956, and to the west defoliation by processionary caterpillar tended to be troublesome.

Plantation ownership was very fragmented (Burdon 2001a), generally in small holdings of individual farmers, but including small communal forests.

3.3 The Choice of Radiata

In those areas where radiata grew well, a number of attributes soon distinguished it from other species being tried that were also growing well. The first attributes to generally receive approving notice were ease of handling and transplanting nursery stock and above all the exceptionally rapid early growth of young trees. Then came appreciation that stands of this pine also produce a large volume of harvestable wood per unit area. It was later understood that radiata's high growth rates reflect its ability to grow whenever temperature and moisture permit growth, that it rapidly occupies the space available to it, and that it subsequently maintains a high rate of bole-wood growth (Burdon and Miller 1992). If the young radiata pines were neglected after very dense stockings were achieved, the stand often showed a remarkable capacity for self-thinning, allowing the eventual survivors to maintain health and growth rate (Burdon 2001a) (Box 3.1).

Box 3.1 Clues to Growth Rate and Adaptation
Radiata has a long opportunistic growing season which evidently accounts for it being the fastest-growing temperate conifer. It thrives in mild, temperate oceanic climates, not necessarily matching closely the native climate, as occur mainly in Chile, New Zealand, Australia, Spain and South Africa, between latitudes 32 and 46 degrees. A small, secondary geographic "window" for satisfactory growth also exists close to the Equator. As a true pine, it has modest soil-fertility demands, so can produce good, even growth throughout quite variable sites.

Compared to, say, coast redwood, radiata was often a difficult, even ugly tree. For example, where about 90% of the redwoods in a moderately dense stand usually exhibited straight boles with relatively small branches, about 90% of the radiata pines grown similarly could, especially on highly productive sites, exhibit forked and/or crooked boles and/or relatively large, steep-angled branches. Thus both enthusiasts and skeptics were surprised by how much value was later recovered from even untended radiata plantations. These substantial values were largely due to some inherent virtues of its wood and the often relatively long clear lengths of bole between **branch clusters**[G] (somewhat erroneously called "whorls"). As planting of radiata increased, it was soon realised that radiata could be established easily on a greater variety of sites than could many of the more-valuable fast-growing species, such as redwood. Furthermore, radiata seeds are held available in the closed cones, are easily collected, store well, and (except for those from Guadalupe Island) germinate rapidly without special pretreatment when planted. The seedling crops grew well in nurseries, usually providing field-ready seedlings in less than a year. If treated properly, these seedlings were transplanted easily, survived well and were relatively tolerant of weed competition.

It was then realised that, in terms of wood quality and volume of wood per hectare, radiata was more valuable and grew faster than many of the other species that planted easily. For example, maritime pine planted similarly well, and was proving more tolerant of infertile sites and high summer temperatures. But it often had poor stem form and did not produce the volume per hectare or quality of logs that radiata did on sites that suited radiata. Thus, radiata was preferred on sites where both species grew well. Near Perth in Western Australia and in South Africa, where climates suited both species, maritime pine was mostly used on soils unfavourable for radiata.

One of radiata's more attractive features was that it responds well to intensive management, despite notable tolerance of neglect. Thinning stands to desired spacing and pruning off lower branches may greatly increase the values recoverable in well-managed plantations. Prescribed thinning involves early removal of the less-desirable trees to give greater growing space to those more desired. The retained crop trees of radiata exhibit little exposure shock following thinning, and rapidly recover growth vigour. Pruning wounds heal quickly, usually with little or no fungal infection.

The combination of these various observations was compelling, and once recognised usually placed radiata pine ahead of all alternative species in those regions where it could be grown well in plantations.

3.3.1 Australia: Ecological Fit and Commercial Acceptance

In the first half of the 20th century, attention was given to both natural stands and plantations of Australia's native conifers as sources of coniferous wood. Although the relatively extensive distribution, good growth and fine qualities of native hoop pine in Queensland kept it in the race, by 1950 production from exotic slash pine

(*P. elliottii*) plantations there exceeded that of natural stands of this native species. There were radiata pine plantations in all Australian states, and it had become the dominant plantation species in all but Western Australia and Queensland (Hall 1951; Scott 1960; Lewis 1991) (Table 3.1; Fig. 3.3).

Table 3.1 Coniferous plantations in Australia (December 1952, ha) (Adapted from Annual Report Forestry and Timber Bureau 1952)

State or territory	Government owned		Private[a]	Total
	Radiata	Other species		
Queensland	200	20,200	0	20,400
New South Wales	12,700	5,600	3,600	21,900
Victoria	13,100	5,500	8,100	26,700
South Australia	34,800	2,800	7,300	44,900
Western Australia	1,300	4,800	N/A[b]	6,100
Tasmania	2,800	100	800	3,700
Australian Capital Territory	5,300	800	0	6,100
Total	7,020	39,800	19,800	129,800

[a]Species details not available, but mainly radiata
[b]Estimated area of radiata was 40 ha

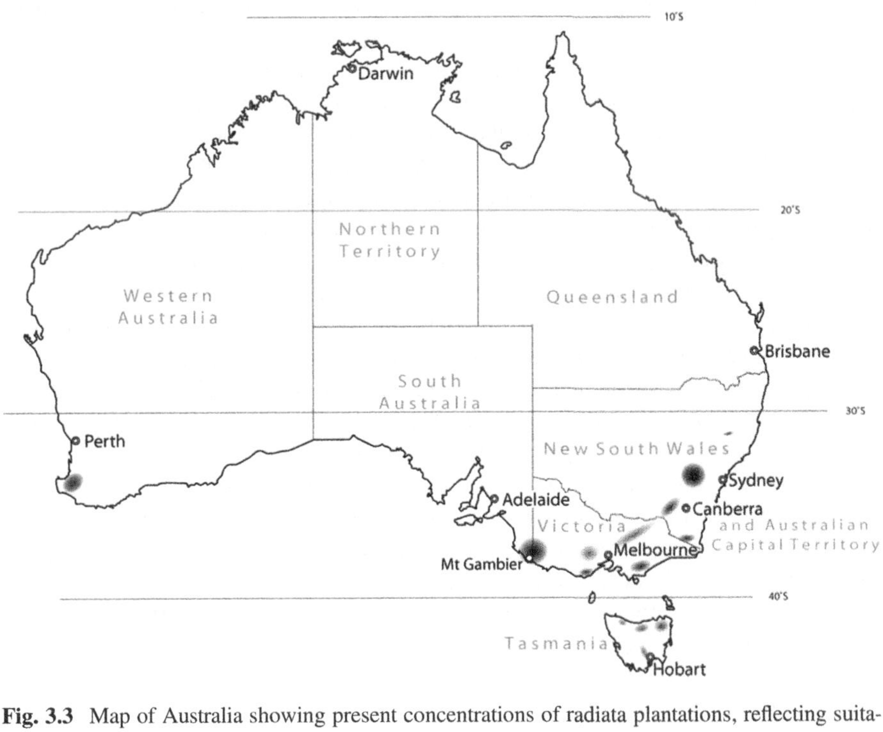

Fig. 3.3 Map of Australia showing present concentrations of radiata plantations, reflecting suitable sites

In some of the regions of Australia where the climate was suitable for radiata pine, problems of "**dieback**"[G] and "spindle stands" were ascribed to soil nutrition in the late 1920s and 1930s. Debilitating phosphorus deficiency was found to be easily corrected, initially by adding blood and bone or, more conveniently, superphosphate. The potential of added zinc emerged in 1936 from a trial established the previous year. Subsequently, both phosphate and zinc were used routinely to successfully correct ill-health and serious bole deformation on suspect sites (Kessell and Stoate 1936, 1938; Stoate 1950; Boardman and McGuire 1990a, b). It was also found that overall growth could often be radically improved by the addition of fertilizer, especially phosphate-based, often where acute nutrient deficiencies were not evident. These management advances greatly increased confidence in radiata as the species of choice in such regions.

3.3.1.1 South-Eastern Australia

Although radiata had been planted earlier in relatively small numbers in and near the native range in California, in Europe, and elsewhere, the plantations begun in the 1880s in South Australia are the oldest extensive radiata *forests* created by planting. By 1901, it was apparent from early species trials and informal plantings that radiata grew well in various parts of south-eastern Australia, and its growth was particularly spectacular in Tasmania. In wood-short South Australia, wood milled in 1902 from early plantings proved to be suitable for packing cases for the fruit industry, and it looked promising for an exceptional variety of other uses. Planting of radiata expanded in the first decade of the 20th century in extensive areas of the south-east of South Australia and, closer to the main market for it, on smaller areas near Adelaide (Lewis 1975; Boardman 1988). Nevertheless, as might be expected, the plantation enterprise in South Australia encountered and had to resolve a series of biological (and thus financial) crises during its first several decades.

By 1900, species trials had led to decisions to plant radiata in regions of appropriate soil and climate, and nurseries had been established to produce seedlings for the plantations. Just as South Australia began ahead of the rest of the world's radiata-growing regions in committing to growing this then-unappreciated species, so this state also led in research into how to grow the species and at how to use its wood (Lewis 1975). Initiatives in research in the prior 25 years to 1900 were followed by initiatives in plantation development. Commercial acceptance of its wood, however, did not come readily, so the South Australian Woods and Forests Department built its own sawmills including seasoning kilns. These mills served to establish important technology and standards in a process that has now extended over a century (Lewis 1975). Research initiatives on how to grow and use radiata were in some cases logically foreseeable, but in others they were reactive, in response to such disconcerting problems of radiata as "dieback" in the 1920s (found to be due to zinc deficiency), and to "spindle stands" in the 1940s (found to be due to low phosphorus).

To the north-east, in New South Wales (NSW) (Grant 1989), radiata had been tried fragmentarily in the late 19th century (Grant 1989; Sect. 2.3.2). The first "**commercial**"[G] plantation attempted was with radiata and maritime pines at Tuncurry (a coastal site, 30 km south of Taree) in 1912, new plantings continuing annually until 1934; the unsatisfactory nature of that locality's sites for radiata pine became progressively apparent and the stands were mercifully killed by fire in 1939. Other early plantings were at Armidale (starting 1915, totalling about 100 ha), Windsor (poor soil, 1917–1934), Wyong (1917–1918, a failure), Eden (1918–1935, a failure), Mathoura (on sandhills, 1918 and intermittently thereafter; small scale), Belanglo and Penrose (near Moss Vale, 1919–, reaching about 2000 ha by 1987), and Lidsdale (1919–, reaching 600 ha by 1987). As of 1918–1919, the total area of all pines (several species) on NSW State Forests was only 675 ha. There was a real problem in the early 20th century of identifying sites suitable for plantations—those close to markets were of poor quality, while better sites on the Tablelands were far from markets (Grant 1989). (The NSW Tablelands region, which includes the Great Dividing Range, extend from the north to the south of the state and adjoin the relatively narrow coastal plains; the boundary on the east is a contour line rising from about 600 m altitude in the south to 950 m in the north, and on the west by a line rising from 600 m in the south to about 750 m in the north (Fig. 3.3)). By the 1930s it was realised that radiata did really well only in the Tablelands region of NSW on sites that combined adequate winter rain, dry summers and adequate or better soils (Scott 1960).

Although the states of Victoria and then New South Wales, in 1968 and 1971 respectively, later surpassed South Australia in area planted to radiata, South Australia still had the most in 1951 (Table 3.1). Some of the plantations in Victoria were near the border with South Australia, in a planting region that crosses state boundaries and is now called the Green Triangle. The main radiata plantings in Victoria, however, were and still are in the eastern part of the state, in Gippsland. In New South Wales, several regions have small areas of radiata plantations, but the main radiata plantings were and still are concentrated towards the south, in the western foothills of the Great Dividing Range.

As was the case elsewhere, learning to use the wood was also important in gaining support for Australian radiata plantations (Carron 1985). In South Australia Lewis (1975), the annual harvest of radiata more than quadrupled between 1938 and 1950. As more radiata wood became available, uses for it extended from predominantly low-grade boxes to a wide range of sawn, round and pulp end-products. This expansion resulted from an impressive commitment to both basic and applied research on the fundamental properties and potential uses of radiata wood (Algar 1988). The Mt Burr sawmill was opened with a gang-saw in 1931; Penola with a bandsaw in 1937 and Nangwarry with a larger band-saw in 1941. The Mt Gambier mill opened in 1958 with the largest capacity in Australia at that time. Kraft paper was made at Botany (NSW) in 1927 (Benjamin et al. 1928). Cardboard production commenced at Millicent (SA) in 1941, and tissue manufacture nearby in 1960.

The fraction of South Australian sawn-timber needs supplied from plantation wood grown in that state increased from 19% to 48% during period from 1938 to 1950. This source of wood was invaluable during the World War II years of 1939–1945 when imports were largely cut off. Radiata contributed much of this increase, and this period was pivotal for gaining general acceptance of wood products from it (Bednall 1956). Following the war, full-scale production runs in Australian, American and UK pulp and paper mills verified radiata's early promise for the pulp and paper industry. It became the first species to be found suitable and used for the full range of paper products from tough kraft bags through newsprint to fine printing and writing papers, using both chemical and groundwood mechanical pulping processes (Dadswell 1961; Sutton 1986).

3.3.1.2 The Edge: Queensland and Western Australia

As with any species, even the regions where radiata grows well contain areas where transitional or different ecological conditions result in marginal or unreliable performance. Such areas, of course, occur near or within the favourable sites in south-eastern Australia, but they are more significant at sites where radiata was tried in Queensland and Western Australia (WA). In all Australian states there are widespread inappropriate conditions where it just cannot be grown successfully. But with the great success of this pine elsewhere, it seemed worth trying, and it is interesting to consider what was attempted in order to adapt radiata to these sites on the edge.

In Queensland, trials of both native and exotic species for plantations were organised by N.W. Jolly by 1916. Because of the importance of hoop pine then being cut from natural stands, there was much interest in planting that species; in 1922 the development of a successful planting tube (a cylindrical plant container) greatly improved establishment results for hoop pine and other species under trial. A review in 1926 perceived little role for radiata in Queensland except possibly in the high-elevation Toowoomba–Stanthorpe area in the south to provide wood for fruit cases (Carron1985). In that region, on the present Passchendaele State Forest, land for farms had been given to veterans of World War I but many were not able to make a living on the poor, shallow granitic soils. The Queensland Forest Service acquired much of the marginal land in the region. Pines native to the south-eastern United States or to Mexico were planted there, as was the increasingly-appreciated radiata. Productivity of radiata on those sites proved to be low (7–8 m^3 ha^{-1} yr^{-1}) compared to yields in more favourable regions (often over 20 m^3 ha^{-1} yr^{-1}). Frequent drought and two fungal problems (*Phytophthora*-caused root rot and *Diplodia*-caused dieback following hail storms) all troubled the average radiata. It was there, and in similar summer-rainfall areas in South Africa, that the interaction of hail and *Diplodia* became most apparent as a factor limiting successful radiata plantations (Rogers 1957; Scott 1960; Clapp 1995).

In WA, following English tradition, most early conifer plantings were on sites that neither had productive native forests nor were good for farms. Radiata seldom did well in such marginal situations. The first trial plantings of pines in WA had been made on coastal dunes near Bunbury in 1897, but they failed (Sect. 2.3.2). By 1916, 240 ha of radiata had been established near Hamel (100 km south of Perth and 10 km inland) and near Ludlow (100 km further south and 25 km inland), but on the poor sandy soils where it was mostly planted growth was again disappointing.

Extensive formal species trials, installed in the 1920s and 1930s, identified maritime pine and radiata as the two conifers most likely to succeed in WA conditions. These trials also better defined which local climates and soil types were best suited to each species. Maritime pine was satisfactory on dryer, less fertile soils and in hotter climates than radiata could tolerate. The Mediterranean-type climate where radiata excelled in WA forms only a relatively small region in the "highlands" of WA's south-west, 30–300 m asl. As in south-eastern Australia, soil quality commonly proved to be a decisive limiting factor (Scott 1960; Bartle and Shea 1989); the addition of appropriate fertilizers would improve the growth of radiata on some sites within that region, but could not overcome all site constraints. North of Perth, in areas of 600–700 mm annual rainfall, radiata had survived well and grown, but suffered serious sunscald on the north-western face of the bole, splitting the bark and degrading the exposed wood (Butcher 1997).

3.3.2 New Zealand

Kauri had been identified in the late 1700s as New Zealand's largest and most valuable conifer, and its wood was prized for many uses. Sawmilling of kauri harvested from native stands reached its peak in 1905, a mere 133 years after the first kauri was felled by Europeans. At about that time, several mobile mills had sawn radiata logs obtained from windbreaks or mixed-species farm plantings, particularly in the Canterbury region. A preview of the radiata era began in 1893 with the sawing of radiata logs from an 1870s plantation (Royal Commission 1913), the first New Zealand harvest of radiata from trees planted with sawn timber in mind. This activity was closely followed by the 1913 report of the Royal Commission on Forestry, urging the development of exotic forest plantations and recognizing the potential of radiata (Simpson 1973).

Chief Forester Matthews of the New Zealand Lands Department, charged with the early state afforestation program from 1896 to 1909, had at first thought radiata to be of little value except as inferior firewood. Thus, when it was included in timber-oriented plantations, it was mostly as a **nurse tree** [G] for species considered more valuable. Ironically, in most situations where it was so used, that underdog species far outperformed the species it was supposed to be nursing. This led to Matthews' belated and grudging acceptance of radiata, noting that it had been "giving the most satisfying results" (Matthews 1905). But of the 25,000,000 forest-tree seedlings planted between 1896 and 1909, only 110,000 were radiata, not quite one-half of one percent of that programme (Sutton 1995).

By 1900, T.W. Adams' trials and arboretum in Canterbury had been leading him and others to increasingly strong conclusions about the relative merits of the different tree species they had been observing there and in other settings. Adams included 50 native species in his plantings at Greendale, where a few grew well but many failed in the droughty Canterbury conditions. He had come to regard the natives as having more of a future as ornamentals than as reliable sources of wood. Even as ornamentals, he considered them "singular rather than beautiful." Among the exotic timber species he trialled, Adams was by then most strongly recommending about five as best suited to the Canterbury region. Based largely its outstanding growth on a wide variety of sites, he had rated radiata second only to Corsican pine (McKelvey 1991).

A.H. Cockayne, a botanist (whose father was a member of the 1913 Royal Commission), was among the first influential New Zealanders to see the future value of radiata as a timber tree. He published these views in 1914, with abundant and effective photographs (Cockayne 1914). In that publication, he reviewed not only the biological qualities of radiata's rapid growth and its adaptedness to many New Zealand sites, but he also surveyed and summarized the uses to which the sawn timber had been successfully put as of that date. For example, in an important and later much-quoted passage, Cockayne wrote: "*Pinus radiata* was first employed in house building some 37 years ago when Mr Cathcart[1] Wason had some 20-year-old trees converted into timber. With this material, several houses and outbuildings were erected in the vicinity of Barr Hill in Canterbury. These are still standing and the timber, which has been kept painted, is still in an excellent state of preservation."

Cockayne also thoroughly and persuasively documented the likely increased need for produce boxes (for shipping fruit, butter, cheese and minor products such as pickles); he projected the imminent shortage of the native kahikatea (earlier called "white pine" by colonists anxious to maintain connections to their Northern Hemisphere roots) then being used for this purpose; he discussed the several qualities of radiata that made it even better suited than kahikatea for produce boxes; he calculated how much radiata plantation would be needed and when; and he analysed the likely profitability of such plantations to their owners and the savings to the users of their plantation-grown radiata wood. He also noted the probability of technological advances such as wood preservatives that would qualify radiata sawn timber for other uses, some of much greater value than boxwood. Perhaps one of the most important people he influenced was his father, Leonard Cockayne, a pioneer plant ecologist whose opinion on matters biological carried much weight (Hegan 1993). While A.H. Cockayne was preparing the manuscript for his 1914

[1]This implies that the trees were planted in 1857, whereas they must have been planted in the 1870s. Yet to mill his own trees, however young, and to use the timber for his own building, were fully consistent with how Wason ran his Barrhill estate (Pawson 1993). Waning fortunes of the settlement, however, make unlikely sawing and building much after 1897. However, there is a record of a North Canterbury farmer, Duncan Rutherford, having milled 21–22-year-old radiata pines in 1893 for building farm sheds (Royal Commission 1913).

paper, Leonard Cockayne was busy as one of the six commissioners on the 1913 Royal Commission on Forestry.

H.A. Goudie was one of the hosts during the 1913 inspections of field sites by the Royal Commission. He conducted their visits to the Waiotapu and Whakarewarewa plantations in central North Island, where radiata was included, revealing favourable comparisons to other candidate species. Adams was one of the commissioners, and he made special efforts to ensure that both the growth and utilization of radiata received the attention of the commission. On several occasions during the 1913 fact-finding trips through New Zealand, Adams departed from the main itinerary to collect additional data on radiata's performance.

The year before he died, Adams was made a Life Member of the New Zealand Forestry League, an organisation dedicated to the effective management of New Zealand's forests. In his 1919 obituary for Adams, his fellow commissioner, L. Cockayne, wrote: "… his researches with regard to Monterey pine (*P. radiata*) as a timber tree can without hesitation be declared the most important advance which forestry has made in New Zealand up to the present time, and one which will eventually add great wealth to the country" (McKelvey 1991).

The 1913 Royal Commission surveyed and evaluated then-extant species trials and plantations. Based on their available 1909 data, radiata had ranked 18th in a list of the 19 species occupying substantial areas in state plantation. Remarkably, in the recommendations for a greatly expanded plantation program, radiata was finally listed as first choice on its new list of promising species. In making this judgement, the Commission noted the wide variety of New Zealand soils on which radiata grew well, the rapidity of its growth, and examples of successful uses of its wood ranging from butter boxes to sawn structural timbers. The report additionally opined that, had radiata been hardy in Europe, its "… admission into forestry practices would have been recognized long ago …." The Commission also was aware of radiata's performance in Australian plantations, and acknowledged its success there as providing some confidence in their recommendation.

Additionally, the Commission specifically recognised the variability of radiata trees in plantations, and recommended that seed be collected from trees "… proved to produce plants of the required character …" (Royal Commission 1913). This latter recommendation was accepted by staff in a few forests, was for many years ignored in most, and was not institutionalized by the NZ Forest Service until 1951.

By 1951 in New Zealand over 200 prospective timber species had been evaluated at one or more sites (Olsen 1996). It was from a combination of formal and informal species-introduction trials that radiata emerged as the clear species of choice for most New Zealand plantation opportunities. During the first half of the 20th century, radiata's wood uses progressed from mostly firewood to boxwood, to sawn timber, veneer and plywood, and beginning in 1937, to pulpwood (Simpson 1973).

Goudie was catalytic in bringing attention to radiata on several occasions, including his influence on the Royal Commission in 1913 and his suggestions leading to Perpetual Forests in 1917. His 1923 pamphlet, *The Remarkable Pine: P. radiata in New Zealand* (Goudie 1923), makes remarkable reading today.

It was particularly timely and influential in creating enthusiasm for radiata in the planting boom of the 1920s and 1930s.

An "Annexure" included in Goudie's 1923 pamphlet described plantations near Matamata, north-west of Rotorua in central North Island. It was written by P.M. Page, Forest Extension Officer; the following quotes from it describe the performance of radiata in the Matamata plantations, and perhaps capture the sense of the day (Page 1923):

> The species planted (in 1873) were *Pinus radiata* (Monterey pine) and *Pinus pinaster* (cluster (or maritime) pine), and the results obtained were greatly in favour of the Monterey pine, which yielded three times the amount of merchantable timber the cluster pine produced in the same period.
>
> The milled timber was sold on the trucks at Matamata in any-sized flitches at 18s per 100 board measure, and much of it was exported to Australia as New Zealand pitch-pine. Large quantities were also shipped to Fiji and other South Sea islands for the manufacture of containers. Several houses have been built at Matamata of timber produced in this plantation, and have given every satisfaction. Houses have also been built of this timber in many parts of the Dominion, and it is being used for building and joinery purposes more extensively each year.
>
> The general verdict now is that this timber is a thousand per cent better than it has ever been given credit for, and the Monterey pine is coming into its own.

As of the 1920s, the few homes and somewhat-more-frequent farm buildings constructed with radiata wood were unusual, and such use was mostly in wood-deficient regions (Sutton 1986). As A.H. Cockayne had suggested, the primary use of sawn radiata wood in the 1920s was for boxes in which to export fruit, cheese and butter. The ease of drying and nailing, good nail-holding, and absence of tainting chemicals, had indeed proved radiata sawn timbers ideal and even improved substitutes for those of native kahikatea, increasingly in short supply. Its use as firewood had established that its wood split readily for that purpose; so it was a pleasant surprise to learn that, compared to Douglas-fir's sawn timbers, radiata's did not split as often in the course of nailing, nor did amateur carpenters bend as many nails while hammering them in.

However, acceptance of radiata's wood for a broad range of uses was hampered by poor natural durability, and by the inferior properties of timbers sawn from the relatively young trees being harvested from quickly-grown plantations. Research has since shown that so-called "juvenile wood", properly termed "**corewood**"[G], comprises the central rings of radiata logs, whether formed in youth in the bottom log or, as the tree matures, in the upper logs, and such wood necessarily predominates in young trees. Several fundamental properties of the wood's cells that govern wood quality progressively improve from the inner corewood to the later rings termed "**outerwood**"[G]. In general, the more rings from the pith (which serves to identify the first year of growth at that point along the bole), the better the wood in the outermost rings. A more-or-less arbitrary level of acceptability for most purposes is attained by about the 15th ring from the centre, this varying somewhat among trees, among logs from the same tree, and among wood properties. These interesting fundamental properties include the length of the wood cells (specifically, of the **tracheids**[G]), the cross-sectional shape and thickness of the cells and their cell walls (affecting the density, or specific gravity, of the wood), the angle of the long-axis of the cells from the vertical (affecting the severity of spiral grain as a defect), and the angle of inclination of the **cellulose microfibrils**[G] which impart structure and strength to the cell walls. These properties of radiata's wood cells in turn affect strength properties of both pulp and solid-wood products, and the **dimensional stability**[G] and **stiffness**[G] of sawn-timber products.

Almost all of these properties are much improved by letting the trees grow longer before cutting them. As the age of the trees being harvested increased from about 20 (when all logs were mostly corewood) toward 35 or more years (when the lower logs were mostly outerwood), the sawn timber from them, on average, got better and better. While the details of causation were not understood at the time, the effects of increasing age at harvest were generally appreciated. Meanwhile additional products were being produced. Reconstituted fibreboards were produced from radiata pulp as softboard for insulation from 1941, and a few years later as hardboard. By 1946, experiments had shown that well-formed radiata logs peeled cleanly and smoothly for veneer products, with fewer drying and distortion problems than experienced with the native species it was replacing (Sutton 1986).

Research and new technologies to further reduce problems associated with corewood came mainly after 1951. They included better drying procedures to improve dimensional stability. Product grading was devised so that excellent sawn timbers would be properly valued and used for demanding purposes and lower-grade sawn timbers would be used for more permissive purposes. Highly effective treatment with various wood preservatives is facilitated by the same anatomical and chemical properties of radiata wood as give it poor natural durability. These properties allowed treated radiata sawn timbers and posts to be generally among the most durable of either treated or untreated woods of all other species. Other pleasant findings included the relative ease of gluing, staining, painting and varnishing radiata wood, and its good fastening properties. Also, logs of young radiata were much more amenable to mechanical pulping than those from various other pine species. This knowledge led to radiata's wood being used in a greater range of products than even its early enthusiasts could have envisioned.

For foresters and sawmillers with European training or experience, a continuing concern damping the rising enthusiasm for radiata was that it grew *too* fast. For Europe's major timber species, Scots pine and Norway spruce, rate of diameter growth was and is a strong inverse indicator of wood quality. It took both research and accumulating experience to show that this need not be the case for radiata, and thus to overcome the widely-held belief that its fast growth meant low-quality wood (Sutton 1995). The association of wide annual rings with low wood quality is not a causal one—the fundamental relationship is that corewood is of lesser quality than outerwood whatever the width of rings in the latter. Thus wide outerwood rings associated with continued rapid growth do not preclude high wood quality.

In 1925, Owen Jones, disappointed with the Victorian Forestry Commission's unwillingness to implement his policy of forestry expansion (he was its Chair), came to supervise and implement the Perpetual Forests plantings. He brought a background of education and experience in Europe, a visit to the native forests at Monterey and experience gained in Australian radiata plantations. It was his leadership and knowledge, and the dedication he inspired in his nursery and woods crews, that carried a financial venture to plantation reality as the Perpetual Forests enterprise grew.

In a 1929 speech to the Science Congress in Auckland, A.R. Entrican, then the Forest Products Engineer for the NZ Forest Service, elaborated on Ellis' vision of two million hectares of planted forest making the country a major exporter of pulp and paper (Simpson 1973). That suggestion, while premature, was prescient. It was, however, based at least in part on work done in 1927 during a stay at the US Forest Products Laboratory in Madison, Wisconsin, where Entrican and local colleagues demonstrated the feasibility of making newsprint entirely from radiata—a world first with pine wood (Orr 2017).

New Zealand's climate differs from that of coastal California, and of course varies widely within the country. Where radiata grew unusually well in New Zealand the climate was usually maritime or even oceanic, characterized by higher humidity and greater rainfall over a greater part of the year than in its native populations. Most tree species that evolved in temperate climates, Douglas-fir for example, are genetically programmed to cease growth in response to environmental cues such as decreasing hours of daylight. This leads to them going dormant in advance of freezing weather, but at the price of foregoing growth on many spring, autumn or even winter days that favour growth. Radiata, while strongly responsive to environmental cues for such things as reproductive-bud initiation, does not go into a defensive winter-dormant condition. Thus, it is capable of growing whenever conditions are favourable. This results in outstanding growth in the conditions prevailing in large areas of New Zealand, but also in greater vulnerability to episodes of severe winter cold. It is, however, surprisingly tolerant of frost during the long growing season. Furthermore, it was increasingly appreciated that radiata could tolerate and even grow well over a wide range of soils and other site variables (Scott 1960). As knowledge of these risks and advantages accumulated during these first five decades of the 20th century, the trade-offs seemed worth the gamble, and radiata was increasingly preferred among the exotic species being planted.

Notably, radiata can do well the entire length of the country—a range of over 12 degrees latitude—except where elevation is excessive, terrain prohibitive, soils unsuitable or rainfall greatly excessive.

During the first planting boom in the 1920s and early 1930s, several conifer species were extensively planted including, after radiata, ponderosa pine, Douglas-fir, Corsican pine and coast redwood. In order to maintain some species balance, a policy decision in 1931 stated that the species mix on state plantations would attempt to reduce the proportion of radiata to about 30%. This required that preference be given to other species, to balance the increasing enthusiasm for radiata pine (Sutton 1984). There was concern that enthusiasm for radiata would lead to too many eggs in one basket, both with respect to the anticipated needs for different kinds of wood, and the possibility of outbreak of destructive pests or pathogens in an extensive monoculture. The percentages by area of exotic plantation of radiata were basically set in New Zealand until the late 1960s by planting up till 1935. In fact, the percentages at 1964 were 49% radiata in State Forest plantations and 62% overall (Poole 1966).

That prescribed limit on the percentage of radiata in the plantations was often exceeded in the early 1930s, due largely to the short planning horizons driven by the need to quickly provide employment for laid-off workers during the Depression. Seedlings of the other species required two or even three years in the nursery, while spring-sown radiata seedlings were field-ready for the next winter planting season. Radiata's rapid growth following planting also allowed less after-care, and post-planting neglect was common in those chaotic economic times (Allsop 1973). Furthermore, other favoured species proved not only slower growing and more limited in their range of uses, they each proved to have their own set of troublesome site-requirements, pests and pathogens (Burdon 2001b). Experience was rapidly accumulating on the use of sawn timber of radiata. In South Australia, a shortage of wood during the 1939–1945 war forced the premature harvest of some radiata timber stands, whereas in New Zealand large-scale sawing had begun with the opening of the Waipa state mill. The resulting experience of the sawn timber brought greater familiarity with it and acceptance of it. By 1950, radiata was still looking increasingly good.

During the 1920s and 1930s, despite some alarms over epidemics of shoot dieback (Burdon 2011), radiata in New Zealand suffered from fewer long-term biological mistakes and problems than most of the other species being planted. Much was learned from those trial-and-error mistakes that were made with it, and such things as insect or disease epidemics had not become limiting. Its reputation benefited from not only its fundamentally good qualities, but also from the increasing skill in growing it being developed by dedicated workers and several truly gifted foresters during those pioneering times.

Radiata comprised about 40% of the state planting areas and over 85% of the private plantings during the first planting boom. Then, the pace of all new planting declined sharply during the mid-1930s through the mid-1940s, as the economic depression deepened and World War II took centre stage (Sutton 1986). This first exotic-tree planting boom ended partly because of the general economic climate of the mid-1930s, and also because both biological and silvicultural mistakes had led to some plantation failures and some substantial financial losses. Many of these occurred where exotic species other than radiata, notably ponderosa pine and coast redwood, had been planted.

The great success of the plantations, particularly those of radiata, paradoxically led to some problems. A temporary over-abundance of harvested wood resulted in low prices and difficulty in marketing trees that were otherwise ready for harvest. This in turn led to a lack of enthusiasm for further plantations, and to reducing the effort devoted to tending those that were already established. One of the self-inflicted institutional problems of the 1930s had its roots in the New Zealand public's perceived right to own inexpensive houses. The state's response to this expectation was to log high-quality, old-growth native trees and then sell their sawn timber at almost giveaway prices. In addition, state and other lending agencies during this period refused to provide loans for houses built with radiata, ignoring the quiet acceptance of its timber for construction in some areas of the country.

These policies not only frustrated efforts to have the increasingly-available radiata sawn timber accepted, it frustrated the Forest Service's repeated efforts to fulfil the 1913 Commission's recommendations and Ellis' goal concerning conservation of the native forests. The continued large-scale logging of native forests, long after 1951, fuelled a powerful public movement for their preservation (Sutton 1986), the Forest Service getting most of the blame.

Despite the combination of financial, political and biological mistakes and problems, by 1956 nearly half of the annual log-volume harvested in New Zealand was from those exotic plantations, and 95% of the exotic species harvested was radiata (Scott 1960).

3.3.3 *South Africa*

The early attention to exotic plantations in South Africa (Fig. 3.4) by both its public and private sectors (Scott 1960; Poynton 1977) is not surprising, given the very limited native timber resources. During the few years following 1875, plantations of radiata had been established at Tokai, 20 km south of Cape Town, using seeds from trees in the Botanic Gardens in that city. Large trees were cut in 1918–1919 from 70 ha of those earliest plantations. By 1922, 38,000 ha of plantations existed; by 1938 South Africa's plantation area had risen to around 140,000 ha, 96,000 of them planted to various conifers (King 1938, 1951); and by 1951 the areas in plantation areas totalled around 1.3 million ha, including 137,000 ha of conifers. The conifers were almost all pines of which radiata would have been around 10% (cf Scott 1960).

Fig. 3.4 Map of South Africa (see text for location of radiata plantations)

The history of planting radiata in South Africa during the first half of the 20th century has four periods. From the beginning of the century until 1910, although plantation forestry with exotic species was well advanced (King 1938, 1951), radiata apparently was not yet much appreciated, and many other pine species were being trialled (Poynton 1977). Then, during 1911–1925, about 12,000 ha were planted to radiata, over a wide area. Many of those seedlings were planted in inappropriate sites and, after stunting on poor soils and *Diplodia* dieback in wrong climates, it fell out of favour for summer-rainfall areas. That disfavour extended by 1935 to its use on certain kinds of uniform-rainfall sites. Although not exploited at the time, propagation of radiata as rooted cuttings was achieved in South Africa in 1938 (Anon 1938a). By 1940, however, enough was known of radiata's limitations, values and requirements so that planting was resumed on increasingly characterised sites in the region of winter-rainfall climate (Scott 1960).

From 1911 to 1924, some and perhaps most of the radiata seed used to establish radiata plantations evidently came from New Zealand. Even in 1925, when radiata was becoming distrusted in South Africa, the New Zealand forest service sent a further 141 kg of radiata seed from the developing land race of its Nelson region to South Africa (Vincent and Dunstan 1989). Although Poynton reported that some of the episodic afforestation effort used seed mostly collected locally in South Africa, some 6400 kg of seed were also imported. Those imports were mostly from New Zealand land races, but smaller shipments are recorded from Europe, Australia and the United States (Poynton 1977). Thus, the South African radiata land races would have been founded by which seed importations came from where. The long-term winners were doubtless enriched by the later, large-scale importations and then by natural and silvicultural selection on the sites where they were grown.

South Africa has a relatively long history of effectively developing exotic plantation forestry, including outstanding work on the silviculture and utilization of radiata. By the 1950s, the combination of soils and climate inappropriate for radiata, typical of most of Africa, appeared to limit radiata to about 50,000 ha in the western and south-eastern portions of the Cape Province. Elsewhere in South Africa, infertile or poorly drained soils limited it where climate is suitable. High summer humidity, particularly when combined with hail and the subsequent shoot dieback caused by the *Diplodia* fungus attacking the hail-induced wounds, rule it out on otherwise appropriate soils. However, by the 1950s, its growth potential, silvicultural properties and timber qualities were recognised as making radiata the preferred species on sites appropriate for it (King 1951). Even on marginal sites, it was often planted in preference to other, better-adapted species in the expectation that the harvest from an indifferent stand of radiata is likely to be better than that from a healthy stand of other species (Scott 1960). In later years, a realisation that readily-corrected phosphorus deficiency was often the limiting problem on many sites has led to a further shift from maritime pine to radiata. The early extensive use of radiata in summer-rainfall areas in both South Africa and Zimbabwe had largely ceased by 1950, and radiata's use in those regions is now confined to quite small, select areas.

Acceptance of sawn timber of radiata came readily in South Africa, where tree form and wood properties were generally good.

3.3.4 Chile

Like Adams in New Zealand and von Mueller in Australia, the elder Arturo Junge became a catalytic advocate of radiata in Chile. After his initial trials, apparently begun in the 1890s, he promoted modest plantings in Chile's Concepción region (Junge 1953). The first substantial planting was by a coal company, which established about 400 ha of radiata pine for possible use as mine timbers near Lota between 1907 and 1912 (Contesse 1987). Those plantations very likely originated directly or secondarily from the California Monterey population (RDB—pers. observation of cones and seeds near Lota).

Arturo Junge's son (or grandson) later recalled an incident a few years after the elder Junge began promoting radiata. Maritime pine had been successfully employed in a massive plantation in the Landes Region south of Bordeaux, France, on sand dunes and a vast area of nearby low-lying sandy areas. Sr Mahuzier, a Chilean neighbour who came from the Landes, was planting maritime pine in the region of Concepción. A good-natured controversy developed between these two advocates of radiata and maritime pine. A third party, General Koermer, suggested they settle the argument in a practical way by giving the city of Concepción plantations of both species on the bare slopes of an eroding hill, Cerro Caracol. This was done in 1893 (Junge 1953). While the outcome of that pine contest is not recorded, it was surely similar to that of the planting of these two species at Matamata in New Zealand (Sect. 2.3.5).

Radiata soon became irrelevant as a supplier of mine timber in coastal Chile, because planted eucalypts have better properties for this purpose. But some people recognised that radiata's rapid growth and long fibre made it suitable for general construction and papermaking. By 1935, it had been planted on 25,000 ha, some in parks and small plantations in various places (Krebs 1973), but mostly on hillsides near Lota (Clapp 1995).

The climate around Concepción is essentially moist and mild Mediterranean, with most rain in the winter, relatively dry summers and summer temperatures that are warm but rarely very hot. Troublesome winds, frosts, snow and hail are all unusual. Thus radiata proved to be especially healthy and vigorous around Concepción (Fig. 3.5, Region VIII). Although many of Chile's plantation-site soils were poorer than those in most New Zealand plantations, much had been learned about appropriate siting during Chile's first round of plantations. As was evident by 1950, compared to New Zealand, the tradeoffs between Chile's more appropriate climate but poorer soils had resulted in similar average levels of productivity in the two countries' radiata plantations (Scott 1960), However, it should be noted that at the time of that mid-century comparison, net productivity in New Zealand had been temporarily depressed by an epidemic of the woodwasp, *Sirex noctilio*. That epidemic has not been repeated in New Zealand, and eventually even proved beneficial in overstocked stands.

Chile began its large-scale plantation program much later than did Australia and New Zealand. Thus, by the time Chileans decided which species to plant

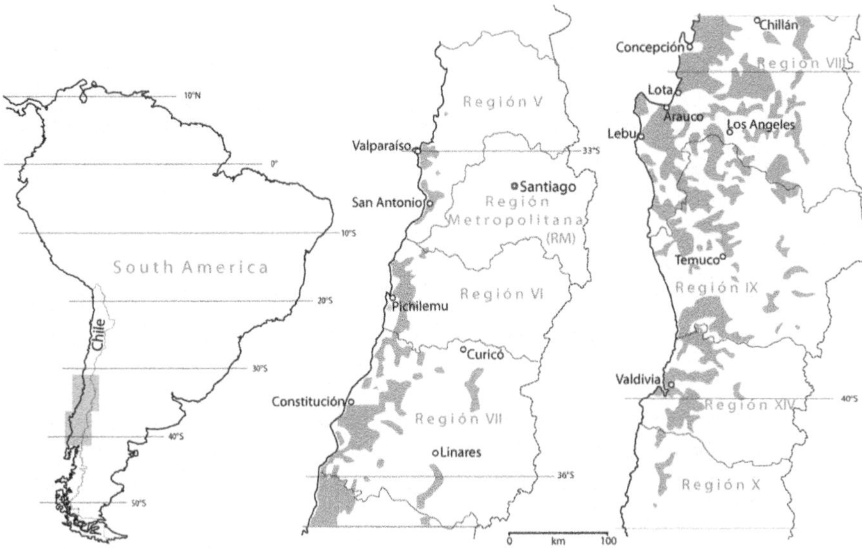

Fig. 3.5 Maps showing current concentrations of radiata in Chile, largely reflecting locations of suitable sites. (From Lewis and Ferguson 1993, permission Ferguson)

more extensively, radiata was already recognised as being among the most promising plantation species on Earth. Chile's commitment to radiata quickly became both absolutely and relatively greater than that of any other country. Large-scale planting began in Concepción province (Región VIII) in about 1935, and soon spread to lesser areas in eight other provinces. In all nine provinces the rate of planting increased through the early 1940s, peaking in different provinces between 1945 and 1951 (Krebs 1973). By 1950 Chile was second only to New Zealand in area planted. Furthermore, radiata made up about 99% of the exotic conifers then being planted in Chile (Scott 1960).

3.3.5 *Spain*

Of the several exotic species tried in the Spanish afforestation program, a eucalypt, blue gum (*E. globulus*), and radiata proved to be the most successful. The main region favoured for growing radiata is on Spain's north-west coast, in the Basque Autonomous Region (Fig. 3.6) between the Cantabrian Mountains and the Bay of Biscay, immediately to the west of the Pyrenees Mountains. The first radiata plantation had been established in 1871, and the pine's excellent growth performance had been noted by 1898 (Michel 2006). Indeed, radiata's unit-area productivity was soon thereafter estimated to be about six times that of Spain's native conifers, attracting both public and private investment. A steady increase in the areas planted began around 1915, and by 1958 about 50,000 ha were growing radiata,

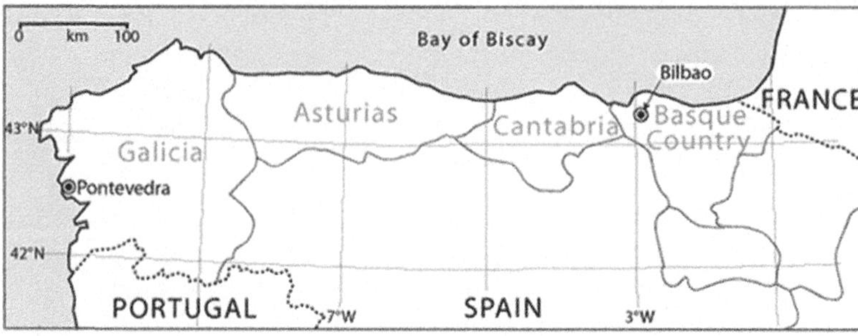

Fig. 3.6 Map of northern Spain (see text for location of radiata plantations)

almost all under 20 years old (Scott 1960; Michel 2006). These early plantations were generally being harvested on short rotations of 16–30 years, depending on the site, and whether the wood was to be used for paper, mine timbers, concrete forms or boxes (Scott 1960).

3.3.6 East and South-Central Africa

The success of radiata as an exotic commercial species in Australia, New Zealand and South Africa in the early 1900s made it a prime candidate for planting elsewhere. The high-altitude grasslands of Kenya, Tanzania and Zimbabwe have fertile soils and, particularly in summer, high rainfall. When plantations were first being established there, those two attributes plus low establishment costs on grassland sites made such areas obvious targets for afforestation, particularly since those regions had no natural conifer forests of any economic consequence. Although a few radiata pines had been introduced in the early 1900s, the main operational plantings were done between 1925 and 1935. By 1940, 3300 ha had been planted in Kenya, mostly above 2500 m elevation; 2500 ha in Zimbabwe between 1700 and 1900 m; and 400 ha in Tanzania (Tanganyika) at about 2000 m (Streets 1962). There was by this time evidence from earlier plantings that radiata grew very fast in the region, and that it produced excellent structural timber (Barnes 1970).

Further plantings were done between about 1000 and 2500 m elevation in Kenya, Tanzania, Uganda and Malawi (Nyasaland) following World War II, commencing about 1946. The seed used in Kenya was generally imported from New Zealand, and that used in Tanzania was from South Australia. Early growth rates were generally excellent. However, early bole malformation caused by boron deficiency and several disease problems (particularly dieback caused by the fungus *Diplodia pinea*) were noted. By 1950, a few thousand additional hectares had been planted in that general region, with plans to continue afforestation on appropriate sites. Predictions of establishing a valuable radiata resource there were cautiously optimistic, given that "*Diplodia* and other pests do not restrict its use" (Scott 1960).

3.3.7 Ecuador

Plantations near the equator that included radiata were established beginning in about 1928, at elevations near 4000 m. The climates there are generally cool and misty, but with several sunny hours per day. There is no dormant season, nor is there an obvious summer. The radiata pines reported on in 1955 had grown modestly by radiata standards, attaining an average height of only 16 m at 27 years of age, but averaging 16 $m^3\ ha^{-1}\ yr^{-1}$ (Scott 1960). At mid-century, it was estimated—optimistically as it now appears—that about a million hectares might be available for planting radiata, Monterey (macrocarpa) cypress and (later) Mexican (white) cedar (*Cupressus lusitanica*).

3.3.8 Elsewhere

As of about 1950, radiata had been tried with varying success in many other places, for example in Brazil, Argentina and Uruguay, at mid-to-high elevations in Peru, Bolivia and Hawaii, in several Mediterranean countries, in the milder parts of Ireland, and in North America in the south-eastern United States, in coastal Oregon, Washington and British Columbia, and throughout much of lower-elevation California. Some of these ventures, trials and experiments were promising and others were, for example in the south-eastern United States, rapid and complete failures.

Between roughly latitudes 10° and 30° on both sides of the equator, very few sites have proved to be suitable for radiata plantations (Burdon 2001a), although isolated trees often fare better. The native populations on Cedros Island and Guadalupe Island are at 28° and 29°, but the trials before 1950 were with mainland-origin stock. Whether radiata from the island populations might narrow this exclusion band has not been explored but it seems unlikely. The mesoclimates that support the pines on Cedros and Guadalupe are seldom if at all closely matched at similar latitudes. At lower altitudes within this 10–30° zone there is often a "damp heat" problem, where radiata pines usually succumb to fungal diseases; the failures often went undocumented after initial good growth was enthusiastically reported. At most other lower-elevation sites, there are frequent lethal excessively hot, dry episodes. At higher altitudes there tends to be either excessive humidity, leading to serious disease problems, or an excessively long dry season. Key nutrients are often deficient in the soils within this latitude band. Nearer the equator, however, as in Ecuador and Kenya, there are relatively small, high-altitude (generally 2500–3500 m) zones where radiata can be satisfactorily grown (Burdon 2001a).

Radiata has been repeatedly and extensively tried in France since 1787. Although the young trees often succeeded for a decade or more, sooner or later a mass of cold continental air moves through France from the north-east, killing radiata pines, eucalypts and other cold-sensitive trees. Radiata has survived these events on the Brittany Peninsula, in the lower foothills of the western Pyrenees Mountains (with less frequent disasters), and close to the coast between these two regions of France. However, the processionary caterpillar is damaging to radiata near the coast (Burdon 2001a). These caterpillars seem to prefer radiata, and they cause their damage by eating many of the tree's needles. These interesting pests move from tree to tree along the ground as a single column of 20–30 rapidly-moving caterpillars, hence the name. However, they spend most of their time eating foliage within spun webbing that protects them from sprayed insecticides and most predators. These caterpillars are therefore hard to control, and that has discouraged extensive commitment to radiata pine in several regions where it otherwise thrives.

In Great Britain, individual radiata pines had survived and even grown into large trees, mainly in the less-frosty south and west, but use of radiata as a plantation crop remained rare. On this densely-inhabited island, land suitable for radiata was and still is generally farmed. Furthermore, a mysterious condition called "yellows" occurred with much tree-to-tree variation, apparently triggered by mild, cloudy weather in late winter. Yellows produces severe needle-cast progressing from the older to the younger foliage, with a consequent reduction in growth. A similar although perhaps different "yellows" problem with radiata was later studied in Tasmania, but no clear causal explanation was discovered (Podger and Wardlaw 1990a, b). Frost is less limiting in Ireland than in Great Britain, and a few plantations of radiata have been established there. However, radiata does not do well on Ireland's common peaty soils, and "yellows" occurs throughout.

Around the Mediterranean Sea, radiata would doubtless have featured in many trial plantings before 1950, for which there is little readily accessible documentation. Systematic documentation of its potential in Italy, by Alberto Eccher, began in the 1960s, but is mainly in Italian. There, radiata pine often starts promisingly, only to perform poorly after canopy closure in the face of summer heat and drought (A. de Phillipis, pers. comm. to RDB 1963). Two insects, the processionary caterpillar and the European pine shoot moth, very often cause severe damage, even in northern Spain. The shoot-moth larvae kill the top one or two metres of the leading shoot, with resulting bole crooks and forks when one or more branches become the new leading shoots. With repeated or continuous outbreaks, the multiply-forked "cabbage pines" become nearly worthless in spite of continued fast growth (pers. observ. WJL 1980). Within stands of such "cabbage pines" are occasional single-boled trees, apparently due to them initiating a single fascicle-origin new leading shoot below the killed top. These could be interesting trees in future breeding programs. Nevertheless, substantial areas of plantation have been established in Sardinia (G. Mughini pers. comm. to RDB 1982), with more-modest areas in Albania (C. Palmberg pers. comm. to RDB 1989), Greece (wjl picture of radiata plantation around Aristotle statue) and along the Sea of Marmara and Black Sea in Turkey.

In the south-eastern United States, a large plantation effort was launched in the 1920s that included millions of radiata seedlings being planted in a single year. Three years later, no radiata had survived (B.J. Zobel pers. comm. to WJL 1992).

They had died so quickly that the causes were never ascertained. But it was clear that some lethal combination of the south-eastern climate and biota ruled radiata out as a viable option for the region.

On the North American west coast, by 1950 or shortly thereafter, modest plantations of radiata were established in the Coos Bay region of Oregon, with smaller trials growing successfully in coastal Washington and British Columbia. T.W. Stoate from Australia spent some time in northern California, and many small trials and plantations were established there and even at lower elevations (generally < 1000 m) in the west-slope forest zone of the northern Sierra Nevada Range as a result of his enthusiasm for and experience with radiata in Australia. It was also commonly used as a landscaping tree in many near-coastal housing developments and parks. Forestry consultants in California found radiata's easy establishment and fast early growth strong selling points for small landowners wishing to replant cutover land. In urban settings, however, those fast-growing pines soon became problems for their neighbouring homes, dropping large branches on roofs and vehicles, and explosively propagating the wildfires that periodically occurred in California. Western gall rust, a pathogen widespread on pines in western North America, soon appeared in many of the young radiata plantations and became epidemic in some. Where it became epidemic, some young pines were killed by it. However, its main effect was to deform and weaken a high percentage of the valuable lower logs, resulting in a quick drop in enthusiasm for plantations in its native country. A common observation by foresters managing the better forest sites in north-coastal California was "Why grow a risky species like Monterey pine, when we can grow safer, more-valuable redwood?"

Thus, by mid-century, the period of species trials and resulting identification of regions and kinds of sites suitable for radiata was drawing to a close. In those regions where radiata was a sufficient success, the basis for the next stage of its domestication was available.

3.4 The Development of Plantation Management

As commitments were made to serious plantation programmes, fine-tuning of available knowledge commenced. After deciding to plant radiata pine in a region, appropriate sites needed to be located. Meanwhile, techniques of storing and germinating its seeds and raising its seedlings were being refined. The site-preparation, planting and aftercare practices that were needed to establish the plantations were being evaluated and modified. Once established and growing, the trees needed to be protected and managed until ready for harvest. Harvesting practices needed improvement, with respect to both efficiency and impacts on the sites—care was necessary to ensure that site resources essential for succeeding crops were conserved. Finally, the harvested wood needed to be turned into useful products, and its potential users needed to be educated in order to win market acceptance. All such developments needed a lot of research.

The forestry sector, be it private or public, had a long tradition of putting a relatively small fraction of gross or net funds into research, compared to agriculture or most other manufacturing enterprises. Sensible people—guided by generally accepted rates of return on investments—probably would have accepted the general opinion that forestry is a second-class enterprise, deserving only low levels of research attention and investment. But, as contrasted to the hunting-and-gathering mind-set often found among most people who exploited native forests for their wood, there often was and is a special quality associated with people devoted to first planting and then raising trees. It is not often described as "sensible." If it succeeds well it may be described as visionary, generally long after those concerned have died.

As the problems of where and how to begin to grow radiata became satisfactorily addressed, important management decisions began to focus on "to what size or age to grow it" and "for what to grow it." It seemed appropriate to grow it for needed and high-value products. Sawn timber and plywood stock became the logical foci of "for what to grow it," with various reconstituted products (pulp for paper, chips for chipboards, with fibreboards and various glued products following later), and ancillary products (cones, bark, mulch, boiler fuel) occurring incidentally to or as residue from timber harvests and milling. Wildlife habitat, recreation, soil and watershed protection, and livestock shelter were sometimes primary management goals but, far more commonly, these were added benefits of establishing forest plantations for the primary purpose of timber production.

Once the "for what?" was reasonably answered, the "for how long" question became more focused. For many tree species, an upper age at harvest had often been determined by biological considerations, such as the age at which losses to insects and diseases become unacceptable (sometimes referred to as a "pathological rotation") or the age at which many members of the species lose vigour and mortality becomes common. Most western North American timber-producing species live two, three or more centuries before serious losses or old age become problems.

Few if any radiata pines reach 200 years of age, although some of the surviving trees on Guadalupe Island may be substantially older. The native mainland-California stands often encounter difficulties after about 60 years, and are typically breaking up by 100 years. By 1950, some of the older radiata plantations, for example on South Island in New Zealand and at Mt Burr in South Australia, were giving evidence of remaining healthy for longer than that, but radiata plantations on some of the warmer and wetter sites were becoming decadent within 60 years (Burdon and Miller 1992). Thus, "where" became an important component of the upper expectation of "how old." But it was also becoming known that radiata on most sites could produce high-quality logs in far fewer than 60 years, often in 35 years or even sooner. In those cases where the primary management objective was wildlife habitat, soil or watershed protection, and/or recreation, then longer-lived species could be and were increasingly judged more suitable than radiata, including plantations of locally-native species. In sharp contrast, although radiata is relatively short-lived for a timber tree, the focus of the "how old" question for timber harvest became the youngest age for satisfactory harvest rather than the oldest to safely harvest.

To decide on the lower age of harvest, managers need to learn how long stands must grow to produce the desired product(s). Once that is determined, then a trade-off is considered between how additional years of growth add to both quantitative and qualitative value of the product(s) and of the other services provided by the plantation, against the cost of the money tied up in the plantation minus costs of re-establishing a new plantation. The nutritional cost of the felling and re-establishment operation also became increasingly relevant to the question of sustaining site productivity (Jolly 1950; Squire and Flynn 1981; Nambiar 1996; O'Hehir and Nambiar 2010). How soon to cut to optimise the values of the products plus various environmental services and the financial returns to investors was not, as of 1957, resolved (Kennedy 1957).

Financial analysts who become involved in these management decisions naturally worry about the cost of money, which pushes forest growers (or their managers) in the direction of reducing harvest ages. Their actual calculations, however, have chronically tended to overlook or under-factor two important aspects: (1) improvements in wood quality that can accrue with increasing tree age, which have tended not to be recognised by timely market signals; and (2) various important public benefits of having forest cover on land.

3.4.1 Nursery and Planting Practice

From the times of the earliest plantations in the previous century, and drawing on experience from horticulture and European forestry, nursery and planting techniques specific to radiata were tried and progressively refined for local conditions (Matthews 1905; Maxwell 1930). The early metrics evaluating nursery and planting performance tended to focus mostly on survival of the planted seedlings. In New Zealand, Goudie's 1923 pamphlet enthusiastically described nursery and planting practices that resulted in radiata pines a metre tall at age three years. This performance was excellent compared to European, North American and other Southern Hemisphere experience at the time and with other species, yet is far below what later became acceptable standards for radiata. Radiata seemed to perform remarkably well both in the nursery and during plantation establishment, and few foresters in the 1920s had any idea how good it could be with even better practices.

When he assumed leadership of the New Zealand Forest Service in 1920, Ellis found that basically European forest-nursery and plantation-establishment practices were being applied. These practices included dense sowing of seeds in the nursery, followed by a labour-intensive transplant within the nursery to grow them on to a relatively large size as "1/1" seedlings over a total of two years (Matthews 1905). These large seedlings were then field-planted at very close spacing (1.2 by 1.2 m) into pre-dug pits. The close spacing in European practice was meant to keep diameter-growth rates low and branch (thus knot) size small. The practice of transplanting in the nursery and producing very large seedlings was used for radiata in Tasmania and Victoria, Australia, particularly for unfenced planting sites where browsing was a significant problem (Eldridge pers. comm. to WJL 1997).

L. McIntosh Ellis introduced machines that, while relatively crude, sowed radiata seeds at wider spacing in the nursery. This gave the germinants a better start so that they could be grown in one season to field-readiness without a nursery transplant. The resulting smaller, field-ready "1/0" seedlings did not require pre-dug pits in the field. Furthermore, he increased prescribed field-spacing, and thus fewer seedlings were planted per hectare. These modified techniques and practices, once established, quartered the cost per hectare of plantation establishment, while still achieving satisfactory per-hectare growth of the resulting plantations.

Ellis' innovations (Roche 1990; Boyd 1992) were crucial to maintaining and even increasing support for the New Zealand planting boom of the 1920–1930s, particularly during the cost-cutting depression years (Allsop 1973). While changes in spacing of planted seeds in the nursery and of seedlings in plantations proved biologically and fiscally sound for radiata, after Ellis departed and the depression stringencies became intense, inevitably there were many corners cut. Most of those changes did not work out so well, to the detriment of some of the plantations then being established. One experiment, however, that did succeed over a significant area was direct drill-sowing of seeds on some nearly-flat, lightly vegetated sites (Roche 1990, Boyd 1992). Where higher-than-expected germination and survival of the seedlings produced very crowded rows of trees, the situation was saved by the species' great capacity for self-thinning.

In Australia and New Zealand the rediscovery (cf Elwes and Henry 1910) of successful propagation of radiata pine by rooting cuttings was reported in 1934 and 1937 (Field 1934; Anon 1937; 1938b; Jacobs 1939). In Canberra, about 2 ha of plantations were established with rooted cuttings in 1938, followed by about 40 ha in 1939. More field trials of such cuttings were also established at Mt Burr, SA, beginning in 1940 (Anon 1939; Pawsey 1967), and elsewhere soon after.

Despite the excessive cost-cutting that sometimes occurred, much excellent plantation establishment occurred during the 1920s and 1930s. New planting, however, largely lapsed, especially during World War II, with much disbandment of nursery and planting crews, with significant loss of hard-won expertise.

3.4.2 Soils and Nutrition

Many plants, at least as young seedlings, have "root-hair" extensions from their outer cells near the root tips, greatly increasing their surface area and assisting them to absorb water and minerals. In contrast, the young, usually white extending roots of conifers are smooth. Those conifer roots, however, typically have close mutualistic associations with several species of fungi that form an external sheath over the outside of the roots and modify them to create distinctive structures called **mycorrhizae**[G] (Atkinson 1983). Through the extensive mycelia of the fungi (thread-like growths extending into the soil), mycorrhizae greatly increase the surface area of absorbing tissue exposed to the soil environment. This feature is particularly significant for the uptake of phosphorus in nutrient-poor soils, where that element is suboptimally present and available unless phosphorus fertiliser has recently been applied; uptake to the roots occurs with the help of their intimate proximity to mycelial surfaces. Improved phosphorus availability usually improves root growth, which in turn enables better uptake of water and other elements.

Where radiata has been planted outside of California, its associated mycorrhizal fungi generally soon appeared in the soil ecosystems, having presumably being introduced with seedlings or soil or even in infected seeds from California. In newly developed nurseries, however, or in any nursery where the soil is routinely fumigated and thus sterilized, introducing mycorrhizal fungi in forest **duff**[G] from healthy pine plantations into the nursery soil, or inoculating the soil with appropriate fungi from laboratory cultures, helps ensure quicker, more-even seedling growth in the nursery and in the field. Despite some Australian reports of mycorrhizal deficiency in the field (Lewis and Ferguson 1993), the only recorded large-scale absence of mycorrhizal fungi was in Kenya (Pudden 1957). Many fungi can form such beneficial associations with radiata pine roots, for example *Hebeloma crustuliniforme* (from Europe and North America, and introduced into Australia), *Suillus* spp. (mostly from northern temperate locations, some species introduced to the Southern Hemisphere), *Inocybe* spp. (widely distributed including Australia), *Amanita muscaria* (native throughout temperate and boreal Northern Hemisphere, widely introduced to the Southern Hemisphere), *Rhizopogon luteolus*, *Rhizopogon rubescens*, *Rhizopogon roseolus* (*Rhizopogon* is widely distributed; *R. luteolus* was deliberately introduced into radiata plantations in Western Australia early in the 20th century), *Laccaria laccata* (from North America), *Endogone flammicorona* (from Europe and North America), *Boletus* spp. (from North America and Europe, widely introduced elsewhere) and *Tuber* spp. (from North America and Europe; cultivated in New Zealand and Australia). In fact, a spontaneous succession of mycorrhizal fungi has since been demonstrated with radiata, different species being dominant at different stages of the host's growth (Chu-Chou 1979).

Dieback problems in Australian radiata plantations were reported in 1907. The dieback of terminal shoots not only stunted growth, but often resulted in deformed boles when one or more surviving branches became the main stem(s), with resulting bole fork or crook. There was a loss of confidence in radiata in the regions of South Australia and Western Australia where the dieback problem was serious, and this caused a drastic reduction in the areas planted to radiata between 1915 and 1922 in those states. By 1925, insects and diseases had been ruled out as the cause of that particular problem, and the disorder had been noted as occurring on only some types of soils. Large-scale planting was then resumed on soil types that supported radiata without such dieback. In 1936, a principal cause was found to be zinc deficiency in Western Australia and, by 1940, trials near Mt Burr demonstrated zinc deficiency to be the cause in South Australia as well (Stoate and Bednall 1957). The problem was then easily avoided by the spraying of zinc sulphate onto the foliage of young trees on suspect sites (Lewis 1975).

Boron deficiency has caused much shoot dieback, especially in young radiata in Chile, New Zealand and some African countries (Burdon 2001a), but it also is generally easily corrected by appropriate fertiliser treatment.

By the mid-1930s, there was concern in parts of Australia (Kessell and Stoate 1936) and New Zealand (Weston 1956) about "spindle-formed" trees. These had low taper and fine branching, both desirable traits, but also exhibited poor vigour associated with short, mildly chlorotic (yellowish) needles that shed a year or so earlier than normal. Low availability of phosphorus in the soil and loss of organic matter from the soil were implicated, particularly following clearfelling of a previous radiata crop. Burning of logging residues often accentuated the problem, and was eventually terminated in South Australia in 1963 (Woods 1980). The prime remedies, however, became use of phosphatic fertiliser (Burdon 2001a) and/or effective control of (particularly) European weeds that aggressively compete for soil nutrients. This latter is often coupled with mechanically fragmenting ("mulching") the post-logging **slash**[G] and mechanically incorporating much of it into upper layer of the soil, thus averting nutrient volatilisation, reducing leaching, and spreading its decomposition and release of its nutrients over longer periods of time than with burning.

Nutrition research, begun in the late 1930s and continued in following decades, has led to the prescribed addition of macro- and micro-nutrients, particularly superphosphate and zinc, and less frequently, nitrogen, potassium, copper and boron. Such treatments were found to be useful on many marginal sites in rescuing plantations where radiata pines were showing serious distress symptoms, as well as increasing the growth and uniformity in many healthy plantations on better sites. By the 1950s, such fertilization and other changes in management practices both maintained and increased the range of sites on which radiata performs successfully as a tree crop in sequential generations (Kessell and Stoate 1938; Stoate 1950, 1953; Anon. 1957; Boardman 1988; Nambiar 1996). Fertilization with phosphorus and nitrogen often improved sites enough to tip the species choice from maritime pine to radiata, especially in South Australia (e.g. Adelaide Hills) and South Africa.

Nutritional problems also often affected radiata in Chile and South Africa (Burdon 2001a), and they were to remain uncorrected there for longer than in Australia and New Zealand. In Chile, large areas had been planted on what proved to be nutrient-deficient sites. Of those sites, particularly on hill slopes, some had suffered patchy erosive loss of topsoil; others, on flat ground or inland dunes, were mainly nutrient-deficient water-borne volcanic ash. In South Africa, the preferred option of growing maritime pine on nutrient-deficient soils probably deflected foresters from addressing some of radiata's nutritional problems.

"True pines" (the 100-odd species of the genus *Pinus*) have relatively low nutrient requirements for healthy growth. Radiata has nutritional demands that are modest, but greater than those of many other pines. This may be due partly to its high growth potential, and partly to its coastal origin, where nutritionally significant inputs of certain elements, notably boron, magnesium and sulphur, come from wind-borne sea spray. (Interestingly, those same salt-laden winds harm and restrict coast redwood and Douglas-fir, radiata's main competitors in its northern native population.)

3.4.3 Genetics of Radiata

Genetics as a field dates back only to the first years of the 20th century, when Gregor Mendel's mid-19th-century published research became known and finally understood. However, observant people had long appreciated, albeit vaguely, that heredity was important even for trees (Zobel and Sprague 1993). For example, a late-17th-century book on agriculture discussing the planting of sugi (*Cryptomeria*) for sawn products in Japan noted "… it is important to select seed of red and handsome trees showing vigorous growth—cuttings must also be collected from young and handsome trees" (Miyazaki 1696). And the Englishman Patrick Matthew (1831) outlined how the principles of natural selection could operate in forest trees, over 20 years before Darwin (1859) published his catalytic book *On the Origin of Species*.

A.H. Cockayne was not only far-sighted in his advocacy of expanded plantings and uses of radiata in New Zealand, he was in 1914 among the first to consider taking advantage of the likely variation its genetic quality. After noting the effect of breeding dairy herds for increased butterfat on the demand for butter boxes, he then extended this principle to the species he had nominated to supply wood for those butter boxes. First, he had noticed, correctly, that "Locally produced seed of *P. radiata* compares more than favourably with that imported from California." That advantage of seeds collected from New Zealand trees was mostly a matter of cost and germination quality, but it would later be shown to have a genetic component as well. Cockayne then stated: "Farmers saving their own seed can, of course, select the trees from which the cones are collected, and in this way may procure those types that are most suitable for their particular locality." He also noted the great variation among radiata trees growing in similar conditions, and the great differences in the value of timber produced from such contrasting trees. He suggested that seeds might be saved from promising trees and, upon milling, the seeds from those whose logs produced the best boards could then be used to establish the next-generation plantations. Cockayne concluded his section on genetics by stating that the development of improved breeds (he called them "pure races," perhaps asking too much in the foreseeable future) would be of more importance in forestry than in other branches of agriculture. He particularly emphasized the effectiveness of selecting among trees growing in similar environments for those that differed favourably in important traits (Cockayne 1914).

3.4.3.1 Early Historical Influences

Transfers of seeds, between countries or regions within countries, was not in those early days intended to be any conscious genetic intervention. Yet, because of the potential influence on the genetic make-up of crops of the recipient growers, and thence on the genetic bases of developing regional land races and later breeding programmes, such transfers may have been of profound significance. During the 1920s through the 1950s, New Zealand was a major supplier of radiata seed to South Africa and at least to Victoria among the Australian states. For instance,

162 kg of radiata seed collected in New Zealand's Nelson region in 1924 was recorded as having been dispatched by the NZ Forest Service for export to South Africa (Sect. 3.3.3). Later, a private seed collector, H.G. Kingsland, is known to have collected and exported (for example to Australia between 1930 and 1950) very large quantities of radiata seed from the Nelson region. A File Note by J.M. Fielding, 18 June 1959, reads: "The son of Mr Kingsland told me today that practically all the seed supplied came from a two-acre patch of pine planted at wide spacing. He knew nothing of the origin of the seed used to plant the small stand." Such a small source, which conceivably had a widespread founding influence on various Australian land races, could have been atypical of the regional Nelson land race. Interestingly, that regional land race evidently has close to 50:50 Año Nuevo and Monterey native-population origins (markedly more Monterey and less Año Nuevo than studied land races in central North Island and the far south of New Zealand—Burdon 1992; Burdon et al. 1997). Even so, it appears that original founding stocks of other countries' land races of radiata generally included a reasonably similar mix of just Año Nuevo and Monterey origins, similar to studies New Zealand land races in general.

In parts of Australia, the rescue of failing plantations and improvements in growth by attention to siting and the addition of macro- and micronutrients were impressive. This led to a widely-held view that site and nutrition were of overriding importance to the success of radiata plantations, as indeed they are. This in turn led to a focus on soil nutrition alone, which in retrospect proved to be myopic. For several decades, the genetic quality of the planting stock was largely ignored. Worse, for an embarrassing number of decades in at least two Australian states a high proportion of the seeds to produce the next crop was reputedly collected from first thinnings, a practice for which details were not usually documented. In some plantations, the trees first thinned were removed systematically (Bednall 1956; Scott 1960), for example by cutting down all of the trees in every third row. Seeds collected from trees felled in such first thinnings would be genetically similar to seeds from the plantation at large. However, it was and is more common to use thinning to concentrate subsequent growth on the trees in the stand with the better-quality boles by selecting the poorer-formed trees for removal. Seeds collected from the trees felled in that more-common thinning practice were generally from the poorer-formed and/or less vigorous trees in the plantations. So, while the average productivity of plantations was being greatly improved by attention to site and nutrition, the genetic quality of some developing Australian land-races was liable to silviculturally-induced decline.

The founding populations of the Western Australian radiata land race(s) reportedly came in several waves, largely from South Australia and New Zealand, and some from New South Wales. Forest managers presumably knew at the time that they were getting first-thinning seeds from South Australia, but in those days they knew little of the developing science of genetics. In a somewhat different route to a similar end as that being taken by the New South Wales forest managers, where the established land-races were affected by **dysgenic selection**^{G} after their founding, the Western Australian land races are thought to have been largely founded by seeds from some of the worst possible maternal parents then available in South Australia and New South Wales.

The two most-important forests established in New Zealand's first planting boom were the Kaingaroa and the Kinleith, adjacent to each other in the central North Island. As that first planting boom rapidly expanded, seeds of radiata were increasingly in short supply. In the mid-1920s, a collector named Oxnam was engaged by the NZ Forest Service to gather available seeds from cone-bearing trees. While based in southern North Island, Oxnam reportedly collected all over North Island (Poole 1997), although there is no record of what proportions of his seed collections came from where. He and other seedsmen gathered many of those seeds from farmers' shelterbelts and woodlots whose felled trees were being milled (Goudie 1923). Many of those farmers' plantings, as well as available trees in parks, probably came or were derived from the seeds Hector had distributed from the Wellington Botanic Garden. The seeds collected for the Forest Service were then used to create the Kaingaroa Forest. The establishment of the Kinleith Forest was similarly accomplished by Perpetual Forests, with radiata seed first collected from nearby plantations at Matamata (Healy 1982). Seedlings for those Matamata plantations had been raised on the estate in 1873, but the source of those founding seeds was not given (Page 1923). When the supply of Matamata seeds became insufficient, they were supplemented with seeds from trees in Pukekura Park in New Plymouth, and then further supplemented with seeds collected in the Canterbury region of South Island (P.F. Olsen pers. comm. to WJL 1997). Those widely-scattered collections are likely to have resulted in broad but somewhat different genetic bases for the Kaingaroa and Kinleith forests, at least within the limits imposed by the then-available land races. However, the needs for inexpensive and abundant seeds often favoured collections from big-branched, squat, heavily cone-bearing trees, a genetic step in the wrong direction and thus a problem continuing among the offspring of those convenient seed parents in the following generations.

Despite the potential for practices of convenience in seed collection to be dysgenic, plantation silviculture in New Zealand led to some genetic improvement in tree form and growth (Burdon 1992), meaning that there was net **eugenic selection**$^{\mathbf{G}}$. The actual impacts of this eugenic selection only became clear from the mid-1960s onwards (Sect. 5.5.4.1), but the processes involved began well before 1950, so they are outlined below. On the other hand, the natural stands in mainland California suffered disturbances (Sect. 2.1.5) since the sampling that founded the New Zealand land races. Suffice to say here that it became clear that New Zealand obtained its founding stocks from the two generally most appropriate parts of radiata's natural range, and that the consequent land-race stocks, while not ideal, were the appropriate genetic base for starting an intensive breeding programme.

At least three activities contributed to such early land-race improvement in New Zealand and elsewhere. One was the breakup of neighbourhood family structure during sequences of seed collection and distribution, nursery sowing and outplanting; that led to more frequent outcrossing among unrelated plantation trees from the same native population and even stronger outcrossing when Monterey-origin trees hybridised with trees from Año Nuevo. A second was the collection of seeds in the plantations and shelterbelts at harvest age when the trees were felled, the poorer trees having been almost eliminated through natural competition or occasionally through selective thinnings. That meant that the seed parents and most pollen parents were more vigorous and were generally better-formed. A third was collection of seeds from the better crop trees in plantations, generally shortly before harvest, something that came much later (Sect. 4.6).

Meanwhile, the genetic quality of particularly the Monterey native population had presumably degraded since the founding seed-collections had been made in the late 19th century, for two likely reasons. One was that some stands in the Monterey population had suffered from continued dysgenic selection as the better trees had been selectively felled and their timber milled for generally local construction, leaving the trees less-suited for timber production to reproduce. The second was that substantial areas that had earlier been cleared for pasture were reinvaded by radiata, often from only scattered remnant parents. That resulted in much larger neighbourhoods of mostly offspring from the same seed parent, and thus even higher levels of average inbreeding among *their* offspring than had been present earlier. Those two processes had also occurred in the Año Nuevo population, but less severely. It was such genetically degraded mid-20th-century seed collections from those two populations that were compared to genetically improved mid-20th-century land races in New Zealand, and those native-population mid-20th-century samples probably performed less well than the 19th-century seed collections that founded the land races would have performed, had they been available for comparison in those mid-20th-century trials.

Thus it seems likely that a combination of relatively recent dysgenic selection in the two sampled native radiata populations, plus little inbreeding and some eugenic selection in managed New Zealand plantations, accounts for the substantial differences in performance in favour of then-current land races observed in mid-century trials. Those trials included essentially-random seedlings from trees in New Zealand plantations and corresponding essentially-random seedlings from recently imported seeds from the native populations.

Those observed differences, mostly in growth and form, have substantial implications for entering source material in current and future breeding programmes in New Zealand. Early on, it was decided that entering additional genetic material from the native populations would be a step backwards from the better source material in the land races. That decision was short-sighted because the gap between the performance of future domesticated breeds and performance on average of trees from the native groves will only increase as domestication of radiata proceeds. But breeders of long-domesticated crops still find it useful to add genetic inputs from native plants after many generations of breeding. Surely, future radiata breeders should find it similarly useful to add desired genetic variation from the Cambria, Cedros and Guadalupe native populations, and from some above-average trees within the conserved Monterey and Año Nuevo native populations. Yet that option was largely ignored for the next 40 years as serious breeding of radiata commenced.

For some of the species other than radiata that were used in New Zealand, however, genetic problems were serious. As one example, some plantations were established in the 1930s using politically-correct seeds of ponderosa pine from Canada rather than biologically-correct seeds of ponderosa pine from California. Importation of seeds from Canada satisfied the Ottawa Agreement of 1932 for Empire preference, making them the correct and perhaps required choice by the political rules of the time (Allsop 1973). The problem was that these Canadian ponderosa pines were adapted to the physical climate of south-western Canada, which made them biologically incorrect for the climates and substantially different summer daylengths of New Zealand. Luckily, radiata did not occur naturally anywhere in the Empire, and its land-race-founding seed importations proved to be well-adapted to many New Zealand conditions.

3.4.3.2 The Beginnings of Purposeful Genetic Interventions

The first formal genetic work done with radiata was in the early 1930s, at the Institute of Forest Genetics in Placerville, California. There, the emphasis was on interspecific hybridization, both as a genetic-improvement initiative and as basic research investigating evolutionary relationships within the genus *Pinus*. Their most successful controlled inter-species crosses were between radiata and knobcone pines (Critchfield 1967); among other things, these *P.* × *attenuradiata* hybrids provided strong biological evidence for a very close evolutionary relationship between the two parent species (G.L. Stebbins pers. comm. to WJL 1965).

In the climate at Placerville, the winters were too cold and summers too hot and dry for radiata to do well. Most radiata pines planted there and in similar Californian locations died or sustained repeated injury. But not all the radiata pines there suffered such early fates. Between the establishment of the Institute as the Eddy Tree Breeding Station in 1928 and the mid-1970s, the fastest-growing and largest pine on the Institute grounds was one radiata pine, one of hundreds of radiata pines planted there soon after 1928. This special tree was finally surpassed in size by local ponderosa and sugar pines in the late 1970s and, at just over 50 years of age, it also died in that too-harsh climate. A cloned copy of it now lives at the University of California Russell Reserve, in a climate kinder but still marginal for radiata.

The *P.* × *attenuradiata* hybrids outgrew the knobcone pines planted with them at the Institute and had better form as well. They also outgrew most of the repeatedly-injured radiata pines that survived for a few decades on that site. This success story was widely proclaimed, and *P.* × *attenuradiata* hybrids were produced at Placerville using different and better knobcone parents. They were then planted in many locations in California, and elsewhere around the world. As with almost any new exotic introduction, there were many failures. But they did well in many places too harsh for radiata and, at least for a few decades, in such locations they typically outperformed most or all other species and hybrids in those later trials.

Early genetic work focused on radiata began in the Southern Hemisphere. In 1933 in South Australia, following A.D. Lindsay's visit to California, C.E. Lane

Poole and G.J. Rodger established trial plantings of native provenances of radiata, which were followed up by similar plantings in 1941.

Meanwhile, in South Africa in 1933, S.P. Sherry collected seeds from 14 young radiata trees that looked very different from each other, which he used to establish an **open-pollinated**[G] (OP) progeny trial at Tokai in 1939. These OP (seed-parent) families soon demonstrated marked differences in average growth rate, stem form, **internode**[G] length, branch size and crown shape, and such characteristics tended to resemble those of the known seed parents (Poynton 1977). Thus there was strong evidence of heritability for a range of radiata's important traits. A marriage of genetics and plant propagation was proposed and encouraged in a 1939 publication by Jacobs (1939), to whom this book is dedicated. In those very early days, he predicted that clonal forestry would be the best way of delivering genetic improvement to operational stands. A more ambitious genetics program was also begun in 1939 at Mt Burr in South Australia, following the appointment of J.M. Fielding in 1938. Fielding was spread thinly over 28 "major" and 26 "minor" trials established by mid-1939, including studies of spacing, pruning, thinning, diseases, nursery practices, planting practices and felling systems Importantly, Fielding also initiated a study of inheritance in progenies of distinctive parents, and established some extremely informative clonal trials. World War II, however, stalled this very fast start.

Following the war, C.K. Pawsey continued the work at Mt Burr and Fielding moved to Canberra. Employing radiata cuttings, Pawsey added clonal studies of "autumn brown top," a drought-induced dieback that proved to be strongly inherited in its incidence. In 1948, controlled pollinations were conducted at Mt Burr to test for full-sib family differences in responses to fertilizer and the development of "spindle form" (Pawsey 1967; Boardman 1988). Meanwhile, in Canberra, Fielding was expanding on the pre-war work by Jacobs. He acquired new samples of seed from the three native mainland populations and established field trials of his new mainland collections in the Australian Capital Territory (ACT) in 1950. He also organised separate trial plantings of seed samples from the island populations of radiata, and provenance trials of the closely related bishop and knobcone pines. In a pioneering move, Fielding also organised searches throughout the ACT) of radiata plantations for above-average "plus trees," and gathered their seeds and cuttings for propagation research and trial plantings. The first large open-pollinated (OP) progeny test was established in 1951 and a **control-pollinated**[G] (CP) test, using a haphazard design due to difficulty of reaching "**flowers**"[G] on mature trees) in 1953 (Fielding 1953, 1961).

In New Zealand, M.H. Bannister began studying variation in radiata. Significantly, he embarked on an open-pollinated progeny trial from 26 radiata seed parents in1949.

3.4.4 Silviculture

As might be expected, forest husbandry (called **silviculture**[G]) practices at first followed from training and practice in Europe. For example, in Spain, radiata plantation management followed the European model of closely spaced planting, frequent light thinning and low pruning that removed the small dead branches off the lower bole (Scott 1960).

The earlier Western Australia plantations similarly employed the European traditions of dense initial stocking and conservative thinning. T.N. (Bill) Stoate recognized that an important effect of more aggressive thinning was to make more water available to the remaining trees.

In South Australia thinning practices were conservative, but there was generally the advantage of easy terrain which facilitated repeated light thinnings. Routine commercial thinning commenced in about 1940, and in that decade second thinnings became routine (Jolly 1950; Boardman 1988).

In the ACT, relatively wide initial spacing (3.6 × 3.6 m) and early pruning were adopted in 1927 because of the absence of a market for small-diameter thinnings. But by 1944 initial spacing had been reduced to 2.4 × 2.4 m (Terrell 1944) to provide scope for selecting better final-crop trees and to supply the evolving market with small-diameter thinnings for case timber and flooring, lining and battens. (In 1939, seasoning and reconditioning facilities had been installed at the government sawmill; in 1943, radiata thinnings provided the only softwood used in ACT.)

A South African, I.J. Craib, broke with European tradition in 1934. He had been working in a context of intense competition among trees for water and nutrients. The upshot was that he advocated somewhat wider-spaced planting and heavier early thinning of pines, including radiata, to increase diameter growth on the crop trees, with early removal of branches (pruning) to improve log quality (Craib 1939, 1947). This proposal was mostly ignored or vigorously opposed by traditional silviculturists of the time (Sutton 1995).

However, in New Zealand, by the 1940s two things were becoming apparent: natural regeneration after felling of radiata could often be excessively dense; and little or no market was foreseen for small-diameter knotty logs removed in so-called "early commercial" thinnings. It was reckoned that if such thinning was to be done, it would be at a net or even total loss. A somewhat more daring departure from European thinking was proposed for New Zealand's Kaingaroa Forest in 1949 (Ure 1949), namely to thin even earlier and much more aggressively than even Craib had advocated. It was proposed that, after a preliminary thinning of the regeneration before age 4, the number of stems per hectare be reduced from about 3000 to 740 when the trees were only 9–12 m tall, at age about 8 or 9. Rather than delaying this thinning operation in a vain effort to recover some financial return, there would be no attempt to use these early thinnings. Instead, they would be cut "to waste," and left to decay and recycle their nutrients on the site. Furthermore, problems with branch and knot size would be sidestepped at that wider post-thinning spacing by pruning the branches off the bottom logs (Ure 1949; Sutton 1984). After a period of misgivings over that proposed regime, an even more radical approach was adopted later on (see Chap. 5).

In Chile, early silvicultural practice followed the European model. However, in the large-scale plantings that began around 1940 very little thinning or pruning were practised.

3.5 Some Alarms

In southern Africa, during the 1920s and early 1930s, large areas of young radiata plantation that had initially survived and grown well failed because of dieback caused by shoot pathogens. This was a pattern later seen in many other subtropical summer-rainfall regions where radiata was tried. It was an extended "moment of truth" for understanding and then accepting radiata's maladapdtation to such climates.

Elsewhere, there was more of an incremental process of learning environmental limits of radiata pine, in terms of rainfall and altitude in relation to exposure and the risk of snow damage. In New Zealand, however, there were some events that caused much alarm when they occurred, but did not necessarily reflect fundamental maladaptation.

At the beginning of the 1930s, large areas of young radiata pine plantation in central North Island of New Zealand were severely affected by shoot dieback caused by the pathogen *Allantophomopsis* (syn. *Phomospsis strobi*) (Burdon 2011). This outbreak, while it caused much permanent bole malformation, has not been repeated. In hindsight, it largely reflected a combination of an extreme weather event and big areas of plantation at a vulnerable age on high-hazard sites, although there appears to have also been a "first-rotation" effect (*op. cit.*).

After World War II, New Zealand had a very large untended resource of radiata, but nowhere near the labour force to cope with the backlog of silvicultural work that had accumulated as workers were called to wartime service. This was a dangerous situation, especially with the three unusually warm, dry summers that followed. Early in 1946, nearly 12,000 ha of radiata plantation, around 15 years old, got burnt north of Lake Taupo in the middle of the North Island (MacArthur 1952; Healy 1982). At the time, the burned-over lands were seen by some of the owner's company management as superfluous. However, the seed shed from the closed cones opened by the conflagration produced profuse regeneration, with up to 2.5 million or so live seedlings per hectare. Any effective thinning was an impossible task, yet the trees eventually "thinned themselves" to produce a very worthwhile crop.

In the following three years there was a bigger alarm, because of the far greater area involved. A European woodwasp (*Sirex noctilio*) ("sirex") had become established in New Zealand in the early 1900s (Bain et al. 2012). A fungus, *Amylostereum areolatum*, is carried by those woodwasps when emerging from infected dead trees and it is introduced with those wasps' eggs into other, generally already-weakened trees. Radiata is extremely sensitive to this fungus, and it is the fungal disease spreading from multiple infections, as an indirect effect of mass attack by sirex, that usually kills the tree (Elliott et al. 1998). After local drying of the wood near the oviposition site on the bole, the sirex eggs hatch. If at about this stage the weakened, multiply-infected tree dies, the fungus continues to spread throughout the dead tree's wood. The sirex larvae bore through the wood, feeding on its symbiotic fungus, and after many months pupate near the surface of the tree's bole (Bedding c.1993).

Fig. 3.7 Radiata stand showing high mortality from the sirex epidemic in New Zealand in the late 1940s. In the most severe cases stocking was reduced to around 125 stems per ha, but remaining trees were almost all well-formed and responded with very fast growth

By the early to mid-1940s, many of the radiata plantations in New Zealand were about 20 years of age and, due largely to worsening economic conditions and then World War II, many had remained unthinned. In each local neighbourhood of eight or so trees, the stronger trees were suppressing and weakening their neighbours. With the extensive unthinned plantations becoming increasingly crowded, sirex populations found an abundance of vulnerable hosts. Helped by some warm, dry summers, they built up to outbreak and then epidemic numbers. Mortality in many stands was dramatic (Fig. 3.7), and the epidemic severely affected over 100,000 ha of radiata plantations on the volcanic plateau of central North Island, with less spectacular damage elsewhere (Burdon and Miller 1992). In some cases the 1500–2500 living pines per hectare were reduced to about 300 or even fewer survivors (Sutton 1984). At the height of that epidemic, it appeared that the radiata plantation programme could be an expensive failure. What made the situation even more alarming was that planning was already underway on the establishment of very large wood-processing plants which would depend for wood on plantations that might be dying.

After the war, in the late 1940s, the sirex crisis in the plantations was addressed on several fronts. Maintaining stands in a healthy condition by timely thinning and other appropriate husbandry reduced the potential for epidemic buildup on weakened hosts. In addition, scientists sponsored warfare within the invaders, namely biological control using three effective parasites of sirex. Two parasitic wasp species were successfully imported and then effectively attacked the sirex larvae. Later, a local species of nematode was found that invaded and colonized the sexual organs of the adult wasps, rendering them infertile. The successful culturing and infective dispersal of those nematodes provided the first example of the successful use of nematodes as biocontrol agents (Bedding and Akhurst 1974;

Bedding c.1993). These silvicultural measures and biological controls were so successful that it is today difficult to find sirex in New Zealand plantations, and those that are there do negligible damage.

In retrospect, this perceived disaster did much good. Sirex attacked mostly the weaker trees in a stand, which fortuitously included a high frequency of those with bole malformations. While the main sirex attack was in progress (Fig. 3.7), it was both widespread and alarming. After it subsided, it became apparent that relatively few valuable trees had been killed. Mostly these woodwasps did a remarkably good job of thinning over 150,000 ha of radiata stands that needed it while the human labour force was tied up on wartime priorities. Indeed, a crusty forest pathologist-entomologist then said that sirex was "the best bloody silviculturist we've ever had." Perhaps about a third of the volume in these overstocked stands was lost but, even in the most seriously attacked stands, a closed tree canopy soon developed and full per-hectare volume growth returned (Sutton 1984).

Perhaps most important, the sirex epidemic refocused attention on the wisdom of having so much of New Zealand's forest resource in plantations of a single species. This insight fostered efforts to find other species that would (1) replace radiata in case of a subsequent event or epidemic that did not turn out so well, and/or (2) provide wood better suited for some important alternative uses, and/or (3) grow well on sites where radiata does not do well. Trialled additional species have been only partly successful thus far, in large measure because radiata grows so well on so many different New Zealand sites, because its wood can be used for so many purposes, and increasingly because genetic improvement was apparently so successful (Burdon and Miller 1992; Burdon 2001a). That final reason is the topic of Chaps. 4 and 5. Indeed, despite the alarms the preferred-species status of radiata in New Zealand remained secure. Later on, a further alarm was actually to accentuate that status.

3.6 Summary of Domestication Progress

Planting booms in the Southern Hemisphere, which were dominated by radiata, led to a major rise in the status of radiata as a commercial timber species. One boom began in New Zealand in the 1920s and another in Chile beginning around 1940, with lesser increases in radiata afforestation in Australia and South Africa. Research into the processing and utilisation of its wood endorsed the findings of initial utilisation efforts, and vindicated the developing massive commitments to growing radiata. This led on to commitments to establish large-scale processing plants.

Unsurprisingly, enthusiasm for radiata led to overreaches in where it was planted. In some cases, notably in Australia where unidentified nutrient deficiencies had threatened total failure of radiata plantations, these deficiencies could be corrected cheaply and straightforwardly after they were diagnosed. In other cases, notably in South Africa where the climate favoured some fungal pathogens, the siting overreach only became evident some years after planting. Those pathogen problems could not be corrected operationally, necessitating the use of other species for afforestation there. In addition, there were some passing scares from transient biotic events that did not recur.

Despite innovative work on plantation establishment, the growing of radiata featured few innovative advances in other aspects of crop management during this era of "passive" domestication, except in the greatly expanded scale on which it was planted as an exotic forest-tree species. Meanwhile, the stage was being set for major advances in tending regimes for several forest-tree species, including radiata. In South Africa pilot work was done for radical refinement of tending regimes, while in Australia and South Africa the potential for intensive radiata breeding programmes was being probed with proposals and early trials.

The utilisation of radiata wood was being researched in Australia and New Zealand, and some processing plants specifically designed to cope with the unfamiliar problems and advantages of the wood were built and came on-line. Governments were heavily involved in the planting and the early wood processing, either directly or through providing financial incentives for private investment. In New Zealand, however, there was independently a major private-company planting boom in the 1920s, followed by private investment in building processing plants.

References

Algar WH (1988) Forestry and forest products. Australian Academy of Technological Sciences and Engineering, Parkville, p 60. Reprinted from (pp 193–252) Technology in Australia 1788–1988 for Australian Forest Development Institute on the occasion of The International Forestry Conference for the Australian Bicentenary 1988

Allsop F (1973) The first fifty years of New Zealand's forest service. Government Printer, Wellington, p 123

Anon (1937) Annual report Commonwealth Forestry Bureau Australia. Government Printer, Canberra, p 15

Anon (1938a) Annual Report for year ended 31st March 1938. Division of Forestry, Union of South Africa, Pretoria, p 35

Anon (1938b) Annual report Commonwealth Forestry Bureau Australia. p 15

Anon (1939) Annual report Commonwealth Forestry Bureau Australia. p 21

Anon (1957) Exotic conifers in Western Australia. Paper to 7th British Commonwealth Forestry Conference, Australia and New Zealand. p 22

Anon (1968) Annual report of New Zealand Forest Service. p 62

Anon (1969) Fifty years of forestry in Western Australia. Forests Department, Western Australia, p 90

Atkinson D (ed.) (1983) Tree root systems and their mycorrhizas. Martinus Nijhoff/Dr W. Junk Publishers, The Hague, p 524

Bain J, Sopow SL, Bulman LS (2012) The sirex woodwasp in New Zealand: history and current status. In: Slippers B, de Groot P, Wingfield MJ (eds) The sirex woodwasp and its fungal symbiont: research and management of a worldwide invertebrate pest. Springer, Dordrecht, pp 167–173

Barnes RD (1970) The prospects of re-establishing *Pinus radiata* as a commercially important species in Rhodesia. S Afr For J 72:17–19

Bartle J, Shea PS (1989) Development of a pulpwood cropping industry in Western Australia. Land Water Res News 1:5–9
Bedding RA (1993) Biological control of *Sirex noctilio* using the nematode *Deladenus siricidicola*. In: Bedding R, Akhurst R, Kaya H (eds) Nematodes and the biological control of insect pests. CSIRO, East Melbourne, pp 11–20
Bedding RA, Akhurst RJ (1974) Use of the nematode *Deladenus siricidicola* in the biological control of *Sirex noctilio* in Australia. J Aust Entomol Soc 13(2):129–135
Bednall BN (1956) Silviculture in relation to management of *Pinus radiata* in South Australia. Aust For 20:30–36
Benjamin LR, Somerville JL, Jeffries RB et al (1928) Kraft pulp and paper from *Pinus insignis*. Bulletin CSIR Melbourne No. 35, p 32
Boardman R (1988) Living on the edge—the development of silviculture in South Australian pine plantations. Aust For 51:135–156
Boardman R, McGuire DO (1990a) The role of zinc in forestry. I. Zinc in forest environments, ecosystems and tree nutrition. For Ecol Manag 37:167–205
Boardman R, McGuire DO (1990b) The role of zinc in forestry. II. Zinc in forest environments, ecosystems and tree nutrition. For Ecol Manag 37:207–218
Boyd J (1992) Pumice and pines: the story of Kaingaroa Forest. GP Publications, Wellington
Burdon RD (1992) Genetic survey of *Pinus radiata*. 9. General discussion and implications for genetic management. N Z J For Sci 22:274–298
Burdon RD (2001a) *Pinus radiata*. In: Last FT (ed.) Ecosystems of the world, Vol. 19, Tree crop ecosystems. Elsevier, Amsterdam, pp 99–161
Burdon RD (2001b) Genetic aspects of risk—species diversification, genetic management and genetic engineering. N Z J For 45(4):20–25
Burdon RD (2011) Shoot dieback epidemics of *Pinus radiata* in New Zealand: is there an endophyte story? N Z J For Sci 41:131–139
Burdon RD (compil.), Miller JT (ed) (1992) Introduced forest trees in New Zealand: recognition, role and seed source. 12. Radiata pine (*Pinus radiata* D.Don). FRI Bulletin 124. New Zealand Forest Research Institute. Rotorua, NZ. p 59
Burdon RD, Broekhuizen P, Zabkiewicz JA (1997) Comparison of native-population and New Zealand land-race samples of *Pinus radiata* using cortical oleoresin monoterpenes, pp 50–56 in Burdon RD and Moore JM (eds) IUFRO '97 Genetics of Radiata Pine. Proc NZFRI-IUFRO Conference 1–4 December and Workshop 5 December, Rotorua, New Zealand. FRI Bulletin 203
Butcher TB (1997) 15 Jan 1997 e-mail to WJL
Carron LT (1985) A history of forestry in Australia. Australian National University Press, Canberra, x + 355 p
Carron LT (1993) Changing nature of federal–state relations in forestry. In: Dargavel J, Feary S (eds) Australia's ever-changing forests. II. CRES, ANU, Canberra, pp 207–239
Chandler WG (1957) Commercial forests from waste scrub lands. Aust For 21:43–47
Chu-Chou M (1979) Mycorrhizal fungi of *Pinus radiata* in New Zealand. Soil Biol Soil Biochem 11:557–562
Clapp RA (1995) The unnatural history of the Monterey pine. Geogr Rev 85(1):1–19
Cockayne AH (1914) The Monterey pine—the great timber-tree of the future. N Z J Agric 8(1):1–26
Collett NG (2010) A history of forestry education in Victoria, 1910–1980. Aust For 73:34–40
Contesse (González) D (1987) Apuntes y consideraciones para la historia del pino radiata en Chile. Boletín de la Academía Chileno de la Historia No. 97a (1986), pp 351–373
Craib IJ (1939) Thinning, pruning and management studies on the main exotic conifers grown in South Africa. South Africa Department of Agriculture and Forestry Science, Pretoria, Bulletin No. 196. p 179
Craib IJ (1947) The silviculture of exotic conifers in South Africa. Paper to 5th British Empire Forestry Conference, GB. p 35

Critchfield WB (1967) Crossability and relationships of the Californian closed cone pines. Silvae Genet 16:89–97

Dadswell HE (1961) Lectures in the national science foundation forest biology workshop. North Carolina State University, Raleigh, and personal communications to WJL during the workshop

Dargavel J (1995) Fashioning Australia's forests. OUP, Australia, p 312

Dargavel J (2008) The zealous conservator: a life of Charles Lane Poole. UWA Press, Crawley, p 252

Darwin C (1859) On the Origin of Species. Various reprinted editions (see http://darwin-online.org.uk)

Donald DGM (1993) Radiata pine in southern Africa. Lewis and Ferguson, p 353–364

Eldridge K, Davidson J, Harwood C et al (1993) Eucalypt domestication and breeding. Clarendon Press, Oxford, p 288

Elliott HJ, Felton KC, Jarman SJ et al (2008) A history of innovation: eighty-five years of research and development at Forestry Tasmania. Forestry Tasmania, Hobart, p vii + 392

Elliott HJ, Ohmart CP, Wylie FR (1998) Insect pests of Australian forests: ecology and management. Inkata Press, Melbourne, p 214

Elwes HJ, Henry A (1910) The trees of Great Britain & Ireland. *Pinus radiata*. Vol V. Privately published in Edinburgh, pp 1079–1084

FAO Secretariat (1958) The Spanish afforestation program. Unasylva 12:3–8

Field JF (1934) Experimental growing of insignis pine from slips. N Z J For 3:185–186

Fielding JM (1953) Variations in Monterey pine. Bulletin, Forestry and Timber Bureau Australia, No. 31. p 43

Fielding JM (1961) Provenances of Monterey and Bishop Pines. Forestry and Timber Bureau, Canberra, Bulletin 10, p 30

Gillbank L (1993) Nineteenth century perceptions of Victorian forests: ideas and concerns of Ferdinand Mueller. In: Dargavel J, Feary S (eds) Australia's ever-changing forests. II. CRES, ANU, Canberra, pp 3–14

Goudie HA (1923) The remarkable pine. *Pinus radiata* in New Zealand. New Zealand forest service circular No. 3. Government Printing Office. Wellington. p 14 (This was revised in 1925 under the title: The insignis pine (the remarkable pine).)

Grant TC (1989) History of forestry in New South Wales 1788 to 1988. The Author, Sydney, p 320

Hall N (1951) The establishment of coniferous plantations in Australia. Aust J Sci 13:164–168

Healy B (1982) A hundred million trees. The story of N.Z. Forest Products Ltd. Hodder and Stoughton, Auckland, p 222

Hegan C (1993) Radiata, the prince of pines. N Z Geogr 20(Oct–Dec):88–114

Hutchins DE (1916) A discussion of Australian forestry. Forest Department, Western Australia, p 434

Jacobs MR (1939) The Vegetative Repropagation of Forest Trees I. Experiments with cuttings of Pinus radiata Don. Forestry and Timber Bureau, Australia, Bulletin 25, p 28

Jolly NW (1950) The thinning of *Pinus radiata* plantations in the south east of South Australia. Bulletin Woods and Forests Department Adelaide. No. 4. p 27

Junge A (1953) Historia del 'pino insigne' en Chile, y el premio 'Arturo Junge Sahr'. Chile Maderero 3(10):4

Kennedy D (1957) The establishment, development and regeneration of exotic species. Technical Paper No. 17, New Zealand Forest Service, Forest Research Institute, Rotorua. p 11

Kessell SL, Stoate TN (1936) Plant nutrients and pine growth. Aust For 1:4–13

Kessell SL, Stoate TN (1938) Pine nutrition No. 50. Bulletin, Forests Department Western, Australia, p 45

King NL (1938) Historical sketch of the development of forestry in South Africa. J S Afr For Assoc 1:4–16

King NL (1951) Tree planting in South Africa. J S Afr For Assoc 21:1–102

Krebs JS (1973) Monterey pine: an introduced species in Chile. PhD dissertation. Department of Geography, University of Colorado. Boulder, CO, USA. p 386

Langlands I, Thomas AJ (1939). Handbook of structural timber design. Technical paper No. 32. Division of Forest Products, CSIR, Melbourne, Australia. p 283

Lewis NB (1975) A hundred years of state forestry—South Australia 1875–1975. Woods and Forests Department, Adelaide, (A Centenary Publication) Bulletin No. 22. p 122

Lewis NB (1991) Management of radiata pine in Australia. In: McKinnell FH, Hopkins ER, Fox JED (eds) Forest management in Australia. Surrey Beatty and Sons, Chipping Norton, pp 190–202

Lewis NB, Ferguson IS (1993) Management of radiata pine. Inkata Press (Butterworth-Heinemann), Melbourne–Sydney, p 404

Lindsay AD (1932) Monterey pine (*Pinus radiata* Don) in its native habitat Bulletin No. 10, Forestry and Timber Bureau, Canberra, p 57

MacArthur RS (1952) Early thinning treatment and growth development in stands of *Pinus radiata* D.Don established at excessively high density. N Z J For 6:333–338

Matthew P (1831) On naval timber and arboriculture. Adam Black, Edinburgh, p 391

Matthews HJ (1905) Tree culture in New Zealand. Government Printer, Wellington, p 358

Maxwell E (1930) Afforestation in southern lands. Whitcombe and Tombs, Auckland, p 315

McKelvey P (1991) Thomas William Adams 1842–1919. Early farm forester. N Z J For 36(2):23–25

McKelvey P (1999) Sand Forests. A Historical Perspective of the Stabilisation and Afforestation of Coastal Sands in New Zealand. Canterbury University Press, Christchurch. p 168

Michel M (2006) El pino radiata en la historia forestal vasca. Sustral: Rev Agropesquera 2006:56–59

Miyazaki Y (1696) A cyclopedia of agriculture. Vol 9. Various trees. agris.fao.org/agris-search/search.do?recordID=US201300185227

Moulds FR (1991) The dynamic forest. Lynedoch Publications, Richmond Victoria, Australia, p 232

(von) Mueller F (1858) On a general introduction of useful plants into Victoria. Trans Philos Inst Vic 2:93–109

Nambiar EKS (1996) Sustained productivity of forests is a continuing challenge to soil science. Soil Sci Soc Am J 60:1629–1642

New Zealand Forest Service (1952) Annual report. p 79

O'Hehir J, Nambiar S (2010) Productivity of successive rotations of *Pinus radiata* plantations in South Australia over a century. For Ecol Manag 259:1857–1869

Olsen P (1996) Presentation to New Zealand Institute of Forestry, Rotorua. 28 Nov 1996, and subsequent December personal communications to WJL

Orr E (2017) Keeping New Zealand green. Steele Roberts Aotearoa Publishers, Wellington, p 234

Page PM (1923) Pine plantations, Matamata estate, Waikato, New Zealand. In: Goudie HA (eds) The remarkable pine: *Pinus radiata* in New Zealand. State Forest Service Circular No. 3. Government Printing Office, Wellington, pp 11–14. p 14

Pawsey CK (1967) Breeding Monterey pine in the south-east of South Australia. MSc thesis, University of Melbourne. p 38 plus 11 supporting articles

Pawson E (1993) Wason, John Cathcart—biography. In dictionary of New Zealand biography (Vol. II). Te Ara—the encyclopedia of New Zealand. <www.teara.govt.nz/en/biographies/2w10> Last accessed 6 July 2012

Podger FD, Wardlaw TJ (1990a) Spring needle-cast of *Pinus radiata* in Tasmania: I. Symptoms, distribution and association with *Cyclaneusma minus*. N Z J For Sci 20:184–205

Podger FD, Wardlaw TJ (1990b) Spring needle-cast of *Pinus radiata* in Tasmania: II. Effects of fertilizers on disease severity, and the impact of the disease on growth. N Z J For Sci 20:206–219

Poole, AL (1966). Forest practices and policy. In McClintock, AH (ed) An encyclopaedia of New Zealand. Te Ara

Poole L (1997) Address at dedication of the horse barn during the 50th anniversary celebration of the New Zealand Forest Research Institute, and personal communications to RDB and WJL following the address

Poynton RJ (1960) Notes on exotic forest trees in South Africa. (2nd edn, revised). Department of Forestry, Government Printer, Pretoria, p 186

Poynton RJ (1977) Tree planting in South Africa. Volume 1. The pines. Department of Forestry, Republic of South Africa, p 576

Poynton RJ (1979) Tree planting in South Africa. Volume 2. The eucalypts. Department of Forestry, Republic of South Africa, p 882

Pudden HH (1957) Exotic forest trees in the Kenya highlands. Government Printer, Nairobi, p 37 (7th British Commonwealth Forestry Conference, Australia and New Zealand)

RBG (Royal Botanic Gardens, Melbourne) (1995) Horticultural Flora of South-eastern Australia. Vol 1. Ferns, Conifers and their Allies. UNSW Press, Sydney. p 361

Restall AA (1964) Sand dune reclamation on Woodhill Forest. N Z J For 9:154–161

Roche MM (1990) History of New Zealand forestry. New Zealand Forestry Corporation in association with GP Books, Wellington, p 466

Rogers LJ (1957) Exotic forest trees in Queensland. Paper to 7th British Forestry Conference, Australia and New Zealand, p 25

Royal Commission (1909) Report on timber and timber-building industries. Appendix to the journals of the house of representatives,1909, session II, H-24. Wellington. New Zealand

Royal Commission (1913) Report on forestry. New Zealand Government Printer, Wellington, p 77 plus p 87 of minutes plus maps and photographs

Scott CW (1960) *Pinus radiata*. FAO forestry and forest products studies No. 14. FAO, Rome, p 328

Simpson TE (1973) Kauri to radiata: origin and expansion of the timber industry of New Zealand. Hodder & Stoughton, Auckland, p 442

Sinclair EK (1991) The spreading tree: a history of APM and Amcor 1844–1989. Allen and Unwin, North Sydney, p 278

Squire RO, Flinn DW (1981) Site disturbance and nutrient economy of plantations with special reference to radiata pine on sands. pp 291–302 in Productivity in perpetuity: proceedings of Australian Forest Nutrition Workshop, 10–14 August, Canberra

Stoate TN (1950) Nutrition of the pine. Bulletin Forestry and Timber Bureau, Australia, No. 30. p 61

Stoate TN (1953) Later disorders in pine stands in South and Western Australia. Leaflet Forestry and Timber Bureau, Australia, No. 66. p 14

Stoate TN, Bednall BH (1957) Disorders in conifer forests. Bulletin, Woods and Forests Department of South Australia No. 11. p 9

Streets RJ (1962) Exotic forest trees in the British Commonwealth. Clarendon Press, Oxford, p 765

Sutton WRJ (1984) New Zealand experience with radiata pine. The HR MacMillan lecture in forestry. University of British Columbia, Vancouver, p 21

Sutton WRJ (1986) Review of the New Zealand radiata pine industry. p 18. in Managing and marketing the changing timber resource. proceedings of the forest products research society meeting, 18–20 March 1986, Fort Worth, TX, USA

Sutton WRJ (1987) The NZ Forest Service contribution to plantation forestry. N Z J For 32(1):24–27

Sutton WRJ (1991) New Zealand radiata pine plantations: an example of sustainability. Plenary paper, Society of American Foresters National Convention, San Francisco, p 5 (metric version)

Sutton WRJ (1995) Field trip and personal communication to WJL

Terrell AK (1944) Forestry operations and ornamental planting in the Australian Capital Territory. Queensl J For 38(1):5–10

Ure J (1949) The natural regeneration of *Pinus radiata* on Kaingaroa forest. N Z J For 6(1): 30–38

Vincent TG, Dunstan JS (1989) Register of commercial seedlots issued by the New Zealand Forest Service. Ministry of Forestry, New Zealand Forest Research Institute, Rotorua, FRI Bulletin 144. p 155

Weston (1956) Fertiliser trials in unthrifty pine plantations at Riverhead forest. N Z J For 8: 35–46

Whitehead PS (1964) Sand dune reclamation in New Zealand. N Z J For 9:146–153

Woods RV (1980) Management of nitrogen in the *P. radiata* plantations of the south east of South Australia. pp 354–367 in Rummery, RA and FJ Hingston (eds) Managing nitrogen economies of natural and man-made forest ecosystems. Proceedings of a workshop at Mandurah, WA. 5–9 October 1980

Zobel BJ, Sprague JR (1993) A forestry revolution. A history of tree improvement in the southern United States. Carolina Academic Press. p 161

Chapter 4
The Beginnings of Genetic Improvement: 1952–1968

4.1 General Context

While genetic improvement is the main theme of this chapter, we first review the context in which the improvement began to be vigorously pursued for growing radiata pine 1951–1968 was a period leading up to major upsurges of new planting in the main grower countries. Areas planted in radiata pine as at 1956, by individual country, are listed in Table 4.1. Meanwhile some major industrial plants processing radiata wood were opened, and more were being planned. Also, further biotic scares occurred in several regions. Since these and other circumstances differed markedly among the main grower countries, we review the context of such events and initiatives by country.

Particularly alarming during the 1960s, red band needle blight, caused by the fungus *Dothistroma pini* ("dothi"), emerged as a fast-spreading and destructive disease of radiata in many areas where it was being grown as an exotic. This is typically troublesome in moist summer conditions, especially with poor air circulation. While dothi took only a few years to appear in different continents, its impact on the status of radiata as a crop varied widely among the different regions where it was being grown.

4.1.1 Australia

By 1951 there were substantial industrial ventures based on processing radiata wood, with a major processing plant of Australian Paper Manufacturers (APM) in the state of Victoria, and three large sawmills and a small pulp mill in south-eastern South Australia (in the Green Triangle) (Lewis 1975). In Queensland, there was active plantation establishment with conifers, but almost entirely with species other

R. Burdon et al., *Domestication of Radiata Pine*, Forestry Sciences,
DOI 10.1007/978-3-319-65018-0_4

Table 4.1 Approximate areas of radiata plantation, by country as at 1956 (Scott 1960)

Country	Area (ha)
New Zealand	228,000
Chile	200,000
Australia	122,400
Spain	51,000
South Africa	21,200
Kenya	3,500
Others	4,000?
Total	~630,000

than radiata. Despite those activities, and plantation forestry with radiata in New South Wales, Western Australia and Tasmania, plus the presence of native forests, most parts of the country were faced with the prospect of increasing dependence on timber imports. Such import demand would be mainly for easily processed and user-friendly softwoods. A response to this situation was led by Dr Max Jacobs, after he became Director General of the Forestry Timber Bureau in Canberra in 1959 (Bevege 2010). Jacobs "oversaw the establishment of the Australia Forestry Council in 1964." Its first major policy initiative led to the Softwood Forestry Agreements (SFA) Act 1967. That and follow-up Acts were important parts of a national commitment to substantially increase Australia's softwood plantation resource and, over most of the areas being committed to plantations, radiata was clearly the species of choice. The later impacts of these initiatives feature in the coming chapters.

Also important during this period, in 1957 the Seventh British Commonwealth Forestry Conference was held in Australia and New Zealand (BCFC 1957). Importantly for tree breeding, one of its recommendations (2/4) strongly urged genetic improvement and underpinning research. This resolution has been widely crediting with prompting the various states to set up their individual tree-improvement programmes, in most of which radiata was central. Also, a Committee Report from that conference envisaged an expanded central forest research organisation, to complement state research arms that were seen as concerned "mainly with practical problems within their boundaries."

While dothi appeared in Australia, especially in parts of New South Wales, it was not sufficiently severe or widespread to cause major concern.

4.1.2 New Zealand

During this 1951–1968 period the situation in New Zealand was quite complicated. Key factors were:

- decisions having been made to establish large industrial complexes that included pulp and paper mills designed to process radiata wood
- nevertheless, a continuing surplus of wood resource

- an increased wariness of biotic risks in single-species forest plantation estates after the sirex epidemic
- intensified demand for land for new pastoral farms;
- awareness of new and continuing fungal disease epidemics or outbreaks in radiata plantations around the world
- but radiata emerging as more than ever the preferred species, becoming increasingly the pre-eminent wood resource in New Zealand.

Decisions had been made by government and private industry to establish two major industrial complexes in central North Island to process wood from the maturing radiata pine plantations there. One belonged to a new company, Tasman Pulp and Paper Co. Ltd (Tasman), which was formed as a consortium of the government and two private companies. The other belonged to New Zealand Forest Products Ltd (NZFP), which featured in Chap. 3 for creating its own massive resource of radiata. There had been talk of a single major wood-processing complex for the volcanic plateau of the central North Island, but a personality clash between the heads of the State Forest Service and NZFP precluded that option (Roche 1990). Both facilities were integrated processing plants, combining sawmilling with pulping and papermaking so as to make good use of logs of widely varying size and quality. The Tasman complex at Kawerau produced newsprint paper, and NZFP at Kinleith a mix of kraft and other papers. The Tasman venture was decided upon by 1951 and the processing plant came on stream at both sites in the mid-1950s (Roche 1990).

Although various sawmills had already been cutting radiata pine and three smaller pulp-based mills had been operating for some years, the opening of these two major industrial complexes had important ramifications. The massive financial investments in processing facilities entailed committing large areas of land to an assured wood supply, with sizeable annual restocking programmes. Assuring continuing supply for the Kawerau mill entailed restocking of forest on challenging sites, where ground-level frost was often a real hazard. Even after successful re-establishment, the silviculture of the second crops proved to be a challenge. In the longer term, however, sustainably grown forest products were increasingly seen as an important diversification from the country's pasture-based export portfolio, in addition to making New Zealand self-sufficient.

Yet despite the size of the new processing facilities, a surplus of radiata growing stock remained. That was partly a consequence of harvest delays occasioned by lack of new investment during World War II, but it was exacerbated by the government continuing to sell high-quality standing native timber at almost giveaway prices, partly in misguided efforts to keep housing cheap and partly to support employment opportunities in sawmilling and construction. Export of unprocessed logs became a means of dealing with the temporary surplus of radiata wood.

Along with that surplus of wood came increased competition for land for pastoral farming. The trace mineral deficiency problem that had plagued livestock on the pumicelands where much of the radiata resource had been planted had been solved and so could be corrected. Farmland was wanted for "rehabilitation" of returned World War II veterans. Moreover, there were zealots in positions of influence who insisted that if land could be farmed with any success then it should be. The land-allocation conflicts, along with the continuing surplus of forest plantation growing stock extended the lull in afforestation (Fig. 3.1). That in turn exacerbated a worrying imbalance in age classes of the plantation estate.

Even though the sirex epidemic had subsided, there remained a wariness of possible future epidemics. The Forest Service engaged a senior forest entomologist from Canada, J.J. de Gryse, to produce a report on forest health (de Gryse 1955). Recommendations included establishing a regular system to monitor forest health, and undertaking fundamental research towards developing a range of control measures for forest diseases and pests. It mentioned a possible future role for genetics, but made no specific mention of breeding for pest or disease resistance. It did, however, reinforce concern over excessive dependence on radiata. That concern spurred interest in increased plantings of a Mexican pine species, *Pinus patula*, and of two species from the south-eastern United States, loblolly pine (*Pinus taeda*) and slash pine (*Pinus elliottii*). Some existing New Zealand stands of these species had suffered much less from sirex (Jackson 1955) and/or could tolerate poorer soils than radiata could. However, those three pine species showed up very disappointingly in local sawing studies, with poor basic wood properties and/or very undesirable defects (R.W.M. Williams pers. comm. to RDB 1959). Moreover, the range of soils where radiata could grow well was then being much extended by aerial topdressing with superphosphate (Weston 1956; Conway 1962), an operation in which New Zealand had developed expertise for pastoral farming.

Another biotic alarm, however, occurred in the mid-1960s, with the appearance of the red band needle blight caused by the fungus *D. pini* ("dothi") (Fig. 4.1). Over large areas of where radiata was growing in New Zealand climatic conditions favour infection by the pathogen. In central North Island, young stands of radiata were quickly and alarmingly defoliated, and very large areas soon became badly affected. Even so, the worst fears were not realised. Aerial spraying with very small amounts of copper fungicide was developed as a dependable control measure (Figs. 4.2 and 4.3), a loss-saving research result that represented a huge payoff for the investment in the research capability of the Forest Research Institute. Spontaneous recovery often occurred after a year or more of alarming levels of defoliation, but far from always. Silvicultural practices that were being adopted served to reduce the disease hazard, mostly by improving air circulation. Moreover, except on sites of extreme disease hazard, stands over 15 years old were essentially unaffected, reflecting "**ontogenetic**"^{G} resistance. Two key "insurance" species which had been planted on a large scale, namely Corsican and ponderosa pines, were much worse affected, and neither showed resistance as they got older. At much the same time, Douglas-fir, the favoured second-choice species after radiata pine, started to be severely affected by Swiss needle cast caused by the fungus *Phaeocryptopus gaeumannii*. Trying to mitigate biotic and other risks by pre-emptive diversification into these other species had incurred considerable opportunity costs, yet in the end it often amounted to "negative insurance," achieving increased biotic vulnerability in return for those costs. Thus, despite

Fig. 4.1 Young stand of radiata in Kaingaroa severely defoliated by dothistroma ("dothi") around 1965, during the initial major outbreak of the disease in New Zealand. Infection spread upwards in crowns, killed foliage being reddish, and a spot-infection pattern was common. Growth was severely affected for several years, although such stands could eventually recover spontaneously without fungicidal spraying

Fig. 4.2 Pilot aerial spraying operation to control dothi

Fig. 4.3 Aerial view of dothi spray trial, showing green foliage in swaths after summer spraying

several alarms and fears for its future role, radiata emerged as more than ever the generally preferred species.

During most of the 1951–1968 period, because new planting was limited, the then-current views on choice of species had little impact on the subsequent species composition of New Zealand's forest estate.

4.1.3 Chile

From around 1935 a planting boom had developed, with radiata the greatly predominant species (Scott 1960). That boom, while it peaked during 1947–1949, declined lowly, in contrast to the abrupt end of the first planting boom in New Zealand. These Chilean plantings were made by numerous landowners, with financial encouragement from governments that were doubtless uncomfortably aware of the country's high dependence on copper for its export base. Individual holdings of plantation, however, were mostly too small to each support major industrial plant, raising the question of how the wood resource could be processed and used.

One locally owned company, CMPC (Companía Manufacturera de Papeles y Cartones) was big enough to develop its own industrial capacity, and opened a pulp mill as early as 1959, at La Laja. For wood from most of the remaining plantations, the government agency Corporación de Fomento de la Producción (CORFO), with the help of the United Nations, engaged in planning the processing and utilisation of the wood that was rapidly coming available (Scott 1960). As a first step in that planning, inventory of growing stock began in 1953–1954. In 1957 a preliminary external report advocated pulping almost the entire plantation wood harvest, as a lack of silviculture led to relatively few higher-value logs.

But a later report, produced in 1957, envisaged about one-third of the harvest being sawn for lumber. Even on the basis of the latter report, major additional pulping capacity was still needed. The United Nations Development Programme (UNDP), engaged along with CORFO in the planning, with continuing inventory of growing stock and surveys of potential industrial processes and processing sites, a process that began towards 1960 and was very active during the 1960s.

For growing, harvesting and wood processing, the United Nations agency FAO was involved, along with Chilean government agencies. Genetic improvement, however, was not considered, let alone implemented, until later (Chap. 5).

As in New Zealand and then in Australia, dothi appeared, being first reported in 1966 (Cobb and Miller 1968), but severe disease was not sufficiently widespread to cause major alarm.

4.1.4 South Africa

By 1951 radiata was one of several favoured forest plantation species, with the limits imposed by its site requirements having become clear during the 1930s (Chap. 3). It was generally the preferred species for producing sawlogs wherever it would perform satisfactorily and reliably. Despite the limits to where it could be grown satisfactorily, the firm recommendation (2/4) concerning genetic improvement from the Seventh British Commonwealth Forestry Conference (BCFC 1957) was meant to provide strong impetus for intensive breeding of radiata pine, which it did. That recommendation was promptly and enthusiastically adopted in South Africa.

4.1.5 Elsewhere

In Spain, planting of radiata proceeded, generally on fragmented land holdings, to be followed with conservative silvicultural regimes. Genetic improvement was not seriously contemplated during the period 1951–1968.

At low latitudes, in various countries, dothi was found to severely restrict or even eliminate radiata as a species option, despite its earlier promise. On many high-altitude sites in these countries, either shoot dieback (generally caused by the fungus *Diplodia pinea* – syn. *Sphaeropsis sapinea*) or droughts were already restricting where radiata pine would do well, and dothi further restricted the plantation range. Indeed, the presence of dothi virtually eliminated radiata for planting in Hawaii (Burdon 2001) and Malawi, Zimbabwe and Zambia in Africa (Poynton 1960, 1977). In Kenya, bigger areas of acceptable sites for planting radiata were to remain, and trees were selected for resistance to the disease, although that did not lead to a vigorous, sustained breeding programme there.

In its native California and adjacent Oregon, some enthusiasm for plantations of radiata developed in the 1950s. But that enthusiasm soon vanished as western gall rust infected, deformed and often killed high percentages of young trees. Then dothi epidemics killed most of the still-juvenile and adolescent survivors in the high-rainfall areas of north-coastal California and south-coastal Oregon. While cultivated radiata remained the species of choice for Christmas trees in drier central and southern California, few or no plantations have been initiated for producing wood in its native country since 1969.

4.2 Introductory Remarks on Breeding

As stated in Chap. 1, the two key planks of domestication of a species are proactive management inputs and genetic improvement, both to better serve some defined human purposes. The management inputs include all stages from producing and then sowing the genetically-improved seeds, producing planting stock, allocating the stock to appropriate sites, site preparation, planting, tending, harvesting and finally changes in processing and utilization. All these aspects need to be managed to take advantage of the known changes in the genetic makeup of the tree crop.

The genetic plank of domesticating a species is founded on choosing or assembling some appropriate base population(s), then selection of desired individuals within the population(s), followed by repeated cycles of intermating, evaluation, selection and so on, to achieve cumulative genetic changes over multiple generations. In some cases hybridisation between or among two or more different populations or even different species can be included.

But whatever genetic improvements are achieved, they need to be delivered in the operational growing stock. Those new trees have to be appropriately deployed and managed in order to reap the full benefits of the genetic improvement. Moreover, and crucially, for genetic changes to represent true improvement, selection must be directed at correctly defined **breeding goals**[G].

The interplay between management inputs and genetic improvement is crucial. The benefits of the two may be simply additive, the sum of the benefits of each individually. They may in some cases be substitutional, being alternative ways of achieving certain ends; for instance, breeding for resistance or chemical control may be alternatives for controlling a troublesome fungal disease. Ideally, and very often, the benefits of management inputs and genetic improvement are synergistic, with combined benefits exceeding the sum of the individual benefits. Such synergisms have long been achieved in animal husbandry and the growing of agricultural crop plants. A classic example is with the dwarf wheats of the "green revolution": biomass production is reallocated from producing superfluous straw to increased grain yield, and fertilizer boosts yields, instead of often causing traditional long-strawed wheats to fall over ("lodge") prior to harvest with loss of potential grain yield. For forest trees, such synergisms likely exist, but so far have generally been on a lesser scale, at least partly because domestication of trees is still in early stages.

Some work on studying and exploiting genetic variation in forest trees had already been done centuries ago (Toda 1974; Burley 2004), and other thoughtful foresters had long ago advocated what we now see as commonsense precautions, such as collecting seeds in appropriate localities from individual trees of good health, vigour and form (Myazaki 1696; Matthew 1831 [Sect. 3.4.3]; Burley 2004; White et al. 2007, Table 1.1). Nevertheless, systematic and intensive genetic improvement of forest trees began quite recently. Beginning the modern era of purposeful breeding, in the 1920s, several organizations and their researchers in eastern North America and Europe seized upon a then-known feature of poplars. Hybrids between some poplar species very often outgrew seedlings or clones of both their parent species, an observation made more acceptable by the success of hybrid corn becoming well known. That attractive phenomenon could be easily exploited, because poplars are routinely produced by raising cuttings in the nursery or setting small unrooted poles in the field, allowing easy testing and mass-production of individual hybrid clones.

In Placerville, California, beginning in the 1920s it was found that closely-related pine species could be successfully hybridized (Duffield 1952). While that gave promise of combining desired properties of different pine species, seed-set was generally low and thus mass production of most of the hybrids was impracticable. The most successful of the hundreds of combinations attempted was the hybrid between radiata and knobcone pines (Sect. 3.4.3.2), and small-to-medium scale operational plantings of this hybrid were established. However, with the difficulties of mass-production of hybrid seed and limited availability of appropriate plantation sites, the hybrid has not gained a significant niche in California. However, it has found use on some sites in adjacent south-central Oregon.

The earliest documented study of variation within a forest tree species is perhaps that of Vilmorin, who in 1820 near Paris planted seeds from 30 provenances of Scots pine and subsequently published the results (Pravdin 1964; Giertych 1991). In Scandinavia, interest in Danish research on tree-to-tree genetic variation within forest-tree species prompted active tree breeding. With Scots pine in the 1940s Sweden mounted the earliest modern program of purposeful within-species genetic improvement of a forest tree. In many R&D (research and development) breeding projects, much of the research came first, and practical development followed from it. For forest trees, the reverse was more common. Forest-tree breeding was frequently launched by borrowing well-tried principles from agricultural crop breeding, rather than being based on well-proven inheritance in the trees. Of even more note institutionally, while most tree-breeding programmes have been operational in nature, they were typically embedded in research organizations, and much of the early forest-genetic research was then conducted within the framework of the operational breeding.

By 1914 in New Zealand, A.H. Cockayne (1914) had foreseen the potential and need for genetic improvement of radiata (Sect. 3.4.3). Radiata pines typically had vigorous early growth, but their health and bole form often left much to be desired, with only a small proportion of wild or early-plantation individuals that were satisfactory in all three respects. But even in those early times, the year-after-year repeatability of characteristics such as annual growth or "**internode**"$^{\mathbf{G}}$ length in individual trees strongly suggested genetic differences that might be used to advantage.

By around 1950, better knowledge of genetic variation in several forest-tree species was becoming available, plantation forestry was increasing, and the distracting impacts of World War II were well over. The stage was thus set for intensive genetic improvement programmes to start in various parts of the world. The countries growing radiata were no exception. Pioneering work on genetic variation in radiata had already started in Australia by the late 1930s (Fielding 1953), and it corroborated some overseas studies, notably those in Scandinavia on other tree species (e.g. Larsen 1956).

While the field of genetics became formally established in the first decade of the 20th century, its principles had been informally applied for centuries in plant and animal husbandry before that (Darwin 1868). Yet, despite the visionary contributions made by pioneers in tree breeding, the forestry profession was often slow to appreciate the potential of purposeful genetic modifications of their future forest crops. As late as the 1960s, genetics was not a required topic for students in professional forestry curricula in many universities. As a result, the genetic quality of the seeds used for plantations was often ignored. Many regeneration foresters in the first half of the 20th century thought they were doing their job professionally and well if they got the right species for the site, and the least expensive seeds of that species were often purchased or collected. There were, however, exceptions, like Cockayne (1914) and Maxwell (1930) in New Zealand, who recommended collecting seed from vigorous, well-formed trees.

Zobel and Sprague's (1993) influential book, *A Forestry Revolution*, identified drought as the "great excuse" often used for plantation failure by foresters throughout the world, even though the seedlings might have been of physiologically poor quality, the site preparation inadequate, the planting done badly, competition control lacking, or the wrong provenance or even the wrong species used. Carl Syrach Larsen's (1956) seminal book, *Genetics in Silviculture*, noted that forestry and horticulture had common beginnings, but that they parted ways in the mid-19th century. While there were advantages to forestry in this divergence, most foresters also turned their backs on the grafting knife, and on the appreciation of heritable variation that is classically exploited in a horticulturist's use of grafts.

Zobel and Sprague went beyond inadequate training and general ignorance of genetics in explaining the use of incorrect provenances or otherwise inappropriate seeds for plantations. They claimed that, unlike most farmers, and most fanciers of ornamental flowers, dogs, cats, racehorses, etc., many foresters were actually antagonistic to the idea that genetic variation in trees might be important, or that it even existed. It is often said that silviculture is "a combination of science and art." It's not uncommon for artists to get prickly when scientists intrude on their domain, and this perhaps explains some of the early antagonism of practising silviculturists to this new science of genetics. Yet neglect of genetic principles in forestry could be much worse than missing opportunities, because various "practices of convenience" could be downright harmful. Wrong provenances (Fig. 4.4), or genetically inferior individuals, can all too easily be the sources of the most abundant and cheapest seeds.

Fig. 4.4 *Pinus ponderosa* provenance trial, Kaingaroa Forest, New Zealand, faster-growing Eldorado (western Sierra Nevada) provenance on right, Mt Lassen (southern Cascades) on left. Provenances from Rocky Mountains (not shown) were much slower still

4.3 Theory

4.3.1 Some Basic Principles and Application to Radiata

Effective domestication of forest trees should be founded on identifying a fairly short list of appropriate species for such effort; indeed, some promisingly executed tree-breeding programmes have been halted because it emerged that they involved insufficiently promising species (e.g. with Scots pine in Britain). In some species, very large geographic differences among natural populations can and do exist in important traits such as growth potential, bole form, foliage oil content and adaptations to substantially different environments. Within such species, it is often crucial to identify appropriate provenances for genetic improvement and possible domestication. As noted in Sect. 3.4.3 for ponderosa pine in New Zealand, various countries have had bad experiences with collecting or importing seeds from inappropriate provenances for use in plantations and developing land races.

With an appropriate provenance or group of provenances identified, it becomes appropriate to embark upon selective breeding, which is based primarily on exploiting genetic variation among individual trees. However, hybridization between provenances, if practicable, may give immediate benefits, while in the longer term it may contribute to the variation among individual trees. For forest trees, with few exceptions, there is abundant genetic variation within provenances that allows considerable changes to be made by selective breeding. Missing attributes, such as frost tolerance or resistance to certain diseases, can sometimes be made good by hybridising with other provenances or even species. Such gains from hybridisation, however, should be built upon selecting superior individuals within the populations or species that are being crossed. Countries or regions that can grow radiata successfully over large areas have generally had little problem in identifying it as a species that could and should be grown. So choosing radiata has followed the orthodox model for domestication (Box 4.1).

Box 4.1 Foundations of Genetic Improvement

Classically, genetic improvement of a tree species starts with choice of one or more appropriate geographic populations, or provenances, followed by selection and breeding of individuals with such provenances. Indeed provenance selection can be almost paramount in some tree species. This initial step, however, was skipped in the case of radiata, largely on the assumption that provenance variation would be a non-issue. While that assumption was wrong, it was actually appropriate to begin radiata breeding by selecting within the existing cultivated stocks.

Much less orthodox, however, was choice of provenance base for domesticating radiata. Despite the five natural populations being discrete, and thus likely to show some important genetic differences, choice of provenance was not seen as a crucial issue when intensive breeding of the species was begun. Thus in Australia, New Zealand and South Africa intensive breeding began with selecting from within land-race material. Despite that decision being based on some misconceptions (Sect. 4.7.1), making an immediate start on intensive breeding proved basically correct. As for tree-to-tree genetic variation, however, there was never serious doubt about its presence and significance.

4.3.2 Concepts and Principles of Quantitative Genetics and Breeding

For many readers the following box may suffice, rather than reading through the subsequent text. Conversely, readers wanting to delve deeper than the following text are referred to Lerner (1958) or the less-detailed and less-specialist coverage by Lerner and Libby (1976) (Box 4.2).

Box 4.2 Scope and Basis for Improvement by Selective Breeding

The scope for breeding for any particular trait is governed by the combination of tree-to-tree variability (often expressed as the coefficient of variation) and the heritability, namely the degree to which the attributes of parents are passed on to their offspring.

The phenotype, namely the individual as it is, reflects in simplest terms the combined influences of its genotype, namely its genetic makeup, and its environment. The degree to which on average the phenotype reflects the genotype is another measure of heritability.

The scope for breeding simultaneously for more than one trait is also governed by the genetic correlations between traits. A genetic correlation between traits means that selecting for only one trait will mean a correlated response in one or more other traits. Favourable genetic correlations make life very easy for the breeder. Adverse genetic correlations severely restrict the genetic improvement simultaneously achievable in the traits concerned. That puts the breeder under pressure to know the comparative economic worth of genetic gain in the various traits. In radiata, for instance, breeding for long lengths between clusters of knots tends strongly to mean larger knots where they do occur.

Even without genetic correlations between traits, simultaneous selection for multiple traits dilutes genetic gain obtainable in individual traits. The breeder therefore needs to know what few traits really matter, and to focus on them.

Given the founding samples of the provenances and their derived land races comprising the base population, efficient breeding is then based on selection among each base population's individuals. **Response to selection**, in any particular trait, is governed by the trait's "**heritability**^{G} multiplied by the **selection differential**^{G} (Fig. 4.5). Heritability is the proportion of the **phenotypic**^{G} variation (the variation among individuals as they are) that is genetic, the balance of the variation being essentially **environmental**^{G}. (Actually, for statistically-valid reasons, variation is typically expressed in terms of the square of its standard deviation, or "variance.") All heritability depends on differences between **alleles**, the alternative forms of genes present at individual **loci**^{G} on chromosomes. The selection differential is the difference between the trait's average in the selected parents from that in the base population. Expressed in standard deviations of the base or candidate populations, the selection differential is termed the **selection intensity**^{G}. Standard deviation is a statistically useful measure of a trait's variability—when it is divided by the population average one gets the "**coefficient of variation**" (CV). The more candidates evaluated per selection made—or the lower the **selection rate**^{G}—the greater the selection differential and thus the selection intensity, but the increase in selection differential is strongly subject to the Law of Diminishing Returns.

For example, if the measured candidate (i.e., base) population was growing an average of 200 cm per year in height, and the parents selected from it were averaging 220 cm per year, the **selection differential** is 20 cm (or 10%). If their

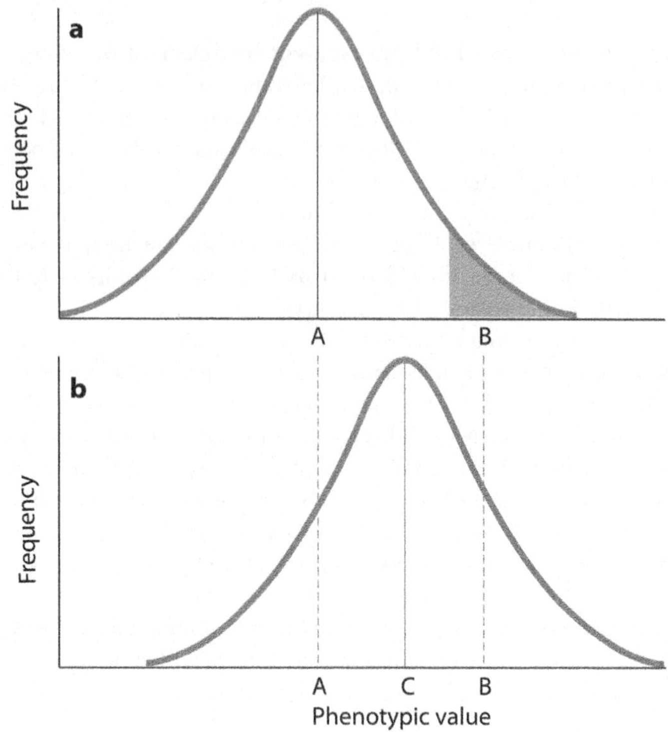

Fig. 4.5 Schematic diagram showing frequency distributions for phenotypic values of a trait in (a) parental candidate population and (b) offspring of selected parents. A denotes mean for the parental candidate population, B the mean for the selected subset parental population (shaded), and C the mean for the offspring of the selected parents. The distance AB denotes selection differential, AC response to selection, and AC/BC the realized heritability. This applies to both seed- and pollen parents being select; with seed parents only being select, AC would be halved. Note that there are now more individuals excelling the selected mean, B, and some of them excel the best of the parental generation. These better offspring provide the basis for ongoing generations of response to forwards selection for the trait(s) under selection

progeny then grew 4% faster on average than the base population, the **realised heritability** of that trait is 0.4. (One can roughly interpret that as: 40% of the observed variation in the candidate population is due to genes affecting height growth additively (see later), while 60% is due to environmental and non-additive genetic differences).

Knowing both heritability and CV gives the breeder a picture of the prospects of altering the trait by various intensities of selection, expected genetic change being proportional to the product of the two as a function of how different the selected parents are from the candidate-population average. The selection differential and selection intensity are often subject to the greatest control by the breeder, although they increase much less than in proportion to the number of candidates evaluated per tree selected. Heritability, standard deviation and coefficient of variation are, strictly speaking, specific properties of the breeding population growing in the environment where it is selected and/or will be deployed.

Scope for improving by selection more than one trait simultaneously is strongly governed by **genetic correlations**^{G} between traits. Genetic change from selecting just for one trait will lead to genetic changes in other traits, in proportion to their respective genetic correlations with the selected trait. Reasons for genetic correlations can be several; the usual one, however, is likely to be **pleiotropy**^{G}, whereby the same gene(s) exert effects on more than one trait. If causing an adverse genetic correlation, pleiotropy is not readily overcome. Favourable genetic correlations make life easier for the breeder, but adverse correlations can pose great difficulties, and have proven to be a serious issue with radiata pine. For instance, an adverse genetic correlation between bole-volume production and wood density will reduce the simultaneous gain achievable in the two traits; and also means that selecting solely for one trait will produce an unwanted change in the other.

Note, though, that even without adverse correlations simultaneous selection for multiple traits will dilute genetic gains achievable in each trait. The breeder thus needs to focus on well-considered improvement of a small number of traits, usually those of known or anticipated high economic importance and/or addressing substantial risk of plantation failure.

For effective selection, the breeder wants to minimize or perhaps statistically remove some of the effects of **environmental variation**^{G}. Often that latter can be crudely done by mentally adjusting candidate-tree measurements or observations for gross effects of site; for example, it may be better to choose a reasonable-looking tree on a site where almost everything looks bad than choosing a good-looking tree where everything else looks nearly as good. But most serious breeding programmes conduct progeny tests, progeny trials and/or clonal tests, which more effectively reduce the "noise" variation that can largely mask the genetic merit (or demerits) of selection candidates. If the site for such a trial or test has been prepared well, and all trees planted well, that will avoid much "environmental noise" variation. Furthermore, unavoidable microsite and similar environmental variation can be coped with by using proper statistical layouts in such tests and trials (e.g. Williams and Matheson 1994). Analyses of data from such field designs provide comparisons that avoid bias, and the size of the trial or test can be preplanned to achieve desired statistical power and thus the desired level of precision. In the process of such analysis and selection, one may also learn much about the inheritance of various traits, including their genetic intercorrelations.

In the long term, breeding programmes want to accumulate genetic changes over generations. That is achievable by recurrent cycles of selection, intercrossing, evaluation, selection, and so on. Such changes accrue from **additive gene effects**^{G}, essentially the effects whereby the average of the offspring tends to be intermediate between the averages of their pollen- and seed parents (whose additive genetic merits can be expressed as their **breeding values**^{G}). This works because the parents' genes have been rearranged in new combinations in the offspring. While the average of new combinations in an additive system approximately equals the average of the two parents, some offspring will be below average and others not only above average, but above the value of the best of the parents. It is those relatively few outstanding offspring that are selected to further advance the next generation. In each generation, the intercrossing produces fresh genetic combinations upon which selection can work to produce individuals that have increasingly accumulated the genes that modify the selected traits toward the desired condition.

Non-additive gene effects[G], whereby offspring tend to depart from intermediacy between their parents, will also exist but in degrees that can vary widely among populations and traits. Such effects include **dominance**[G] of genes at individual loci, and interactions between gene effects at different loci (**epistasis**[G]). These can be used to advantage in some situations, although in multi-generation breeding using additive genetic effects such non-additive effects do not accumulate over generations. They can be expressed upon hybridization between species, but have been very widely exploited following hybridization between highly-inbred lines within species of crop plants, something that is hard to achieve with long generation times such as trees typically have. Such specific genetic combinations producing the desired expression of important traits can, however, be captured in full by testing and selecting clones, which then have the advantage of preserving those specific desirable combinations and thus avoiding segregation that produces unwanted genetic variability. Since non-additive gene effects are difficult to exploit sexually, that gives an advantage to easily cloned species like poplars, where cuttings are the propagation method of choice.

Additive genetic variance is the numerator for **narrow-sense heritability**, which applies to many situations of propagation by seed, and thus to cumulative genetic changes achievable over generations of breeding. Both additive and non-additive genetic variances, however, form the numerator for **broad-sense heritability**, which is relevant for selection of clones (Fig. 4.6) for propagation and deployment.

Fig. 4.6 Three rows of radiata trees representing different clones showing clonal consistency of branching pattern and other features. Heavy, steep-angle branching (right, accentuated by reduced competition) contrasts with slender, more regular branching (left); these are among the classic "Jacobs" clones (Fielding 1953), planted in 1938–39, as at 1967 (Photos AGB)

Genetic diversity is generally desired in both the breeding population and across the plantation estate, and it can be crucial in several ways:

- as the basis for continued genetic gain for a few target traits, or for a menu of many different traits
- to avoid inbreeding and its ill-effects in the deployed crop trees
- to spread risks of massive loss to pest or pathogen epidemics and to unusual physical events such as drought, or extreme heat or cold
- to accommodate future changes in markets and biotic and abiotic factors by being able to effectively pursue changed breeding goals.

If a breeding programme has to serve a substantial range of plantation environments, **genotype-by-environment interactions** can be very important. Such interactions occur within a species when the performance of some breeds, families or clones relative to others differs among environments. In extreme cases, selected breeds, families or clones that are "superior" in one environment may be average or even "inferior" elsewhere. For example, a radiata clone that grows over more days of the year may grow relatively well in a permissive environment but be seriously damaged in environments with greater extremes of temperature or rainfall, while a more "conservative" clone will avoid such damage from extremes in the latter environments but miss growth opportunities in more-permissive environments. Separate breeding programmes may be needed for sufficiently different environments, which becomes very expensive. Even without such "regionalisation" of breeding programmes, it is possible deploy families or clones on planting sites where they perform to best advantage.

While really extreme cases of genotype-site interaction are uncommon in radiata, interaction is sufficient in parts of Australia to pose challenges for the breeder (e.g. Raymond 2011). The most conspicuous and specific source of interaction in radiata has been P deficiency (Fielding and Brown 1961; Burdon 1971; Johnson and Burdon 1990), which is prevalent in Australia and South Africa and regional in New Zealand (Burdon 2001), but this deficiency is now been routinely corrected with superphosphate or rock phosphate.

4.4 The Scandinavian School of Tree Breeding

While genetic information about forest trees was beginning to accumulate in the first decades of the 20th century and before (Pravdin 1964), and a few hybrids had been attempted as a method of radically altering available species, few programmes had tried to breed better trees within a species or population. True, as early as 1879, M. Kienitz urged German foresters to improve their native species rather than attempt forest "improvement" by bringing in exotics (Larsen 1956). That did not, however, deflect interest from exotics. Because very different successes and failures were being obtained in plantations with different lots of imported seeds, of either true exotics or widely distributed native species, attention became focused on heredity towards 1900, but at the provenance level.

Substantial attention was also given to interspecific hybrids during the first decades of the 20th century. This largely involved poplars, in which some interspecies hybrids were easily propagated as cuttings and were substantially more vigorous than seedlings or clones of the parent species. This early hybrid-poplar work had been undertaken in both Europe and eastern USA. Similar work on pines in California had almost no practical outcome there, largely because of the difficulties of mass production.

Selection of desirable clones had been practised with sugi (*Cryptomeria*) for centuries in Japan, earlier with Chinese-fir (*Cunninghamia*), and even earlier with willows and poplars in the Mediterranean region, but such clonal selection did not involve tree breeding as such. As hoped for in most subsequent clonal forestry, it favoured well-adapted, useful clones that remained relatively easy to mass-propagate as cuttings, as experience accumulated over time (Ahuja and Libby 1993).

Adapting within-species breeding to forest-tree species was pioneered in northern Europe and particularly in Scandinavia during the fourth and fifth decades of the 20th century (Zobel and Sprague 1993). Early in the 20th century's fifth decade, a society for "practical forest-tree breeding" had been formed in Sweden. Then, while much of the world occupied itself with killing and survival during World War II, Swedish tree breeding continued to develop. Following the war, as foresters and others turned to peacetime pursuits, they found that Sweden had established a well-supported institutional programme of forest-tree breeding. Åke Gustafsson, an eminent geneticist with a background in cereal breeding, in 1946 took the leadership of its forest-genetics section and associated it with his laboratory in Stockholm. A 1946 book by Bertil Lindquist, *Forest Genetics and the Practice of Silviculture*, also available in German translation (Lindquist 1948), had significant impact on many of the foresters and forest managers who read it (Zobel and Sprague 1993).

The writings of Carl Syrach Larsen, Director of the Danish National Arboretum, had begun to attract attention in the 1930s. Both in the arboretum and in pictures in his book, Larsen made use of vegetative propagation (by grafting and/or rooting cuttings) to create "tree shows." The differences between contrasting groups of clonally replicated trees provided powerful evidence to convince not only uninformed foresters but skeptics as well of the important influence that genetics exerts on growth, form and other important attributes of trees.

Larsen did his work in Denmark on trees adapted to high latitudes. He was, however, very aware of radiata—and its developing economic importance and its impressive genetic variability. In the preface to his book, he stated:

> Today one must travel to the most northerly parts of New Zealand to find some small, still untouched, forests of kauri. To see these forests, beautiful with magnificent knot-free stems of that tree, is reward for the journey. Yet in its way it also gives one a thrill to behold hundreds of thousands of acres of Monterey pine in the district around Rotorua. This is perhaps the world's largest and boldest planting scheme with a single species, undertaken in a short period of years… by means of this remarkable tree with an insignificant natural range in California there have been created here by man's efforts very great resources, which are even now already ripe for harvesting.

In Fig. 17b of his book, he used a picture of 14-year-old trees of radiata pine, rooted as juvenile cuttings in 1930 and planted in Palmerston North. These share the facing page with Fig. 17a, a picture of the famous "Higo-measa" sugi clone in Japan. One wonders if Larsen would be surprised by the importance of this upstart radiata in the late 20th century's Japanese wood market.

With the English translation of Larsen's *Genetics in Silviculture* book in 1956, a much larger forestry readership became familiar with the possibilities offered by the application of genetics to silviculture, and consequently several programmes then began to explore techniques of tree-breeding and propagation. Around 1950 and thereafter, the Scandinavian programme of tree breeding was studied and adopted by many new programmes of tree breeding throughout the world. It is fairly simple in concept, which led to widespread understanding and acceptance. It consists of four steps:

1. Select trees at a high intensity (originally the "best" tree in a large area, perhaps the "best" tree in a million), choosing trees of outstanding size, stem form, health and other desired attributes. Such individuals, pending verification of genetic "superiority," are termed **plus trees**^{G}.
2. Assemble those trees (usually by **grafting**^{G} **scions**—buds or shoots—removed from them) in a common area, called a **seed orchard**. Increase their numbers, and thus their multiplying power, by grafting many copies of each tree onto (usually) seedling rootstocks. This was usually done by putting one of each selected clone into each orchard block, each orchard having 20, 40 or even 100 orchard blocks. Typically, copies ("**ramets**"^{G}) of about 50–100 selected trees would be grafted into the orchard. According to plan, the various clones would all interpollinate each other.
3. Test the genetic worth of these selected parents by growing their offspring in formal **progeny tests**^{G} (usually common-garden plantings). Sometimes the offspring would be the result of controlled pollinations of two or more selected parents, but more frequently in the early stages of a tree-breeding program, open-pollinated (OP) seeds from the original tree (the **ortet**^{G} would be used. (This latter approach invoked the hopeful assumption that the average genetic value of the male parents in each neighbourhood of each ortet, and thus of its seeds, would approximate the overall population average.) The purpose of such progeny tests is ranking the parents for their **breeding values**^{G} for a series of important traits. Meanwhile, back at the seed orchard, the grafted parents are beginning to produce male and female "**flowers**"^{G} (Fig. 4.7), and thus to produce large numbers of seeds.
4. After the progeny tests have reached a size and age when their important traits can be evaluated, the lower-ranked parents are removed, or "rogued," from the seed orchard. Finally, after one or more rounds of such roguing, only the best 15–20 parents remain, and the average genetic quality of the seeds from the rogued seed orchard is consequently raised even further. Alternatively, seed orchards could be established afresh, without the lower-ranked parents. Either way, the selection involved in progeny-testing followed by roguing came in

Fig. 4.7 "Flower" structures in radiata: (a) pollen-bearing (male) strobili ("catkins"), and (b) female strobili (conelets), before pollination. Note different arrangements and locations on trees, giving physical separation that helps favour outbreeding

some programmes to be called "1.5-generation" or "backwards" selection, implying further genetic improvement in the current generation but markedly less improvement than anticipated when the best of the offspring are bred to produce the next generation.

This model for tree breeding was implemented on a national scale in Sweden. It was also adopted in three high-profile breeding programmes with southern pines in south-eastern USA. The first was headed by Bruce Zobel, based in Texas, under a cooperative arrangement between forest industry companies, the Texas Forest Service and Texas A & M University. After a disagreement with the Texas Forest Service, Zobel left to head a similar programme based at North Carolina State University (NCSU); that programme, with the strong quantitative genetics base at NCSU, became a key training ground for many forest geneticists/tree breeders. A third such programme was based at University of Florida. The Scandinavian model was also the basis for radiata pine breeding programmes set up in New Zealand, Australia and (via NCSU training) South Africa.

However, programmes following the Scandinavian model were liable to develop a weakness in seeing Steps 1–4 as very intense initial selection of a few elite parents to quickly achieve an early end-point, rather than as a very early stage of long-term domestication. For long-term breeding, extending over multiple generations, much larger numbers of unrelated parents, and thence greater genetic diversity in the founding ("base") breeding population, are likely to be crucial.

A leading role in developing that insight was played in the 1960s by Gene Namkoong, a U.S. Forest Service Pioneer Scientist stationed at North Carolina State University. Specifically, he and several colleagues (Namkoong et al. 1966) had noted that, (1) for the reselection of clones for roguing first-generation seed orchards and (2) to continue further selections into the indefinite future while avoiding inbreeding, much larger initial numbers of parents selected from base populations were needed in order to offer really worthwhile initial and continuing genetic changes. Wide variation in heritability among important traits made the large numbers especially important. That work was followed up by Shelbourne (1969) for the context of New Zealand's breeding programme. The impact of those works was one of several important developments that define the end of the period covered by this chapter.

4.5 Institutional Settings

During the 1951–1968 period reviewed in this chapter, just three countries had become actively involved in intensive genetic improvement of radiata: Australia, New Zealand and South Africa. Nevertheless, important early research was done in USA, specifically California where the species comes from, so it will be outlined first.

4.5.1 USA

Pioneering work was done in California, intensively exploring hybrids between pine species. This work was instigated by Thomas G. Eddy, a forward-thinking President of the Port Blakeley Timber Company. Eddy engaged as a consultant the famous plant breeder, Luther Burbank, who advised on the research plans and searched for an appropriate site and climate in which to do it. He finally chose a location at Placerville, California, where the Eddy Tree Breeding Station was established in 1925. After the planting of a breeding arboretum that successfully grew most of the 100-odd species of pines, and beginning the hybridising work that shed light on their evolutionary relationships, the Great Depression imposed financial constraints and it converted to a non-profit organization. Eddy gifted the site, arboretum and buildings to "the people" in 1935. Its research was continued by the US Forest Service, as the first Institute of Forest Genetics (IFG), through its Pacific Southwest Forest & Range Experiment Station, located 200 km distant in Berkeley. IFG Service used the site and buildings as its field station, and retained the name Eddy Arboretum.

Among the species studied was radiata, which proved highly crossable with its close relative, knobcone pine, and extremely difficult to hybridize with another apparently close relative, bishop pine (*Pinus muricata*) (Critchfield 1967). While experimental plantings of many species-pair hybrids were made and some showed some promising features, most had no practical impact on forest practice. Some low-level research effort on radiata continued—notably the independent research of D.T. MacDougal (1938) on its growth behavior within its native range.

In 1961, the University of California, Berkeley, became involved through appointing one of us (WJL) to its faculty. Because of work already done in New Zealand and Australia, WJL first developed a strong interest in radiata as a model species for basic research. As he came to realize its importance as a renewable resource, he broadened that interest to its performance as a plantation species, and remained for many years a key contact and collaborator for radiata breeders in Australia and New Zealand.

4.5.2 Australia

As of 1950, Australia had a federal organization, the Forestry and Timber Bureau, based in Canberra, that both conducted research and managed national forest resource and trade databases. Its operational and territorial responsibilities were mostly confined to the Australian Capital Territory, which despite its total size of only about 230,000 ha already had substantial plantation forest. In both Canberra and at a cooperative research outpost near Mt Gambier in South Australia, important silviculture and genetic studies on radiata had been conducted from the late 1930s. Public ownership and management of forests, however, was almost all vested in the various states, most of which sooner or later mounted their own tree breeding programmes, on occasions using plus trees obtained by exchange as well as local selections. APM Forests, formed in 1951, quickly became a large private-sector forest owner and welcomed the involvement of federal personnel in their tree-breeding programme in Victoria; Ken Eldridge took up that role in 1958. The company was also active in a variety of other aspects of plantation forest science, including soils and nutrition, silviculture, stand management and inventory, and harvesting. A sawmiller in the same region, Wilbur Saxton, developed an interest in tree breeding and established a successful seed orchard; subsequently, from about 2005, this operation has been managed by Seed Energy (http://www.seedenergy.com/).

Other institutional changes occurred over time—the (federal) Forestry and Timber Bureau's research activity in 1962 grew into the (Australian) Forest Research Institute, which in 1975 was absorbed into the Commonwealth Scientific and Industrial Research Organisation (CSIRO) as the Division of Forest Research. One important consequence of the establishment of the Forest Research Institute was a substantial increase in research personnel and facilities. Another development was the formation of the Australian Forestry Council in 1964 (Bevege 2010), which aimed at coordinating work among state and federal organisations. Significantly, its Research Working Group (RWG) No. 1, formed in 1967 (Shepherd 1977), was devoted to genetics. Importantly, New Zealand tree breeders soon participated as observers and presenters at the RWG1 technical sessions, beginning in 1970. Relations among the various organisations' breeders were typically cordial, but effective coordination of breeding work generally remained constrained by the emphatically independent nature of the states' forest services. One significant cooperative outcome, however, was the eventual formation of the Southern Tree Breeding Association (STBA) (http://stba.com.au) by some of the organisations participating in the working group (Sect. 5.3.1).

Underpinning the breeding programmes was a continuing or intensified commitment to expanding plantations particularly of radiata pine in southern Australian states, although Western Australia also endorsed maritime pine for certain important site types where it survived and grew better than radiata (Butcher 2007). Queensland, despite having begun active breeding of radiata and other species in the late 1940s (Wu et al. 2007), lacked any large area of sites suitable for radiata and so concentrated on other species.

In parallel with research related to growing and later breeding radiata pine was research on wood properties and processing, conducted in Melbourne by the CSIRO Division of Forest Products which had been formed in 1928 (Preston 1968). In collaboration with Canberra researchers, this was the scene of important early research on the inheritance of various wood properties (e.g. Dadswell et al. 1961; Nicholls et al. 1964). (Studies of timber engineering and wood structure were later recognised in awards of the Marcus Wallenberg Prize to Robert Leicester in 2000 and to Robert Evans in 2001 (http://www.mwp.org/prizewinners.cfm)).

4.5.3 New Zealand

In the few years before 1951, some individual enthusiasts had identified a few apparent super-trees of radiata, with a view to their potential for later selective breeding. Meanwhile, Martin Bannister, who had joined the Botany Division of the Department of Scientific and Industrial Research (DSIR) in 1948, was assigned to conduct basic research on variation in the radiata by the Director of the Division, Lindsay Poole, a forester.

The big institutional commitment to improving radiata occurred, however, in 1951 with the appointment of Ib Thulin, a Danish pupil of Syrach Larsen's, to the fledgling Forest Research Institute (FRI) within the New Zealand Forest Service (NZFS) (Burdon et al. 2008). With strong support of A.R. Entrican, then the NZFS head, Thulin soon embarked on a programme of intensive selection and description of radiata plus trees in New Zealand land-race plantations (Figs. 4.8 and 4.9) complemented by collecting open-pollinated seeds from those trees, archiving the selected seed parents as grafts, and commencing a controlled-crossing programme among many of these archived selections. Planting clonal seed orchards was begun at two different sites in 1957–58, and two more sites in the 1960s. Thulin employed the Scandinavian model, using grafts of very few intensively selected parent clones, which were expected to interpollinate by uncontrolled wind-blown pollen from within the confines of the orchards.

In the meantime, Martin Bannister had been transferred in 1952 from DSIR to FRI to start fundamental research on base populations of radiata. Besides studying

Fig. 4.8 An early plus tree selection ("PR7") in New Zealand. Growing in the open in the Nelson region, it was just under 60 m tall. Despite its impressive appearance, its progeny performed disappointingly in progeny trials, a common pattern among plus trees elected on open-grown performance

two of the uncertain number of New Zealand land races, he variously acquired, organized or collected essentially random seed-parent samples of the five native populations as they then existed. Seed samples for the Californian mainland populations came from the cone collection made by Margot Forde (1964a) for her PhD research at Davis, California, seed from Guadalupe Island trees was provided by Dr Reid

Fig. 4.9 Another early New Zealand plus tree selection ("Clone 55"), many years after selection and grafting into clonal archive. Its outstanding growth rate and resistance to foliage diseases were consistently exhibited in its progeny, as was its low wood density. It made a major early contribution to the breeding programmes in New Zealand and in more than one Australian state

Moran, and was supplemented by some material collected by Libby, Bannister and Linhart (1968), who also made a systematic collection on Cedros Island. All collections were from essentially random seed parents, save for avoiding obviously poor trees, although some seed from superior trees on both islands was also collected and used for planting at the University of California Russell Reserve in Lafayette. The seven near-random population samples were grown intimately together in duplicate common-garden research trials which constituted the Genetic Survey experiment planted in New Zealand during 1964–1967 (Burdon et al. 1992b).

Of great importance, from 1951 until the mid-1980s, was that all the tree-improvement work, while conducted from FRI in Rotorua, was done within a single national forest service. The resulting programmatic consistency later proved to have both advantages and weaknesses. Also, FRI was unusual in embracing on-site forest-products research in addition to forestry research, a feature that was exploited to advantage. A less favourable feature, however, was that the forests where many of the field operations had to be done were generally managed on site by forest rangers who were often (but by no means always) suspicious of university-trained FRI foresters and scientists. On the other hand, some rangers were recruited to the tree-breeding programme within which they gave sterling service.

In 1960, the Forest Research Institute came under a formally designated Director of Research, Dennis Richardson, who saw to it that three highly qualified scientists (Geoff Sweet, Rowland Burdon and Tony Shelbourne) were added to the force working on the genetics and breeding of radiata. While Thulin's experience in Denmark brought useful familiarity with the Scandinavian school's programme, these three independently-trained scientists were able to think outside of that box and identify some of its long-term weaknesses. By 1968, much had occurred towards integrating the applied breeding programme and the more fundamental research on the native populations and two local land-race samples of radiata. Under Thulin's vigorous leadership, by 1960 the breeding programme was well underway, with the first seed orchard producing its first operational-level collection of seed in 1968.

On another front, and of great significance for the next chapter, Richardson in 1962 recruited R.T. Fenton from within the Forest Service to research the economics of the silviculture and sawmilling of radiata.

4.5.4 South Africa

In South Africa, breeding of several species of forest trees started in 1953, about the time most other modern programmes got started, although genetic research on radiata had begun before 1940 (Poynton 1960). However, breeding of radiata was a relative latecomer, intensive selection commencing in 1962. Several classical Scandinavian-style seed orchards were established during 1972–1978, ranging in size from 6 to 17 ha, the later ones containing up to 240 plus-tree clones (van Wyk and Roeder 1978).

4.6 Interim Genetic Improvement Measures

As genetics became appreciated within some sectors of the forestry profession in the early 1950s, and pioneering breeding programmes got underway in important species in many countries, the benefits of intensive breeding were clearly going to take some years to meet demands for planned and active planting programmes. Interim genetic-improvement measures were therefore adopted.

In radiata, these early measures usually first involved finding trees in long-established plantations nearing harvest age that were substantially better than their near neighbours in health, size and bole form. These were used as seed parents, made more practical because, in the absence of fire, intense summer heat or cone-boring insects, radiata pines can store viable seeds in closed cones on the tree for up to seven years or even longer. Such seed-parent selection had four consequences, of which at least one was important in various plantation forestry programmes, and all four were important in Australia and New Zealand.

- *First*, collecting seeds from selective first thinnings was stopped, thus eliminating the dysgenic effects of that otherwise-convenient practice.
- *Second*, half of the genes of that next generation came from those well-above-average seed parents. The genes comprising the remaining half were those of the pollen parents which, while less superior than those of the seed parents, were still superior to the average of the trees planted because of inferior trees becoming suppressed, dying naturally or being removed in thinnings. The overall genetic effect on the resulting plantations was clearly positive and worthwhile.
- *Third*, since seeds for establishing plantations were no longer collected from isolated trees or shelterbelts, as was "conveniently" done to establish the early plantations, various undesirable effects of inbreeding were averted.
- *Fourth*, the effort devoted to finding the better trees in plantations and collecting cones from them focused the attention of field foresters and other forest workers on the newly recognised importance of genetics. This paved the way for easier acceptance of the products of intensive breeding as they became available.

In the early stages of waiting for seed orchards to meet all the seed requirements in New Zealand, crews worked just in advance of clearfell logging operations to produce "felling select" seed collections (Vincent and Dunstan 1989). They would find about 10–25 of the best trees per hectare, fell them, and strip them of their cone crop. This worked especially well in cooler, moister climates where cones remained closed on the trees for long periods.

Later, in 1968, the New Zealand cone-collection crews started moving into somewhat younger (post-1945) stands where logging was not imminent, to produce **climbing select**[G] collections. There they would climb about 25 marked trees per hectare that had been selected as the best phenotypes, and collect all cones that could be reached.

In Australia, Tasmanian foresters could similarly collect seeds of radiata from selected trees, a practice introduced in the mid-1950s (Elliott et al. 2008). In hotter parts of Australia, however, cones generally shed seed soon after they ripen.

In 1966, Eldridge (1966) reported on the development of **seed production areas**[G], initiated in nine-year-old stands in Gippsland, Victoria. These were routine plantations, except that they were heavily thinned to 250 trees per ha and fertilised, with a view to enhancing both cone production by the remaining trees and the genetic quality of the seeds produced. The accumulated cones were collected following clear-felling some five years after thinning. That treatment resulted in a five-fold increase in cone production per unit area, and was continued on some stands in Gippsland into the 1970s. By the mid-1970s, however, about half of APM Forests' seed requirement was produced from its seed orchards; the seed production areas in commercial plantations provided the balance (pers. comm. Henry Lieshout 6 June 2012).

Dedicated seed-production stands, however, never became popular with radiata, probably because of the ease of collecting seed from select individuals within routine commercial stands.

Another early 1960s approach to obtaining interim genetic upgrading of a forest was tentatively explored, namely grafting plus-tree scions onto a sprinkling of seedlings among natural regeneration in New Zealand, in the hope that these grafted trees would contribute substantially to the harvested crop of trees (Sweet 1963); Fielding tried a similar approach in Canberra. Despite some initial promise, the technique was abandoned because of both cost and the disappointing performance of the grafts which became increasingly evident during the 1960s. Thus the relatively slower growth of the grafted trees compared to their seedling neighbours made this early attempt a clear failure. By the time maturation state was better understood in radiata (Libby et al. 1972; Sweet 1973; Tufuor and Libby 1973; Libby and Hood 1976; Hood and Libby 1978), the mainstream option of planting genetically improved offspring was clearly better than upgrading naturally regenerated or planted forests by grafting.

4.7 Intensive Breeding

4.7.1 Background

It was a remarkable paradox that, while radiata was in key respects such an outstanding species, there were also glaring needs for improvement of several of its problem traits.

The first step in an operational tree-breeding programme was often plus-tree selection in genetically unimproved plantation stands, and the early radiata breeding programmes followed this pioneering protocol. The effectiveness of such selection is improved if that selection is done in stands that include some of the best available genetic material.

For the tree breeders it was reasonable to assume that two or more preceding generations of cultivation as an exotic would have allowed radiata to evolve better adaptation to local conditions, by some combination of natural and silvicultural selection, and this assumption was at least implicit in how all the early radiata breeding programmes began. Less appreciated at the time were the pressures of dysgenic selection that had been operating in native stands in California where "highgrading" or "creaming" meant selective removal of the best and straightest trees, which also tended to make land-race stands genetically superior.

Another widely-held belief was that radiata was confined to three nearby populations on similar parts of the Californian mainland coast; although Guadalupe Island pine had also been classified as a part of radiata, little was known about it, and Cedros Island pine was generally thought to be a population of bishop pine (*P. muricata*). An associated leap-of-faith was that provenance variation among the native radiata populations was, as in various other pines of limited distribution, relatively minor and thus not a substantial issue. This assumption was also convenient, in the sense of justifying an immediate start on intensive breeding.

From the beginning, however, this assumption was suspect. Not only had Guadalupe Island pine been classified as part of radiata, but the belief that Cedros Island pine was a population of bishop pine was in question. Moreover, some early attention had been paid to the possibility of provenance variation in radiata. Seeds had been acquired from the native Cambria population in the 1930s for trial plantings at Mt Burr (Pawsey 1967) and Australian Capital Territory (Fielding 1961a). Poor performance of the Cambria trees presumably dampened interest in radiata provenances; even so, experimental plantings of trees from the island populations were made in the 1950s (Fielding 1961b). In addition, *Pinus × attenuradiata* hybrids from the Institute of Forest Genetics, Placerville, California, were imported in 1954 for testing and inclusion in a range-wide provenance trial of *Pinus attenuata*, a move that was copied in New Zealand.

In New Zealand, collections of radiata seed from Monterey and Cambria were used to establish very limited provenance trials in 1955 (Shelbourne et al. 1979). In 1958 a few seeds from Guadalupe Island were obtained (Bannister 1958a) and used for a pilot planting. As time went on, it became increasingly clear that the two island populations were indeed radiata, meaning that its geographic range was considerable. Moreover, the possibility of important differences even among the mainland populations was increasingly recognised.

Thus it was that in the early 1960s, Bannister assembled a comprehensive seed collection for a systematic study of genetic variation across the entire natural range of radiata (Sects 4.5.3 and 4.7.6). This stemmed from the classic paper by Bannister (1959), who proposed that in order to realise the full genetic potential of radiata, breeders would need to tap into genetic material from the entire natural range.

In New Zealand, Australia and South Africa, breeding duly began with parents from land-race stands, which were evidently derived from seed imports from just the Año Nuevo and Monterey native populations. Later research (Sect. 5.5.4.1) showed that, by good luck, the Año Nuevo and Monterey populations are the two best pre-adapted to most sites where radiata is grown commercially. Indeed, this element of luck probably helped in the widespread acceptance of the species. For the long term, however, the entire range of genetic variation available in the species will be of interest.

Embarking on intensive breeding of radiata also assumed worthwhile heritability for the highly variable traits of interest. Visual clues, such as year-to-year up-the-bole repeatability of differences between individual trees in stem-form and branching characteristics, plus consistency within and variation between clones (Fielding 1953), were very reassuring. Even so, the choice of traits for future selection awaited verification and refinement, with evidence coming in from some pioneering trials of clones or open-pollinated seedling progenies, and then from progeny trials and progeny tests of plus trees. Verification duly came, but still occasioned some rethinking about refining traits to select for and how best to do the selection.

For curiosity-driven basic research, an ideal is to use the base populations that provide the starting points for breeding programmes. However, doing this basic research independently of ongoing breeding work can unnecessarily duplicate much of the cost and effort. Moreover, unless very intensive selection has previously altered the current breeding population(s) for one or more highly heritable traits, theory indicates that progenies of select material can not only give reasonably reliable estimates of genetic parameters such as heritabilities and genetic correlations, but such estimates are often more appropriate for the populations being domesticated than are estimates appropriate for the unselected native populations. In New Zealand especially, genetic research on radiata pine was done on both base-population and select material, which eventually verified these theoretical expectations (Burdon 1992). Accordingly, not only breeding but basic research efforts have come to be concentrated largely on select breeding populations (Box 4.3).

Box 4.3 Some Concepts Relating to Traits Involved in Breeding

- Breeding-goal traits represent those traits in which genetic improvement is sought, a breeding goal being the level(s) of genetic improvement sought in the trait or group of traits in question. In radiata pine, such traits can include stem volume production, freedom from malformation, timber stiffness, or resistance to certain diseases.
- Selection traits are traits that are conveniently evaluated, of good heritability, and strongly correlated genetically with one or more breeding-goal traits. In radiata pine, selecting for numerous, closely-spaced clusters of even-sized branches has served to pursue a breeding goal involving reduced knot size and susceptibility to malformation.
- Measurement traits are more immediate but conveniently measured proxies for breeding-goal traits. In radiata pine acoustic velocity (rate of travel of acoustic impulses) is such a proxy for wood stiffness, while resistance to penetrometers or torsiometers has been used as a proxy for wood density.
- An ideotype represents a set of characteristics, for example a certain crown configuration, that serve to achieve a certain breeding goal, e.g. increased stemwood production per hectare.

A breed, in the context of improving radiata pine, represents individuals that are all selected in pursuit of a specific breeding goal, differentiated breeds being based on comparative economic worth of different traits varying according to a set of circumstances, e.g. site categories to be planted and/or anticipated end-uses.

4.7.2 Traits Chosen for Early-Generation Selection

In tree-breeding programmes, whatever the species, selection of plus trees generally met several prerequisites. The trees had to be healthy, dominant to superdominant (i.e., at or above the general crown canopy), straight-boled and free of forking, with at least reasonably light and regular branching.

At different times and among different radiata breeding programmes, however, selection criteria varied with perceived priorities of what needed to be improved in the first generations of selective breeding. New Zealand's centralized programme is a case in point. On an important part of its forest estate, the volcanic plateau of the North Island, typical plantation trees' form was a big problem, with high rates of forking, many crooked boles, and too-frequent large, steep-angled branches. (While these phenomena represented a problem, they had the benefit of representing clear expression of important genetic differences.) For that region in particular, great emphasis was placed on bole straightness, and trees with regularly-spaced small-diameter flat-angled (that is, near 90° from the bole) branches were rigorously chosen. Another requirement was driven by pursuing needs for timber for cladding purposes ("weatherboards" on buildings), which required a surface that could be painted satisfactorily. Intergrown knots, produced by live or recently dead branches were acceptable, but loose knots or knotholes severely degraded the boards. Similarly, "cone holes" left in the bole wood by the stems of radiata cones on the main bole were troublesome, so trees with cones low on the bole were avoided during plus-tree selection in the 1950s. As it turned out, the specification of relatively small flat-angled branches led to choosing trees that produced multiple evenly-spaced clusters of branches per year on the bole, with relatively short distances between clusters (frequently called short "internodes") and termed "multinodal."

Only two wood properties (specific gravity and fibre length) were mooted early as likely-important selection criteria in New Zealand. Early plus-tree selections were diligently assayed for these properties in the hope greater wood-density and longer tracheids might usefully enhance stiffness and strength of sawn timbers (Burdon et al. 2008). Similar studies were undertaken in Australia, where spiral grain attracted additional attention as affecting dimensional stability and consequently straightness of sawn timber (unpublished reports by Nicholls & Dadswell, cited in Eldridge and Matthews 1977). Industry inputs in these countries, however, gave no clear call for genetic improvement of wood properties. Thus it was difficult for the breeders to engage on wood properties which were then expensive to evaluate on large numbers of selection candidates, especially if improving the properties genetically would be at the expense of desired changes in other traits (Burdon 2010). Moreover, wood-processing technologists were more comfortable working to solve problems of wood quality than trying to define what wood properties should be improved genetically, and in what ways (op. cit.). So little if any selection was done at the time for wood properties. Indeed, one early plus tree, the celebrated Clone NZ55, whose progeny showed outstanding vigour and health, was selected despite the progeny's very low wood density. That clone was made widely available to other breeding programmes, and its genes were doubtless heavily represented in early seed-orchard outturns. Fortuitously, it was to emerge later that this parent conferred a nice property for mechanical pulping, namely requiring much less energy.

A special case of trait priority was Queensland, where the focus of selection was resistance to the shoot pathogen, *D. pinea* (D.G. Nikles, pers. comm. to RDB 2013). However, because of limited success in breeding for resistance and the small area of radiata pine grown in Queensland, this selective breeding programme was later abandoned.

4.7.3 Regionalisation?

In the 1950s in New Zealand and Australia, it was considered unwise to mount a single breeding programme to serve all areas where radiata was widely grown. Some families that were well-adapted to some areas might be less or even poorly adapted to others, which would call for separate regional breeding lines and programmes. In Australia, breeding programmes serving individual states would have been based on some land-race differentiation and have led to breeding-goal regionalization in response to regional site conditions and different political goals and institutions. However, structuring of breeding programmes according to state boundaries often did not fit the obvious bases for regionalization. Straddling state boundaries were some major areas growing radiata that were obvious natural regions on the bases of climate and soils and served essentially the same markets. On the other hand, states' tree breeders would accept exchange of plus-tree clones, notably some New Zealand ones.

In South Africa, radiata had become confined to a restricted zone of essentially Mediterranean climate, so only one programme was developed.

In New Zealand, however, one agency was breeding for a 12° range of latitude and substantial climatic and edaphic differences, as well as regional variations in harmful pests and pathogens. Plus trees were therefore initially selected for orchards to serve specifically each of three different regions: North Island and Nelson (northern South Island); Canterbury (eastern central South Island); and Southland-Otago (far south) in the hope that such regionalization would be sufficient for almost all radiata sites. Surprisingly, genetic gain trials which were established later on failed to show any justification for even this coarse regionalization. One contributing factor may be that, unlike many conifers, radiata's timing of growth initiation and cessation are evidently little affected by absolute daylength but rather by whether days are shortening or lengthening (Burdon 1994).

4.7.4 Clonal Archives

Part of the Scandinavian breeding model was the establishment of clonal archives, typically as grafts of the plus trees, as a prerequisite for establishing clonal seed orchards. Grafting of radiata quickly proved to be relatively easy with tip-cleft grafting (Thulin 1957) adopted as most convenient, although other grafting techniques also work well. The archives serve several purposes:

- repositories of the founding and later-generation selected clones (Fig. 4.10)
- a means of examining candidate selections more closely in "tree shows"
- providing convenient sources of material for propagating into seed orchards
- allowing convenient controlled pollinations among the select clones (Fig. 4.11).

Tree-show performance, while useful for revealing genetic differences in some traits, was much less reliable for other traits than progeny or clonal performance in well-designed tests. The archived clones did reveal problems like early graft incompatibility of particular clones that might otherwise have been forwarded to seed orchards only to sicken and die as their graft unions failed.

An early New Zealand clonal archive in Rotorua was the scene of pilot work on grafting radiata onto rootstocks of other pine species, in attempts to achieve

Fig. 4.10 Ib Thulin, in 1983 just before retirement, beside graft taken in 1953 from New Zealand's "clone 55." Note the thicker, more deeply fissured bark on the rootstock

Fig. 4.11 Early clonal archive of radiata plus trees at Rotorua, New Zealand, showing isolation bags shortly before controlled pollinations

dwarfing of the grafts for easier breeding and cone collection, and/or boosting of flowering. These efforts were unsuccessful in achieving those goals and were not formally reported. Yet successful grafting was achieved on rootstocks of a notable range of species (Fig. 4.12). Many of these two-species grafted trees survived for several decades in spite of the graft unions being of surprisingly unequal size, the radiata scion often growing increasingly larger in diameter than the rootstock. By contrast, rootstocks of radiata would become much thicker than the scions of other species that were grafted onto it (Box 4.4).

Box 4.4 The Seed Orchard Concept
Seed orchards are effectively a means of bringing together superior parents that may be widely separated in the forest so that they can interpollinate, thereby mating the best with the best to produce genetically improved commercial seed. With radiata, producing seed orchards almost always entails, at least initially, taking grafts from intensively select parents. Progeny testing can reveal which of the originally selected parents are truly superior, so the remainder can be "rogued" in thinning or re-establishing the orchards.

4.7.5 Seed Orchards

In setting up Scandinavian-style clonal seed orchards, much store was set by two facts: that radiata of any age would graft readily on to seedling rootstocks, and that the scions of the grafts showed maturation characteristic of the ages of the ortets (or the previously grafted clonal material) providing the scions. Grafting was far easier than rooting cuttings from sexually mature trees (Fielding 1954, 1957), but grafts

Fig. 4.12 Graft of a southern pine (from USA) scion on a radiata rootstock, showing the faster growth of radiata. Radiata was grafted on to rootstocks of various other pine species, in unsuccessful efforts to reduce its vigour and/or boost seed production

often later proved prone to delayed incompatibility whereby they mysteriously died (Fig. 4.13), and this was sometimes very troublesome (Burdon et al. 2008). Maturation, or "physiological ageing", is manifested in radiata by the propagules developing quite different foliage and general growth habit compared with seedlings of the same size and age from planting. In this context, however, maturation had the big attraction of allowing almost immediate flowering; but this advantage later proved less decisive than originally hoped, because less time is gained from maturation in achieving abundant pollen production than in the onset of seed production.

Siting is always a major consideration for establishing seed orchards, and some early seed-orchard sites proved to be much less than ideal. For ease of management, flat or gentle terrain and proximity to some headquarters were frequent criteria. For open-pollinated seed orchards, isolation from unwanted pollen blowing in from

Fig. 4.13 A case of delayed graft incompatibility in radiata, showing characteristic ribbing of the wood surface that is revealed by removing bark. External signs were progressive yellowing and thinning of crown before eventual death

unimproved stands or contrasting breeds is increasingly important as tree breeding progresses. In seeking isolation, however, some bad mistakes were made in New Zealand. Areas that were physically distant (1 km or more) from other radiata were chosen, two in Kaingaroa Forest (Fig. 4.14) in central North Island and one in Gwavas Forest in Hawke's Bay (Fig. 4.15). However, those in Kaingaroa were where other tree species had been planted because the sites were thought to be marginal for radiata pine, and they could be far from optimal for radiata's growth and sexual performance. Moreover, the pollen isolation achieved was limited, because pollen from extensive radiata plantations often blows in visible clouds over long distances, crossing even very wide isolation zones. Later, however, heavy pollen production within maturing orchards could, by its sheer density, proportionally reduce external pollen contamination.

To serve the South Island of New Zealand, a radiata orchard site was acquired in 1965 at Amberley, near Christchurch (Fig 4.16), and it proved excellent. Well isolated, it was relatively dry (annual precipitation about 650 mm), not frosty, and close to the sea. Indeed, it set a benchmark for orchard siting. Especially high pollen and cone production was subsequently noted on radiata pines close to the sea in both California and New Zealand, and is now an important guideline for identifying preferred seed-orchard sites.

As in New Zealand, Australia had mixed fortunes in choice of radiata seed-orchard sites. Of the Australian orchards, at least three became successful sources of seed: Tallaganda in New South Wales (established to serve both New South Wales and the Australian Capital Territory) and about 50 km from the coast; Saxton's, on a private estate about 15 km from the coast in south-eastern Victoria, and Mt Schank, near Mt Gambier and about 12 km from the coast in South Australia (Pederick and Brown 1976). In New South Wales, orchard planting

Fig. 4.14 Early clonal seed orchard planting in Kaingaroa Forest, New Zealand, showing wide spacing between grafts. Affected by several problems, including root rots, frost kill of conelets near ground level, very heavy early pollen contamination and delayed graft incompatibility of radiata, it eventually produced well, but trees could not be topped to control their height without severely reducing seed production

began at three other sites, in 1957, 1961 and 1966 respectively (Johnson et al. 2008); 31 additional ha were planted by 1967. Other areas by that same year were Tallaganda (NSW + ACT) 9 ha; South Australia forest service 35 ha; Tasmania forest service 4 ha; Victorian private sector 31 ha, forest service 30 ha; and Western Australia forest service 12 ha. The early Australian orchards generally contained significant proportions of clones donated from New Zealand.

Fig. 4.15 Part of Gwavas Seed Orchard, planted in 1966, as at 1991, giving good seed production

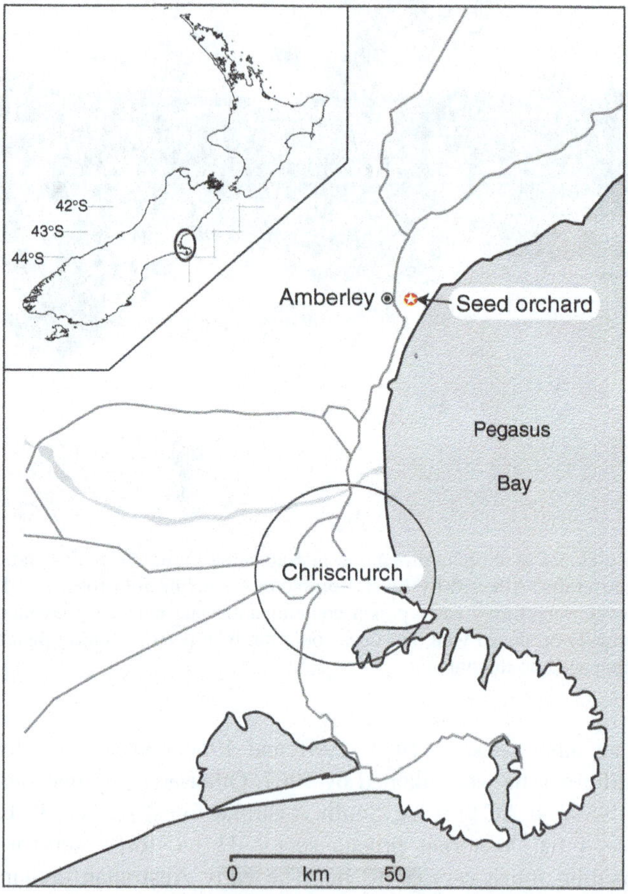

Fig. 4.16 Map showing location of Amberley seed orchard in New Zealand

The summer of 1967–1968 was particularly hot and dry in parts of Australia, and provided clear evidence of the importance of orchard siting: apart from high mortality at sites such as Canberra, cone yields were reduced (Pederick and Brown 1976), as had been noted in earlier droughts in South Australia (Pawsey 1960). Further evidence of the benefit of maritime influence on seed yields in radiata seed orchards was the continued high seed yields of certain early Australian orchards, in accord with later New Zealand experience.

Other problems: Siting was not the only problem with early radiata seed orchards. The early practice of grafting onto container-grown rootstocks often led to those trees later falling over because of root strangulation. Control of weeds and pests like bark beetles posed some unexpected challenges, which included loss of grafts from ringbarking by the beetles. And there was the insidious problem of delayed graft incompatibility, which for reasons unknown has been more troublesome on some sites than on others. It often killed grafts as they were due to commence heavy seed production, after thinnings had been made to boost seed production on the retained trees. Furthermore, the surviving grafts grew inconveniently tall, but topping them to a convenient height greatly reduced seed production in the short term.

For various reasons early seeds from the seed orchards were often small, leading to slow initial growth of the seedlings in the nursery. Nurserymen, who had erroneously expected genetic improvement to show up in the nursery beds, were generally disappointed, whereas it was actually their task to make good any effects of a slow start on the planting stock.

Yet another problem was posed in New Zealand, by quarantine restrictions. Alarm over red-band needle blight (caused by "dothi") led to prohibition of the movement of grafts and cuttings from most of the North Island to the South Island. After the regional South Island seed-orchard units were abandoned in favour of a national set of clones, this prohibition left the Amberley orchard without many of the plus-tree clones needed to serve the main seed-production needs of the country. Actually, the quarantine restrictions were futile, because dothi very soon spread to where, and only where, it would thrive, which did not include anywhere near Amberley in eastern South Island.

4.7.6 Genetic Trials and Mating Designs in Breeding Programmes

Genetic trials, were they **progeny trials**^{G} or clonal trials, served several early purposes:

- Evaluation and ranking of candidate parents in the 1950s and beyond, following the Scandinavian model. **Progeny tests**^{G} were particularly appropriate for ranking the plus-tree parent clones, and then for roguing the lower-rankled clones from seed orchards, in **backwards selection**^{G}.
- Estimation of genetic parameters (Sect. 4.3.2) for various observed traits (notably variances and heritabilities and intercorrelations among those traits). Those early

estimates were used to verify prospects and likely values of genetic improvements of candidate traits or combinations of traits, and to guide the breeder on how to select effectively for those traits and thus for more-useful trees.

- Providing progeny trials designed to identify a promising set of offspring-generation trees to serve as parents of the next generation. Such **forwards selection** concentrates the favourable new genetic combinations that occur among the progeny, according to the efficiency of evaluation made possible by the trial design.

How these various purposes are served, in the cases of genetic-architecture, progeny- or clonal trials, depends on the mating design used, namely the combinations in which parents are intermated. Actually, no one mating design can serve all purposes with optimal efficiency (Burdon and Shelbourne 1971; White et al. 2007).

The quickest and cheapest mating design is generally what Nature gives, namely open pollination. Collecting naturally pollinated seeds from the plus-tree ortets was very convenient with radiata. However, the identity of the individual pollen parents was of course unknowable at the time.

Open-pollinated (OP) families can be complex and highly variable in respect of relatedness for family members, especially in natural stands of forest trees, making it difficult to evaluate the genetic merit of parents and individual offspring. If, however, they approximate well to half-sib families, with very large numbers of unrelated pollen parents, they provide excellent information on the genetic merit of seed parents. For breeding programmes with radiata, successful use of OP families thus depended largely on the various widely-scattered plus-tree seed parents having been all pollinated by a broad and genetically similar mix of pollen. Thankfully, that seems to have been largely true in plantations of radiata—and not too badly wrong even in native stands in mainland California (Burdon et al. 1992a). Anyway, in those early years, OP progeny tests were generally established as a matter of routine. That was done in order to begin the first progeny tests with less-expensive, quickly-available seedlings, and then to be able to rogue the first seed orchards. It was, however, still an act of faith based on the then unverified hope that the different sets of fathers producing the OP families would not seriously invalidate the plus-tree rankings and trait heritability estimates.

The alternative of using control-pollinated progeny in such early trials could have provided much more accurate genetic information, but this is only assured if each candidate tree is involved in crosses with a number of others in a rigorously chosen mating design. Making controlled pollinations on the plus-tree ortets, however, was laborious and dangerous, although it was undertaken in Canberra for a few years. Such pollinations (Fig. 4.17) entail much more expense than collecting OP cones from the select trees, and incur delays of at least two years while the newly-pollinated cones mature. Controlled pollinations done on ramets in breeding orchards (or archives) for that initial generation is much easier, less expensive and safer than if done on ortets in the field, However, they entailed even greater delays from waiting for the grafted ramets of the plus trees to become established and grow to function sexually. Luckily, the OP progeny served surprisingly well

Fig. 4.17 Close-up view of controlled pollination in radiata, showing isolation bags and injection of pollen over receptive conelets

to both rank the genetic merit of the plus-tree seed parents and explore the inheritance of various traits. The heritabilities that they revealed are based on **additive** gene effects, which are the basis for cumulative genetic gains with continuing breeding for those traits in subsequent generations.

Such OP progenies had their limitations, though. The pollen parents were genetically random and unknown, which is a major disadvantages for efficient "forwards" selection of individual offspring for the next generation. And, with only seed parents known, such trials tended to give relatively weak "signal-to-noise" ratios in both evaluating parents and ascertaining the inheritance of traits the trials were very big and thus very expensive. Furthermore, such progeny trials told nothing about the role of **non-additive** gene effects, whereby offspring tend to depart from intermediacy between seed parents and (usually unknown) pollen parents.

Accordingly, by or in the 1960s tree breeders set about making controlled pollinations, ensuring not only that the pollen parents as well as seed parents were plus trees, but also allowing clearer pedigrees to be developed to the benefit of future more-intensive breeding. By mating in judiciously chosen mating designs, additional information could be obtained on inheritance of traits, and with pollen parents known to be plus trees a better genetic base could be provided for selecting in future generations.

As forest-tree breeding progressed beyond the Scandinavian model, largely led by Zobel's North Carolina Cooperative, a major mis-step in choosing a mating design was often made in forest-tree breeding programmes, including the main radiata programmes. Consultation with chicken breeders produced a recommendation to use the four-tester (or North Carolina II) design, in which four of the very best-rated roosters were each crossed sequentially with every one of a set of well-above-average hens. That design, however, was based on the ability of a rooster to sire far more chickens than a hen could be mother to, especially within any convenient period; this is quite a different situation from the large numbers of independent cones that a radiata tree can simultaneously produce, along with the possibility of using a single "male" as a pollen source for each of almost any numbers of cones. Having half the genes in the next generation of selection candidates come from only four parents was no great drawback for estimating additive and non-additive components of heritability and for ranking the parents for "backwards" selection, but it is highly undesirable for "forwards" selection (Burdon and Shelbourne 1971), narrowing the genetic base and accelerating the future onset of inbreeding.

Four-tester mating designs were dutifully implemented for small numbers of parents selected in the 1950s in Australia and New Zealand, but in the latter most of the seeds thus produced were destroyed in a coolstore disaster in early 1966. But, as we will review in the next chapter, that was a blessing in disguise.

As a supplement to controlled pair-crossing some controlled self-pollinations were made, to verify the likely effects of severe inbreeding (M.H. Bannister pers. comm. to RDB 1965; Pawsey 1964; Wilcox 1983; Kumar 2004). Inbreeding depression proved to be serious in radiata, especially for seed set and growth rate, but extremely variable among different parents' selfed progenies.

Not all progeny trials involved plus-trees, however. Some open-pollinated trials were established with parents that were chosen effectively at random or else for some distinctive features such as contrasting branching patterns (e.g. Bannister and Vine 1981 and preceding papers in the series). For studying the natural range of variation of the species, in common-garden experiments comparing natural populations and some land races (e.g. Burdon 1992; Burdon et al. 1992a), the OP seed parents or clones (Libby and Hood 1976; Hood and Libby 1978), were essentially random samples from each population.

Clonal trials, which had started to be established in Australia before 1951, were used largely to study inheritance—not only of growth and tree form (Fielding 1953) but also of wood properties (e.g. Dadswell et al. 1961). In 1955, a small clonal trial was planted on four New Zealand sites (Burdon 1971, 1976; Burdon and Harris 1973), using rooted nursery cuttings that had been produced from young seedlings. The clonal trials could provide information complementing that from progeny tests, were more precise in estimating the proportions of non-genetic ("environmental") and total genetic ("additive" plus "non-additive") contributions to variation in each trait evaluated, and/or needed fewer trial trees for equivalent confidence in the estimates, and if site-replicated had a special advantage for probing genotype-by-environment interactions involving differential responses of different individual clones to different environments. Notably, those early New Zealand clonal trials confirmed substantial genetic variation in tolerance of phosphorus deficiency as driving strong clone-by-site interactions, as had been previously observed in Australia (Fielding and Brown 1961). A clonal population-architecture trial in northern California was to show not only substantial differences in growth and form among clones from near-neighbour ortets and within mainland populations, but also differences at the local-clonal and population levels in cold and animal damage (Hood and Libby 1980).

By 1968, a picture was beginning to emerge of the overall genetic architecture of radiata, namely of the proportions of total genetic variation for various traits among the three (later all five) native populations, among stands occupying contrasting environments within each population, among OP families or clones within stands, and among OP sibs or clones within families. In addition, the heritabilities and genetic intercorrelations for an increasing number of studied traits were being estimated, although much of the story was then still unpublished. The two island populations, while both taxonomically belonging within radiata, were found to be quite distinct from each other and from those on the mainland. Among the mainland populations, Cambria was the most distinct, but generally performing worst in the New Zealand, Australia and California trials for most studied traits of anticipated importance. By this time it was becoming clear that not only were all five native populations members of the radiata pine species, but each is at least appreciably different from the other four in many distinguishing and important traits.

4.8 Vegetative Propagation

By the 1950s, most forestry people professionally growing trees as a wood crop knew that people who grow trees for their apple or almond or other fruit or nut crops had long abandoned planting seedlings in their orchards. The advantages of uniform reliable clones were too great to commit their orchard lands to seedlings, which would belatedly produce fruit or nuts of uncertain quality and unwanted variability.

The Scandinavian tree-breeding model had followed suit. It favoured clonal orchards to produce seeds, not for eating them but to create forest plantations from them. As in horticulture at the time, most early forestry seed orchards then relied on grafting to propagate their clones. Early trials employing grafted trees to produce wood were initiated by Sweet in New Zealand and Fielding in Australia but this soon proved costly and impractical (Sect. 4.6, final para.). So attention during 1951–1968 turned to trials and research using other methods of cloning forest trees. Species such as poplars and willows were known to be easily propagated by cuttings, but most others, including most conifers, were proving to be more difficult.

Knowing that radiata could be propagated directly by setting and rooting cuttings from young seedlings, and quite readily in a good, well-run nursery, it was hoped that this could be a means of mass-propagation of genetically improved planting stock. Cuttings from well-characterised proven clones (that is, used in clonal forestry) would not only be genetically superior; they would also lack the unwanted variation resulting from genetic segregation that always occurs in producing seedlings. As a nice bonus, if the cuttings are taken from young seedling-origin trees, that is, in a late-juvenile or adolescent maturation state, their resulting clonal trees tend to have much better tree form than the original seedlings (e.g.

Fielding 1970). Another bonus, which emerged later on, can be near-immunity to such juvenile diseases as western gall rust (caused by the fungus *Endocronartium harknessii*) (Zagory and Libby 1985).

As from the 1950s Fielding in Australia had become concerned with the untidy growth habit of radiata seedlings in the "adolescent" phase of maturation, with adverse effects on log form. Experience in the Northern Hemisphere with some other pines and some spruces had shown that it was far harder to produce good, field-ready cuttings from older trees than from young seedlings. Yet Fielding was so concerned about radiata's early tree form that he set out to overcome the difficulties of producing plantable cuttings of post-juvenile material, in order to enjoy the advantages of both clonal uniformity and better tree form associated with maturation. The quest for such clonal/maturation-state solutions for radiata was begun in Australia (Fielding 1970; Libby et al. 1972), and was followed in California (Libby and Conkle 1966) and in 1966 in New Zealand (Wilcox et al. 1976), beginning in 1962 and 1966 respectively. It was to falter because of continuing difficulties of maintaining clones in a maturation state that allowed easy and cheap propagation (Libby et al. 1972; Burdon et al. 2008). Even so, the production of cuttings, for more than one purpose, was to be revived later on, and then *in-vitro* technologies were developed as further options for vegetative propagation.

4.9 Other Studies

Complementing unreported work on controlled self-pollination, beginning in the early 1950s, Bannister (1965) studied natural inbreeding in radiata, where it was colonizing ground outside plantations. Isolated seedlings in open ground, while free of competing trees, were likely to be largely self-pollinating, and the offspring that sprang up around them would still be pollinated largely by close relatives. Despite the inbreeding entailed, radiata pine was well able to invade essentially open ground.

The progeny and clonal trials described above were mostly evaluated for quantitative-trait variation. Most of the traits initially chosen for study, mainly height and diameter growth, form, adaptedness, pest and pathogen susceptibility, and nutrient requirements, show continuous variation. That suggested that these traits were each controlled by moderate-to-large numbers of genes, at several-to-many loci (sites on chromosomes), each exerting relatively small effects. An exception arose in the offspring of "fastigiate" individuals, in which their unusual growth habit (upright and repeatedly forked), small cone size and anomalous location of seed cones (typical of location and arrangement of pollen catkins on normal trees); a progeny trial established in New Zealand in 1958 confirmed the role of a recessive allele at a single locus. Chlorophyll defects, often expressed in inbreeding studies as albino seedlings in self-families, were also apparently due to single recessive genes.

Some work in herbaria helped confirm the taxonomic place of the island populations within the species (Bannister 1958a), and to describe some spontaneous hybrids of radiata pine with knobcone pine (*P. attenuata*) (Bannister 1958b, c) that had occurred in Canterbury, New Zealand. Studies of constituents of their turpentines helped characterise taxonomic and evolutionary relationships among the closed-cone pine species and their hybrids (Forde 1964b), helped confirm the species and varietal status of the radiata island populations (Bannister and McDonald 1983), and pointed to which native populations contributed to the different New Zealand land races then being studied (Bannister et al. 1962). While driven purely by scientific curiosity, such work provided information (Blight et al. 1964) that, among other things, led to the building of a profitable turpentine recovery plant in New Zealand.

Notable among other work was a study by Bannister (1962a) that characterised the bewildering variability of branching patterns in radiata within New Zealand.

4.10 Programme Reviews

While Australia continued to hold regular meetings on its tree breeding programmes following the formation of the Australian Forestry Council in 1964, neither New Zealand nor South Africa had a close counterpart. True, there were periodic Research Advisory Committee meetings in New Zealand, but these were effectively public-relations exercises during an era of essentially stable public funding, although they served that purpose well.

In 1965, however, "The Improvement of Radiata Pine Symposium" was held in Rotorua at the instigation of Dennis Richardson. It addressed the entire New Zealand tree-breeding programme, with particular focus on radiata (Burdon and Thulin 1966). The 40 participants included 10 local research scientists and breeders, 26 growing, processing and utilization specialists from the forestry sector, and four overseas visitors. Among these latter visitors was Bruce Zobel, who had hosted a month-long international working conference on forest biology and tree breeding in North Carolina in 1960.

Topics at the Rotorua symposium covered or at least included what breeding had already been done in New Zealand, what was known of forest-tree genetics and breeding technology, and what traits most needed to be improved in radiata. Held in the context of the recent appearance and uncertain impact of red-band needle blight in New Zealand, there was a general if cautious endorsement of both the developing cooperation with the pathology programme in New Zealand and the breeding programme as it stood. Industry participants did make an emphatic call for the short-"internode" or "multinodal" branching habit, with its benefits of better general tree form and easier pruning, rather than for "long-internode" (or "uninodal") trees that could provide knot-free recuttings of lumber without pruning but would have larger branches and somewhat poorer tree form. Calls for genetic improvement of wood properties, however, were muted or absent, despite the promise that their excellent heritabilities gave for rapid and needed improvement; there was both uncertainty as to exactly what wood properties were wanted and a general unwillingness to sacrifice potential genetic gains for volume growth rate by selecting for wood properties.

Yet, within three years after that meeting, the structure of the New Zealand breeding programme was to be radically rethought, so much so that it was largely started afresh.

4.11 Summary of Domestication Progress

Intensive genetic improvement programmes for radiata were launched from the 1950s, in New Zealand, Australia and South Africa. An unorthodox feature of the programmes, which was essentially vindicated, was to base them on existing landrace stocks, without waiting for research on variation among native provenances. Launching the tree-breeding programmes was a major step in domestication, although the beginnings of these programmes in general entailed steep learning curves for the breeders, and delivered few or no harvested genetic gain during the 1951–68 period. Wide-ranging research was embarked upon to underpin tree-breeding programmes, including much not within the actual breeding programmes. The "Swedish"or "Scandinavian" model for tree breeding, based on clonal seed orchards, was widely adopted and was the basis for many of the early tree-breeding programmes. However, this model and features borrowed from animal-breeding programmes came under increasing scrutiny in the late 1960s.

In silviculture, re-establishment of harvested areas posed some challenges, for both restocking and safeguarding site productivity, but these were surmountable. **Natural regeneration** was often used on felled areas, but posed problems for controlling stocking and stand tending.

New plantings proceeded apace in Chile, and had taken off in northern Spain, with Australia and New Zealand gearing up for fresh **planting booms**.

A major biotic **alarm** occurred, with the pathogen *D. pini* appearing in various countries, including the three main growers of radiata. While this eliminated radiata as a species option in some countries, solving the problem that it posed actually strengthened the status of radiata research and radiata itself as the preferred species in New Zealand. Aerial spraying of a cheap copper compound was developed as an acceptable control option, a very significant development for commercial forestry. A similar development was aerial topdressing of established plantations with fertiliser.

Clonal systems were explored in pursuit of expected advantages over growing seedlings, but without **real** operational success with radiata.

More rigorous examination of the economics of commercial forestry was undertaken for radiata. **Initially**, it demonstrated that it was a commercially competitive land use, capable of giving a fully acceptable rate of return on investment. Such research was then extended for optimising establishment and tending regimes.

References

Ahuja MR, Libby WJ (eds) (1993) Clonal forestry. Vol. 1. Genetics and biotechnology (15 chapters, p ix + 277). Vol. 2. Conservation and application (p 240). Springer-Verlag, Heidelberg

Bannister MH (1958a) Specimens of two pine trees from Guadalupe Island, Mexico. N Z J For 7:81–87

Bannister MH (1958b) Evidence of hybridization between *Pinus attenuata* and *Pinus radiata* in New Zealand. 2. Variation in samples of two-year-old *Pinus attenuata*, *Pinus radiata* and their hybrids. Trans R Soc N Z 85:217–225

Bannister MH (1958c) Variation in samples of two-year-old *Pinus attenuata*, *Pinus radiata* and their hybrids. Trans R Soc N Z 85:227–236

Bannister MH (1959) Artificial selection and *Pinus radiata*. N Z J For 8:69–90

Bannister MH (1962a) Some variations in the growth pattern of *Pinus radiata* in New Zealand. N Z J Sci 5:342–370

Bannister MH (1962b) Variation of turpentine composition in five population samples of *Pinus radiata*. N Z J Sci 5:486–495

Bannister MH (1965a) Variation in the breeding system of *Pinus radiata*. In: Baker HG and Stebbins GL (ed) The genetics of colonising species. Academic Press. pp 353–374

Bannister MH, McDonald IRC (1983) Turpentine composition of the pines of Guadalupe and Cedros Islands, Baja California. N Z J Bot 21:373–377

Bannister MH, Vine MH (1981) An early progeny trial in *Pinus radiata*. 4. Wood density. N Z J For Sci 11:221–243

Bannister MH, Williams AL, McDonald IRC et al (1962) Variation of turpentine composition of five population samples of *Pinus radiata*. N Z J Sci 5:486–495

BCFC (1957) Seventh British Commonwealth Forestry Conference, Australia and New Zealand: summary report, resolutions and reports of committees. Commonwealth Government Printer, Canberra, p 83

Bevege DI (2010) The challenge of change in a climate of uncertainty: looking back to the future through the prism of Max Jacobs' forestry career. Aust For 73:140–155

Blight MM, Brewerton HV, McDonald IRC et al (1964) β-pinene content of the *Pinus radiata* population of Kaingaroa Forest. N Z J Sci 7:457–459

Burdon RD (1971) Clonal repeatabilities and clone-site interactions in *Pinus radiata*. Silvae Genet 20:33–39

Burdon RD (1976) Foliar macronutrient concentrations and foliage retention in radiata pine on four sites. N Z J For Sci 5:250–259

Burdon RD (1992) Genetic survey of *Pinus radiata*. 9. General discussion and implications for genetic management. N Z J For Sci 22:274–298

Burdon RD (1994) Annual growth stages for height and diameter in *Pinus radiata*. N Z J For Sci 24:11–17

Burdon RD (2001) *Pinus radiata*. In: Last FT (ed) Ecosystems of the world, Vol. 19, Tree crop ecosystems. Elsevier, Amsterdam, pp 99–161

Burdon RD (2010) Wood properties and genetic improvement of radiata pine. N Z J For 55 (2):22–27

Burdon RD, Harris JM (1973) Wood density in radiata pine clones on four different sites. N Z J For Sci 3:286–303

Burdon RD, Shelbourne CJA (1971) Breeding populations for recurrent selection: conflicts and possible solutions. N Z J For Sci 1:174–193

Burdon RD (compil.), Thulin IJ (ed) (1966) The improvement of *Pinus radiata*. Proceedings, FRI symposium No. 6, Rotorua, November 1965. New Zealand Forest Service, Forest Research Institute, Rotorua, p 92

Burdon RD, Bannister MH, Madgwick HAI et al (1992a) Genetic survey of *Pinus radiata*. 1: introduction, description of experiment, and basic methodology. N Z J For Sci 22:119–137

Burdon RD, Bannister MH, Low CB (1992b) Genetic survey of *Pinus radiata*. 2: population comparisons for growth rate, disease resistance and morphology. N Z J For Sci 22:138–159

Burdon RD, Carson MJ, Shelbourne CJA (2008) Achievements in forest tree genetic improvement in Australia and New Zealand 10: *Pinus radiata* in New Zealand. Aust For 71:263–279

Burley J (2004) A historical overview of forest tree improvement. In: Burley J, Evans J, Youngquist J (eds) Encyclopedia of forest sciences. Elsevier Academic Press, Oxford and San Diego, pp 1532–1538

Butcher TB (2007) Achievements in forest tree genetic improvement in Australia and New Zealand. 7: maritime pine and Brutian pine tree improvement programs in Western Australia. Aust For 70:141–151

Cobb FW, Miller DR (1968) Hosts and geographic distribution of *Schirria pini*—the cause of red band needle blight in California. J For 66:930–933

Cockayne AH (1914) The Monterey pine—the great timber-tree of the future. The [N Z] J Agric 8(1):1–26

Conway MJ (1962) Aerial application of phosphate fertilisers to radiata pine forests in New Zealand. Empire Forestry Review 41(3):234–245

Critchfield WB (1967) Crossability and relationships of the Californian closed cone pines. Silvae Genet 16:89–97

Dadswell HE, Fielding JM, Nicholls JWP et al (1961) Tree-to-tree variations and the gross heritability of wood characteristics of *Pinus radiata*. Tappi 44:174–179

Darwin CR (1868) The variation of plants and animals under domestication; vol. 1, p viii + 411; vol. 2, p vii + 486. John Murray, London (More recent editions available)

de Gryse JJ (1955) Forest pathology in New Zealand. New Zealand Forest Service, Bulletin No. 11. p vi + 62

Duffield JW (1952) Relationships and species hybridization in the genus *Pinus*. Z Forstgenet Forstpflanzenzücht 1:93–100

Eldridge KG (1966) A seed production experiment with *Pinus radiata*. Aust For 30:43–56

Eldridge KG, Matthews CWF (1977) Bibliography of Australian tree breeding and forest genetics to 1975. Divisional Report No. 1, CSIRO Division of Forest Research, Canberra

Elliott HJ, Felton KC, Jarman SJ et al (2008) A history of innovation: eighty-five years of research and development at Forestry Tasmania. Forestry Tasmania, Hobart, p vii + 392

Fielding JM (1953) Variations in Monterey pine. Bulletin, Forestry and Timber Bureau Australia, Canberra, No. 31. p 43

Fielding JM (1954) Methods of raising Monterey pine from cuttings in open beds. Bulletin, Forestry and Timber Bureau Australia, Canberra, No. 32. p 29

Fielding JM (1957) The breeding of Monterey pine in the Australian Capital Territory. In: Paper to 7th British Commonwealth Forestry Conference, Australia and New Zealand, p 19

Fielding, JM (1961a) Provenances of Monterey and bishop pines. Forestry and Timber Bureau, Canberra, Bulletin No. 38. p 30

Fielding JM (1961b) The pines of Cedros Island. Aust For 24:62–65

Fielding JM (1970) Trees grown from cuttings compared with trees grown from seed (*Pinus radiata* D.Don). Silvae Genet 19:54–63

Fielding JM, Brown AG (1961) Tree-to-tree variations in the health and some effect of superphosphate on the growth and development of Monterey pine on a low quality site. Forestry and Timber Bureau, Australia, Leaflet No. 79. p 19

Forde MB (1964a) Variation in native populations of *Pinus radiata*. 1. Sampling methods and branch characters. N Z J Bot 2:213–236

Forde MB (1964b) Inheritance of turpentine composition in *Pinus attenuata x radiata* hybrids. New Zeal J Bot 2(1):53–59

Giertych M (1991) Provenance variation in growth and phenology. In: Giertych M, Mátyás C (eds) Genetics of Scots pine. Elsevier, Amsterdam, The Netherlands, pp 87–101

Hood JV, Libby WJ (1978) Continuing effects of maturation state in radiata pine and a general maturation model. In: Hughes KW, Henke R, Constantin M (eds) Proceedings, international symposium on "propagation of higher plants through tissue culture." USDOE Conf. 7804111, pp 220–232

Hood JV, Libby WJ (1980) A clonal study of intraspecific variability in radiata pine. I. Cold and animal damage. Aust For Res 10:9–20

Jackson DS (1955) The *Pinus radiata/Sirex noctilio* relationship at Rotoehu Forest. N Z J For 7 (2):26–41

Johnson GR, Burdon RD (1990) Family–site interaction in *Pinus radiata*: implications for progeny testing strategy and regionalised breeding in New Zealand. Silvae Genet 39:55–62

Johnson IG, Cotterill IM, Raymond CA et al (2008) Half a century of *Pinus radiata* tree improvement in New South Wales. N Z J For 52(4):7–13

Kumar S (2004) Effect of selfing on various economic traits in *Pinus radiata* and some implications for breeding strategy. For Sci 50:571–578

Larsen CS (1956) Genetics in silviculture (trans: Anderson ML). Oliver and Boyd, London. pp 224

Lerner IM (1958) The genetic basis of selection. John Wiley & Sons, NY, Chapman & Hall, London, p 298

Lerner IM, Libby WJ (1976) Heredity, evolution and society, 2nd edn. W.H. Freeman & Co, San Francisco, p 431

Lewis, N.B. 1975. A hundred years of state forestry—South Australia 1875–1975. Woods and Forests Department, Adelaide. (A Centenary Publication) Bulletin No. 22. p 122

Libby WJ, Conkle MT (1966) Effects of auxin treatment, tree age, tree vigor and cold storage on rooting young Monterey pine. For Sci 12:484–505

Libby WJ, Hood JV (1976) Juvenility in hedged radiata pine. Acta Hortic 59:91–98

Libby WJ, Bannister MH, Linhart YB (1968) The pines of Cedros and Guadalupe islands. J For 66:846–853

Libby WJ, Brown AG, Fielding JM (1972) Effects of hedging radiata pine on production, rooting and early growth of cuttings. N Z J For Sci 2:263–283

Lindqvist B (1948) Genetics in Swedish forestry practice [Translation of Lindquist (1946)]. Chronica Botanica, Waltham, MA

MacDougal DT (1938) Tree growth. Chronica Botanica, Leiden, p 240

Matthew P (1831) On naval timber and arboriculture. Adam Black, Edinburgh, p 391

Maxwell E (1930) Afforestation in southern lands. Whitcombe and Tombs, Auckland, New Zealand, p 315

Miyazaki Y (1696) A cyclopedia of agriculture. Volume 9. Various trees. agris.fao.org/agris-search/search.do?recordID=US201300185227

Namkoong G, Snyder EB, Stonecypher RW (1966) Heritability and gain concepts for evaluating breeding systems and seedling orchards. Silvae Genet 15:76–84

Nicholls JWP, Dadswell HE, Fielding JM (1964) The heritability of wood characteristics of *Pinus radiata*. Silvae Genet 13:68–71

Pawsey CK (1960) Cone production reduced, apparently by drought, in the south-east of South Australia. Aust For 24:74–75

Pawsey, CK (1964) Inbreeding of radiata pine (*Pinus radiata* D.Don). Forestry and Timber Bureau, Australia, Leaflet No. 89. p 31

Pawsey CK (1967) Breeding Monterey pine in the south-east of South Australia. MSc thesis, University of Melbourne. p 38 plus 11 supporting articles

Pederick LA, Brown AG (1976) Seed production in radiata pine seed orchards in Australia. Aust For 40:144–150

Poynton RJ (1960) Notes on exotic forest trees in South Africa, 2nd edn. revised. Department of Forestry, Government Printer, Pretoria, p 186

Poynton RJ (1977) Tree planting in South Africa. Volume 1. the pines. Department of Forestry, Republic of South Africa, p 576

Pravdin LF (1964) Scots pine. Variation, intraspecific taxonomy and selection. Translated from Russian by R. Karshon in 1969. Israel Program for Scientific Translations Ltd. Annarbor Humphrey Science Publishers Ltd., London, p iv + 208

Preston S (1968) Division of forest products—a brief history. CSIRO Forest Products Newsletter No. 350: 4–8

Raymond CA (2011) Genotype by environment interaction for *Pinus radiata* in New South Wales, Australia. Tree Genet Genomes 7:819–833

Roche MM (1990) History of New Zealand forestry. New Zealand Forestry Corporation in association with GP Books, Wellington, p 466

Scott CW (1960) *Pinus radiata*. FAO Forestry and Forest Products Studies No. 14. Rome, p 328

Shelbourne CJA (1969) Tree breeding methods. New Zealand Forest Service, Forest Research Institute, Rotorua, Technical Paper No. 55. p 43

Shelbourne CJA, Burdon RD, Bannister MH et al (1979) Choosing the best provenances of radiata pine for different sites in New Zealand. N Z J For 24:288–300

Shepherd KR (1977) Forest tree breeding in Australia. Divisional report 2. CSIRO Division of Forest Research, Canberra, p 29

Sweet GB (1963) A pilot trial: Grafting on to natural regeneration of *Pinus radiata*. New Zealand Forest Service, Forest Research Institute, Rotorua, Forest Tree Improvement Report No. 13 (unpubl.)

Sweet GB (1973) The effect of maturation on the growth and form of vegetative propagules of radiata pine. N Z J For Sci 3:191–210

Thulin IJ (1957) Application of tree breeding to New Zealand forestry. New Zealand Forest Service, Forest Research Institute Technical Paper No. 22. p 12

Toda R (1974) Vegetative propagation in relation to Japanese forest tree improvement. N Z J For Sci 4:410–417

Tufuor K, Libby WJ (1973) First-lift pruning times of radiata pine seedlings and rooted cuttings in a small California experiment. N Z J For 18:124–132

van Wyk G, Roeder KR (1978) The status of tree breeding in South Africa. South Afr For J 107 (1):54–59

Vincent TG, Dunstan JS (1989) Register of commercial seedlots issued by the New Zealand Forest Service. Ministry of Forestry, New Zealand Forest Research Institute, Rotorua, FRI Bulletin 144. p 155

Weston (1956) Fertiliser trials in unthrifty pine plantations at Riverhead Forest. N Z J For 8:35–46

White TL, Adams WT, Neale DB (2007) Forest genetics. CABI Publishing, Wallingford and Cambridge, MA, p x + 682

Wilcox MD (1983) Inbreeding depression and variances estimated from self- and cross-pollinated families of *Pinus radiata*. Silvae Genet 32:89–96

Wilcox MD, Thulin IJ, Vincent TG (1976) Selection of *Pinus radiata* clones for planting from cuttings. N Z J For 21:239–247

Williams ER, Matheson AC (1994) Experimental design and analysis for use in tree improvement. CSIRO, Victoria, Australia, p 174

Wu HX, Eldridge KG, Matheson AC et al (2007) Achievements in forest tree improvement in Australia and New Zealand 8: Successful introduction and breeding of radiata pine in Australia. Aust For 70:215–225

Zagory D, Libby WJ (1985) Maturation-related resistance of radiata pine to western gall rust. Phytopathology 75:1443–1447

Zobel BJ, Sprague JR (1993) A forestry revolution. A history of tree improvement in the Southern United States. Carolina Academic Press, (Reprint and online version available) p 161

Chapter 5
Development of the Management Concept While Tree Improvement Shifts Gears: 1969–1983

5.1 Introduction

In 1967, Australia had just made its first operational collections of seed-orchard seeds, and such a collection was made a year later in New Zealand. The year 1968 also began a new era for breeding radiata in New Zealand, with the selection afresh of vastly more plus trees from New Zealand land-race plantations. This latter advance was based on Gene Namkoong's contributions to breeding theory appropriate for forest trees, and Tony Shelbourne's strategic planning and implementation of that theory focused on radiata in New Zealand. The resulting update of the breeding strategy greatly increased the numbers of plus trees, and thence the scope for reselecting outstanding parents on progeny performance. That also increased the available genetic diversity in the New Zealand base breeding population.

At about the same time, important developments were taking place on other fronts, making the late 1960s even more important as radiata domestication accelerated. Within New Zealand, radical new silvicultural regimes were being developed, and soon were implemented. The so-called "Direct Sawlog" (or "Direct") regime included much wider initial spacing at planting, earlier thinning to remove the apparently poorer trees and achieve even wider spacing, and aggressive pruning, all focused on more quickly producing larger and more-valuable final-crop trees. This early selection and spacing of the crop trees, plus pruning boles to about six metres height, was meant to achieve larger, higher-quality butt logs, while genetic improvement was counted on to defend the quality and value of unpruned upper logs. Nursery and field establishment practices were soon to be revolutionised (Maclaren 1993; Mead 2013), which had the additional benefit of allowing much better yields of planting stock from the scarce, genetically improved seed coming from the early seed orchards.

Also occurring in New Zealand, a National Development Conference was held in 1969; it endorsed the view that New Zealand needed a major expansion of its plantation forestry estate (Fig. 3.1), both for profit and for broadening the

R. Burdon et al., *Domestication of Radiata Pine*, Forestry Sciences,
DOI 10.1007/978-3-319-65018-0_5

country's narrow base of export products, and that became part of government policy (Roche 1990). This roughly coincided with the lifting of price controls on sawn timber which had encouraged profligate use of the fast-dwindling native timber resource. In Australia, at about the same time (from 1966), central government had been persuaded to encourage new forest plantings to reduce dependence on imported softwood products (Johnson et al. 2008; Bevege 2010). In both countries, expansion of plantation forests thereby became much more active during the next few years. In several Australian states the new plantings were increasingly being concentrated on sites where previous experience indicated radiata would do well, often leading to the clearing of eucalypt forest for the plantations. Such clearing of native forest led to community opposition (e.g. Routley and Routley, 1975), and in the 1980s in Victoria, for example, such clearing was phased out (Lavery 1988) and new plantings to expand the forest estate were mostly on former farmland. Fertility on these former farming sites had often been built up through pastoral practices such as the cultivation of legumes and the addition of superphosphate. By the 1980s this factor, together with increased application of fertilisers (e.g. Boardman 1988), better weed control and a trend to lower stocking and shorter rotations (e.g. Lavery 1988) combined to present cultural environments often biologically and physically very different from those in which the first-generation seed-orchard trees had been selected.

In Chile, major developments in radiata breeding, silviculture and plantation management came later, and involved both institutions and policies. After the military coup of 1973, central government renewed vigorous promotion of new forest planting, mindful of the country's extreme dependence on copper exports. Breeding of radiata began in the 1970s, and greatly benefitted from hard-won experience in New Zealand, at the time freely shared by FRI. Large processing plants were established in Chile during the early 1970s, with a heavy emphasis on pulping, to deal with the big resources of Chilean radiata wood becoming available. By comparison, such industrial developments were modest in Australia, New Zealand and South Africa.

Many important developments leading to more-intensive breeding of radiata in Australia and New Zealand followed increased support by central governments for radiata breeding in those countries. Operationally, those breeding programmes had by 1983 advanced into second-generation selections among partially-pedigreed offspring of the early plus-tree selections and progeny tests. Also, by 1983 the outturn from New Zealand's seed orchards was close to meeting the entire national planting requirements, which it did in 1986. However, progress towards that goal was abruptly checked in South Australia by disastrous fires early in 1983, which consumed not only some seed orchards but also major stocks of seed in storage. The resulting shortfall was greatly exacerbated by the urgent need to restock large areas of burned forest. Some four million seedlings, supplemented by half a million rooted cuttings from orchard seed, were raised per annum, resulting in the largest use of vegetatively-propagated radiata to that time.

By the mid-1970s, the quantitative genetic architecture of radiata was becoming much better known, for a wide range of traits. Variation among provenances was much better characterized, and information on several genetic parameters, notably heritabilities and genetic correlations, was increasingly solid. The new information on provenance variation prompted further larger collections of seeds from all five native populations in 1978, in order to boost available genetic resources. The collecting expeditions were sponsored and staffed by Australia and New Zealand, with additional financial support from the US National Science Foundation and important local collaborations in Mexico and California (Eldridge 1978). Besides making additional founder collections of native seeds, the participants added to previous observations of the character and condition of the native radiata pines. Among those observations was a reminder that the open-grown pines on Guadalupe Island, while exposed to repeated gale-force winds, had remarkably straight boles, unlike similarly exposed mainland radiata pines.

Moving beyond the first generation of intensive breeding provided much of the context for developing tree-breeding strategy worldwide, as a discipline in its own right. Breeding strategies were advanced while addressing several issues of propagation and silviculture of radiata, presented in detail later.

Importantly, constructing a more robust breeding strategy classically involved a hierarchical organization of breeding material (Libby 1973): from top-ranked material for immediate commercial deployment, underpinned the breeding population in which intensive genetic gain was pursued over successive generations, which in turn are underpinned by genetically-diverse collections of land-race and native-population base samples at the lowest level of the hierarchy. Controlled pollination among parents in appropriate mating designs was another key plank of the evolving strategies. Underpinning the developments in breeding strategy were developments in quantitative breeding methodology (Burdon et al. 2008).

Better field experimental designs were generally being adopted, often to accommodate greatly enlarged breeding populations and the large areas of ground needed for genetic tests of candidate trees. Selection procedures were made more objective and efficient, with improved laboratory and field assays of wood properties, and computer-based statistical analyses of the floods of data being generated. Furthermore, analyses of genotype-by-site interactions were developed and included in setting tree-breeding goals and implementing strategies during the time period of this chapter.

Much effort was devoted to improving breeding technology. Controlled-pollination techniques were improved (Burdon et al. 2008). Despite seed orchards eventually delivering genetically improved seeds, various problems with seed orchards became clearer, and were vigorously researched. In the course of seed-orchard research, new technologies were developed for rooting cuttings taken from grafts of mature trees that were at risk of delayed graft incompatibility. Such cuttings were then used in seed orchards.

Technology for vegetative propagation has applications well beyond breeding technology, since mass-production of cuttings from pilot quantities of genetically elite seed is a means of quickly and massively delivering genetic improvements in growing stock. Indeed, acquiring cuttings from serial propagation of young seedlings and/or from juvenile hedges (Libby et al. 1972) was being developed for such mass-propagation, while the foundations were laid for future production of **plantlets**[G] by *in-vitro* tissue culture.

The need arose to demonstrate and quantify genetic improvements, partly for political purposes to secure industry support and continued or increased funding, and partly for better projections of future forest yields and log characteristics. For all this, better information than then available was needed. It had been assumed that comparing the genetically improved trees with unimproved "control" trees in properly replicated progeny trials would suffice, but questions were raised as to how much improvement would be expressed in operational plantations where differences between select and control material would not be amplified by competition effects. The impetus for better information first came in Australia, where J.D. Ovington of the Australian National University had queried the scale of genetic gain likely from breeding programs. New Zealand followed soon after, establishing a network of genetic-gain comparison trials (Sect. 5.5.7) on a range of sites, the first of which were planted in 1978. Progress was subsequently made in modelling performance, towards harvest age, of the genetically improved stock (Carson et al. 1999).

Meanwhile, well before 1983, the understanding of where radiata could be grown successfully had advanced, and we next review (Sect. 5.2) that enhanced understanding. Moreover, the improvements in nursery, field-establishment and management practices were continuing to extend the range of sites where radiata could be grown successfully (Sect. 5.4.2).

On the global scene, Australia hosted at Canberra the Third World Consultation on Forest Tree Breeding, under the auspices of FAO and IUFRO (International Union of Forest Research Organisations). Breeding radiata featured prominently in the proceedings, and a IUFRO Working Party on Breeding Radiata Pine became reactivated as a vehicle for exchanging information and ideas.

5.2 Where Which Species Will Perform Well

It was good to learn the range of sites and conditions where radiata would grow well. It was even better to learn where it would remain preferred over other valuable species that could also grow well there. The earlier, often informal species-introduction trials, current and predicted markets, and desires for some diversification, drove decisions to plant radiata and/or some alternative species. Between 1951 and the 1970s, the picture of all the then-contesting species' site tolerances changed, largely through the advent of some fungal pathogens, but partly through developments in management practice for radiata (Sect. 5.4). It was found that, relative to such attractive alternatives as some acacias, Douglas-fir, some eucalypts, some various other pines, and coast redwood, radiata was usually preferred in Chile and New Zealand. Radiata was widely preferred in south-eastern Australia but less so farther west and north. In South Africa, Spain and some other Mediterranean countries (Italy, Albania, Greece and Turkey) climates were locally congenial but suitable soils were often even more localised. In California itself, sites that permitted rapid tree growth and good health in radiata proved very localized.

An alarming biotic event had been the spread of the needle-cast pathogen *Dothistroma pini*. Known earlier only as a nuisance pathogen in Christmas-tree growing in USA, it spread to many parts of the world during the late 1950s and the 1960s (Gibson 1972). It effectively eliminated radiata as a species option on many high-altitude sites in East Africa and Hawaii where it had previously grown well (Burdon 2001). In New Zealand there were large areas where the future of radiata came into question. A multi-pronged genetic response included: collecting and stockpiling large amounts of seed of *Pinus patula* (Mexican weeping pine) which was known to be resistant, attempting to cross radiata with *P. patula*, selecting and cloning healthy radiata pines growing among heavily infected neighbours, and importing radiata clones selected for resistance in Kenya. However, the immediate practical solution lay in FRI developing an effective and quite cheap protocol for aerial spraying with copper compounds that controlled the disease (Kininmonth 1997b; Mead 2013). Because radiata's susceptibility typically gave way to effective "ontogenetic" resistance before about age 15 years, such protective spraying could usually be discontinued after only two or three applications. This solution for those plantations at risk maintained confidence in the species among growers. Ironically, after the initial scare, radiata became the preferred species in New Zealand more than ever, because other pines (particularly Corsican and ponderosa) that had been planted widely remained highly susceptible even as mature trees. After appearing in both Australia and Chile, dothistroma was not perceived there as such a big threat, because severe disease was much more localized and did not significantly affect radiata in its most important plantation-growing areas.

Decisive advances in management inputs occurred in both establishment practice and use of fertilizers. Producing more-vigorous and better-conditioned nursery stock, in conjunction with better site-preparation practice, allowed radiata to be grown over large areas in New Zealand that were subject to severe ground-level frosts and had previously been planted with more-frost-hardy pines. Nutrient deficiencies had been recognised at a range of sites in Australia in the 1930s; applications of zinc and phosphorus produced strong responses on sites where these elements were limiting factors (e.g. Stoate 1950). By the 1980s, nutritional constraints and opportunities for benefits—especially those involving nitrogen—on the old soils typical of Australia were much better understood (Australian Forestry Council 1981).

In New Zealand correction of phosphorus deficiency had been developed during the 1950s (Weston 1956, 1962). The existence of an agricultural aviation industry topdressing hill-country pastures with superphosphate made application to forestry straightforward. As a result, radiata could thus be grown over a much larger area in the far north of the country, and plantings there were a significant component of accelerated new planting from the early 1970s. Also in New Zealand, boron deficiency as a cause of shoot dieback and consequent stem malformation was recognized as a local problem (Stone and Will 1965), but correction by application of boron was usually straightforward. Often, however, the practical limits for growing radiata commercially in New Zealand were and are set by the terrain rather than by climate, soil fertility and/or distance to ports or local markets.

In Chile, multiple nutrient deficiencies were evident in radiata on the degraded Coast Range soils that had been widely afforested. In addition, boron deficiency symptoms were often severe on radiata pines planted in weathered granite or river-borne volcanic ash, causing much bole malformation (Schlatter and Gerding 1985). These nutrient deficiencies seldom prevented people from planting and growing radiata on such soils, but more prompt correction of known deficiencies would often have greatly improved the value of their harvested trees.

In South Africa, radiata plantations, after some ambitious and ill-fated siting, became confined almost entirely to a strip with winter-rainfall climates near the south coast extending westwards from slightly east of George. However, until widespread phosphorus deficiency was recognised as affecting radiata (Donald and Glen 1974; Donald et al. 1987), maritime pine was preferred for the less fertile sites in this zone.

In Spain, new planting continued along the north coast, mainly near the western end of the Pyrenees in the Basque Country. Despite frost-kill of some 20,000 ha of radiata plantation in the province of Guipuzcoa (E of Lat. 2°W) in 1956, new planting had accelerated sharply from around 1954 till 1973, by when the area of radiata in the region had increased from around 60,000 to 160,000 ha (Michel 2006).

Near the Equator in South America (Ecuador and Colombia) considerable plantings of radiata have since been made (P.G. Cannon, pers. comm. to RDB 1993), presumably using seed from Chile. While radiata does tolerably on these high-elevation sites (Miller 1974), interestingly, as of the end of the 20th century, several provenances of the related bishop (muricata) pine were looking better (pers. obs. WJL in 2001).

Geographic ranges of radiata plantations in the main grower countries are summarised in Table 5.1.

Table 5.1 Approximate latitudinal and altitudinal ranges of main areas of radiata plantations, by country (After Burdon 2001)

Country	Latitude (°)	Location	Altitude (m a.s.l.)
New Zealand	34½*–46½	North	0–700
		Far south	0–300
Chile	35–40		0–800
Australia	33–43½*	New South Wales	400–1,100
		W. Australia	0–300*
		Victoria & S. Australia	0–700
		Tasmania	0–400
Spain	42½–43½*		0–400
South Africa	33–35*		0–1,000

*Range limited by extent of landmass or lack of elevated topography

5.3 Institutional Developments

During 1969 to 1983, actual institutional changes in Australia, New Zealand and South Africa were generally modest, although important developments occurred affecting the growing of radiata. Institutional changes in Chile, however, were dramatic, with political turmoil followed by a military coup with its bloody aftermath. For the forestry sector, along with the rest of the country, it was a turbulent period, but in the latter half there were important developments in the growing, utilization and breeding of radiata.

5.3.1 Australia

The rate of new planting of radiata in Australia, with government support, increased markedly during the period (Table 5.2). This challenged breeders and seed-orchard managers to supply more genetically-improved seed. In view of the lead time needed to increase supply, the seed orchards coped well with the challenge. Moreover, by the end of 1982 it was shown to be operationally practicable to greatly extend an inadequate supply of high-quality seeds by taking and rooting cuttings from young seedlings.

In addition to these public-sector activities, planting by private-sector corporate entities was significant in these years (Table 5.3). The largest areas of private radiata pine plantations were in Victoria, New South Wales and Tasmania. In eastern Victoria, APM Forests had an active seed-orchard programme. Two other private-sector entities participated in efforts to improve the genetic quality of radiata: Saxton's in Gippsland—who developed and operated a seed orchard—and Francis Clark in NSW who (not far from Canberra) propagated and sold rooted cuttings of selected families he acquired from the CSIRO breeding programme. Saxton's orchard was eventually to come under the management of Seed Energy using genetic material under licence from the Southern Tree Breeding Association (STBA; see below).

Table 5.2 New areas of radiata planted annually in Australia (ha)*

Year	Forest service							Private	Total
	NSW	Victoria	Queensland	South Australia	Western Australia	Tasmania	ACT		
1965	3,400	2,200	60	2,110	500	730	0	2,350	11,350
1968	6,040	4,610	100	2,710	800	1,760	210	..	16,230
1969	8,310	2,450	110	2,410	430	1,880	100	..	15,690
1983	5,320	2,070	0	1,890	1,380	900	0	12,400 (1982)	23,960

*These figures, derived mainly from annual reports, are indicative because a few of the smaller numbers include replanting of small felled or burnt areas, and the reported timing of inventories within calendar years was not consistent

.. = not available

Table 5.3 New areas of conifer plantations established in 1975 in Australia (ha)*

Sector	State or territory							Total
	NSW	Victoria	Queensland	South Australia	Western Australia	Tasmania	ACT	
Private	1,837	5,175	625	515	900	1,687	0	10,739
Public	7,143	4,297	5,457	3,042	2,218	1,795	553	24,505
Total	8,980	9,472	6,082	3,557	3,118	3,482	553	35,244

*See footnote to Table 5.2. Also, most of this planting was of radiata; the exceptions were in Queensland (where almost all was other species) and in Western Australia (where about one-third was radiata)

The first meeting of the Australian Forestry Council's Research Working Group No. 1, Forest Genetics, took place in 1967, and subsequent meetings were held at intervals of two to three years, usually with observers from New Zealand. (Other working groups, for example Soils and Nutrition, concurrently became active and contributed to the improved management of radiata plantations). Information and genetic material (mostly plus-tree pollen and scions) was shared among members, and opportunities to further this cooperation were explored through a sub-set of group members. As the period drew to a close, extensive but unstructured collaboration that had evolved culminated (in 1982) in the formation of the Southern Tree Breeding Association. Based in Mt Gambier, the STBA was an "incorporated non-profit organization funded by industry in the "Green Triangle" region of south-eastern SA and western Victoria"; it "was established with the assistance of CSIRO for two companies [Softwood Holdings Ltd and Sapfor] [growing radiata] and the state government of SA." This move was intended to mitigate some organizational difficulties and permitted some economies of scale, but could not eliminate the downsides of geographic and organisational dispersal of the national breeding effort with radiata. Straddling state borders as it did, the STBA was a major development that later became a national organisation serving about half the radiata estate and breeding of species additional to radiata (Wu et al. 2007; http://www.stba.com.au/page/home). Among its first responsibilities under David Boomsma was the management of the seed orchards established by the South Australian Woods and Forests Department.

In 1975 the Australian federal government transferred the Forest Research Institute of the Forestry and Timber Bureau to the Commonwealth Scientific and Industrial Research Organisation (CSIRO), paving the way for the appointment of additional specialist staff in Canberra and Mt Gambier to undertake research and development related to forest genetics and tree breeding.

The appointment of J.J. Landsberg in 1981 as the second Chief of the CSIRO Division of Forest Research led to a surge in ecophysiological research on radiata; close collaboration later followed with New Zealand tree physiologists led by David Rook. An important outcome of this work was the development of growth models that drew together a wide range of research results on water availability, nutrition, stand conditions and other environmental factors (e.g. Landsberg et al. 1988).

5.3.2 New Zealand

In New Zealand, the main institutional development during this period was the commitment of Tasman Forestry Ltd (Tasman) to radiata breeding, led by John Gleed, who had been their nurseryman. The company established its own seed orchard, using parents selected by FRI and then progeny-tested by FRI near Tasman plantations. Located on a warm, low-altitude site subject to sea breezes, that orchard produced seeds relatively quickly and in abundance.

The then-biggest private forestry company, NZ Forest Products Ltd (NZFP), made substantial forays into its own tree-breeding and seed-orchard development, specific to their lands and goals. Compared with the Tasman initiatives under Gleed, however, NZFP commitments were inconsistent, and were thus without lasting impact either in advancing their own local radiata breeds or on public and professional recognition for producing genetic information.

FRI remained overwhelmingly the dominant player with respect to radiata's domestication in New Zealand, and perhaps the world. Besides continually improving understanding of radiata's potential on various New Zealand sites, and of how broadly given families or clones could be deployed over a range of contrasting sites, FRI aggressively continued genetics research, practical breeding, and trying various management practices.

In 1970, FRI hosted a second Symposium on Pruning and Thinning Practice (James et al. 1970; Sutton 1970), its first symposium on these topics having been held in 1963. This second symposium soon proved far more influential in respect of silvicultural regimes. For genetic improvement, one of the papers (Shelbourne 1970) importantly foreshadowed the concept of radiata breeds and clones differentiated on technical properties of the trees and likely end-products rather than on regional adaptation (Sect. 5.5.2). Later, FRI hosted two International Union of Forest Research Organizations (IUFRO) meetings, one in 1973 on Vegetative Propagation (mainly focused on radiata) and the other, in 1982, on Breeding Radiata Pine. These latter two meetings both produced and synthesised ideas and data on breeding and management of radiata, thus supporting and progressing the early domestication of radiata both locally and worldwide.

In 1970, a largely-personal initiative was conceived and launched by the NC-State-trained geneticist Tony Shelbourne of FRI. He induced geneticists and tree breeders in Australia, France, Kenya, New Zealand, South Africa and the US to contribute select radiata material (largely as single-parent open-pollinated (OP) progenies from seed orchards) for a series of International Gene Pool trials. Within Australia, eight breeding programmes contributed seeds of their plus-tree lines, and the US added some OP families from apparent plus trees in the five native populations. Research plantings were made in five countries, although only three of those countries (Australia, New Zealand and South Africa) produced reported results (Burdon et al. 1997). Some of the findings indicated where parents could be selected or tested effectively for what sets of sites; this involved how well progeny differences were resolved on the various sites and how consistent progeny rankings were between pairs of sites, which is valuable information. However, the project was conceived primarily to share valuable breeding material among programmes in different countries. Many selections were later made from these trials for incorporation in local breeding programmes.

In 1982 a new Propagation and Early Growth (PEG) research unit was created at FRI, which was separate from the Genetics and Tree Improvement (GTI) unit. This removed responsibility for more-effective propagation from geneticists and breeders. While it led to some communication gaps, it helped bring to better enlist the expertise of tree physiologists and nursery research personnel.

5.3.3 South Africa

The national Department of Forestry, which had initiated radiata breeding in the early 1960s, became in 1980 the Directorate of Forestry within the newly created Department of Water Affairs, Forestry and Environmental Affairs. While the Directorate retained responsibility for operational breeding of radiata, which it ran from Saasveld, Stellenbosch University over time became increasingly involved in silvicultural research on this species.

5.3.4 Chile

Purposeful tree breeding started much later in Chile than in Australia, New Zealand and South Africa, despite the large size of Chile's plantation estate of radiata. In the early 1970s, plus trees were intensively selected under the direction of Norman Smith, a Chilean forester of Instituto Forestal, the central government's forest research organization. Seeds were collected from these trees and extracted, but the fledgling breeding programme fell victim to political upheaval leading up to and following the military coup in 1973 and the subsequent exodus of almost all of Instituto Forestal staff. While some experienced staff were laid off or redeployed, many moved pre-emptively into other employment. Later, in 1978, some plus trees that had been chosen by Norman Smith could be relocated, but the seed that had originally been collected from all of them was mixed together before sowing in the nursery, precluding any effective progeny testing.

During this period, the University of Chile became involved in research on genetics of radiata, and that research also met with disasters resulting directly from political upheaval. During the unrest leading up to the military coup, Jochen Kummerow, a professor at the university, was raising seedlings in a greenhouse from seed provided by Libby from the five native-provenance native populations for replicating provenance-progeny trials in the US, New Zealand and Australia. Kummerow got beaten up by striking students, and the seedlings perished from lack of watering while he was recovering in hospital. Martin Bannister from New Zealand's FRI visited the university during the unrest in 1973. With Dr Harald

Schmidt as his host, they collected cones from vigorous well-formed trees in a large number of Chilean radiata plantations, intending to share the seeds between Chile and New Zealand. While such seeds would have been of great value for research and breeding, they just disappeared during the turmoil.

Well before the political upheavals, moves had been made by the Chilean government to develop manufacturing industries to process wood from the radiata plantations. FAO personnel had been active in Chile from the mid-1950s (Scott 1960) in surveying the young radiata resource with the intention of guiding the establishment of strategically located pulpmills. Even so, one big Chilean-owned company, CMPC, had already set up its own industrial plant, dominated by a kraft pulpmill that had opened on a modest scale in 1959, while a smaller company had established a **hardboard**[G] mill. However, there were large areas of radiata plantation, in varyingly fragmented ownership, for which no provision had been made to process harvested wood. There was much consolidation of forest ownership, which then facilitated the supply of wood for the mills. Despite the disruptions caused by political upheavals and much nationalization of forests and businesses, three pulpmills were established by 1977. Sawmills had been established, but on nowhere near the scale needed to deal with the increasing volume of wood that was becoming available for harvest.

After the military coup in 1973, the new government soon vigorously renewed strong financial encouragement of additional afforestation. This was prompted by a combination of factors: a desire to diversity Chile's export markets from the extreme exposure to the copper market; an appreciation of the comparative advantages of growing wood on the lower-quality land; a response to high unemployment; and a need to rehabilitate large areas of badly-degraded farmland. (Ironically, this was massive state assistance on the part of a right-wing government that a few years later was true to its political colours with vigorous privatization.) In response, planting rates boomed, and radiata was chosen for most of the plantations. Genetic improvement of radiata resumed within a new institutional structure that was founded in 1976. It followed the pioneering 1960s model of the North Carolina State University (NCSU) /Industry Tree Improvement Cooperative that was operating with loblolly pine, but was called a "Convenio" ("agreement") because "Cooperativa" ("cooperative") was seen by the right-wing military government as socialist. The model was set up with the advice from Bruce Zobel, Director of the NCSU Cooperative. The founding Director of the Convenio was Roberto Delmastro, a former graduate student of Zobel's and a faculty member of the Forestry School of Universidad Austral de Chile at Valdivia.

In the Convenio, each member company agreed to have its own clonal seed orchard, with the Chilean forest service (CONAF, the Corporación Nacional Forestal) operating as a member sponsoring two of the orchards. Compared with the NCSU Cooperative, the membership of the Convenio was more heterogeneous. Members of the former were typically vertically integrated companies, growing much of the wood that they processed, with individual company sizes largely governed by the economies of scale of kraft pulpmills. In the Convenio, the sizes of companies and the nature of their operations were more varied.

In 1978, Rowland Burdon, after 14 years' experience in New Zealand's FRI, served for two months as FAO Consultant to the Convenio (Burdon 1978), checking plus-tree selection, propagation practice and orchard siting, and then reviewing the overall direction of the programme. Significantly, much had been learned from the early and often bad New Zealand and Australian experiences of orchard siting, and Convenio members thus avoided repeating some of those early mistakes. The concept of the breeding cooperative entailed each member company assuming operational responsibility for its own breeding programme, with technical guidance from Director Delmastro at Universidad Austral. The two CONAF orchards were established to provide seeds to serve forest owners outside the Convenio. Advantages included a sense of ownership of their own breeding programmes and spurring of competition among the companies. The emphasis on individual ownership, however, did lead to a lack of national coordination in managing the genetic material.

Following up on Burdon's visit, a party of six personnel from within the Chilean Convenio, mostly young tree breeders, was sent on a short tour to New Zealand by FAO in 1979.

With the prospect of intensive genetic improvement and with increasing technical contact with New Zealand, other aspects of growing radiata began to change in Chile. Nursery practice there had been driven by considerations of cost and the pious belief that first-year survival after planting was a sufficient measure of planting success. Yet with the genetic improvement yielding initially limited amounts of very valuable seed, obtaining more and better-quality planting stock from the available seed became very attractive.

Thinning and pruning in Chile had usually been minimal at best, and typically low rates of natural mortality from inter-stem competition (despite radiata's good assertion of crown dominance) meant that thinning in order to harvest stems before they died was not a pressing issue. Nevertheless, interest increased in active tending to improve economic returns. Various study tours and consultancies continued between New Zealand and Chile through 1983 and beyond.

5.4 The Context of Intensifying Management Practices

5.4.1 Introduction

For radiata, we have addressed in Chap. 4 early work on the second of the two main planks of domestication, namely genetic improvement. The early development of the first plank, namely the technology of growing the species and efficient crop management, has been traced in Chap. 3. Ideally, these two planks are not just interactive but strongly synergistic, and both should be aimed at likely future markets. Some major changes in practices for growing radiata came after the late 1960s, and these were largely centred in New Zealand. In some respects the early genetic improvement and new growing practices meshed well, but in other respects they did not (Sects. 5.4.4.6 and 6.3.2.2), creating a fresh challenge to better integrate the two domains.

As with the ongoing genetic improvement, learning how to grow radiata as a plantation crop progressively better has been important, both in itself and as a path-breaking exercise in the development of late 20th-century plantation forestry with other species as well. The process was not always smooth and orderly, because traditional approaches often proved wanting, as did certain of the much-vaunted remedies (Sects. 5.4.4.6 and 6.3.2.2).

After large-scale harvesting began, restocking clear-felled areas was often a vexing task. It progressed from management of natural regeneration, through some use of direct seeding, and then usually to a commitment to immediate post-harvest planting. It was only after trial-and-error with restocking practices that planting became accepted or even mandated practice. In support of this, the growing of field-ready planting-stock was much improved by better and more intensive nursery practices, and the tending regimes for plantations were greatly intensified (Box 5.1).

Box 5.1 Regeneration Systems

Natural regeneration by seed can work spectacularly for radiata. However, both it and broadcast seeding give poor control over stocking and limited control over genetic quality. Therefore, planting of nursery stock has become almost universally preferred for establishing plantations of radiata. To this end, the raising and planting of nursery stock have been greatly refined.

5.4.2 Crop Establishment

5.4.2.1 Seed Collection

Radiata is one of the easiest of all commercial tree species for seed collection, even though its cone-bearing trees are often tall. The cones generally each contain over a hundred seeds, and can remain unopened on the tree for years (Fig. 5.1). In cool, moist climates, the trees can accumulate seed crops year after year, unlike many tree species in which good seed crops occur only occasionally and then need to be collected during narrow time windows between when the seeds are ripe enough and when they fall and scatter. Thus, while some of the seeds may be released by some cones opening in summer heat, radiata's closed cones typically provide a "seed bank" ready for release by fire or after climatic damage creates gaps in stands. While the cones remain closed, the seeds therein remain viable almost indefinitely, and even after extraction they store remarkably well (Shrestha et al. 1985).

This ease of seed collection is also of some help to breeders undertaking controlled pollinations. However, there is typically more than a two-year delay in

Fig. 5.1 Accumulation of ripe, unopened cones on a radiata graft within a hedged seed orchard

most pine species from pollination to availability of fully-ripened seeds, a drawback compared to the less-than-a-year delay in such species as redwood and Douglas-fir. Indeed, keeping that delay to only two years for radiata can depend on "curing" still-green cones picked early in order to accelerate ripening but retain full viability of its seeds (Rimbawanto et al. 1988).

5.4.2.2 Natural Regeneration

The accumulated seed crops on felled trees often allow natural restocking of clear-felled radiata plantations. After felling, the cones scattered in the slash can open readily to shed their seeds, especially with summer felling. Many of those seeds germinate successfully, especially if logging activities mix the seeds into the soil. This initially inexpensive option, however, has proved a mixed blessing. The genetic merit of the resulting crop was limited by that of the parent stand, although there were some benefits from earlier natural selection and if selection imposed by thinning of the crop was eugenic. Moreover, stocking and spacing proved hard to control. Stocking was often excessive and/or uneven (Fig. 5.2). Overstocking may require expensive thinning, and gaps may require infill planting. Lack of such follow-up management resulted in uneven diameter growth and variable knot sizes, which often led to reduced crop values.

At the other extreme, natural regeneration was sometimes inadequate, leaving areas of unstocked ground and/or trees that developed poor form. Reasons for inadequate regeneration were various, including excessive weed growth, felling the parent crop during winter (meaning that seeds do not get mixed with soil when cones open), burning the logging debris, and/or too much seed predation by birds and/or mice.

Fig. 5.2 Aerial view of part of Kaingaroa Forest, New Zealand, showing patchiness of natural regeneration of radiata occurring after clearfelling operations. Such natural regeneration is no longer relied upon, and is now discouraged

These problems with natural regeneration of plantations, in which regeneration often was over-dense, too sparse or unacceptably patchy, have occurred in all the countries that are the main growers of radiata as their plantation managers attempted to rely on natural regeneration.

5.4.2.3 Direct Seeding

Failing an assurance of good natural regeneration, managers sometimes spread supplemental seeds on the ground (called "broadcast sowing") in order to achieve restocking. This was sometimes done from aircraft following plantation harvest. Such broadcast sowing still left some problems of variable stocking. It also required the acquisition and use of large numbers of seeds per successful seedling, which militated against selection for better genetic quality.

It has proved possible to use drill-sowing of radiata seeds on open grassland or in other light vegetation, which was done successfully around 1930 over a significant area to help create the Kaingaroa Forest in New Zealand. This technique has also been used, but on site-prepared ground, with maritime pine in the sandy Landes region of south-western France. With broadcast and drill-sowing of seeds, follow-up measures are needed to achieve even spacing of seedlings at an appropriate stocking. In the 1960s, following aerial sowing of winter-felled compartments in Kaingaroa Forest, problems of uneven stocking persisted despite such follow-up. Because of these problems, and the desirability of making efficient use of more-expensive seed from select parents, aerial sowing was discontinued in New Zealand after 1971 (Levack 1973).

5.4.2.4 Nursery Practices and Transplanting

For the most part, radiata seeds germinate readily. Their germination can be hastened and made more uniform by stratification, which involves keeping seeds chilled and moist before sowing. After germination, growth in the nursery is typically rapid such that, in many climates, field-ready seedlings can be easily produced within a single year (Fig. 5.3).

By 1983 it was clear that radiata could be propagated on a large scale by rooting cuttings from very young seedlings and/or from low-hedged seedlings or still-juvenile cuttings (Fig. 5.4). While not as easy as with such species as willows or poplars, it was proving much easier than effectively rooting cuttings from the great majority of conifers. The many prospective advantages of being able to operationally deploy excellent, well-characterized clones (Libby & Rauter 1984) were prompting development of rooting technology and knowledge in Australia, California and New Zealand for radiata. Operational implementation with radiata, however, did not prove to be straightforward (Sects. 5.5.6; 6.8.2).

Seedlings and cuttings of radiata are easy to transplant into the field, even if bare-rooted. Provided the site is well prepared, the planting is done well and the weather is favourable, close to 100% survival can often be achieved.

Even so, it is desirable to get the planting stock conditioned, to help assure good survival even in adverse conditions, and good growth from the outset. An important advance achieved from research in New Zealand during this period was repeated undercutting of the roots in the nursery beds, to encourage fibrous multiple-root development and to harden off the tops so the propagules can resist some desiccation shortly after field planting. Such treatment is inclusively termed "wrenching," and begins by undercutting the tap roots of seedlings or sinker roots of cuttings.

Fig. 5.3 Nursery beds of radiata seedlings, showing green apical tufts of primary (juvenile) needles. With time and increasing height the shoot tips transition into brown, sealed buds

Fig. 5.4 Stem cuttings of radiata in nursery bed 10 months after collection from stool beds

It became appreciated that good initial field survival of planted propagules is not enough. Instead, rapid and uniform growth was sought right from the first year in the field, which required even-sized, robust and well-conditioned nursery stock. This was vigorously addressed by the New Zealand Forest Research Institute, especially during the late 1960s and 1970s (Chavasse 1969a, 1969b, 1973, 1976, 1981), these developments being encapsulated by Maclaren (1988) (see also Mead 2013). Better precision-sowing of individual seeds was developed as a basic step in obtaining seedlings of the right size and robustness. To condition the planting stock, and control planting-stock size, the traditional treatment of wrenching was much refined by repeated mechanical undercutting of roots and trimming of long lateral roots.

Good planting is of course also needed, not just to ensure survival and immediate growth. Getting the root system well configured and distributing the roots well in the soil can be crucial to avoiding young seedlings "toppling," particularly on windy sites with heavy soils. Such toppling results in swept and leaning stems that drastically reduce the value of the resulting logs (Fig. 5.5). Surprisingly, cuttings are much less susceptible to toppling, but cuttings-production technology was by 1983 not yet sufficiently advanced to exploit that feature. Meanwhile, much effort was directed at refining the planting techniques for seedlings, still the favoured propagule type as of 1983.

Improved nursery and planting practices, particularly the developing cuttings technology, were worthwhile in their own right, through reducing costs of subsequent tending, assuring better yields and log quality, and even making for slightly earlier harvest. In addition, these developments effectively complemented genetic improvement, with such vegetative multiplication allowing much more efficient use of scarce seeds of top genetic quality.

Fig. 5.5 Result of post-planting tree toppling, showing persistent butt deformation that greatly reduces tree value. Toppling can result from any or all of poor practice in the nursery, poor planting, soft and sticky soils and failure to consolidate cultivated soil

5.4.2.5 Site Preparation and Weed Control

For reasons indicated above, planting had by the late 1960s become very much the preferred option for both afforestation and replanting with radiata. Preparing ground for either operation serves one or more of three main purposes: controlling and/or killing unwanted "weed" vegetation (Fig. 5.6), including much that survived logging or previous land uses; ameliorating the soil; and mitigating harsh microclimates (Fig. 5.7). Clearing vegetation can take various forms, such as burning or crushing the logging slash, felling or crushing scrub and unwanted small trees, burning bracken fern, and/or use of herbicides to control various types of weed vegetation (Fig. 5.8). The soil can be ameliorated by such measures as ploughing or ripping to control vegetation and/or break up hardpans, and by additions of fertilizers. The near-ground microclimate can be improved for newly planted trees, for instance by creating raised soil beds, and/or by reduction or elimination of vegetation to allow better cold-air drainage and reduce the extent of surfaces that create radiation frosts.

Radiata's early rapid growth and its tolerances of some drought and of relatively low soil fertility mean that super-intensive control of competing vegetation is not

Fig. 5.6 Site preparation for planting achieved by mechanical clearing of existing vegetation, but involving some undesirable scalping of topsoil

Fig 5.7 Site preparation using a controlled burnoff of logging debris. While producing a clean site, often reducing frost hazard by improving cold-air drainage, this practice has been widely abandoned. On some soils it causes site deterioration, and it is always objectionable to the public. An alternative for controlling weed growth is use of desiccant sprays

Fig. 5.8 Spot spraying of grass before planting radiata, to avert competition from pasture grasses

required in its establishment phase. Even so, control of competing vegetation is sometimes needed after planting, to give the young trees enough light or to reduce competition for water and nutrients. Being tolerant of various herbicides, radiata in some countries featured in the development of mechanised (often aerial) post-planting applications of herbicides.

Intensive site preparation and follow-up weed control did not begin in the plantations of radiata. For growing certain tropical tree species (including hoop pine in Australia), where weed growth was typically luxuriant and aggressive, such practices were already traditional and intensive. The adoption of such practices for radiata, however, is of note for several reasons including the following three as examples: while not essential for achieving survival, the economic advantages of a rapid take-off were recognized as important; mechanisation of some of these operations was developed to a new level; and use of better selective herbicides (that killed most vegetation but did not harm radiata) came to the fore.

For re-establishing plantations after harvest, site preparation involved some different issues which centre around the use of fire. Burning of logging slash to clear sites for planting had several attractions, including low cost per hectare. In addition it can achieve one or several of: precluding unwanted natural regeneration, facilitating replanting, preventing outbreaks of insect pests that damage or kill young trees, and ameliorating the microclimate by allowing better cold-air drainage. However, in South Australia, slash burning was used largely to preclude over-dense regeneration. On those nutrient-poor soils, burning volatilized some nutrients and released others in a single pulse, allowing most to escape the site. The consequent loss of humus and already-scarce nutrients led to serious site degradation during the 1950s and 1960s. This degradation was halted and strongly reversed by abandoning slash burning, instead mechanically shattering and partially burying the slash, thus allowing a much slower release of nutrients that were mostly captured by the growing roots of the new trees. The sites were then improved by supplementing the on-site nutrients with applied fertilisers (Nambiar and O'Hehir 2010; O'Hehir and Nambiar 2010). On New Zealand's generally more nutrient-rich sites, slash burning had been widely adopted during the 1970s without the same ill-effects, but was later discontinued on environmental grounds, with public objections to the smoke and concerns over the eventual impacts on soils. In Chile, a common practice developed following slash burning that allowed farmers to grow wheat crops for a few years on the ash beds among the recently-planted trees; this shared the costs of site preparation and averted competition from certain weeds, but was to the further detriment of soils that had already been degraded before afforestation. Locally there, burning of logging debris has been observed by RDB to cause severe soil deterioration.

5.4.2.6 Spacing

Traditionally, based largely on earlier European practice, plantation forestry had usually prescribed planting trees at a much greater density than needed for the final crop that was to be harvested in a clearfelling. With radiata there have been at least two main reasons for this traditional practice: (1) for producing high-quality sawn timber, the dense early stocking gave control of branch size—and thence knot sizes in timber—especially in the lower logs that provide most of the financial value in a tree to be harvested for sawn timber; and (2) more importantly on some site types than others, dense initial stocking gave the forester a safety margin to allow for thinning out high percentages of trees with malformed boles or poor vigour, leaving the better-quality and more vigorous trees to grow on to harvest. In addition, suppression of weed growth was often a bonus from dense planting. However, adoption of very dense planting to allow for post-planting mortality is unsatisfactory; it does not address any underlying establishment problem, and usually leads to patches of both overstocking and understocking. Also, while not directly related to very dense initial stocking, the harvesting of thinnings was seen as a means of securing the advantages of early financial returns but, as we shall see later, the financial benefits depend on favourable cost and price structures at the time of thinning.

5.4.3 Use of Fertilisers

By the 1950s, substantial areas of radiata plantation in both New Zealand and Australia were growing poorly, even after correction of the dramatic problems caused by deficiencies of zinc (Sect. 5.2) (Fig. 5.9). In New Zealand, from about Lat. 37½°S northwards, many stands were extremely unthrifty. Phosphorus deficiency was diagnosed, and could be effectively corrected by aerial topdressing with superphosphate (Weston 1956, 1962; Conway 1962; Ballard 1977), that becoming routine practice on such sites by the early 1960s (Fig. 5.10). Application of phosphatic and other fertiliser, including nitrogenous ones, also expanded in Australia, notably on

Fig. 5.9 Five nursery seedlings showing zinc deficiency compared with one with deficiency corrected by spraying with zinc sulphate

Fig. 5.10 Response of radiata to phosphatic fertliser on strongly P-deficient site. The tree on the left is severely deficient; on right this deficiency has been corrected

phosphate-poor sites in New South Wales, the ACT and parts of South Australia (e.g. Boardman 1988; Horne 1988). That input, together with weed control, was sometimes sufficient to make radiata an economic crop on sites otherwise regarded as unsuitable for the species (Boardman 1988). Understanding of the interactions of fertilizer application with weed management, with thinning and with stand age led to appreciable gains in volume production (Australian Forestry Council 1981). The increased use of former "improved" pastureland for plantations was found to give rise to a new set of problems—soil nitrogen levels perhaps two orders of magnitude greater than those of unimproved pasture resulted in imbalances with several other nutrient elements (Horne 1988), plus ill-effects on tree form and wood properties.

In both South Africa and Chile, where nutrient deficiencies were often manifest in radiata, use of fertiliser applications came much later. South Africa continued to rely on maritime pine on its less-fertile sites within the Mediterranean climate zone, while in Chile the deficiencies tended to remain uncorrected in radiata plantations until after 1983.

5.4.4 Thinning and Pruning Regimes

Thinning commercial forest stands was long-established practice, aimed at removing cull stems and concentrating growth on the better trees to produce logs that are more valuable on account of both their quality and size. Pruning, in the form of removing lower branches, can be complementary to thinning. Low pruning, up to a height of around 2 m, has often been practiced to improve access and/or reduce fire hazard, with production of some clear, knot-free timber a bonus. Far less common had been pruning up to a full log length of 5–6 m to obtain much bigger yields of clear timber. Since cost-effectively producing such clearwood depends greatly on diameter growth after pruning, thinning and pruning practices need to be closely integrated.

The evolution of thinning regimes and the frequent adoption of pruning are important parts of our domestication story. With radiata, the application of the traditional approach of repeated thinnings, and harvesting the thinnings, had not been straightforward. Indeed, there are some intriguing paradoxes. Left untouched, dense radiata stands can progressively thin themselves from thousands of small young trees per hectare to about 200 or fewer large survivors as the stands mature. Luckily, those survivors tend to be better-formed trees, with low rates of forking and other malformation. Their near neighbours therefore become suppressed (overtopped) and, depending on the environment, those increasingly weakened trees readily suffer mortality due to natural causes. The trees that achieve local dominance thus create enough space for themselves to eventually make the diameter growth that is needed to produce reasonably valuable logs. That is what happened in the huge plantings created in New Zealand during 1925–1935, to a lesser extent in the Chilean plantings done mainly during the 1940s, and locally in Australia. Such strong assertion of dominance in radiata by a minority of trees is unusual among tree species. Nevertheless, radiata is still highly responsive to silvicultural intervention in the form of thinnings and pruning, but margins for error in the timing and execution are often small.

In New Zealand during 1938–1945, because of the Second World War, there had been neither the manpower nor the money available to do widespread thinning. A high proportion of the suppressed trees usually were destined to die in New Zealand, even without the additional mortality due to a late-1940s outbreak of a woodwasp that killed huge numbers of weakened radiata pines. In Chile, with its generally less natural stem mortality, a high proportion of suppressed stems could typically be harvested in the course of early clearfellings—where markets could be found for very small logs.

5.4.4.1 Some Problems with Commercial Thinning

By the mid-1960s, it was already known that, despite radiata's potential for self-thinning, its health and growth vigour are highly responsive to thinning and other interventions. An immediate problem, though, was that the financial returns from the typically small trees harvested in thinnings usually did not cover the costs of planning, felling, logging and transporting the numerous small logs produced. In New Zealand, A.R. Entrican, who during the 1920s and early 1930s had organised research on the utilisation of wood from radiata plantations (and later became head of the Forest Service), envisioned the thinnings strictly as pulpwood. But even when pulpmills opened and created markets for pulpwood, it was soon clear that the price and cost structures associated with most thinnings made their harvesting unprofitable. This was seen as partly because, even after the timber from the self-thinned early radiata plantations came on stream, the government (through the State Forest Service) continued to sell high-quality standing timber from native forests at artificially low prices, in the name of providing easily affordable houses (Orr 2017). That practice effectively ignored a key goal of the 1920s plantations, which was to have their wood substitute for native woods, in order to allow more native forests to be retained unharvested. In addition, mechanised harvesting of thinnings was still primitive, and not yet cost-efficient. Actually, some forest managers did just accept that young thinnings would not be profitable, and left the small felled trees to rot on the ground, often seeing that as a penalty of having planted too many trees in the first place. This issue prompted radical solutions (Kininmonth 1997a), which we will trace below (see Sect. 5.4.4.3).

In Australia, where wood supplies were often scarce and the terrain often easier, harvesting of radiata thinnings tended to be less of a technical and commercial challenge. On easy terrain, with the lower rates of stem malformation associated with Australia's lower average site fertility, a much cheaper and simpler approach to thinning would often work. In it, thinning out whole rows of trees could be done with little adverse effect on the quality of the final crop. The removal, for example, of all trees in every seventh row ("outrows") provided convenient access throughout the stands for modified highway trucks ("crane trucks"). Also, the timber markets often allowed small harvested logs to often be profitably sawn, for example for making boxes for fruit, as well as being used for pulp in some locations.

In Chile, where labour costs were much lower and bullock teams were available, with little tree mortality from suppression, it was common to make a profit on harvesting the considerable volumes of small, closely-spaced trees that comprised thinnings. Alternatively, unthinned plantation stands were often clearfelled quite young, meaning very little loss of wood to natural stem mortality.

It was in New Zealand, where competition-induced tree mortality was often high, that the conundrum of what to do about thinning was a pressing issue, and where a radical approach was developed. There are several threads to the story, in which a few foresters and scientists played key roles.

5.4.4.2 Some Lessons from South Africa

Long before the issue of thinning regimes for radiata was recognised as being so problematic in New Zealand, some ground-breaking research had been done in South Africa by I.J. Craib. A research forester with a doctorate from Yale University, Craib worked for the South African Department of Forestry. In the 1920s and 1930s, he was working with several fast-growing tree species including radiata, in a climate characterised by long dry periods. Among his studies was a series of spacing and thinning trials with the radiata. The results led him to prescribe planting at much wider initial spacings and heavier thinnings than previously used. Where they were adopted in South Africa, his prescriptions were a great success. Significantly, he cautioned strongly against letting thinning operations fall behind prescribed schedules, because that would lead to delays in logs reaching millable size, which would badly affected the projected financial profitability of South African plantations.

5.4.4.3 Towards Radical Changes in New Zealand Practice

In New Zealand, from the late 1940s, forester John Ure was overseeing the restocking and subsequent tending of large areas following clearfelling, after the harvesting of the 1920s plantations accelerated sharply in Kaingaroa Forest. He was mindful of experience with overcrowded stands planted before 1940, which were suffering heavy mortality from attack by the wood wasp *Sirex noctilio*. At the same time, he was often faced with over-dense natural regeneration, especially where stands had been clearfelled during summer. In response, he prescribed very early thinning of the natural post-harvest regeneration, to be followed by relatively few and heavy thinnings later on (Ure 1949) (Fig. 5.11), in the expectation that the later thinning could be harvested profitably.

Despite Ure's hopes, attempts at commercial thinning remained stubbornly unprofitable in most parts of New Zealand. The scene was set for a more radical approach (Kininmonth 1997a). In that, three key players were, in approximate historical order, C.H. Brown, E.H. Bunn and R.T. Fenton[1].

[1]The remainder of this section is heavily based on personal recollections of one of us (RDB) who was in contact with Fenton and Bunn during the time period.

Fig. 5.11 Early trial tending natural regeneration in Kaingaroa Forest, New Zealand, showing heavy early thinning and low pruning

Brown was a forester responsible for areas in the far south of New Zealand where young plantations were at ages when decisions had to be made about their tending. He was well-acquainted with problems of trying to thin relatively old, over-crowded stands of radiata. Later on, he also noted the disastrously poor grades of sawn timber from a Corsican pine stand that had been thinned heavily to allow the crop trees free growth, but the resulting large lower logs had received only very belated pruning. In 1956 he therefore decided to focus thinning efforts on his young stands, and he realised he needed to address the questions of when and how to prune. In taking the initiative, he was even prepared to physically do experimental tending himself. In tackling the younger stands, thinning heavily and doing first pruning before age 10, his aim was to keep the "knotty core" of logs small so the butt logs would contain a thick outer zone of clearwood.

Bunn, to whom Fenton reported while researching thinning regimes, had had some important prior experience. In 1951, while waiting to attend the Forestry School in Canberra, Bunn was working at the Waipa State Sawmill in Rotorua and was given the task of tracking logs in the mill to identify categories that should be pulped rather than sawn. The result was a big surprise, in that he found that the large, rough logs from stand-margin trees tended to produce better grades of sawn timber than did the smaller, finer-branched logs from trees in the stand interiors. Another experience may have helped shape his outlook. At a New Zealand Institute of Foresters Annual Conference in 1960, which Bunn attended, J.H. Beekhuis was presenting research results from sample plots of radiata thinned to several regimes spanning what had been the range of accepted practice when he started out. Beekhuis was rewarded with a short but withering verbal blast from A.P. Thomson, who later became Director-General of the Forest Service. Thomson pointed out that the New Zealand Forest Service had already adopted heavier thinning regimes than any studied by Beekhuis. When Bunn later oversaw FRI's subsequent silvicultural research, he was guided by two strong, related principles: "be prepared for the unexpected," and "try a range of treatments extending beyond the bounds of any anticipated practice."

Those principles led Bunn to strongly support Fenton, and to exhort Fenton's juniors on those principles—at great length and sometimes late into the night. Fenton had been recruited in early 1962 by the Forest Research Institute, Rotorua, to study the Economics of Silviculture. He had previously worked in a sawmill in the same region as where Brown was working. He was acquainted with Brown's investigations, and shared with Brown experience of the results of very delayed pruning together with heavy thinning of Corsican pine.

Crucially, Fenton was not one to flinch from following an argument to its logical conclusion. He headed a team that worked on a broad research front. They relied largely on sawmill studies of sawn-timber grades from radiata stands that had received known treatments. New thinning and pruning studies were begun, log outturns from alternative tending regimes were projected from existing yield tables, and costs and returns for the alternative regimes were used for financial analyses.

Unavoidably, some important assumptions had to be made. A supply of good-length clear (knot-free) sawn timber of radiata was essentially non-existent at the time, while abundant clear sawn timber was still available from native forests in New Zealand. Thus there was no real market information on the value of clear radiata timber. On the other hand, large premiums for clear versus knotty timber of other pines were commonplace overseas, so it was logical for Fenton to assume that a similar premium would eventually prevail for clear timber of radiata. Since there was no real impetus from the wood-processing and building industries, the whole exercise stood as an example of technology-push rather than demand-pull.

On the silvicultural front, one notable trial, in northern Kaingaroa Forest, included a treatment thinning in one step to final-crop stocking of 200 stems per hectare, combined with immediate pruning of the trees left. The outcome was remarkable in that the resulting stand developed with good tree form, without unduly heavy branching.

Preceding their work on alternative radiata growing regimes was a detailed simulation study comparing the economics of developing an area of land into forest plantation versus pasture. That study had been based on accepted silvicultural and pastoral-development practices at the time, along with the then-current cost and price structures. But, once it was completed, Fenton and his assistants turned to re-examining the economics of alternative silvicultural regimes in search of ones that would be more financially profitable than the existing ones. This took the economic study of industrial forestry to a new level.

Certain of their modelled findings were clear-cut, and some of them were very surprising, especially to foresters of classical silvicultural training. Having adopted reasonable assumptions, their models indicated that afforestation with radiata could be financially competitive with the alternative of developing all the available land for pastoral farming, often projecting an internal rate of return from afforestation with radiata of over 7%. Their modelling also predicted that pruning of butt-logs to produce clear sawn timber could be highly rewarding if combined with sufficiently early and heavy thinning. In this connection, the second log, from 5–10 m up the tree from stump level, posed a problem, being too high to prune cost-effectively, even with economies of scale over large areas. Yet, those "second logs" were modelled as too large to treat as just a low-value by-product of producing high-quality butt logs. This meant that while improving the size and quality was largely in the hands of culturists, offsetting adverse effects of the lower stockings on branch size and general stem quality in the second and higher logs would be up to the breeders.

Perhaps most interesting, Fenton and his team found that delaying thinning until the trees were big and old enough to make the thinning immediately profitable could severely compromise the overall profitability of the crop. Such a delay in thinning reduced the average diameter growth in the remaining crop, in turn delaying the final harvest. Commercial thinning of older larger trees could also lead to substantial damage to the final-crop trees, further reducing the volume of wood that could be harvested at low cost per unit of generally more-valuable wood. In other words, waiting until a thinning could immediately return more than the thinning operation cost carried a major, if far from intuitively obvious, opportunity cost.

Accordingly, Fenton (1972) proposed the "Direct Sawlog" ("Direct") regime. It entailed more than one non-commercial thinning, leaving the thinnings to rot on the ground, and accepting significant sacrifice in total wood yield per hectare. The financial and productivity costs were justified on the expectation that the regime achieved greater full-rotation profitability with a major contribution from better-quality sawn timbers and veneers. Such a regime was indeed radical, but it underscored with much stronger data Craib's key tenet, namely that thinnings should not be delayed in hope of an immediate net profit on thinning.

The Fenton team's recommendations focused on getting the course of stand development firmly set during its first 10 years from planting, which is before post-thinning wind damage becomes a significant risk. All operations, including site preparation, spacing of initial planting, beneficial fertilization, thinning, pruning, and protection from competition and damage, were focused on the production of valuable sawlogs. They recommended using pre-commercial thinning pre-emptively rather than as a source of pulpwood, rejecting any notion that zero-revenue thinning should be a last resort in the absence of any profitable market for thinnings. That made pulpwood, mostly from upper logs, almost entirely a by-product of growing for solid-wood products. Despite much initial scepticism, the science-based data looked compelling, and the Direct regime became widely adopted. Often, however, the practices of Fenton's enthusiastic followers in New Zealand became even more extreme than prescribed in the Direct regime.

Actually, the results of Fenton's studies did not entail an unconditional proscription of commercial thinnings. Rather, those studies set a framework for identifying the various conditions that would have to be met for the option of commercial thinning to be truly economic, in the sense of contributing positively to the profitability of the crop over the whole rotation (Box 5.2).

Box 5.2 Thinning Issues
Thinning has traditionally been practised in forest plantations, providing some early wood yields, often harvesting trees that might die and rot before the final harvest, and concentrating wood production on the best stems that form the final crop. However, delaying thinnings until the extraction can be done at a profit can affect the economics of the final crop unacceptably, making it appropriate to thin early without harvesting the thinnings. Moreover, returns from pruning depend heavily on the "crop" trees having enough space in which to put on clear, knot-free timber after pruning. In New Zealand a combination of heavy, non-commercial thinnings and intensive pruning was widely adopted. Often, however, it was to compromise unduly—and more than anticipated—both the yields and wood quality of radiata.

At least three other factors helped adoption of Fenton's more radical approach. Firstly, with genetic improvement, many fewer trees needed to be planted per hectare in order to guarantee a final crop that would be composed almost entirely of trees of good form (Figs 5.12 and 5.13). Secondly, it was accepted that the benefits of pruning would be enhanced by stems being straighter, which was another consequence of genetic improvement. Thirdly, there came a realization that the early stockings needed to allow for stem malformation (even in genetically improved stock), in order to assure a well-formed final crop, would be less than had been feared. This was because mild stem malformation, in the form of forks or **ramicorn**[G] branches was found to be very often self-correcting, although it would still cause some degrade inside the log. Some of the early stands thinned to Ure's prescriptions were found a few years later to have disturbing percentages of malformed tops which, because the self-correction rate among malformations was not yet appreciated, caused alarm at the time over aggressive thinning regimes. Importantly, the promulgation and later acceptance of the Direct Sawlog regime came when a second radiata planting boom was gathering momentum in New Zealand.

An offshoot of the move towards radical silviculture in New Zealand was an agroforestry movement in New Zealand (Maclaren 1988). This movement, while arising in parallel with a worldwide agroforestry movement, was mostly prompted by local factors. These factors included the disenchantment with commercial thinning, and much disenchantment with poor profitability of pastoral farming on large areas of hill country. A solution proposed was to convert from pure pasture to agroforestry by planting trees into problem pastures at low initial stockings. That would let sheep graze the pasture during the several years between when the young trees were large enough to avoid serious damage from stock until when the pasture grasses were suppressed by canopy closure. That way, "pasturage" would

Fig. 5.12 Tree stocking trial with the Direct Sawlog regime, all thinning done without waiting until thinnings were worth harvesting

provide an early financial yield in place of commercial thinnings. Where radiata was the tree species of choice, it was important that the young radiata pines proved to not be seriously vulnerable to damage by grazing stock. Furthermore, ex-pasture sites often gave very rapid tree growth because of the enhanced soil fertility that resulted from relatively intensive pastoral farming.

Neil Barr was a visionary and charismatic early farm forester of great enthusiasm, energy and affability, and he helped found the New Zealand Farm Forestry Association. Barr was a vigorous proponent of pasturage as a yield in place of thinnings, and was much emboldened after a team of workers pruned a young stand far more severely than he had intended, but with excellent results. Barr's farm, however, was in a locality north of Auckland where problems with tree form in radiata tended to be minimal. Many early recruits to agroforestry visited that farm, which perhaps led to some over-optimism.

Fig. 5.13 Stand grown under Direct Sawlog regime ready for harvest; early thinning to low final-crop stocking was accompanied by pruning

Fig. 5.14 Early stage of agroforestry trial with radiata, established in pasture with the idea that pasture could provide an intermediate yield in lieu of thinnings, prior to canopy closure

Major impetus for agroforestry research in New Zealand undoubtedly stemmed from a personal friendship between Bunn and Barr. Within FRI and with Bunn's blessing and support, Leith Knowles led early research efforts on the agroforestry option (Fig. 5.14), joined by personnel from the Ministry/Department of Agriculture. In the following three decades, Knowles made the results of that research broadly available, and joined Barr in strongly advocating the apparent advantages of agroforestry. The availability of genetically improved radiata with better stem form, while not specifically directed to uses in agroforestry, seemed to fit nicely the requirements of such widely-spaced farm forests.

Roughly paralleling refinements in establishing agroforestry blocks came refinements in the treatment of radiata shelterbelts to produce high-quality timber. Historically, shelterbelts had often been grown for producing firewood as well as providing shelter. This had often entailed periodic topping to control shelterbelt height, which also better maintained continuity of shelter and made the trees resistant to windthrow. Where the trees had not been topped they could produce respectable sawlogs. As radiata became increasingly accepted as a sawtimber, farmers became increasingly interested in producing quality butt logs material in shelterbelts. This led to a profusion of designs and management regimes for shelterbelts (Maclaren 1993; Mead 2013) (Fig. 5.15). For instance, some of the trees could be pruned for clearwood production, while intervening trees could be trimmed to produce continuing low shelter below pruning height; alternatively, the intervening trees were often of other, slower-growing species.

Fig. 5.15 Sophisticated shelterbelt grown on Canterbury Plains, New Zealand, by P. Smail, a prominent farm forester. The large trees are radiata, alternate trees being pruned for producing clear timber and left unpruned for providing continuous shelter. The smaller trees are deodar cedar, to give longer-term shelter from the ground upwards

5.4.4.4 Practice in Other Countries

Outside New Zealand, there was not the same rapid adoption of radical regimes, such as Fenton's Direct Sawlog regime, for a variety of reasons.

In Australia, there is typically easier terrain, which facilitated financially-viable commercial thinning; often less natural stem mortality from within-stand competition; an Australian focus on producing structural rather than appearance-grade timber; and often a declared management objective of maximising the volume of wood harvested per unit area (e.g. Jolly 1950). As it turned out, there often were problems with stands that had suffered as managers waited until a commercial thinning could be profitable. Silvicultural regimes did vary according to factors such as site characteristics, management policies, and availability and lucrativeness of outlets for young, small-diameter logs (e.g. Boardman 1988; Horne 1988; Lavery 1988). Even so, the different stand-management regimes adopted (Lewis and Ferguson 1993) have tended to be "incremental" modifications of classic, textbook practice. Thus much less pruning was practised, stockings tended to be kept higher and rotations were generally longer than in New Zealand. Exceptions included heavier thinning on low-rainfall, sandy sites in Western Australia (on which sites, however, maritime pine tended to be preferred over radiata), and in Tasmania where more aggressive tending regimes were sometimes adopted for radiata (Elliott et al. 2008).

In Chile, there had been some harvesting of thinnings in the early-1900s plantings, which were largely located near coal mines. However, in the large plantings that had been made later, especially those planted in the 1940s, very little silviculture was practised. Young trees in some areas had been pruned, but most stands had been left unthinned, as had happened in the stands established in New Zealand during the planting boom of 1925–1935. For utilisation of the wood, a modest-sized pulpmill had been commissioned in 1959, but this was later enlarged with three more pulpmills coming on stream during the early to mid-1970s as the large 1940s plantations matured. Compared with New Zealand, and even with Australia, between 1969 and 1983 Chile presented a very different context for silvicultural regimes to evolve in.

This Chilean context during 1969–1983 included a greater focus on efficient pulpwood recovery (associated with increasing pulping capacity), a high level of vertical integration between forest growing and wood processing; lower costs of harvesting of thinnings, less natural mortality in stands; and felling of stands at somewhat younger ages and thus producing smaller sizes than in New Zealand and Australia. While some stands had been pruned for clearwood production, silviculture has overall been more conservative than in New Zealand (Mead 2010, 2013). Compared with New Zealand, initial stockings in Chile were still typically higher, and pruning, when it was done, was less aggressive, as were thinnings. This was helped by natural, competition-induced tree mortality being comparatively low in Chile (Fenton 1979) (Fig. 5.16). Some Chilean radiata crops were even being grown exclusively for pulpwood, depending on the nature of the crops that then could be grown satisfactorily on the particular sites and the locations in relation to the pulpmills.

Fig. 5.16 Unthinned radiata stand in Chile, some 20 years old, showing relatively low natural stem mortality characteristic of Chile. Note the contrasting frequencies of branch clusters between the tree in the left centre foreground and one in the far right foreground (Photo W. Sutton)

5.4.4.5 Longer-Term Follow-up

Following on from Fenton's work in New Zealand, one of his team, W.R.J. Sutton, vigorously promoted the production of clearwood by the silvicultural regimes developed by the team. The assumptions driving the rationale at the time were two-fold: (1) such clearwood sawn timber would soon command much better prices than knotty structural timber; and (2) growers of radiata in several countries, especially those in New Zealand, would continue to enjoy a big comparative advantage for growing such clearwood timber because of radiata's outstanding growth rates and suitability as a plantation species.

Later, in 1979–1982, Sutton was charged by Bunn with leading a NZ Forest Service Task Force to model the growth and outturns of radiata stands under various site conditions and growing regimes. That team effort produced a software package, SILMOD, which was released in 1983. An important development supporting plantation forestry, SILMOD became a widely used decision-aid for forest planners and plantation managers. The SILMOD software has been updated and extended over the years, being superseded in NZ first by STANDPAK in 1997 (Kininmonth 1997b), and then by "forecaster" under the Scion software development company ATLAS. Such data-based decision-aids have been another leading-edge development in both plantation forestry and the beginnings of forest-tree domestication worldwide.

5.4.4.6 Tending Regimes in Perspective

Revisions of tending regimes were influenced not only by information on how different regimes had performed or, with increasing experience with them, how they would perform. They were also based on perceptions of what sites would be available for planting and what genetic improvement would achieve. Even with accumulating data on performance, making some important assumptions was still unavoidable. While some of those assumptions were explicit and consciously made, others were being made unwittingly. Some of the assumptions later proved to be wrong. This led to some disappointments with the performance of tending regimes adopted, and also with some of the selection criteria used in breeding. Moreover, these disappointments revealed some misalignments between silvicultural practice and the selective breeding. Details will be covered later, but the misalignments that later emerged are previewed briefly here.

By 1983 in New Zealand, accumulating experiences with the more aggressive Direct Sawlog and Agroforestry regimes were starting to reveal previously unrecognized limitations in what could be achieved by selective breeding, and to expose how genetic improvement of wood properties could be far more crucial than had been anticipated under such regimes (Sects. 5.4.4.6, 6.3.2.2, 6.7.2.4 and 6.10.1).

In Australia some misalignment between breeding and plantation siting had occurred, despite less aggressive silvicultural regimes. That was because early plus-tree selection there had mostly been done in heavily stocked stands on quite infertile sites. Yet the resulting seed orchards often went on to serve plantings on sites where fertility had been elevated by prior pastoral farming or by fertilising the plantations. That exposed serious problems with tree form (Birk 1990, 1991) when the offspring of those orchards were grown on such sites. As a result, progenies of NZ clones that had been selected on higher-rainfall and more-fertile sites were later found to be better suited to many NSW sites than offspring of NSW selections.

In Chile, the early preferences of harvesting for pulpwood had contributed to less aggressive silvicultural regimes than became fashionable in New Zealand. There the breeding programme did not pursue the option of breeding for longer lengths of knot-free timber without needing to prune (namely for "long-internode" trees), which seemed regrettable, because such trees appeared to be satisfactory for a wide range of Chilean sites (Burdon 1978). However, the eventual arrival of the European pine shoot moth (Sect. 6.3.1) would likely have nullified that advantage, because such trees are affected worse by leader damage.

5.5 Genetic Improvement

5.5.1 *Operational Breeding Activities*

We first outline major developments in operational breeding during the period 1969–1983 before later addressing their technical aspects in more depth. Some of the ongoing or routine activities, such as continuing seed-orchard establishment and replacement that did not entail important changes or advances, are not automatically reviewed.

Advancement through the first two generations of breeding had initially been relatively slow, partly through some false starts, such that selection and testing of genetically improved radiata pines had by 1983 seldom if at all proceeded past a second generation.

5.5.1.1 Australia

At the beginning of the period of interest in this chapter, a review of seed procurement for radiata plantations concluded that "seed collection is typically made from trees felled in routine thinning operations [of] older stands [in which] one or two thinnings would have already been carried out, thus removing the poorer fraction of the stand ..." During the period, seed production from first-generation seed orchards increased greatly, although production was variable between orchards and from year to year (Pederick and Brown, 1976). In some planting programmes, deficits in the supplies of orchard seed were offset by collections from clear-felled trees in heavily-thinned plantations and, in some cases, from specially-treated seed-production stands.

By the early 1970s the South Australian and Queensland forest services were then and subsequently able to use only improved planting stock from their own breeding programmes for all their radiata afforestation and reforestation (Wu et al. 2007). Cone collection, however, became difficult as the trees in established orchards increased in height. Pollarding (cutting back the upper crowns) was explored as a means of both increasing seed yield and keeping cones sufficiently accessible for collection, but neither goal was thus achieved (Matheson and Willcocks 1976; Pederick and Brown 1976). It became evident that the fiscally-useful life of radiata trees in seed orchards did not extend much beyond 20 years. The consequent decline in cost-effective seed-harvest had not been fully anticipated in at least some of Australia's radiata breeding programmes, with the result that some shortfalls in supply occurred during the 1970s and 1980s before new orchards came into production (e.g. Johnson et al. 2008). Nevertheless, by 1983 half of Australia's state forest service radiata planting programs were based on orchard seed only, and in others the fraction ranged between 70% and 82%.

As in New Zealand, it became clear that choice of site was a critical factor for a successful orchard. The radiata orchard at Mt Schank near Mt Gambier (SA) proved to be on an excellent site, and it provided notably high seed yields as well as good information on favourable orchard-site characteristics for radiata in Australia: temperate (maritime) rather than continental temperatures and a favourable soil moisture regime. At some locations with previously-established seed orchards, seed yields were substantially increased, and costs per seed reduced, by the application of fertiliser and/or irrigation. In some orchards, foraging by parrots (cockatoos) substantially reduced seed yields.

Using information from progeny tests, most of the radiata clonal seed orchards established during the 1950s–1970s had by 1983 been only lightly culled (less than 30% of the parent clones removed). The need to replace old orchards to facilitate seed collection also afforded the opportunity to use advanced-generation clones in them; by the early 1980s three of Australia's eight radiata breeding programmes were able to plant new orchards with second-generation clones, and most of the others were preparing to do so. This step was possible because early progeny tests attained sufficient age to permit assessment, and capacity to process the extensive data from them was greatly improved by better access to computers and evolving software for managing data and assessing breeding values, and for managing adverse genetic correlations (e.g. Cotterill and Jackson 1981; Dean et al. 1983). Their breeding objectives maintained an emphasis on improving growth, branching habit and bole form, although tolerance to drought was added in a secondary program in South Australia and resistance to diseases was sought in several programmes (Ivković et al. 2010); and APM recognised the importance of wood density in Gippsland, adding its improvement as a breeding objective. An Australia-wide genotype-by-environment interaction study established in 1969 and 1970 did not reveal interactions large enough of be operational importance among the genotypes and sites studied (Matheson and Raymond 1984).

Work in Australia (and New Zealand) was stimulated by field tours and discussions during and after the previously noted Third World Consultation on Forest Tree Breeding, held in Canberra in 1977. That meeting was sponsored jointly by the Food and Agriculture Organization of the United Nations (FAO) and the International Union of Forestry Research Organizations (IUFRO). A series of consultations and ongoing interactions through the working parties of IUFRO, both before and continuing after the 1977 Canberra meeting, has made important contributions to the development and conduct both of tree breeding and relevant international cooperation over the half-century from 1960 to 2010.

As in New Zealand, reviews of strategies for advanced selection led to the conclusion that most of the radiata breeding programmes had too few selections available as the founding parents for long-term breeding. Remedial steps included additional selections in then-established Australian plantations not previously evaluated and acquisition of seeds from other programmes, for adding to breeding populations.

The Woods and Forests Department of South Australia's radiata breeding programme was by 1969 being led by Bob Boardman and assisted by Forestry and Timber Bureau/CSIRO staff at Mt Gambier. At the commencement of the period considered in this chapter, that programme followed the basic Scandinavian model, but in the last third of the time a new round of selection in plantations added 627 first-generation plus trees (selected in unimproved land-race stands) to those already on hand to better provide diversity for later generations of breeding. Observations of the value to breeding programmes of good planning, and the greater availability of the information necessary for more complex work such as index selection (Cotterill and Dean 1990), led increasingly to the preparation of formal strategies and plans for genetic management that could be discussed, critically reviewed and periodically revised (Cotterill 1984)—a desirable practice that has been often emulated elsewhere.

Unfortunately, some of the serious forest fires in February 1983 (Sect. 5.1), in the Green Triangle and Adelaide Hills in South Australia, destroyed the major radiata seed orchard in the former area and other breeding resources, including progeny tests and clonal archives of plus trees. Thereby, it set back the breeding programme in the region by more than five years.

In Victoria the Forests Commission had hoped that by the end of the period there would be sufficient orchard seed from its breeding work, led by Leon Pederick, for the whole annual planting programme. Unfortunately, a 30-ha first-generation orchard at Campaspe—one-third of the commission's orchard estate—was burnt early in 1983. Accordingly, replacement seed was collected from old stands clearfelled because of fire damage, and was supplemented by large-scale multiplication of nursery cuttings of small quantities of seed-orchard seed. APM Forests in East Gippsland, primarily interested in pulp production, in collaboration with CSIRO wrestled with the negative correlation between volume production and wood density (Dean et al. 1983). Controlled pollinations were made between members of high-volume families and of high-density families, including selections from the Guadalupe provenance. Under the leadership of Mike Hall and John Cameron, they also maintained their seed-orchard programme, and a broad range of research related to vegetative propagation and stand management.

Seed for the very large New South Wales radiata planting programme was mostly derived from three seed orchards—Tallaganda (4.4 ha established 1957–1973), Green Hills near Tumut (17 ha, 1961–1969) and Vulcan, near Bathurst (67 ha, 1966–1972) (Johnson et al. 2008). Progeny tests of first-generation selections were assessed in 1977–1978, and the data used for culling the Green Hills and part of the Vulcan orchards soon afterwards. A new round of selection in land-race stands, during1980–1986, yielded 347 additional first-generation plus trees. Also, progeny trials seven years or more in age provided a further 240 select trees for the breeding population. Between 1979 and 1982, four large progeny and provenance trials, including the then-recent collections from natural populations, were planted within the state.

Queensland's commitment to breeding radiata continued, albeit at a low level because of the limited prospects of developing a large radiata estate in that subtropical, summer-rainfall environment. Trials including first-generation New Zealand progenies (planted from the late 1950s) and NSW Tallaganda (1970) progenies indicated that growth of progenies of trees selected locally was relatively inferior, probably a consequence of the nature of the local founding plantation population. Although no radiata seed orchards were reported by Queensland in a national survey in 1976 (Shepherd 1977), "local improved seed was available in the 1960s" and an orchard was planted in 1981 (Wu et al. 2007). Eventually, however, Queensland abandoned its own radiata breeding programme.

In Tasmania, the state forest service and a private company (APPM Ltd) established four orchards totaling 18 ha by 1976, in general following the Scandinavian first-generation protocols. Then, in the early 1980s, seedlings from more than 250 open-pollinated (mainly) and pair-cross families from programmes in Australia and New Zealand were planted in progeny trials. These were established to provide for selecting trees to produce a greatly expanded and more diverse base breeding population, to provide performance data for informing future breeding, and in the expectation that at least some of the included families would prove to be sufficiently adapted to Tasmanian conditions for near-future within-family selection, amplification and deployment (Elliott et al. 1995).

In Western Australia the Forests Department began planting a radiata seed orchard in 1965, using a minimal Scandinavian protocol because of uncertainty about the role of radiata there. Establishment of a second orchard, at Manjimup, was completed in 1972. While awaiting seed production to commence in this second orchard, seed was obtained from a heavily-thinned and fertilised 40-year-old stand by felling. As elsewhere, early selections in WA for seed orchards were on the basis of vigour and stem form. In 1978, however, the programme took the distinctive step of researching and breeding for improving resistance to the troublesome root pathogen *Phytophthora cinnamomi* (Butcher et al. 1984). Not only was within-population resistance (actually, tolerance) identified, but the previously neglected Cambria population was shown to include a high percentage of tolerant trees. Testing of Cambria trees was extended with a view to their inclusion in the breeding population (Wu et al. 2007).

A further round of seed collections of radiata, from ecologically-diverse natural stands across the entire natural range of the species, was undertaken in 1978, primarily to serve Australia and New Zealand (Sect. 5.5.4.1 below; Eldridge 1978).

5.5.1.2 New Zealand

A major expansion in the breeding population was begun in 1968 with selecting afresh an intended 600 plus trees in unimproved stands 12–18 years old (Shelbourne 1986; Burdon et al. 2008). For obtaining so many plus trees, the intensiveness of selection was reduced to the best tree in each 1.2 ha. Establishing and assessing three huge resulting open-pollinated (OP) progeny trials planted in 1969 was a major commitment for the FRI tree breeders. Those trials were established from cones collected from each of 588 new plus trees at the time the trees were found and selected in 1968 (Shelbourne et al. 1986; Burdon et al. 2008), resulting in around 30,000 OP offspring trees at each of three sites. Several purposes were served: testing the plus-tree parents, estimation of genetic parameters (variances, heritabilities, genetic correlations between traits, and the magnitude of genotype-site interactions), and providing a basis for forwards selection among the offspring. In the event, forwards selection was confined almost entirely to just one of the three sites, with the nearest and best-quality progeny trial. First assessment of the three trials was done in 1975, five years after planting out year-old seedlings, giving some research information, an early growth-and-form basis for an initial culling from those 600 parent clones in archives/orchards, and a very early selection of individual offspring for further testing. Of the 588 parent clones (a few were not grafted successfully), 250 were provisionally incorporated into fresh seed-orchard plantings on the basis of that very early assessment (Shelbourne 1986; Vincent 1986). Taking increment cores from progeny, those 250 parent clones were evaluated for corewood density, with about the lowest 50% culled on this basis (Shelbourne 1986).

As for the OP offspring, they had non-select male parents that were essentially random and virtually all different from each other. However, the very large number of half-pedigreed families with a total of many thousands of offspring provided a major advantage for belatedly including a new and high-heritability trait for selection, namely wood density. Previous work had already established that the high heritability of corewood density ($\geq$0.5) was substantially greater than heritability for growth (ca. 0.2) (Burdon 1992). Accordingly, among offspring provisionally selected on both family- and individual-tree performance for growth and form, those with above-average corewood density were chosen. Those higher-density offspring trees were used for new seed-orchard plantings, and for intercrossing to produce the next generation of what was thus a new breeding population, to serve as a candidate population for a new generation of breeding (Shelbourne et al. 1986).

One hundred such "forwards" selections were made in 1975 and these seedling trees were then used immediately for controlled crossing *in situ* (that is, where each tree stood in the progeny trials). This was a laborious operation, but it saved time: in retrospect it proved to be in a very fortuitous time window when most trees were sexually mature but had not grown too tall. Two years later, seeds were extracted from these cones, germinated, and planted the following year in full-sib progeny trials. In the next few years, however, many of those 100 selections became badly affected by a needle-cast disease caused by *Cyclaneusma minus* (Fig. 5.17), despite good health at the time of selection. Accordingly, in 1980 nearly 200 additional trees were chosen in the same type of selection from one of those 1969 OP progeny trials, but with heavy emphasis on crown health and less emphasis on wood density. Those 200 selections were archived as grafts and later used for controlled crossing.

By 1980, largely in order to reduce evaluation costs, wood density was becoming assessed indirectly, more quickly but with lower accuracy, first by torsiometer but later by penetrometer tools applied directly to the candidate-tree boles.

A concern existed among breeders that the 1968 selections, which had explicitly favoured the short-internode branching habit, would preclude the option of obtaining good lengths of clear timber between knot clusters in unpruned logs. So in 1970 and 1973, a total of 140 plus trees was selected for a long-internode branching habit from within the same base population, and OP progeny trials were similarly established from them. In effect, a separate breed was thereby created.

Fig. 5.17 Cyclaneusma needle cast affecting several trees in the left foreground in a young radiata stand near Rotorua, New Zealand. This condition can severely affect growth rate, and resistance is now routinely sought in selection in New Zealand

Further controlled crosses were made around 1980, choosing parents on the basis of accumulated progeny-test information. Those crosses involved several series of select parents, and progeny trials were duly established.

In addition, several other sets of field selections were made for different special purposes (Sect. 5.5.2), but the programmes were later dropped because the effort and resources needed later proved on various counts to be unwarranted.

Assessments continued on other trials that had been established before 1969, yielding mainly research information on quantitative inheritance of a wide range of traits (e.g. Burdon 1992). That included encouraging evidence of worthwhile heritable variation in resistance to *Dothistroma* (Wilcox 1983).

5.5.1.3 Chile

The ill-fated breeding programme of Instituto Forestal (Sect. 5.3.4) involved selection in the early 1970s of some 200 plus trees from land-race stands over a wide area. This followed much the same protocols as followed in New Zealand in 1968, opting for fast-growing, straight-boled, small-branched trees with a short-internode branching habit. Clones were archived, and Instituto Forestal collaborated with one company that wanted to establish its own seed orchard. Unfortunately the practices that worked with initial archiving failed in trying to establish the orchard, so the grafts almost all perished. The OP seeds collected from those trees were stored for a while but, in the confusion of the time, were then all mixed together for sowing in a nursery.

Soon after the establishment of the breeding Convenio in 1976, eight member-company seed-orchard programmes were launched, with some others soon to follow.

In the Convenio, each member company was planned to establish its own clonal seed orchard, each orchard containing 40–44 plus-tree clones. Those clones were selected independently, where possible from each company's own local plantations, but on essentially the same criteria as the Instituto Forestal's previous selections (and the 1968 New Zealand selections). Importantly, much was learned from New Zealand's less fortunate experiences of orchard siting, with similar siting mistakes thus avoided in Chile.

A disadvantage, or at least difference from developments in New Zealand, of having separate independent company programmes, was that management of radiata's genetic material was not integrated across the country's radiata estate. This persisted despite counsel (Burdon 1978) that research be done on how well plus-tree progenies of each company performed on other companies' lands compared with on its own lands. But Zobel, who had originally been consulted, may have feared that most or all companies could be tempted to concentrate on a very few top-ranking parent clones at the cost of reducing genetic variation that would be needed for the future; anyway, he advised the Chileans to keep the companies' breeding programs and individual base-populations separate.

The Chilean focus on only the short-internode branching habit, while arguable, was defensible, since there was then no established market for clear sawn timbers in or from Chile. However, the long-internode trees in Chile often had far fewer tree-form problems than those on New Zealand's more fertile sites.

5.5.1.4 South Africa

Although South Africa's radiata breeding programme remained in place after 1969, based at Saasveld, the levels of financial and breeder commitments to it were limited. This reflected the fact that radiata was only one of five important pine species in South Africa's production-forestry sector, and it was grown over much smaller areas than in New Zealand, Chile or Australia. Moreover, interest was starting to increase in genetic improvement of eucalypts, with more financing and staff allocated to the eucalypt programmes. While South Africa followed New Zealand and Australia after 1968 in using many more clones in new seed-orchard plantings (van Wyk and Roeder 1978), no development of central importance to radiata's domestication occurred there between 1969 and 1983.

5.5.2 Structuring and Reshaping of Breeding Programmes

Radiata breeding programmes, in the countries where it was grown widely, tended to be divided into geographic units. With time, however, some geographic divisions were revised, while specialization according the breeding objectives began.

In Australia, the politically separate radiata breeding programmes of the various states imposed some degree of regionalization, however fortuitous or biologically imperfect that diversification might be in terms of rational and efficient domestication. On the other hand, the more lucrative markets for radiata logs in Australia remained quite narrowly based, being very predominantly used for light structural sawn timber. While several abundant eucalypt species (Eldridge et al. 1993) provided sawn timbers with strength and durability that excelled radiata timber, those timbers posed many more problems with sawing and drying than softwood timbers do. Those considerations contributed strongly to the national policy decision to greatly expand the softwood timber resource.

In South Africa, growing radiata was confined to more fertile sites in a restricted climatic zone in southern Cape Province. As in Australia, the more lucrative markets for radiata included light structural timbers for which it was the preferred pine species.

In Chile, the organization of radiata breeding within the Convenio involved company-programme independence, which led to a high degree of regionalization. Nevertheless, all those different radiata breeders employed similar foci on selection for traits important for light structural timbers despite the heavy emphasis on pulp production.

In New Zealand, the wide geographic scope of FRI's single radiata breeding programme had initially prompted internal regionalization of plus-tree selection and clonal composition of seed orchards. However, by 1983 this particular regionalisation appeared to be groundless for radiata. Data from plus-tree progeny tests replicated in very different parts of the country gave similar rankings for the same sets of progenies with respect to early growth and bole form. In other words, there was limited genotype-by-site interaction in those few traits and, as of 1983, separate breeding programmes and seed orchards were thus deemed "not needed" for either the North or South Islands.

On the other hand, New Zealand tree breeders had been uneasy about the exclusive focus on a "short-internode" branching habit, hence the decision to mount a smaller "side programme" based on selecting long-internode plus trees (Sect. 5.5.1.2). True, the high heritability of branching pattern, and favourable genetic correlations of more and shorter with growth rate and general tree form, had led to selecting for that branching pattern as a means of not only controlling branch size but also of improving other aspects of bole form (in a classic case of **indirect selection**$^{\mathbf{G}}$). Yet given an established market for **clearcuttings**$^{\mathbf{G}}$ between defects in sawn timber, a continuing supply of long-internode logs produced without pruning was attractive. Long-internode genotypes would serve the option of being able to obtain useful knot-free lengths cut out from between clusters of knots in unpruned logs. Field observations of the long-internode trees that occurred in most plantations indicated that there were some sites, usually with lower fertility, where the long-internode trees would not incur serious bole-form problems.

Accordingly, in 1970 and again in1973, plus trees were selected specifically for a long-internode branching pattern, in addition to good growth, bole form and

Fig. 5.18 Radiata trees showing contrasting branching habits: "short-internode" (left) and "long-internode" (right). The former was pursued automatically in the mainstream breeding programme in New Zealand, despite giving only short lengths of clear timber between knot clusters, while the latter was pursued in a side programme to provide good lengths of clear timber without pruning, despite disadvantages in early growth and tree form (From Burdon 2004, permission Elsevier)

general health. Of the 140 such selections, 104 were selected in 1970 in northern Kaingaroa Forest, in the same general areas where the 1968 plus-tree selections favouring short-internode trees had been made. Fewer long-internode than short-internode plus trees could be found per unit area because of a combination of two factors: trees of the appropriate branching pattern were less common, and among them fewer had both high vigour and superior bole form. The remaining 37 such selections came from individual long-internode offspring within controlled crosses among plus trees selected in the 1950s. The long-internode selections and their offspring came to be recognized as a **Long Internode (LI) breed**, with a differentiated breeding goal that was primarily defined by the long-internode trait (Figs. 5.18 and 5.19). Clonal archives of these LI plus trees, and an LI seed-orchard and LI progeny tests were established. Strangely, no demand for seed from that orchard ever materialised.

The first-generation plus tress and their offspring in the mainstream (short-internode) FRI breeding programme then came to be designated the **Growth and Form (GF)** breed. A desire to acquire greater resistance to dothistroma (red-band) needle blight led to creation of a **Dothistroma Resistant (DR)** breed. That was

Fig. 5.19 Open pollinated offspring of long-internode radiata trees in a 15-year-old field progeny trial on a New Zealand site of moderate growth rate

first developed as an offshoot of the GF breed although, in the FRI programme, dothistroma resistance later became part of its mainstream breeding goal rather than the basis of a specialized breed. Some selections for dothistroma resistance, imported from Kenya, did prove to have superior resistance, but had not been selected for bole form and were therefore not integrated into the FRI breeding programme.

Two other specialized selection programmes were embarked upon experimentally in New Zealand, but both were abandoned, largely because the needs served later proved to be insufficient to justify the effort.

In New Zealand, some experimental selections were made in 1969 and 1970 for resistance to shoot dieback caused by *Diplodia pinea* (syn. *Sphaeropsis sapinea*). Despite indications of success (Burdon et al. 1982), there were no significant recurrence of the disease outbreak that prompted the selection for resistance (Burdon 2011), so the programme lapsed.

Provisional selections for frost resistance were made in 1973 in New Zealand, following a severe unseasonal frost. But, despite supporting evidence of genetic variation in New Zealand trials (Menzies et al. 1987), and strong evidence of heritability in California trials (Hood and Libby 1980), silvicultural research proved that the New Zealand frost problem could be overcome just by good nursery and establishment practices. In northern California, plantings of radiata on cold sites were abandoned due to the western gall rust epidemic raging in earlier radiata plantations there.

In Queensland, Australia, where hailstone damage to radiata provided entrance wounds for diplodia-associated dieback, the forest service had before 1950 selected for resistance to the disease, but this programme also lapsed, largely because of a combination of limited response to selection and the modest scale on which radiata was being grown in that state (Wu et al. 2007).

5.5.3 Tree-Breeding Strategy and Quantitative Methodology

The experience and theory concerning need for greater genetic diversity in the breeding lines that led to largely re-starting the New Zealand breeding programme in 1968 was a key prompt for forest tree breeding worldwide, with many species, to adopt a more strategic approach to forest tree breeding. Given the long time-frames involved, it was all too easy for tree breeders to find themselves "painted into a corner" through making strategic mistakes, having to essentially start afresh after many years, in order to assure good continued progress.

While early delivery of usefully-improved planting stock was needed in order to satisfy the accountants as well as foresters' demands for improved planting stock, thoughtful tree breeders were by the late 1960s increasingly considering how to provide for continued genetic improvement in the long term, and how to prepare for enforced changes in breeding goals. Changes in goals could come from various causes, such as political or managerial fiats, different markets, new wood-processing technologies, better financial or genetic information, or crises resulting from new and serious pests or diseases. For the short term, using the best available information on heritabilities and other genetic parameters, breeders needed to predict magnitudes and net values of genetic gains under alternative mating designs, progeny testing systems and selection scenarios. The need for such projections was driven not only by breeder curiosity and for better fiscal efficiency, but to satisfy people in both industry and government that they would get value for money spent on tree breeding. As was documented by Shelbourne (1969), such projections followed on from the theoretical quantitative genetics work of Namkoong and his American colleagues (Namkoong et al. 1966). For the longer term, early choices of mating design and overall population structure were typically crucial to laying the foundations for future genetic changes in the service of effective domestication. Choice among mating designs was complicated by the fact that no one design can efficiently serve all the various purposes of these designs (Burdon and Shelbourne 1971; Zobel and Talbert 1984). This led to the suggestion that a combination of mating designs may be best despite additional costs (Burdon and Shelbourne 1971).

By the early 1960s, North Carolina State University (NC State) at Raleigh had become internationally recognised as a powerhouse for both quantitative genetics and experimental statistics. Zobel's industry-university-government tree breeding cooperative was then well-established on the Raleigh campus as a model operational programme with loblolly pine, and it served as a dominant training ground for both students and experienced tree breeders from around the world.

Several mating designs had been developed in service of the ongoing domestication of corn (maize) or chickens. The so-called **NC-1** was a hierarchical design for corn, which like pines is monoecious, bearing separate male and female organs on the same plant. NC-1 nests groups of three or four (preferably random and unselected) plants serving as "females," the plants in each group mated to a single different "male" from an independent set of plants. This produced groups of full-sibs with half-sib relationships through the males in common, useful for estimating additive genetic, some non-additive genetic, and environmental components of variation, which were in turn useful for estimating narrow-sense (additive) heritabilities and parts of broad-sense (total) heritabilities.

The so-called **diallel** design was also available, in which small groups of (usually 8–10 random) plants are all mated to each other serving as both males and females, including selfs. From this theoretically elegant design, narrow-sense heritabilities, inbreeding effects, maternal effects and (if the sibs are cloned) broad-sense heritabilites can be accurately estimated, although for groups containing few parent plants. It was a difficult design to complete, and few forest-tree programmes adopted it. (The Norway spruce programme at Aas, Norway, managed to complete and analyse three of them.) For a decade from 1976, under the auspices of CSIRO and STBA, 21 sets of 6 × 6 half-diallel crosses were made among radiata selections in Australia by RWG 1 members to provide second-generation selections for seed orchards and information to assist strategy development.)

The so-called **NC-2** design, an elongated factorial, employed very few (usually only four or five) tester "males," each mated to all candidate plants as "females." It achieved, through subsequent progeny testing, a fair and accurate evaluation of those candidate plants as parents for backwards selection. But it is terrible for forwards selection, because half of the next generations' genes come from those four or five sexy tester "males." Both the NC-1 and NC-2 designs were initially adopted by Zobel's programme for loblolly pine, and by many or most of the students trained in it. Both designs worked well in the first generation of domestication of various tree species, but both were inadequate for forwards selection in subsequent generations. In hindsight, it was adapted to the fact that a rooster could sire vast numbers of offspring compared with the finite numbers of eggs that individual hens could lay. It was certainly not designed to capitalise on the fact that a monoecious conifer could, especially after clonal propagation, produce huge numbers of seeds. While controlled crossing of seed parents with single pollen parents was possible, wind pollination meant that seeds of a particular parent could be sired by a virtually random collection of pollen parents from a large population—a phenomenon of great importance and convenience for the breeder.

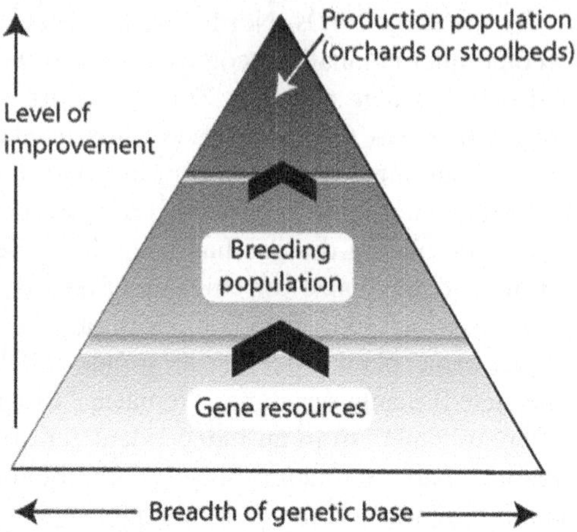

Fig. 5.20 Schematic diagram illustrating the trade-off between level of genetic improvement and genetic diversity in a population "hierarchy."

Later, a **single-pair** mating design was developed that anticipates forwards selection through progeny trials. It mates half of the selected plus trees as "females," each with a different plus tree of the other selected half as a "male." It thereby optimises genetic diversity and value for selection among the offspring to produce the following generation and provide a genetically broad pedigreed base for later generations (Libby 1973).

During a year-long 1971 sabbatical visit by Libby to New Zealand, the concept of organizing genetic material into a **hierarchy of populations** was considered, much discussed, and crystallized (Libby 1973). Basically it accommodated the trade-off between large single-generation genetic change achieved through stringent selection of only a few top-ranked trees to be parents of the entire next generation, on the one hand, and genetic diversity that is left largely intact by maintaining much larger breeding populations with lower selection intensity per generation, on the other. The trade-off is illustrated schematically in Fig. 5.20.

At the top of the proposed hierarchy is the **production population**, typically stringently-selected and then cloned plus trees (which give rise to **deployment populations**). Relatively few stringently-selected tens of clones may serve as replicated parents in seed orchards. The same clones, or perhaps their stringently-selected and cloned next-generation offspring, may be deployed either generally, or (usually later following testing) specifically to different kinds of sites as well-characterised clones.

Underpinning that is the **breeding population**, which may contain a few hundreds of less-intensely-selected individual trees or clones; in it, recurrent cycles of selection and intermating produce cumulative genetic changes over generations.

At the bottom of such a population hierarchy are the **gene resources**, thousands of largely unimproved trees containing almost all of the species' genetic diversity. That is the repository of the programme's foundation genetic diversity, which may be conserved *in situ* as natural populations, in dedicated plantations, or in archives that may be augmented over time by additions from native populations, land races and other breeding programmes not previously sampled and included.

With time, an inclusive concept of a breeding strategy developed. Embracing all aspects of breeding, it involved devising a set of measures that are designed, in conjunction, to ensure near-optimal outcomes, in both the short and long terms, despite the many biological and market uncertainties (cf Shelbourne et al. 1986). As thus conceived, no one measure or feature is to be considered in isolation from the rest of the total "package." For any one aspect of a strategy various options may exist, but the option chosen may dictate the nature of one or more other features of the strategy, so that it is hopefully fail-safe.

Implementation of tree-breeding strategy is much helped by refining methodologies for estimating genetic parameters and for evaluating selection candidates. Both parameter estimation and candidate selection involved choices of trial sites, and among various field designs and methods for data analyses and then actual selection. Experimental designs were refined to achieve greater operational simplicity yet more precise information, while making allowance for limitations in both computers and data-analysis programmes.

Incomplete-block field designs had been used for large progeny tests in Australia from the 1950s (experience and developments in trial design and data analysis were collated by Williams and Matheson 1994). In sharp contrast, for large progeny trials in New Zealand, the sets-in-replicates (set-in-reps) field design (Schutz and Cockerham 1966) was adapted for testing the progeny of the 588 plus-trees selected in 1968. In that case, the progenies were grouped into essentially random sets of about 30 OP families, to create manageable subunits in the hope (generally realized) that differences in average genetic values among sets would be negligible. A further move was made to non-contiguous plots (Libby and Cockerham 1980). However, assigning one individual per family (or clone) per rep/set block within the sets-in-reps field design allowed smaller experimental subunits (blocks) and more-precise comparisons among selection candidates. That also made large datasets easier to analyse, at least with the then-existing computing facilities. It became common to include in progeny trials or clonal trials some of material that represented alternative seed sources, such as standard mixes of OP families or well-characterised "standard" clones, as repeatable "controls" that were then compared with the candidate progenies or clones both within the trials and among different trials.

With a very large breeding population's progeny in trials, promising candidate trees were seldom side-by-side for the tree breeder to compare them simultaneously. Furthermore, selection needed to be done for several traits for each candidate tree, with traits such as wood properties not readily observable in the field. Initial selections therefore needed to be based on data records. Adapting developments in animal breeding, Mike Wilcox in New Zealand pioneered the use of index selection in forest trees during the mid-1970s. However, the University of California at Davis' professor of plant breeding, Bob Allard, emphasized to his students that final selection should always be done by experienced practitioners observing the initially-selected candidates in the field. Among other things, those practitioners could see and then include in their final selection decisions unusually favourable expression of traits that had not been included in the computer's selection index (Allard 1960). Such field checks of computer-selected trees were duly made with radiata. They often eliminated individuals or clones with drawbacks that were not included in the index calculation.

Genotype-by-site interactions, and the possible need to create regional subdivisions of breeding programmes, remained a concern at FRI. It was addressed by Shelbourne (1972) and by the methodology developed by Burdon (1977, 1979) in the 1970s for both analyzing data and accommodating interactions in selection. In New Zealand, interactions involving changes in ranking of candidates for individual traits remained surprisingly minor. An exception was in performance on phosphorus-deficient clays compared with elsewhere, but even those interactions did not necessarily call for regionalized breeding (Johnson and Burdon 1990). In Australia, however, important genotype-by-site interactions proved prevalent, but without obvious patterns with respect to widely diverse climates and soil (Matheson and Raymond 1984), although the effects high fertility of some ex-pasture sites on tree form were clearly a source of difficulty for deployment decisions (Horne 1988; Lavery 1988).

In developing breeding strategy for radiata, continuing interactions with overseas personnel was important. Prominent among the influences and collaborators in Australia and New Zealand were Libby of University of California Berkeley, Namkoong of the US Forest Service who was stationed at North Carolina State University, and Hans van Buijtenen of Texas A and M University. In Chile, Zobel played a key initial role, followed by input from Burdon.

5.5.4 Quantitative Genetic Architecture

Intensive breeding of radiata had begun with very little definite knowledge of either the patterns of variation or of the inheritance of important traits among and within its populations. The first systematic comparison of radiata from the entire natural range of the species was the "Genetic Survey" experiment initiated by Martin Bannister and followed through observations and analyses by Burdon (Burdon 1992; Burdon et al. 1992a). That experiment, however, both depended and built on a study by Margot Forde during 1959–1961 of *in-situ* variation in the three Californian populations (Forde 1964). For that study, she had made essentially random cone collections from ecologically representative sites, and those cones supplied the seed for the mainland population samples in the Genetic Survey. Visits by Bannister and Libby in 1964 to the other two native populations, on the islands, completed cone collection from natural populations for the Genetic Survey, supplementing an earlier collection by Dr Reid Moran on Guadalupe Island and collecting afresh on Cedros Island (Libby et al. 1968). For experimental controls, cones of near-random tree samples were taken from two regional

New Zealand land-race populations; one was near Nelson in northern South Island, and the other in Kaingaroa Forest in central North Island.

Planting was done during 1964–1967 in Kaingaroa Forest, New Zealand, on two contrasting sites, one being markedly warmer, somewhat more fertile and less exposed than the other. The basic structure was 50 OP progenies of each of the five natural populations and two New Zealand ones, with some variations in order to accommodate delays in acquiring some seed. A clonally replicated component was added to the 1967 plantings for providing additional genetic information. The experiment was supplemented by three small provenance trials elsewhere in New Zealand using spare planting stock of the same natural populations.

The field design contained a pioneering adaptation of a field design (Libby and Cockerham 1980) intended to allow a series of two systematic 50% thinnings, while leaving equal numbers of each of the 350 progenies).

The FRI Genetic Survey experiment, however, had been preceded by two earlier radiata provenance trials and stimulated a third. One, near Canberra in Australia and planted mainly in 1950, compared samples of the three Californian mainland populations (Fielding 1961a). It surprisingly showed the Cambria population to be the least adapted to that part of Australia, which had seemed more similar to the Cambria environment than to that of the two more-northern populations. The second, established in 1955 on two New Zealand sites (Shelbourne et al. 1979), included only two bulk seedlots from Monterey, one such seedlot from Cambria, and two New Zealand single-parent OP progenies. Limited though it was, it provided a helpful complement to the Genetic Survey experiment.

Following on Forde's study and with her help, cuttings were collected in 1962 from 12 naturally-established randomly-located seedlings in each of 12 ecologically-contrasting sites in each of the three mainland populations. These cuttings were subsequently rooted (Libby and Conkle 1966), with four per clone used to establish two genetic-conservation *cum* breeding archives near Berkeley in 1963, whence a fresh cycle of cuttings were planted in northern California in a clonal genetic-architecture trial in 1970. The archive at the University of California's Russell Reserve, in Lafayette, CA, still existed in 2014, but the other was removed to make way for corn-genetics experiments in 2010. Following damage from a hard freeze (Hood and Libby 1980), a washed-out bridge, an outbreak of dothistroma needle blight and then a western gall rust epidemic on radiata in California, the cloned genetic-architecture trial was abandoned after assessments up till 1978 (Guinon et al. 1982).

5.5.4.1 Population Comparisons and Action Taken

Differences among the population samples in the Genetic Survey plantings became evident from very early on (Burdon and Bannister 1973; Burdon 1992; Burdon et al. 1992b) (Table 5.4), reinforcing some of the earlier findings in Australia by Fielding

Table 5.4 Comparisons among native provenances for selected traits (After Burdon 2001)

Trait/Date of information	Año Nuevo	Monterey	Cambria	Guadalupe	Cedros
Growth potential	+	+	+	–	– –
Ease of transplanting	+	◦	–	+(+)	– –
Stem straightness	–	–	◦	++	◦
Corewood density	–	–	–(–)	++	++
Resistance to/tolerance of					
Abiotic factors					
Frost	++	+	–	◦*	– –
B deficiency	+	+	+	– –	– –
P deficiency (1997)	–	++	++	•	•
Soil salinity (1982)	◦	+	++	– –	–
Drought (1995)	– –	–	◦?	+	++
Snow damage (1994)	++	(–)	– –	•	•
Fungal diseases					
Dothistroma blight	++	++	– –	◦	– –
Cyclaneusma needle cast	+	++	– –	◦	– –
Diplodia dieback	++	++	– –	– –	•
Phytophthora root (1984)	– –	+	++	•	•
Western gall rust (1986)	◦	– –	– –	++	+
Animal pests					
Deer/rabbit browse	◦	◦	◦	–	+
Pine woolly aphid	+	+	–	– –	+

*Resistance to winter cold evidently superior

Pluses denote superiority relative to average, minuses inferiority, ◦ roughly average, • no convincing data. Entries based on first substantial published report, in 1973, unless later date stated

(1961a, b). Young seedlings and (later) saplings from the two island populations proved to be distinctive from each other in both appearance and very early growth rates, being similar mostly in having their mature needles typically in pairs rather than threes. Evidently lacking a history of adaptation to frequent light fires (Stephens and Libby 2006), island-population seedlings take much longer to develop the thick, fissured, fire-resistant bark on the lower bole characteristic of mainland, and particularly Monterey, seedling-origin trees; in this and some other respects, adolescent seedlings from the islands resemble grafts or cuttings derived from mature trees of the mainland and land-race populations. The Guadalupe samples were appreciably slower growing than those from the mainland populations, due in part, but only in part (S. Kennedy and C.B. Low unpubl.), to those open-pollinated families having greater levels of inbreeding (Burdon and Bannister 1973; Moran et al. 1988a); their greater inbreeding was doubtless as a result of few nearby pollen parents in Guadalupe's linear relict native population, with those few neighbours mostly living near the ridgeline at right angles to the prevailing wind. Seedlings from Cedros Island, despite less inbreeding, were initially by far the slowest-growing of the five native-population samples. Interestingly, these growth differences were not exhibited in early inter-population hybrids between island- and mainland-origin parents. Browsing animals in New Zealand, while preferring Guadalupe seedlings, tended to shun Cedros seedlings, a contrast in palatability that fitted with the biotic situations on the two islands—Cedros Island has long had a small deer species, in contrast to the historic lack of browsing mammals on Guadalupe Island. Seedlings from both island populations, however, were more prone to suffering from boron deficiency than those from the three mainland populations.

Differences in growth potential among the mainland populations (reflected in growth rates without disease or adverse soil conditions) were not evident. Many other trait differences among mainland populations are less compared with corresponding differences between mainland and island populations and between the two island populations. The mainland populations samples, however, differed markedly in average cone size and seed weight, Cambria having the largest and Monterey the smallest. There were subtle differences in several elements of young trees' appearance, allowing an experienced observer to assign almost every individual to a population on appearance alone. Cambria, however, is the most distinctive mainland populations at various ages, and differences in appearance between Año Nuevo and Monterey were the least obvious. In New Zealand trials, Cambria trees have been much more susceptible to needle-cast diseases (cf Ades and Simpson 1997) and to shoot dieback attributed to *D. pinea*.

An interesting feature is variation among provenances in persistence of juvenile characteristics, which was first expressed clearly in how quickly green apical tufts gave way to brown, sealed buds. Among the mainland provenances, Cambria-origin trees maintain juvenility the longest, followed by Monterey and then Año Nuevo, with Guadalupe showing the fastest acquisition of adult characteristics. Cedros was similar to Año Nuevo in the rate of this phase change, although slower growth of Cedros made rigorous comparison difficult. These provenance differences in rate of phase change, however, have implications for resistance to diseases that tend to affect trees mainly in either the juvenile or the adult phase. This was borne out in provenance comparisons of resistance to western gall rust, which affects juvenile radiata, provenance differences in susceptibility reflecting comparative persistence of juvenility (Old et al. 1986).

The cloned genetic-architecture trial in northern California revealed differences among mainland populations in frost tolerance, with Año Nuevo the most tolerant, and in susceptibilities to animal damage (Hood and Libby 1980). The among-population differences in growth potential were negligible, although the Cambria sample became less thrifty with time (Guinon et al. 1982). These results roughly paralleled those from the New Zealand Genetic Survey experiment.

All populations, however, showed strong tree-to-tree variability in traits such as cone size and shape, and branching pattern, the comparative importance of the genetic and environmental components of tree-to-tree variation depending on the trait. Thus, despite some substantial differences among population averages, there was much overlap in variation for most observed traits. All told, the patterns of variation observed among the populations were complex and multidimensional, such that the pattern of resemblances between pairs of populations differed according to the trait(s) considered.

By 1983 it had become clear that not only were all five native populations members of the radiata species, but that each is different from the other four in many distinguishing and important traits (Table 5.4). This in turn supported the theory that radiata had long existed as a number of small migrating populations, rather than the current populations being survivors of a much larger continuous population (Millar 1999). This had implications as well for identifying the contributions of individual native populations in founding the land races developing in their new host countries. It had also became increasingly clear that the two sampled and analysed New Zealand land-races in the Genetic Survey experiment New Zealand most resembled the Año Nuevo population sample (in the Kaingaroa land race), or some combination of parents from Año Nuevo and Monterey (in the Nelson land race). Those two land-race samples were also proving to be clearly the best-adapted, in terms of overall survival, early growth and health, to the specific environments of the several provenance and genetic-architecture trials in New Zealand (NZ) and Australia. They were followed by the Año Nuevo and Monterey samples.

Independent evidence, from turpentine analyses, also indicated that the Kaingaroa land-race sample (from central North Island, NZ) had originated predominantly from Año Nuevo parents (Bannister 1962); and further, that this sample was broadly representative of radiata in the whole Kaingaroa Forest (Blight et al. 1964). Thus the Año Nuevo population was heavily over-represented in the Kaingaroa land-race ancestry, especially in relation to the relative size of the area that population occupies in California. The Nelson land-race sample (from northern South Island, NZ) later appeared to have a greater proportion of Monterey-origin ancestry, close to 50% (Burdon 1992). In Australia, land-race samples were deemed, on the basis of isozyme results (Moran and Bell 1987) to have come from similar ancestry to that of the NZ samples; but evidence of the proportions from Monterey and Año Nuevo was weak. There was no evidence, historical, circumstantial, biochemical or observational, of any contributions to any NZ or Australian land races from the Cambria or two island populations.

Notably, the two land-race samples all grew appreciably faster, by ca 10% in average height, in the Genetic Survey experiment than any of the five native-population samples. This growth difference was probably attributable, at least in part, to the large plantations from which the land races were sampled not being subject to "neighbourhood inbreeding." Such inbreeding no doubt occurred to some degree in the natural stands, due to their population structures whereby near neighbours tended to be relatives. Also, the two New Zealand land-race samples exhibited substantially better bole form than did the samples of their two native progenitor populations. That commercially-important difference is likely due to their divergent histories, since the mid-19th century, of mildly eugenic silviculture in the evolution of New Zealand land races and dysgenic selective logging in the mainland native populations, particularly in the Monterey Peninsula part of the Monterey population. Interestingly, the Guadalupe population, with no historic or current evidence of human management, exhibited straighter stems than did the two land-race samples in those trials.

The pre-1983 provenance comparisons in New Zealand were mainly confined to one region, namely the volcanic soils of the Kaingaroa Forest in central North Island. However, there was some additional evidence (Shelbourne et al. 1979) that suggested that infusion of genes from additional Monterey-population parents into future breeds or clones could help confer better adaptation to some other New Zealand environments, or even to future circumstances arising in New Zealand.

In the light of the latest information then available, an expedition was mounted in 1978 to collect additional OP seeds, afresh and systematically, from better-than-average parents in all the five native populations (Eldridge 1978). Various factors drove the decision. It was by then clear that at least some land-race stocks in Australia and New Zealand had been founded by seeds acquired from a only small portion of the species' natural range. Even from the identified two founding populations, their contributed genetic base was probably unbalanced geographically; furthermore, the numbers of their founder parents were unknown and possibly small. There is some circumstantial historic evidence that many or even most of the seeds shipped to New Zealand to create the early radiata plantings were gathered from trees felled to supply fuel for rendering whale oil by shore-whalers, known to then be operating from the coasts of both the Monterey and Año Nuevo populations. Seeds from parents that might have been more tolerant of the soil conditions, higher and/or lower temperatures, less fog, and other features of sites further inland, could well have been under-represented or even unrepresented in founding the existing land races. Nowhere in New Zealand, Australia, Chile, South Africa, California or any other country had the various provenances, particularly samples of their sub-populations, been tested on the wide range of potential radiata sites that would clearly reveal genotypes with such special adaptations.

There was an option of making controlled crosses among parents of known origins in the then-available provenance trials and archives in New Zealand and Australia, in order to acquire pedigreed pure-population seeds for additional trials and to maintain those native populations *ex situ* for future breeding. That would have assured future pedigree-known pure-population status, in contrast to the seeds compromised by pollen contamination from the widespread radiata plantations if OP families from existing and future provenance trials and archives were used instead. Moreover, in the provenance collections for the Genetic Survey experiment, cones had been carefully collected from essentially random parents, and a new collection could concentrate on parents with better phenotypes. Furthermore, it was decided by the leadership in Australia that a new collecting expedition could expand radiata's genetic resources available for breeding and additional trials, making good the incomplete and patchy representation of the natural range of the species in land-race stocks and offering possible advantages in site tolerances, bole form, disease resistance and wood density. In addition, during the collecting, members of the expedition could observe and learn more about the different ecological characteristics of the native stands being sampled. Meanwhile, the Guadalupe population was under imminent threat of extinction, due to the hungry goats, while much attrition was occurring in New Zealand's *ex-situ* Cedros plantings in trials and archives.

The new 1978 collection was planned to favour parent trees, in terms of size, health and bole form, so few of them if any would have been included in the earlier quasi-random collection, except on Guadalupe Island.

A major event for radiata's future domestication, the collecting expedition was not a budgeted part of operational breeding in the 1970s. Instead, it was done by a joint Australian–New Zealand expedition, operating with local assistance in California and Mexico, and additional funding from the US National Science Foundation, FAO and several California Christmas-tree growers. The California growers financially supported and partly staffed the expedition's collecting trips to the two Mexican islands, as then-ongoing research (Old et al. 1986) was showing the island populations to be much more resistant to the western-gall-rust epidemic then devastating the extensive radiata Christmas-tree farms in California.

With few exceptions, large numbers of seeds were successfully collected from each of around 520 trees in the mainland populations and in lesser quantities from 99 trees in the island populations (Eldridge pers. comm. to WJL 1997). Those seeds were shared mainly by Australia, New Zealand and, in the case of the island collections, Mexico, with small amounts going to other countries, including Chile (Jayawickrama and Balocchi 1993), South Africa (Falkenhagen 1991) and Greece (Matziris 1995). In both New Zealand and Australia large networks of provenance trials were quickly established, especially of mainland material (Eldridge pers. comm. to WJL 1997), in order to obtain comprehensive profiles of the populations' and sub-populations' site tolerances. In addition, substantial gene-resource plantings were made in both New Zealand and Australia.

5.5.4.2 Within-Population Variation

Observed variation in putatively adaptive traits among the samples from different sites with contrasting ecological conditions within each natural population was, as of 1983, generally minor in the various provenance-trial environments in New Zealand and Australia. That was particularly surprising in the case of the samples from the large Monterey population, where there are major among-site differences in soil type, elevation, fog-days, maximum and minimum temperatures, aspect, and distance from salt-laden winds. Accepting this early evidence, the FRI breeders' main interest continued to centre upon general tree-to-tree variation in several growth-and-form traits, their coefficients of variation (standard deviations divided by means), narrow-sense (additive) heritabilities, and genetic correlations among traits.

Analyses of data, by trait and by population sample, from the two large Genetic Survey trials provided support for the following findings from small earlier trials, namely that traits of interest differ widely in both coefficients of variation (CVs) and narrow-sense heritabilities (Burdon 1992) (Table 5.5). Growth traits, especially

Table 5.5 General importance and indicative information on variability and inheritance of selected individual traits in radiata, as basically known by 1983 but often confirmed later (Mainly after Burdon 2001)

Trait	Economic importance	Status[a]	Coefficient of variation (%)[b]	Heritability	Importance of[c]	
					Non-additive gene effects	Genotype-site interaction[d]
Height growth	High	No	12	~0.2	••	••
Diameter at breast height	Very high	Sel	15	~0.2	•••	•••
Stem volume	Very high	BG	30	~0.2	•••	•••
Straightness	High	BG	–	0.4	••	••
Branching frequency	High indirect	Sel	20–30	0.45	◦	•
Branch size	High	Inc	–	0.2	•	•
Branch angle	High	Inc	–	<0.2	•	•
Wood density	Can be high	Sp				
Cores/discs	"	Sp	7–8	≥0.6	◦	◦
Indirect	"	Sel	–	0.3–0.4	◦	◦
Fibre length	Now low	No	10	>0.5	◦?	◦?
Grain spirality	Significant	Sp	>50	>0.5	?	•••
Cone dimensions	None	No	12–18	>0.5	◦?	•
Cone volume/seed weight	None	No	20–35	>0.5	◦?	•
Resistance to						
Dothistroma	Locally high	Sp	–	0.3	◦	◦
Cyclaneusma	Variable	Sp	–	0.1–0.35	◦	◦
Phytophthora	Locally high	Sp	–	0.8	◦?	◦?
Frost	Limited	No	–	<0.2	?	?
Wind damage	Variable	No	–	0.05	?	?
Leader dieback	Variable	Inc	–	0.05	?	?

[a]BG denotes key breeding-goal trait; Sel – Key selection trait for indirect breeding-goal selection; Inc – often addressed incidentally by selection trait(s); - Sp – addressed in specific situations; No – not generally addressed in breeding
[b]Tree-to-tree, if there is a meaningful measure
[c]◦Denotes almost always unimportant, to ••• denoting can be very important
[d]Involving rank changes among environments

bole diameters at standard "breast height" (1.4 m), tended to show high CVs (often 15–20%) but fairly low heritabilities (typically ~0.2). In contrast, wood density showed a much lower CV (~8%) but much higher heritabilities (~0.5–0.7). Internode length (or its approximate inverse, frequency of branch clusters) proved not only unusually variable (CV ~20–30%), but also strongly heritable (~0.5) if measured accurately. Stem straightness, with indeterminate CV, showed lowish observed heritability (0.1–0.3) partly because of the need for visual ratings. Heritability of forking, which can be highly dependent on presence or absence of leader-damaging agents such as brush-tailed possums or diplodia dieback, was estimated as very low (0.05–0.1). These different estimates were generally consistent for each trait analysed across all native population samples, land-race samples, and even across estimates from hierarchies of plus-tree offspring (except where intensive selection had already occurred for branch-cluster frequency). As well as exhibiting low heritabilities, even within sites, bole diameter was later shown to show relatively high genotype-by-environment interactions and non-additive gene effects (Sect. 6.7.2.3). Of largely academic and taxonomic interest, cone traits were very variable among trees and evidently highly heritable (Burdon 1992).

Taking account of some additional trials (e.g. Burdon 1971), diameter growth showed important genotype-by-site interactions, especially when the trials were replicated across contrasting phosphate-deficient and nutritionally balanced sites. Some strong genetic intercorrelations were emerging, long "internodes" between branch clusters tending to be associated with larger, steeper-angled branches, more frequent forking and ramicorn branches, and somewhat more bole crookedness. The cloned genetic-architecture study in California (Guinon et al. 1982) showed broad-sense heritabilities there of around 0.15–0.2 for diameter growth and some tree-form traits, but much lower for incidence of forking (~0.05).

For various between-trait genetic correlations, analyses of data from the Genetic Survey generally supported and refined previous weak estimates (Burdon 1992, Table 3). A set of favourable genetic correlations among branch-cluster frequency (viewing high as desirable), early growth rate, bole straightness and freedom from forking were confirmed. So, too, was the adverse genetic correlation between growth rate and wood density. Looking at different growth-rate variables, however, the story is more complicated (Burdon and Low 1992). The genetic correlation between height growth and wood density was less adverse than that between bole diameter and wood density. In practical terms, that means that trees with a tall, slender growth habit tend to have greater wood density than ones with a short, fat growth habit.

Interestingly, genetic intercorrelations among most traits, as with heritabilities, were generally consistent among the different population samples. That suggested that those correlations are at least partly due to pleiotropy, with at least some of the genes affecting both of the genetically-correlated traits. Where pleiotropy generates adverse genetic correlations, breeders are faced with unavoidable constraints on the amount of desired genetic change that is simultaneously obtainable for such adversely-correlated traits. However, all is not lost when two important but adversely-correlated traits are identified. In the usual case, where the genetic correlation is <1.00, breeders can clonally test and select "correlation breakers" which show some departure from the adverse relationship because of pleiotropy between traits being less than 100%. Trees with favourable combinations of genes affecting only one or the other of adversely correlated traits can be identified, accumulated by breeding (more difficult), and clonally deployed (Libby and Rauter 1984).

5.5.5 Breeding and Seed-Orchard Technology

Problems with seed orchards abounded, including delayed graft incompatibility, poor production of **female strobili**[G], abortion of conelets after pollination, and depredations by various animal pests.

For reasons still unknown, incidence of delayed graft incompatibility varied a lot with site. This incompatibility typically killed many grafts about 10 years after planting, when those ramets were just coming into full seed production. This would frequently be preceded by one or two years of super-abundant "stress crops" of both pollen and female strobili, followed by decline and then death of the tree as translocation of water and nutrients was ceasing at the failing graft union. Quests were conducted for rootstocks with better and more-general graft compatibility, with some success with Douglas-fir. But no "universal recipient" rootstock for radiata was found that was graft-compatible with a wide range of parent scions. An alternative solution was to root cuttings that were at a maturation state near or at sexual competence, usefully aided by "ringbarking" twigs on the donor tree (a graft prior to onset of incompatibility) several months before their removal as cuttings. It was a laborious solution. The efficiency of this option was somewhat improved by hedging late-adolescent ramets of desired parents that were in clonal archives, to produce the numbers of late-adolescent or mature cuttings required for planting in new or expanded seed orchards.

The rapid height growth of radiata grafts and cuttings in the seed orchards made managing those parents and collecting their cones difficult (Fig. 5.21). Also, large differences among the parent clones in fecundity, especially in viable conelet production but also in pollen production, were soon evident. This meant that the "effective number" of parent clones producing the wind-pollinated seeds in an

Fig. 5.21 Pollarding trial in Gwavas Seed Orchard, Hawke's Bay, New Zealand. Aimed at preventing grafts becoming unmanageably tall, this practice reduced seed yields unacceptably. Recourse was made to climbing and/or use of hydraulic hoists

orchard, a measure of genetic diversity in the offspring generation, was often far less than the "census number" of different clones in the orchard. Once recognized, this "effective number" could be brought to a desired level by adjusting the ratios of the different clones in the orchard, favouring the less-fecund in order to boost their representation in seed parentage.

Bark beetles, if sufficiently plentiful, could girdle recently planted grafts, although this was a minor and temporary problem. Birds, particularly chaffinches, ate strobili-containing buds. In New Zealand, brush tailed possums would eat pollen catkins, and subsequently ate immature seed cones. And in Australia, cockatoos all too frequently ripped open nearly-mature cones, playfully and annoyingly ate only one seed, and wantonly discarded the rest. In general these problems could be dealt with, but often that took much time and effort.

In New Zealand, these problems with managing seed orchards spurred much research at FRI, led by Geoff Sweet on several fronts. Graft incompatibility was researched, helped by Bob Kellison during a year-long sabbatical visit from North Carolina State, trying remedies used for loblolly pine, without achieving any clear solution. However, enhanced initiation of both male and female strobili proved possible with the application of gibberellins, specifically GA 4/7 (Ross et al. 1984). Although the most responsive clones were the ones that already flowered best, thus maintaining or increasing fecundity differences, this became and has remained operational practice. Various crown-shaping regimes were tried, analogous to pruning fruit trees, in order to achieve less expensive controlled pollination and more convenient cone collection (e.g. Sweet and Krugman 1978). The FRI orchard-pruning regimes were subsequently adopted in Western Australia, under the leadership of Trevor Butcher, and modified as HAPSOs (Hedged Artificially Pollinated Seed Orchards; Wu et al. 2007) (Fig. 5.22).

Fig. 5.22 Hedged Artificially Pollinated Seed Orchard (HAPSO) in Western Australia, Pollen being blown onto receptive conelets with the help of an air compressor. Further control of pollination was achieved by spraying maleic hydrazide to suppress pollen catkin growth on recipient ramets (Photo T. Butcher)

Given the problems with clonal seed orchards, in 1977 the proposal was made by Burdon (unpubl.) in New Zealand to establish ESSOs (Extensive Seedling Seed Orchards). Controlled crosses could be made among a large number, say 50, of top-ranked clones. As needed, the resulting seedlings could be multiplied in nurseries by rooting several juvenile cuttings from each of them. The resulting cuttings could then be planted in areas managed to produce both good quality logs and seed. While expected genetic gain would be less than from a clonal seed orchard, this was seen as a low-cost back-up for the clonal seed orchards. Proof-of-concept plantings were duly made, but they never produced harvested seed, partly because demand for genetically improved seed was being increasingly met by mature-clone orchards, and partly because possums in uncontrolled numbers were allowed to destroy almost all the growing conelets.

Despite the various problems with early clonal seed orchards, they eventually succeeded in providing abundant seed with major genetic improvement in growth and form. Indeed, it was soon after 1983 that New Zealand's seed orchards were able to meet the radiata planting needs for the entire country (Vincent 1986). Not only that, it was then hoped to export considerable quantities of their seed, of which the progeny of at least some of the parent clones were known to perform well in Australia.

Ironically, the findings of much good and painstaking research were overridden by the realization that correct choice of orchard site could avoid several of the main seed-orchard problems. The early preoccupation with pollen isolation (Fig. 5.23) often led to choosing inherently unsuitable sites for seed production, without always achieving effective isolation. The present-day ideal for a New Zealand radiata seed orchard is a sunny, largely frost-free site, with modest but sufficient rainfall and good drainage, located close to the sea. This important latter

Fig. 5.23 View of Tallaganda Seed Orchard, in NSW near Canberra in an isolated montane location surrounded by eucalypt forests. Such sites were used for a number of seed orchards in NSW, ACT and Victoria, but seed yields were poor or modest compared with those on more maritime sites (Pederick and Brown 1976) (Photo AGB)

criterion was suggested by profuse cone production very near the coast in both planted and natural stands in California, and noted in planted coastal radiata pines in New Zealand. It then was observed to be the case in the already established Amberley seed orchard, located 1–2 km inland from New Zealand's east coast, 20–30 m above sea level about 40 km north of Christchurch. There, the ramets did not grow unduly tall, delayed graft incompatibility was not seriously troublesome, conelet production was early and plentiful, clonal variation in fecundity was less, conelet abortion was minor, and the number of filled seeds per cone was high. Indeed, this orchard was soon seen as the benchmark for orchard performance. The Chilean breeding programme, in particular, benefited greatly from knowledge of the site characteristics and orchard performance at Amberley.

5.5.6 Mass Propagation by Vegetative Means

In the early stages of active breeding of radiata, vegetative propagation in the form of grafting had been widely used as a breeding tool to establish clonal archives and thence clonal seed orchards. When graft incompatibility had not been effectively overcome by around 1970, research attention was increasingly refocused on techniques for using sexually-competent cuttings for establishing additional orchard areas (Box 5.3).

Box 5.3 Applications of Vegetative Propagation
Vegetative propagation by grafting was originally used operationally with radiata for establishing clonal archives as a preliminary to establishing seed orchards. Grafts behave like the tops of the mature trees from which they are usually taken, and in radiata are subject to troublesome delayed incompatibility. An alternative, which works very well starting with young seedlings, is to set and root cuttings, but in-vitro methods have since been developed for vegetative propagation of radiata. Vegetative propagation from an individual produces ramets of a clone that are genetically identical to each other, ensuring faithful reproduction of a genotype, but generating no new genetic variation.

Small quantities of seed can be "extended" by vegetative multiplication, typically to produce large numbers of planting stock from pilot quantities of seed from the latest and very best clonal selections. This is widely done operationally, with the help of measures that delay maturation, maturation being a process whereby cuttings become progressively harder to root as the trees become bigger and older.

However, using vegetative propagation on a massive scale, as a tool to deliver genetically improved planting stock to radiata plantations, was first attempted in the early 1960s by J.M. Fielding, trying to build on propagation research in Australia and New Zealand during the 1930s. Importantly, Fielding hoped to achieve thereby a bonus in tree form by avoiding the untidy juvenile growth habit of seedlings. Large-scale propagation of post-juvenile material, however, soon proved impracticable, and mass vegetative propagation for producing plantations took another two decades to attain operational feasibility.

Prior to the 1970s, for "good" cuttings, breeders and propagators wanted shoot terminals that were growing in full light, thus producing cuttings that were in good physiological condition. In closed-canopy stands, those were branch terminals in the upper crown. That seemed *such* a good idea at the time, yet it was brutally counterproductive for rooting such cuttings to produce plenty of cheap and vigorous planting stock. What was not generally realized then was that successful rooting and good growth of cuttings depended on the material being in both good physiological condition and in an early maturation state. By the mid-1960s enough was known from propagation studies with radiata, and from similar studies of other conifers elsewhere, to discourage breeders from trying to mass-propagate plus trees nearing rotation age as cuttings (Fig. 5.24). Instead, they tried working with cuttings from promising seedling-origin trees at adolescent maturation states, namely trees young enough for their cuttings to root readily, yet large enough to show bole form of at least their butt logs, and old enough to give some confidence in evaluations of their early health, growth rate and some other important traits including wood density. That approach failed because, by the time the better performers among those clones could be confidently identified in clonal tests, the ortets and even their ramets in test had become too mature to repropagate at acceptable cost (Wilcox et al. 1976).

This problem was increasingly understood (Sweet 1973; Tufuor 1973; Greenwood and Hutchison 1993) as due largely to "meristem maturation," which for a time was misleadingly called "physiological ageing." Acceptance by many

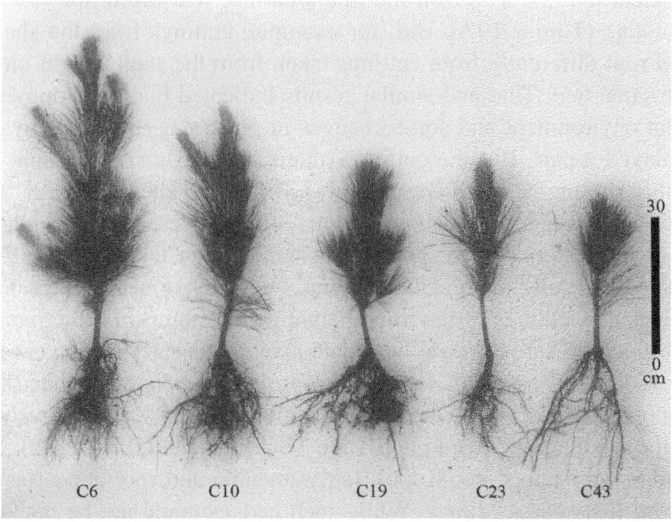

Fig. 5.24 Cuttings taken from trees aged 6, 10, 19, 23 and 43 years, two years after setting and before planting out. These illustrate the progressive decline in vigour which is matched by increasing difficulty of rooting

of that latter term had led to generally ineffective attempts at physiological solutions, such as changes in nutrition or applications of auxins and other plant-growth regulators. Furthermore, it was increasingly appreciated that a tree does not have all its meristems at the same maturation state at the same time (Tufuor 1973). Indeed, the most-mature meristems are in or near the terminal bud, while meristems lower on the tree are at various less-mature maturation states. Vigorous upper shoots on a sizeable tree may be in splendid condition, but they are the most mature and thus very difficult to root as cuttings.

Carefully observing seedling-origin trees at different chronological ages led to better understandings of the maturation process. A very young seedling of radiata looks greatly different from terminal shoots on a mature tree. Young seedlings have soft, paleish-green juvenile foliage dominated by primary needles. That juvenile appearance changes as the seedling grows larger, and genetically-programmed changes in its meristems' DNA-regulation progress occur. Secondary needles grouped mainly in threes (or twos in the island populations) appear as dormant short shoots (fascicles), and brown, sealed buds appear instead of green tufts of foliage at shoot tips. Other changes occur with time as the sapling gets even older and taller. Among those changes, the branching pattern becomes more regular, producing a "tidier" growth habit, pollen- and cone production begin as sexual competence develops, and the lateral meristems (cambia) produce longer and coarser tracheids. So increasing difficulty of rooting cuttings is but one of a whole suite of changes accompanying increasing maturation.

Later, carefully observing replicate cuttings produced from different parts of the same donor tree helped sort out changes in those various traits caused by meristem maturation from more ephemeral changes that could be influenced by altering the environment and/or physiology of the propagule. In general, it was becoming clear that the lower on the tree a cutting was taken, the less mature its maturation state (Tufuor 1973. But, for example, cuttings from the shady side of the tree did root differently from cuttings taken from the same height on the sunny side of the same tree. That and similar results indicated that developmental genetics, ambient environment and some changes in physiology imposed by the propagator all played a part. But the cutting's maturation state seemed most important for efficiently rooting it, and it was the most difficult to alter.

A solution was suggested by observations of tree species (such as redwood) that can produce sprouts or coppice shoots from stumps after felling. Even if the logged trees were old and fully mature, such stump sprouts have nearly the same juvenile characteristics as seedlings of the same height as the stumps. Furthermore, the tops of such fresh stump sprouts seem be at late-juvenile or early-adolescent maturation states. For radiata, a promising solution was to keep seedlings intended to serve as cutting-donors hedged close to the ground (Libby et al. 1972), effectively anticipating scientific findings on maturation state (Bolstad and Libby 1982). Year after year, the regrowth shoots would look fairly juvenile and root almost as readily as cuttings from first-year seedlings. While such hedges could not be maintained in a near-juvenile state indefinitely, this approach had enormous practical applications by the 1990s.

It may be noted that differing maturation states can contribute substantially to **C-effects**^{G} (or c-effects) within clones (Burdon and Shelbourne 1974), which mimic unwanted genetic variation and represent a form of **epigenetic**^{G} effects.

In parallel with ongoing research on rooting radiata cuttings for planting stock, research began at FRI on its propagation in the laboratory using tissue-culture to produce "test tube babies." This was a new avenue of propagation technology, aimed at being able to apply a technology to radiata that was already being developed for a number of plant and animal species, including humans. During the 1970s, under Roger Cameron and then Geoff Sweet at the New Zealand FRI, Kathy Horgan pioneered and led the development of tissue-culture propagation of radiata, operational use of which was vigorously pursued after 1983.

By 1969, the tree breeders at FRI had identified two strategic foci: (1) acquiring and managing radiata's genetic diversity for short- and long-term genetic changes; and (2) researching a short-term operational goal of developing vegetative propagation, particularly rooting and raising mature-phase cuttings so as to make clonal seed orchards work reliably. Developing moderately large-scale vegetative propagation of planting stock for those orchards soon involved an increased emphasis on nursery research, to more reliably grow such stock in a field-ready state. Meanwhile and in parallel, research on growing very large numbers of more robust field-ready seedlings of radiata, destined for large-scale plantations during a broadened planting-time window, had largely come to fruition.

Balanced against the bad effects of maturation on propagability and early growth rate were the prospective benefits of planting late-juvenile and adolescent cuttings (and later, tissue-culture plantlets) on tree form, and their resistance to such things as toppling in heavy wet soils and to diseases of juvenile radiata. It was with this in mind that in 1982 Bunn established a new research unit at FRI, Propagation and Early Growth (PEG), under Mike Menzies (Sect. 5.3.2). It was tasked to address the interplay between available and developing vegetative-propagation techniques and the important differences in growth and bole form developed by various maturation states of radiata cuttings, but even by unrooted cuttings set directly in plantations, and perhaps by plantlets beginning to come from propagation *in vitro*. Of major interest was how maturation of conifers, particularly in radiata, could be effectively and efficiently controlled, and how exactly maturation state affected propagability, growth rate and tree form. Such knowledge would then support more beneficial deployment of radiata propagules at maturation states beyond early-juvenile.

Early efforts to mass propagate cuttings from tree 5–7 years old, while giving promising early results (Fig. 5.25), proved impractical. It emerged that deploying cuttings having 2–3 years' greater initial maturation state, compared to that of seedlings, could substantially benefit tree form (Menzies et al. 1988) and greatly reduce severity of diseases such as western gall rust (Old et al. 1986), without seriously affecting their propagability and early growth (Fig. 5.26). As the promise of bringing "somatic" embryogenesis on line became a longer-term goal, 1980s PEG research nevertheless continued to refine the production of plantlets *in-vitro* by tissue-culture, transitioning the small, delicate plantlets from the conditions inside a culture dish to greenhouse and then to outdoor conditions, growing them in

Fig. 5.25 Tasman Forestry Ltd's oldest stand of cuttings, taken from 7-year-old trees, in June 1986, 16 years after planting. The trees feature good bole form and thin bark, reflecting the age of donor trees and choice of donors with "short-internode" branching pattern; and similarity of habit and cones, a consequence of many of the trees belonging to the same clones

Fig. 5.26 Comparison of seedlings (left) and cuttings taken from 4-year-old trees (right), the latter showing a less bushy, more open crown habit. Ageing of donor trees can improve tree form sooner than it depresses growth rate, but production of cuttings tends to become more expensive as tree form improves

nurseries to become field-ready plantlets, and researching their subsequent growth. Later on, towards 1990, PEG was ready to begin research on somatic embryogenesis, which was aimed at eventually producing field-ready "emblings" and even encapsulated artificial seeds.

Despite the logical links between GTI and PEG, they were for nearly 10 years administratively placed in separate divisions of FRI, a situation that much impeded communication and collaboration toward common goals.

5.5.7 Demonstrating and Quantifying Genetic Changes

Proving, quantifying and financially valuing genetic improvement are important both politically and technically. Politically, this is needed to justify ongoing financial support, not only for continuing the radiata breeding programme but also planned breeding programmes in other species, such as Douglas-fir, cypresses and eucalypts. Technically, several purposes are served. Genetic changes achieved after one or two generations of breeding could be expected to affect harvest yields and some elements of log quality, particularly bole straightness. Information on such effects is needed for planning future harvests and thinking ahead to marketing. Hard data from fair trials of new breeds allows fine-tuning and revision of models predicting future yields and log outturns. Demonstrated genetic changes also have implications for management and silvicultural regimes, and allow knowledge-based changes in initial spacing, in thinning and pruning regimes, and even in future harvest ages.

Some conflicts exist between meeting the political needs and the technical needs, despite both being served by demonstrating genetic changes. Side-by-side plots of genetically improved breeds and unimproved control seedlots (representative of alternative genetic stocks for planting) often showcase dramatic differences and are thus politically effective. However, such paired comparisons tend to be achieved at the expense of the proper, fair experimental replication that is needed to provide reliable data that are convincing to scientists and "hands-on" managers and provide a sound base for future planning.

Early progeny trials typically contained "routine" seedlots, which were generally genetically unimproved seedlings, as "controls." Typically, the controls performed as had been observed in plantations before, and most of the select progenies duly exhibited improvement of the selected traits, which reassured at least the tree breeders. But in Australia, by the mid-1970s, political pressures were coming for more rigorous demonstration of genetic gains. This prompted such analyses and a preliminary paper given at the 1977 3rd World Consultation on Forest Tree Breeding (Sect. 5.1). That paper reported modest but variable increases in growth rates and straighter boles, based on the rigorous analyses of data from three young trials (age 5½–7 years). Later, at ages 10–12 years, further analyses (Eldridge 1982) indicated that such gains were by then more consistent, and thus more certain. Those later analyses indicated around 20% greater average stem volume and much better tree form in seed-orchard seedlots compared to the controls.

In New Zealand, some of the very early progeny trials, both replicated and unreplicated, showed some OP families with, on average, dramatically straighter stems, lighter and more regular branching, and much less malformation. Such manifest gains in these important tree-form traits fully convinced foresters of the benefits of breeding. Even so, it was deemed appropriate to set up special genetic-gain trials (Burdon et al. 2008), to quantify genetic changes obtainable from select-tree seed collections, seedlots from the various regional seed orchards, and from seedlots designed to simulate the outturns of future seed-orchards, and to predict the performance of future plantations to be established using those genetically-improved seeds.

In 1978 and 1979, two major series of multi-site Genetic Gain trials were established. Each year, two types of trial were planted:

1. "Small plots," using six-tree row-plots, well replicated and incorporating a range of control lots, aimed at giving precise rankings among seedlots. These, however, are prone to competitive effects between adjacent rows of seedlots, thus generally exaggerating growth differences among the seedlots.
2. "Large plots," 6 trees × 6 rows giving 36 trees per plot, with fewer controls and less replication, but designed to better simulate crop-level rather than individual-tree performance. Because the outer rows and columns are so subject to competitive effects of the neighbouring plots, data from only the inner 16 trees were used to give a measure of stand-level performance of each genetic entry.

Those early trials, and subsequent ones, were meant to demonstrate and quantify the genetic gains in traits of interest. Their data could form the bases both for financially evaluating such gains and for more persuasive, honest and effective marketing of genetically improved seed (Vincent 1987), and for projecting the harvest yields (Carson et al. 1999) and possible product-grade outturns of the resulting plantations.

Nurserymen often expected to see genetic improvement showing as better-looking, faster-growing seedlings in their nurseries, only to be disappointed. Instead, it turned out that genetic improvement in radiata starts to be expressed clearly in the field only after about five years. On the other hand, for various reasons the seedlings from seed orchards often made comparatively slow starts, which could be made good into meeting planting-stock specifications by skilled nursery management. Despite this, economic benefits of genetic improvement could and often did accrue from the time of planting onwards. For many plantations, foresters decided that fewer trees needed to be planted per hectare, because less provision had to be made for thinning out malformed trees. Pruning better-formed trees was cheaper and the payoffs from pruning were greater with straighter boles. And, in addition to increases in total stemwood per hectare and various changes in log quality, fewer forked and other malformed trees made for safer logging and less logging waste. Finally, at that time it was believed that genetic improvements would be able to offset the adverse effects on overall crop quality associated with silvicultural regimes designed to reduce effective growing costs by such measures as fertilizer use, heavier thinnings and shorter rotations.

5.6 Summary of Domestication Progress

Booms of new planting of radiata developed in New Zealand and Australia, and the planting boom continued in Chile despite a lull associated with political upheavals. Thus the total area of radiata plantation increased substantially. Intensive tree breeding spread to Chile, where tree breeders were able to take advantage of hard-won experience in Australia and New Zealand.

Comparative financial analyses of alternative silvicultural regimes for radiata in New Zealand led to another major advance in radiata plantation forestry. A key finding was that, in most local conditions, commercial harvesting of thinnings was generally uneconomic, because that reduced the overall profitability of the crop. That led to widespread adoption within New Zealand of the unconventional Direct Sawlog regime, which featured wide early spacing, and early and heavy thinnings coordinated with aggressive pruning of lower branches, but with no harvesting of thinnings. As follow-up, stand-modelling software was being developed to predict log and product outturns from alternative growing regimes on various site categories, an activity that has remained the subject of continuing refinement.

Together with developing the Direct Sawlog regime was the development in New Zealand of silvo-pastoral systems for converting pasture land into radiata plantations. While such systems were used there for converting large areas of pasture into forest plantation, the benefits of grazing for interim financial yields were generally disappointing. Tending systems were also being refined for radiata shelterbelts that could provide high-quality timber.

Increasing awareness of the need to assure continued genetic gains led radiata breeders to develop multi-generation breeding strategies, an important contribution to the relatively new discipline of tree breeding. An immediate result was the expansion of many tree-breeding programmes, often by adding large numbers of fresh selections from land races or natural stands, to supplement breeding lines based on too few plus trees. It also led to elaborate population structuring for the genetic material, to avoid deploying inbred material and to better assure future genetic gains while capturing large immediate gains.

By 1983, seed orchards were delivering genetic gains in almost all the radiata planting stock for New Zealand and South Africa and in much of that for Australia. Some of those gains were already giving financial payoffs by saving some important growing costs. However, some downsides of the combination of genetically improved stock and new silvicultural regimes, in the form of greater than expected declines in the wood quality of the harvested crops, were not yet evident.

Intensive research to improve practices in the nursery and in field establishment produced advances in husbandry which were important both in themselves and for capitalising on radiata's genetic improvement. Thereby, better (less wasteful) use could be made of expensive seeds and their promised genetic gains.

Early findings comparing native provenances of radiata prompted a major fresh round of seed collection in all five of radiata's natural populations. This was mounted by Australia and New Zealand, with support from US agencies and FAO. This was to ensure the availability for breeders of the full range of radiata's natural variability, and recognised its international importance as a resource. The seed was used for establishing numerous provenance trials and a number of gene-resource plantings.

Research on vegetative propagation was vigorously pursued. Some was for improving seed-orchard systems, for which novel options were studied, but orchard siting emerged as a pre-eminent factor. Other propagation research achieved successful mass multiplication of seedlings as nursery cuttings, and *in-vitro* multiplication of seedling embryos into field-fit plantlets. Such multiplication systems offered accelerated capture of genetic gain in the quality of planting stock.

Early attempts to develop clonal forestry with radiata, for better tree form, more genetic gain and crop uniformity, had failed, principally due to the phenomenon of maturation. However, some progress in controlling maturation was achieved by the practice of hedging donor plants for cuttings. The various influences of increasing maturation on propagule performance became better characterised.

Genetic-gain trials began to be planted, to achieve rigorous demonstration of genetic gains and to provide a basis for tracking the expression of genetic gain through the life of the crop.

References

Ades PK, Simpson JA (1997) Variation in susceptibility to *Dothistroma* needle blight among provenances of *Pinus radiata* var. *radiata*. Silvae Genet 40:6–13

Allard RW (1960) Principles of plant breeding. Wiley, New York, p 485

Australian Forestry Council (1981) Proceedings Australian forest nutrition workshop [2nd]: Productivity in perpetuity. Research working group 3 (soils and nutrition). ACT, Canberra, p 366

Ballard R (compil.) (1977) Use of fertilisers in New Zealand forestry. New Zealand Forest Service, Forest Research Institute, Rotorua, (compil.) Symposium No. 19. p 322

Bannister MH (1962) Variation of turpentine composition in five population samples of *Pinus radiata*. N Z J Sci 5:486–495

Bevege DI (2010) The challenge of change in a climate of uncertainty: looking back to the future through the prism of Max Jacobs' forestry career. Aust For 73:140–155

Birk EM (1990) Poor tree form of *Pinus radiata* D.Don on former pasture sites in New South Wales. Aust For 53:104–112

Birk EM (1991) Stem and branch form of 20-year-old radiata pine in relation to previous land use. Aust For 54:30–39

Blight MM, Brewerton HV, McDonald IRC et al (1964) β-pinene content of the *Pinus radiata* population of Kaingaroa Forest. N Z J Sci 7:457–459

Boardman R (1988) Living on the edge—the development of silviculture in South Australian pine plantations. Aust For 51:135–156

Bolstad PV, Libby WJ (1982) Comparisons of radiata pine cuttings of hedge and tree-form origin after seven growing seasons. Silvae Genet 31:9–13

Burdon RD (1971) Clonal repeatabilities and clone-site interactions in *Pinus radiata*. Silvae Genet 20:33–39

Burdon RD (1977) Genetic correlation as a concept for studying genotype–environment interaction in forest tree breeding. Silvae Genet 26:168–175

Burdon R (1978) Mejoramiento genetico ([Forest] Genetic Improvement [in Chile]). Project Working Document, FAO Project CHI/7. p 81

Burdon RD (1979) Generalisation of multi-trait selection indices using information from several sites. N Z J For Sci 9:145–152

Burdon RD (1992) Genetic survey of *Pinus radiata*. 9. General discussion and implications for genetic management. N Z J For Sci 22:274–298

Burdon RD (2001) *Pinus radiata*. In: Last FT (ed.) Ecosystems of the World, Vol. 19, Tree Crop Ecosystems. Elsevier, Amsterdam, pp 99–161

Burdon RD (2004) Genetics of *Pinus radiata*. In: Burley J, Evans J, Youngquist JA (eds) Encyclopedia of forest sciences vol. 3. Elsevier Academic Press, Oxford and San Diego, pp 1507–1516

Burdon RD (2011) Shoot dieback epidemics of *Pinus radiata* in New Zealand: is there an endophyte story? N Z J For Sci 41:131–139

Burdon RD, Bannister MH (1973) Provenances of *Pinus radiata*: their early performance and silvicultural potential. N Z J For 18:217–232

Burdon RD, Hong SO, Shelbourne CJA et al. (1997) International gene pool experiments in *Pinus radiata*: patterns of genotype–site interaction. N Z J For Sci 27:101–125

Burdon RD, Low CB (1992) Genetic survey of *Pinus radiata*. 6. Wood properties: variation, heritabilities and interrelationships with other traits. N Z J For Sci 22:228–245

Burdon RD, Shelbourne CJA (1971) Breeding populations for recurrent selection: conflicts and possible solutions. N Z J For Sci 1:174–193

Burdon RD, Shelbourne CJA (1974) Use of vegetative propagules for obtaining genetic information. N Z J For Sci 4:418–425

Burdon RD, Bannister MH, Madgwick HAI et al (1992a) Genetic survey of *Pinus radiata*. 1: Introduction, description of experiment, and basic methodology. N Z J For Sci 22:119–137

Burdon RD, Bannister MH, Low CB (1992b) Genetic survey of *Pinus radiata*. 2: Population comparisons for growth rate, morphology and disease resistance. N Z J For Sci 22:138

Burdon RD, Carson MJ, Shelbourne CJA (2008) Achievements in forest tree genetic improvement in Australia and New Zealand 10: *Pinus radiata* in New Zealand. Aust For 71:263–279

Burdon RD, Chou CKS, Currie D (1982) Response to inoculation with *Diplodia pinea* in progenies of apparently resistant trees of *Pinus radiata*. Australia Plant Pathol 11(4):37–39

Butcher TB, Stukely MJC, Chester GN (1984) Genetic variation in resistance of *Pinus radiata* to *Phytophthora cinnamomi*. For Ecol Manag 8:197–220

Carson SD, García OP, Hayes JD (1999) Realized gain and prediction of yield with genetically improved *Pinus radiata* in New Zealand. For Sci 45:186–200

Chavasse CGR (compil. & ed.) (1969a) Forest nursery and establishment practice in New Zealand. In: Proceedings, New Zealand Forest Service, Forest Research Institute, Rotorua, Symposium no. 9. p 214

Chavasse CGR (compil. & ed.) (1969b) Land preparation for forestry in New Zealand. In: Proceedings, New Zealand Forest Service, Forest Research Institute, Rotorua, Symposium no. 11. p 205

Chavasse CGR (compil.) (1973) Mechanisation of nursery production, forest establishment and tending in New Zealand. In: Proceedings, New Zealand Forest Service, Forest Research Institute, Rotorua, Symposium no. 13. part 1, p 239; part 2, p 119

Chavasse CGR (compil.) (1976) The use of herbicides in forestry in New Zealand. In: Proceedings, New Zealand Forest Service, Forest Research Institute, Rotorua, Symposium No. 18. p 322

Chavasse CGR (compil. & ed) (1981) Forest nursery and establishment practice in New Zealand. In: Proceedings, New Zealand Forest Service, Forest Research Institute, Rotorua, Symposium no. 22. Part 1, nursery, p 261; Part 2, establishment, p 330

Conway MJ (1962) Aerial application of phosphate fertilisers to radiata pine forests in New Zealand. Emp For Rev 41(3):235–245

Cotterill PP (1984) A plan for breeding radiata pine. Silvae Genet 33:84–90

Cotterill PP, Dean CA (1990) Successful tree breeding with index selection. CSIRO, Melbourne, p 80

Cotterill PP, Jackson N (1981) Index selection with restrictions in tree breeding. Silvae Genet 30:106–108

Dean CA, Cotterill PP, Cameron JN (1983) Genetic parameters and gains expected from multiple-trait selection of radiata pine in eastern Victoria. Aust For Res 13:271–278

Donald DGM, Glen LM (1974) The response of *Pinus radiata* and *Pinus pinaster* to N, P and K fertilizers applied at planting. South Afr For J 91(1):19–28

Donald DGM, Lange PW, Schutz CJ et al. (1987) The application of fertilisers to pines in Southern Africa. South Afr For J 141(1):53–62

Eldridge KG (1978) Seed collections in California 1978–. Annual Report 1977–1978. CSIRO Division of Forest Research, Canberra, pp 9–17

Eldridge KG (1982) Genetic improvements from a radiata pine seed orchard. N Z J For Sci 12:404–411

Eldridge K, Davidson J, Harwood C et al. (1993) Eucalypt domestication and breeding. Clarendon Press, Oxford, UK, p 288

Elliott HJ, Felton KC, Jarman SJ et al. (2008) A history of innovation: eighty-five years of research and development in Forestry Tasmania. Forestry Tasmania, Hobart. p 392

Falkenhagen ER (1991) Provenance variation in *Pinus radiata* at six sites in South Africa. Silvae Genet 40:41–50

Fenton RT (1972) Economics of radiata pine for sawlog production. N Z J For Sci 2:313–347

Fenton RT (1979) Comparisons of Chilean and New Zealand radiata pine plantations. N Z J For 24:13–24

Fielding, JM (1961a) Provenances of Monterey and bishop pines. Forestry and Timber Bureau, Canberra, Bulletin No. 38. p 30

Fielding JM (1961b) The pines of Cedros Island. Aust For 24:62–65

Forde MB (1964) Variation in native populations of *Pinus radiata*. 1. Sampling methods and branch characters. N Z J Bot 2:213–236

Gibson IAS (1972) Dothistroma blight of *Pinus radiata*. Annu Rev Phytopathol 10:51–72

Greenwood MS, Hutchison KW (1993) Maturation as a developmental process. In: Ahuja MR, Libby WJ (eds) Clonal forestry I, Genetics and biotechnology. Springer-Verlag, Heidelberg, pp 14–33

Guinon M, Hood JV, Libby WJ (1982) A clonal study of intraspecific variability in radiata pine II. Growth and form. Aust For Res 12:191–201

Hood JV, Libby WJ (1980) A clonal study of intraspecific variability in radiata pine. I. Cold and animal damage. Aust For Res 10:9–20

Horne R (1988) Philosophy and practice of *Pinus radiata* plantation silviculture in New South Wales. Aust For 51:157–170

Ivković M, Wu HX, Kumar S (2010) Bioeconomic modelling as a method for determining economic weights for optimal multiple-trait selection. Silvae Genet 59:77–90

James RN, Sutton WRJ, Tustin JR (compil.) Tustin JR, Bunn EH (eds) (1970) Pruning and thinning practice, vol. 1: proceedings. New Zealand Forest Service, Forest Research Institute, Rotorua, Symposium No. 12. p 188

Jayawickrama KJS, Balocchi C (1993) Growth and form of provenances of *Pinus radiata* (radiata pine) in Chile. Aust For 56:172–178

Johnson GR, Burdon RD (1990) Family–site interaction on *Pinus radiata*: implications for progeny testing strategy and regionalised breeding in New Zealand. Silvae Genet 39:55–62

Johnson IG, Cotterill IM, Raymond CA et al. (2008) Half a century of *Pinus radiata* tree improvement in New South Wales. N Z J For 52(4):7–13

Jolly NW (1950) The thinning of *Pinus radiata* plantations in the south east of South Australia. Bulletin Woods and Forests Department, Adelaide. no. 4. p 27

Kininmonth J (1997a) The development of modern silviculture in New Zealand. N Z [J] For 41 (4):17–19. 22–23

Kininmonth JA (1997b) A history of forestry research in New Zealand: commemorating 50 years of research at the Forest Research Institute (FRI) (1947–1997). New Zealand Forest Research Institute, Rotorua, p 271

Landsberg JJ, Thompson WA, McMurtrie RE (1988) Development and use of a complex model of growth of radiata pine. In Leech, JW et al, (eds) Modelling trees, stands and forests. University of Melbourne, Australia. pp 257–265

Lavery PB (1988) *Pinus radiata* in Victoria—a review of public-sector plantation forestry. Aust For 51:171–184

Levack HH (1973) The Kaingaroa air sowing era, 1960–1971. N Z J For 18:104–108

Lewis NB, Ferguson IS (1993) Management of radiata pine. Inkata Press (Butterworth-Heinemann), Melbourne–Sydney, Australia, p 404

Libby WJ (1973) Domestication strategies for forest trees. Can J For Res 3:265–276

Libby WJ, Cockerham CC (1980) Random non-contiguous plots in interlocking field layouts. Silvae Genet 29:183–190

Libby WJ, Conkle MT (1966) Effects of auxin treatment, tree age, tree vigor and cold storage on rooting young Monterey pine. For Sci 12:484–505

Libby WJ, Rauter RM (1984) Advantages of clonal forestry. For Chronicle 60(3):145–149

Libby WJ, Bannister MH, Linhart YB (1968) The pines of Cedros and Guadalupe Islands. J For 66:846–853

Libby WJ, Brown AG, Fielding JM (1972) Effects of hedging radiata pine on production, rooting and early growth of cuttings. N Z J For Sci 2:263–283

Maclaren [J]P (ed) (1988) Proceedings of the agroforestry symposium, Rotorua, 24–27 November 1986. New Zealand Ministry of Forestry, Rotorua, FRI Bulletin no. 139. p 299

Maclaren JP (1993) Radiata pine growers' manual. New Zealand Forest Research Institute, Rotorua, Bulletin no. 184. p 140

Matheson AC, Raymond CA (1984) The impact of genotype × environment interactions on Australian *Pinus radiata* breeding programs. Aust For Res 14:11–35

Matheson AC, Willcocks KW (1976) Seed yield in radiata pine seed orchard following pollarding. N Z J For Sci 6:14–18

Matziris DI (1995) Provenance variation of *Pinus radiata* grown in Greece. Silvae Genet 44: 88–96

Mead D (2010) Radiata pine silviculture in Chile. N Z Tree Grow 31(2):15–17

Mead DJ (2013) Sustainable management of *Pinus radiata* plantations. FAO Forestry Paper No. 170. FAO, Rome, p xv + 246

Menzies MI, Aimers JP, Whitehouse LJ (eds) (1988) Proceedings, workshop on growing Radiata pine from cuttings, 5–7 May 1986, Rotorua, New Zealand. Ministry of Forestry, FRI Bulletin no. 135. p 197

Menzies MI, Burdon RD, Holden DG et al (1987) Family variation and potential for genetic gain in frost resistance of *Pinus radiata*. New For 3:171–186

Michel M (2006) El pino radiata en la historia forestal vasca. Sustral: Revis Agropesquera 2006:56–59

Millar CI (1999) Evolution and biogeography of *Pinus radiata*, with a proposed revision of its Quaternary history. N Z J For Sci 29:335–365

Miller AD (1974) *Pinus radiata* in Ecuador, its ecology and growth. Unasylva 26:30–34

Moran GF, Bell JC (1987) The origin and genetic diversity of *Pinus radiata* in Australia. Theor Appl Genet 73:616–622

Moran GF, Bell JC, Matheson AC (1988a) The genetic structure and levels of inbreeding in a *Pinus radiata* D.Don seed orchard. Silvae Genet 29:190–193

Moran GF, Eldridge KG, Bell JC (1988b) The genetic structure and the conservation of the five natural populations of *Pinus radiata*. Can J For Res 18:506–514

Nambiar S, O'Hehir J (2010) Increasing production from pine plantations in the Green Triangle. Aust For Grow 33(2):35–37. [Winter 2010]

Namkoong G, Snyder EB, Stonecypher RW (1966) Heritability and gain concepts for evaluating breeding systems and seedling orchards. Silvae Genet 15:76–84

O'Hehir J, Nambiar S (2010) Productivity of successive rotations of *Pinus radiata* plantations in South Australia over a century. For Ecol Manag 259:1857–1869

Old KM, Libby WJ, Russell JH et al. (1986) Genetic variability in susceptibility of *Pinus radiata* to western gall rust. Silvae Genet 35:145–149

Orr E (2017) Keeping New Zealand green. Steele Roberts Aotearoa Publishers, Wellington, p 253

Pederick LA, Brown AG (1976) Seed production in radiata pine seed orchards in Australia. Aust For 40:144–150

Rimbawanto A, Coolbear P, Firth A (1988) Artificial ripening of prematurely harvested cones of New Zealand *Pinus radiata* and its effect on seed quality. N Z J For Sci 18:149–160

Roche MM (1990) History of New Zealand forestry. New Zealand Forestry Corporation in association with GP Books, Wellington, p 466

Ross SD, Bollmannn MP, Pharis RP et al. (1984) Gibberellin 4/7 and the promotion of flowering in *Pinus radiata*. Plant Physiol 76:326–330

Routley R, Routley V (1975) The fight for the forests: the takeover of Australian forests for pines, wood chips, and intensive forestry, 3rd edn. Research School of Social Sciences, Australian National University, Canberra, p 407

Schlatter J, Gerding V (1985) Deficiencia de boro en plantaciones de *Pinus radiata* en Chile I. Distribución y origen. Bosque 6:24–31

Schutz WM, Cockerham CC (1966) The effect of field blocking on gain from selection. Biometrics 22:843–863

Scott CW (1960) *Pinus radiata*. FAO Forestry and Forest Products Studies No. 14. FAO, Rome, p 328

Shelbourne CJA (1969) Tree breeding methods. New Zealand Forest Service, Forest Research Institute, Rotorua, Technical Paper no. 55. p 43

Shelbourne CJA (1970) Genetic improvement in different tree characters of radiata pine and the consequences for silviculture and utilization. In: Sutton, pp 44–58.

Shelbourne CJA (1972) Genotype–environment interaction: its study and its implications in forest tree improvement. Paper B-1. In: Proceedings, IUFRO Genetics-SABRAO joint symposia for advancement of forest tree breeding, Tokyo, October 1972. p 27

Shelbourne CJA (1986) Historical introduction. Chapter I. In: Shelbourne et al, pp 1–6

Shelbourne CJA, Burdon RD, Bannister MH et al. (1979) Choosing the best provenances of radiata pine for different sites in New Zealand. N Z J For 24:288–300

Shelbourne CJA, Burdon RD, Carson SD, et al. (1986) Development plan for radiata pine breeding. New Zealand Forest Service, Forest Research Institute, Rotorua. p 142

Shepherd KR (1977) Forest tree breeding in Australia. Divisional report 2. CSIRO Division of Forest Research, Canberra, p 29

Shrestha KB, Shepherd KR, Turnbull JW (1985) Controlled atmosphere storage for *Pinus radiata* seed. Commonw For Rev 62(2):141–150

Stephens SL, Libby WJ (2006) Anthropogenic fire and bark thickness in coastal and island pine populations from Alta and Baja California. J Biogeogr 33:648–652

Stoate TN (1950) Nutrition of the pine. Bulletin Forestry and Timber Bureau, Australia, No. 30. p 61

Stone EL, Will GM (1965) Boron deficiency in *Pinus radiata* and *P. pinaster*. For Sci 11:425–433

Sutton WRJ (ed.) (1970) Pruning and thinning practice, vol. 2: papers. In: Proceedings, New Zealand Forest Service, Forest Research Institute, Rotorua, symposium no. 12. p 250

Sweet GB (1973) The effect of maturation on the growth and form of vegetative propagules of radiata pine. N Z J For Sci 3:191–210

Sweet GB, Krugman SL (1978) Flowering and seed production problems—a new concept of seed orchards. Paper 4/2. In: Brown, AG, Palmberg CM (compil.) Proceedings of the 3rd world consultation on forest tree breeding, 21–26 March 1977, Canberra, Australia. CSIRO, Canberra, pp 749–759

Tufuor K (1973) Comparative growth performance of seedlings and vegetative propagules of *Pinus radiata* (D.Don) and *Sequoia sempervirens* (D.Don Endl.). University of California Berkeley, Berkeley, PhD dissertation

Ure J (1949) The natural regeneration of *Pinus radiata* on Kaingaroa Forest. N Z J For 6(1):30–38
van Wyk G, Roeder KR (1978) The status of tree breeding in South Africa. South Afr For J 107 (1):54–59
Vincent TG (1986) Seed orchards. In: Shelbourne et al, pp 62–97
Vincent TG (1987) Certification scheme for forest tree seed and planting stock. Ministry of Forestry, Rotorua, FRI Bulletin no. 134. p 17
Weston GC (1956) Fertiliser trials in unthrifty pine plantations at Riverhead Forest. N Z J For 8:35–46
Weston GC (compil. and ed.) (1962) The use of fertilizers in New Zealand forestry. In: Proceedings, New Zealand Forest Service, Forest Research Institute, Rotorua, symposium no. 1. p 133
Wilcox MD, Thulin IJ, Vincent TG (1976) Selection of *Pinus radiata* clones for planting from cuttings. N Z J For 21:239–247
Wilcox MD (1983) Genetic variation and inheritance of resistance to Dothistroma needle blight in *Pinus radiata*. N Z J For Sci 12:14–35
Williams ER, Matheson AC (1994) Experimental design and analysis for use in tree improvement. CSIRO, Victoria, Australia, pp 174
Wu HX, Eldridge KG, Matheson AC et al (2007) Achievements on forest tree improvement in Australia and New Zealand 8: successful introduction and breeding of radiata pine in Australia. Aust For 70:215–225
Zobel B, Talbert J (1984) Applied forest tree improvement. John Wiley & Sons, New York, p 505

Chapter 6
Towards More Precise Genetic Control 1984–1997

6.1 Introduction

George Orwell's futuristic *1984* discussed various kinds of control, including genetic. Perhaps it is just coincidence that the next period of radiata's domestication seems to begin in about that year. This 1984–1997 period saw new levels of genetic control not only developing but also being demanded for radiata.

In retrospect, it is moot as to how much these advances in radiata's domestication were driven by technological capabilities as they became available, or by the demand for them. Surely the research and experience with available breeding and propagation technologies in 1969–1983 were important, among other things pointing to operational strategies for both **family forestry**^{G} and clonal forestry. But biological changes from the previous breeding cycle had also been convincingly demonstrated, and the mixed benefits and setbacks were just becoming clear by 1984. These combined to create major changes in the expectations, attitudes and demands of forest managers and corporate executives contemplating new plantations of radiata.

During 1969–1983, many forest managers abandoned their earlier requirement that the new planting stock, be it seedlings or cuttings, must not cost more at the nursery gate than seedlings from cheaply collected seeds coming from just anywhere. Rather, by 1984 they wanted propagules that had either proof or good promise of better benefit/cost ratios over their entire enterprises, through the nursery and plantations to the final products. They accepted that the predicted near-term and future genetic improvements would contribute importantly to that goal, as would greater uniformity and predictability of performance. Moreover, they were willing to pay a substantial premium for such brave new propagules.

Information was accumulating on the genetic gains already achieved in the first two cycles of breeding of radiata. As the research sample of breed progeny grew older and larger, and as additional traits were evaluated, progressively better information on radiata breeds was mostly coming from the specially designed trials in

R. Burdon et al., *Domestication of Radiata Pine*, Forestry Sciences,
DOI 10.1007/978-3-319-65018-0_6

Australia and New Zealand, and the information fed into a seed- and clone certification scheme. That scheme was first codified as a series of easy-to-comprehend descriptors for seed and planting stock being released in New Zealand for sale. It greatly facilitated the acceptance of, and even catalysed demand for, genetically improved planting stocks among growers and wood-products industries.

Such acceptance and demands were following similar breeding and trial results with several forest species in many other countries. A few examples from a much longer list involved loblolly pine in the south-eastern US, eucalypt hybrids in coastal Brazil, Chinese-fir in China, and some spruce species in Canada and Europe, although improved material was marketed with less innovative descriptors than used for radiata. This acceptance often entailed an institutional willingness to compete and pay much more extra for assured genetic superiority, rather than just minimizing propagule costs.

Intensifying the domestication of radiata, however, was not a smooth process. In New Zealand it occurred during a period of great political and institutional changes, and wildly fluctuating rates of afforestation. In addition, technical perceptions and possibilities were changing rapidly. Genetic improvement in several wood properties was finally emerging as a pressing need. A better understanding was coming of which properties needed to be improved, and advances were coming in the assay technology needed for selecting for those properties. Meanwhile, major advances were being achieved in plant propagation and DNA technology. In 1997, a second international conference was held specifically on genetics and improvement of radiata (Burdon and Moore 1997), a date that is an appropriate end-point for this chapter.

Preceding that meeting, the year 1996 had seen the onset of a steep decline in New Zealand's short-lived 1990s planting boom, as some East Asian economies went into recession and importers could access other sources of wood. Major changes were also occurring in New Zealand's forestry and forest-research institutions. In Chile, Australia and Spain, however, such institutions were comparatively stable through the 1990s.

6.2 Institutions

As noted in our previous chapters, it was the institutions and some key people associated with them that made things happen as forestry was moving further to growing and harvesting trees as a domesticated crop. For the domestication of radiata during 1984–1997, the formalization of cooperative institutions was a major development, with increasing support from and involvement of the forestry companies while support from some governments dwindled.

During the period of this chapter there was much institutional turbulence, with rapid changes in corporate structures and forest ownership, especially in New Zealand. Towards the end, in 1996, came a financial crisis affecting south-east Asia. Advances in gene technology created additional strains on the collaborative relationships that had worked well in the past, and contributed to a cultural and funding divide between operational tree breeders and biotechnology researchers. All told, the scene was set for something of a wild ride from 1998 onwards.

6.2.1 History and Usefulness of Cooperatives

"Tree-improvement" cooperatives had begun in the southern United States in the early 1950s, at a time when intensive tree breeding was based largely on theory and promise (Zobel and Sprague 1993). Those early cooperatives developed and evolved as genetic knowledge of several forest species advanced. At the outset, member companies were largely of similar size, which centred on the economies of scale of kraft pulpmills. At that time in the US, those mills had a pattern of vertical integration, whereby the processing plant and the forest that supplied most or all of its wood were usually under the same ownership. The companies themselves had limited specialized expertise in genetics, in breeding, and in nursery practices for raising propagules other than seedlings. The cooperatives mostly depended for this expertise on universities, which ran the secretariats. Underpinning the success of the cooperatives was a large element of "gentlemen's agreement" among the members in committing to paying subscriptions and sharing genetic material and intellectual property. With time, however, as corporate structures and products became more varied, and member companies diverged in their levels of commitment to incorporating new technical options in their wood-growing and general forest practices, cooperatives experienced increasing internal tensions and conflicts of goals.

In sharp contrast, the continuation of breeding and other aspects of domesticating radiata through formal cooperative organizations in Australia and New Zealand began only after much breeding had already been accomplished. Furthermore, the genetic gains already secured were becoming evident, although the downsides of the combination of breeding and new tending regimes were also becoming evident in New Zealand. It became clear that inputs from industry and other end-users would be to their advantage, with a goal of improving rather than inadvertently degrading the qualities of the wood and wood products they would eventually need to sell.

While better understanding of the financial advantages and costs of current and likely and proven genetic changes helped, other factors complicated the formation of cooperatives. Various prospective member organizations in New Zealand were of widely disparate size, and some already had commitments to their own proprietary technology. In late-20th-century New Zealand, unlike the mutual sharing in the early southern US cooperatives, guarding "intellectual property" led to secrecy and much wasteful duplication of effort and delays in progress. This appeared particularly counterproductive when some of those organizations later merged or acquired each other. Also, the prospect of paying directly for breeding work was nostalgically unwelcome to those who had previously simply paid as taxpayers for government institutions to do the work.

6.2.2 *Institutional Changes in Individual Countries*

6.2.2.1 Australia

In Australia, the Southern Tree Breeding Association (STBA) had been created in 1982 as a breeding cooperative, largely to correct the irrationality of state borders cutting across what was a very well defined natural region with sites favourable to radiata plantations (Boomsma 1997). It centred around the "Green Triangle" straddling western Victoria and south-easternmost South Australia (SA), but for administrative efficiency came to include the rest of those two states in terms of their forest services and their main private forest owners. STBA absorbed the collaborative arrangement that had operated between CSIRO (Commonwealth Scientific and Industrial Research Organisation) at Mt Gambier, SA, and APM (Australian Paper Manufacturers Ltd) in eastern Victoria. Members also joined from Tasmania and Western Australia, although the latter withdrew in 1993 after five years' membership. The STBA was launched under the direction/management of David Boomsma of the Woods and Forests Department of SA, a soil scientist but with a strong personal interest in genetic improvement. Later, the STBA was reconstituted under a Board of Directors and a CEO.

Disastrous forest fires in the Green Triangle in early 1983 (Keeves and Douglas 1983) occurred very soon after the inception of the STBA. Those fires destroyed two of the region's three seed orchards and simultaneously created a huge demand for propagules to restock 21,000 ha of burnt plantations. Thus, the fires gave much urgency to the operation of the STBA, driven by the hope it could quickly provide better, more valuable trees to replace those burnt.

6.2.2.2 New Zealand

From 1987 to 1996 several major institutional changes occurred (Burdon et al. 2008) that are highly relevant to our story.

The New Zealand Radiata Pine Breeding Cooperative (NZRPBC) arose out of a huge institutional upheaval that centred around the dissolution of the Forest Service on 1 April 1987. State plantation forests were transferred to a newly created and fully commercial New Zealand Forestry Corporation (FC), still under government ownership. Meanwhile, FRI remained largely intact but became the largest part of a rump organization, the Ministry of Forestry. As such, FRI personnel had no automatic access to land for new genetic tests and breeding operations in the now-FC forests, although existing tests and breeding material, and access to the same, were nominally subject to covenant protection.

The previously state-owned seed orchards were placed under a small subsidiary of the FC called PROSEED. Significantly, although PROSEED was initially run by Peter Bolton, a nurseryman, its operation was confined to producing and trading forest-tree seeds. It had neither ownership nor control of nurseries to produce planting stocks, but instead had to competitively market its seeds to them. Thus, it became an important player in producing, advertising and selling the improved radiata seedlots resulting from breeding by FRI and later by NZRPBC.

The NZRPBC, which took several months of fraught negotiations to set up, assumed charge of New Zealand's breeding material of radiata, with John Gleed of Tasman Forestry Ltd as Chairman and Randy Johnson of FRI as technical leader. Its breeding resources became what the NZRPBC would support financially. That included: land-race base populations owned by member companies and which could provide new plus-tree selections; select clones in archives; unsown seed from breeding activities; trees in field trials of select families and clones; and highly-selected, well-tested and clonally replicated parents transferred to PROSEED'S seed orchards and already contributing to the planting stock being offered to clients for production plantations (Shelbourne et al. 1986; Vincent 1993). It did not, however, include any of the random or field-selected material from the costly Forde, Bannister and Eldridge collections from the five native populations. That genetic material, along with genetic trials of other tree species, was almost all in ex-Forest Service forests, but was subject to covenant protection. Any felling of it would be subject to FRI consent, but such protection could fail and sometimes did.

Another major change soon occurred, in 1992, when FRI was converted into a Crown Research Institute (CRI) operating as a commercial company. CRIs were conceived with the idea of allowing them to borrow and thus incur debt, in order to allow them to operate more flexibly. Direct government funding was incrementally decreased, although FRI projects were then free to compete for "contestable" government research grants. Two practical outcomes were (1) an intensifying need during a transition period for CRIs to raise much of their own funding, and (2) pressures on CRIs to sequester and guard "intellectual property" in order to secure future revenue streams through patenting and/or licensing. This departed sharply from FRI's recent obligation, under the Official Information Act of 1982, to make its scientific discoveries and knowledge available in the public domain. This new culture of secrecy soon became a source of major tension and frustration as industry organizations intensified sequestering of their own intellectual property, including important advances made in breeding, testing and propagating radiata. At the same time, recent pre-1992 release of information was often placing FRI at a commercial disadvantage.

Soon after FRI became a CRI, an important positive change brought propagation projects and staff into the same group as those doing breeding and basic genetic-research projects, and added new gene-technology research projects and staff, all in FRI's newly-constituted Biotechnology Portfolio.

Two specialised FRI–Industry research groups associated with genetic improvement of radiata were soon formed, but these groups were short-lived.

FRI, while expected to raise increasing proportions of its revenue from the forestry sector, continued to do most of New Zealand's basic research in that sector, and to maintain lesser breeding programmes in other forest-tree species. The basic genetic research on radiata now included DNA-based studies aimed at enhancing "forwards" selection within a new generation "marker-assisted selection," and implementing direct insertion or deletion of single genes in otherwise intact genotypes ("genetic engineering"). FRI's involvement with problems and operations of seed orchards decreased as the orchards were taken over by PROSEED, or newly-established by such corporations as Tasman (later within Fletcher Challenge Corporation) and New Zealand Forest Products (later Carter Holt Harvey Ltd). However, FRI retained full responsibility for developing and operating the certification system for tree seeds and planting stock of all forest species in New Zealand and, most importantly, for the radiata breeds, seedlots and clones being developed and marketed. The development and evolution of that system is a story in itself (Sect. 6.6.4.1).

Further institutional changes occurred in New Zealand during this period. In 1990, the Forestry Commission of New South Wales (NSW), a major forest owner, opted to join the NZRPBC instead of the STBA (Burdon et al. 2008; Johnson et al. 2008). That decision was based on the climatic similarities and parallel disease problems between many New Zealand and NSW sites, on some technical problems with NSW seed orchards, and perhaps most important, on the good performance of many New Zealand select radiata progenies in NSW trials.

Other changes in the NZRPBC stemmed from a series of major changes in ownership of New Zealand's plantation forests. Member-company changes and additions followed from privatisation in 1992 of over 50% of the state-owned commercial forests. The state-owned New Zealand Forestry Corporation, which then morphed into the Forestry Corporation of New Zealand (FCNZ), was no longer by far the largest forest owner among the Cooperative membership; some other forest-owner members become larger, and some new corporate forest owners eventually joined as members. Further major change of forest ownership occurred in 1996, when most of the remaining state-owned (FCNZ) plantation forest located inland from the Bay of Plenty were sold by the state to Fletcher Challenge Corporation (FCC). FCC had much earlier acquired Tasman Forestry Ltd, which had been a founder member of the NZRPBC. This in turn brought a major change in the power structure within the NZRPBC (subsequently RPBC).

Among private corporations, New Zealand Forest Products Ltd (NZFP) and Tasman Forestry Ltd (TFL) engaged in their own intellectual property ventures associated with genetic improvement. TFL had established its own successful seed orchard. John Gleed then led TFL's heavy investment in nursery operations and laboratory-based biotechnology. Based at Te Teko in the Bay of Plenty, its Centre for Advanced Biotechnology later operated within FCC. One of us (WJL) was engaged there for 6 months per year in an advisory and planning role during 1992–1996, with responsibilities for clonal testing, clonal selection, characterisation of clones and "multiclonal varieties," clone deployment, and maintenance of genetic diversity in plantations.

6.2.2.3 Chile

In Chile, their breeding cooperative had by 1991 overcome political ideology by calling itself a "Cooperativa" instead of a "Convenio." Despite some major changes in ownership of some member companies up until the 1980s, the cooperative continued operating in the same format, but under some changes of Director. Anyway, its seed orchards were generally unusually successful in producing improved seeds of its developing breeds during 1984–1997. However, very little was done to integrate the various breeding programmes into a national scheme of genetic management for radiata; specifically, there was little or no research then to characterise the patterns of possible genotype-by-site interactions among contrasting sites in Chile's plantation regions.

In 1990 the Arauco group of companies, the largest in the forestry sector, set up its own biotechnology company, Bioforest. It was led by Claudio Balocchi, an ex-student of Roberto Delmastro at University Austral, PhD graduate from NC State University and ex-Director of the Cooperativa.

During this chapter's 1984–97 period, the Chilean Cooperativa expanded its membership from the original seven companies plus the Chilean forest service (CONAF) to 15 full members. By the end of 1994, despite no solid evidence for or against important genotype-by-region interactions, some members were operating more than one region-specific breeding programme (C.J.A. Shelbourne, pers. comm. 1995).

6.2.2.4 Other Countries

Around 1990, Spain engaged in a radiata breeding programme, within the Basque Autonomous Region, under the auspices of the Basque regional government. It was on a modest scale and focused strongly on producing increased wood volume. Operational breeding was hampered by extremely fragmented forest ownership, and difficulty of finding good seed-orchard and nursery sites within the region. In 1992, but largely independently, another radiata breeding programme was begun in the region of Galicia (Codesido and Fernández-López 2008, 2009), in north-western Spain, under the auspices of its regional government.

In South Africa there were several changes affecting its principal tree-breeding agency (Liebenberg et al. 2004). In 1984 the Directorate of Forestry was renamed the South African Forestry Research Institute (SAFRI), and in 1990 that was incorporated in S. Africa's Council of Scientific and Industrial Research (CSIR) as the Division of Forestry Science and Technology (FORESTEK). One important result of that incorporation was awarding FORESTEK intellectual property rights over all of the country's select genetic material of several species of forest trees, including archived and seed-orchard clones of radiata. Subsequently, in 1997, FORESTEK was merged into the CSIR Division of Water, Forestry and Environment (Environmentek). In 1991, when the FORESTEK professional workforce was overcommitted with the breeding programme's responsibilities, Tony Shelbourne from New Zealand served as an external consultant for tree breeding in South Africa. On the silvicultural front, Stellenbosch University had become heavily involved in mineral nutrition of radiata (for example Donald and Glen 1974; Donald et al. 1987; de Ronde and Donald 1993).

In 1993, all state plantation forests were commercialised under a state-owned company, SAFCOL, which then wanted to operate its own breeding programmes. An acrimonious dispute over pricing by FORESTEK for use of its intellectual property in forest genetics led to litigation and that dispute severely hampered much of the subsequent tree-breeding effort.

6.3 Developments in Planting and Silviculture

6.3.1 Planting Activity

Planting of radiata fluctuated strongly during this period, for somewhat different reasons among the various grower countries.

In Australia, the state of New South Wales led an expansion in hectares planted, it becoming the state with the largest area of radiata plantation (Table 6.1). Some of the expansion included afforesting ex-pasture sites, driven by a decline in the profitability of grazing. There the accumulated soil fertility was associated with far worse tree form (Birk 1990; Birk et al. 1993) than was usual on ex-forest sites in NSW, and thus the ineffectiveness of previous selection for tree form on ex-forest sites by the NSW breeding-programme was brutally exposed. The occasional individual nursery operation, for example one operated by Francis Clarke that was based on cuttings with enough maturation to substantially improve bole form, did address this problem on farm sites (Sect. 6.9).

In New Zealand, the fluctuations were particularly severe (Fig. 3.1). New planting declined sharply from around 50,000 ha/annum in 1984, ending a planting boom that built up in the 1970s. This drop came after a new government axed forestry encouragement grants, bottoming out at under 15,000 ha/annum in 1991 and 1992. Immediately afterwards, a new and even bigger planting boom began, driven largely by a sharp rise in export log and timber prices and a decline in pastoral export prices. That boom peaked at around 100,000 ha/annum, but then

Table 6.1 Areas ('000 ha) of radiata pine plantation in Australia, by state and ownership class, as at March 1989 (Lavery 1990)

State	Area		
	State	Private	Total
New South Wales	165	67	232
Victoria	67	103	170
South Australia	66	26	92
Tasmania	39	31	70
Western Australia	38	16	54
Australian Capital Territory	14	0	14
Queensland	2	1	3
Total	391	244	635

started declining steeply around 1996, following declining log export prices associated with economic recession in some east Asian countries and increased availability of some alternative wood sources. Significantly, much of the planting during the latter boom was on ex-pasture sites, with major participation by many small-scale owners, who collectively and as a result then owned a substantial portion of the national plantation estate. There were two groups of these small-scale owners.

One group was mostly composed of "farm foresters," namely farm owners who continued to plant new stands of trees on their farms, often on problem areas such as erosive hillsides. Their NZ Farm Forestry Association published its own journal, the *NZ Tree Grower*, and held frequent local and annual national meetings to exchange information. The members were often tree enthusiasts, and they planted native species and several exotics in addition to radiata.

The second group, "Partnerships" whose partners were largely concerned with retirement incomes, were mostly organised by several active consulting-forester organizations. These organisations would invite investors to buy shares in proposed plantations. Once a venture was funded (usually with 20 shares), the consulting organisation would buy, lease or share-crop an area of land and plant it, usually to radiata. Unlike most of the farm foresters, the share-owners did not own the land, nor do much hands-on management of the resulting plantation. Their investments had some liquidity, as their shares could be sold at any time, sometimes for speculative profit, sometimes for other reasons.

Both groups shared a strong interest in the new seedlots and clones emerging from RPBC activity and Fletcher Challenge's cloning operation at Te Teko. They were excellent clients as a result, providing both markets and feedback for the new seed stocks and tested and characterised clones.

In Chile, new planting continued at around 70,000 ha/annum until 1993 (Toro and Gessel 1999) (see also Table 6.2), then dropped over several years to around

Table 6.2 The radiata resource, Chile, as of 1990 and 1996 (From Lisboa 1993 and 1996 figures from FAO *ex* Instituto Forestal statistics)

Region	Lat. (~°S)	Area (ha)		Standing volume (million m^3)	
		1990	1996	1990	1996
IV	33	–	656	–	–
V	33–34	23,025	15,853	3.8	2.0
RM	34	966	1,001	0.1	0.2
VI	34–35	57,617	59,715	6.1	7.9
VII	35–36	280,467	326,422	23.6	35.0
VIII	36–38	570,735	642,705	84.2	103.4
IX	37–39	220,731	224,672	25.8	35.5
X	39–41	89,752	116,017	8.9	14.3
Total		1,243,293	1,387,041	152.5	198.3

30,000 ha/annum (Instituto Forestal 2011). At least four factors contributed to the drop, although their relative contributions seem unclear:

1. A government-funded encouragement scheme for afforestation that had been initiated in the mid-1970s ended in 1992, although it was almost immediately revived in much the same form;
2. A resurgence of planting eucalypts in the country, species that better fit with Chile's heavy emphasis on pulp production and its heavy reliance on stone and cement for residential construction, thus making structural timbers of pine relatively less attractive;
3. The arrival of the European pine shoot moth, *Rhyacionia buoliana* which was first detected in Chile in 1985 (Lanfranco et al. 1991). Having badly affected radiata in parts of Europe and Asia Minor, this pest rightly caused caution until its effect could be evaluated in Chile. Having evidently come by way of Argentina, the moth spread from Chile's south, producing crown and bole malformation in many of the affected radiata trees that led to log degrade (Fig. 6.1) and reduced wood recovery at harvest.
4. Arson attributed to native Americans with grievances over land ownership.

Actually, areas of Chile's radiata plantation estate at any one time tend to be less than cumulative areas of reported new plantings, for reasons that are not fully clear.

In South Africa, research on phosphatic fertilisers (for example Payn and de Ronde 1988; de Ronde and Donald 1993) confirmed that the productivity of radiata could be raised substantially on many sites by applying such fertilisers, where it could then become preferred over the less nutrient-demanding maritime pine. However, the pitch canker fungus, *Fusarium subglutinans* f.sp. *pini*, was recorded in South Africa in 1990 (Viljoen and Wingfield 1994), and was subsequently

Fig. 6.1 Persistent malformation caused by tip moth damage on a radiata pine in Chile (Photo G. Hosking)

Table 6.3 Areas of radiata plantation (ha) in Spain as at 1975 and 1995, by region and ownership class (MM Ruiz Urrestarazu pers. comm. 1998)

Autonomous community (i.e. Region, W to E)	Public		Private		Total	
	1975	1995	1975	1995	1975	1995
Galicia	25,717	18,930	7,210	14,684	32,927	33,614
Asturias	17,825	12,488	8,202	13,705	26,027	26,193
Cantabrica	12,397	14,079	2,565	3,363	14,962	17,442
Basque Country	21,646	20,302	140,673	129,897	162,319	150,199
Others (Navarra most important)	8,002	8,670	2,551	2,446	10,553	11,115
Total	85,587	74,469	161,201	164,094	246,788	238,563

becoming troublesome with radiata (Wingfield et al. 1999; Mitchell et al. 2011; J. Mather pers. comm. 2011). At first, pitch canker's main impact lay in poor survival of recently-planted seedlings, resulting from infection of those seedlings in nurseries. But in California, it had by 1997 proved capable of reducing growth on trees of all sizes, malforming crowns as branches died, and even killing radiata saplings over several years of repeated reinfections (Mitchell et al. 2011). South Africans were thus rightly cautious about committing more plantation area to radiata, among the most susceptible of conifers to pitch canker, until its longer-term impact became clear.

In the Basque Autonomous Region of Spain, the area of radiata plantation had stabilised around 1973 at about 160,000 ha, amid some environmental concerns. It was reduced slightly by fires in 1989, followed by some recovery (Michel 2006; see also Table 6.3). Expansion, however, was checked by many farmers preferring to grow short-rotation crops of *Eucalyptus nitens* which is more frost-tolerant than *Eucalyptus globulus*. In the geographically separated region of Galicia, in the north-western corner of the country, new planting of radiata accelerated under incentives to retire marginal farmland from farming, its plantation area reaching 82,000 ha by 2008 (Codesido and Fernández-López 2009).

Thus the worldwide expansion of radiata as a plantation crop largely stalled by 1997, although more intensive domestication continued in New Zealand and began in Chile.

6.3.2 Silviculture

6.3.2.1 Overview

In the previous chapter (Sect. 5.4) we covered revolutionary developments in tending regimes and nursery practices, which were mainly spearheaded by and from FRI for radiata in New Zealand. These developments were aired in a 1989 international conference held at FRI (James and Tarlton 1990), and encapsulated in the *Radiata Pine Growers' Manual* by Maclaren (1993).

Nursery practices for radiata had been researched and refined by FRI in New Zealand, allowing production of more uniform and more robust planting stock that not only survived outplanting almost 100% but usually resumed vigorous growth shortly after outplanting. Largely by their engaging some trained New Zealand nurserymen, that nursery technology became adopted in Chile, where the criterion of successful nursery practice had been no more than good post-planting survival.

In south-eastern Australia, the loss of plantations and some seed orchards to fires in early 1983 posed special challenges to produce very large quantities of genetically acceptable planting stock in a short time. That prompted variably successful work on vegetative propagation, using nursery cuttings to multiply planting stock from small quantities of remaining genetically superior seed.

In New Zealand, a similar pursuit of such propagation technology was already continuing, both to capture genetic gain sooner, and to reap greater benefits from more precise deployment of planting stock. This included the technologies to yield hundreds or even thousands of plantable trees from a single valuable pedigreed seed. The technologies for such **vegetative multiplication**^{G} were developed at FRI and then developed further and massively applied at Tasman's nearby Centre for Advanced Forest Biotechnology at Te Teko.

Control of weed growth in young plantations had been developed considerably, with an increasing awareness of the importance of root competition for water and nutrients. With increased trade and travel, new exotic weeds were arriving in all of Earth's countries. Burning of logging debris, which helped control some weeds, was becoming unpopular. New herbicides were becoming available, and public pressures were growing to outlaw some traditional herbicides. Accordingly, active herbicide research continued, and new herbicides were adopted. Not all weed species are readily controlled by herbicides, and some were found to be amenable to biocontrol, particularly by insects imported from the host weeds' native ranges. Overall, research on weed control was accepted as an ongoing process.

The new Direct Sawlog regime for tending young radiata stands in New Zealand prescribed lower initial stockings and earlier and heavier "thinning to waste" (logs were rarely removed and milled, but were usually left with their needles and branches to decay and recycle their contained nutrients on-site). It also combined the thinning with pruning of the retained "crop" trees in order to improve their clearwood yields. For several reasons (Sect. 5.4.4) New Zealand developed and vigorously implemented this regime, whereas other radiata-growing countries took a less aggressive approach. Not surprisingly, it was in New Zealand that the problems with the regime were encountered first and most strongly.

After initial development of the Direct Sawlog regime, however, came the development of variable-lift pruning, which entailed repeatedly and frequently pruning each bole to a constant prescribed current diameter, rather than to a specified height (Koehler 1984) (Fig. 6.2). That overcame the problem of pruning to a constant height above ground on trees of different heights, exacerbating size differences among neighbouring trees associated with different amounts of live crown retained. Importantly, it better controlled the variation in knotty-core diameters and in amount of valuable clearwood outside the knotty core.

Fig. 6.2 Variable-lift pruning in Tikitere agroforestry trial near Rotorua, New Zealand. Trees in this case were all pruned to half height, rather than to a fixed height, to avoid exaggerated differences in tree size. A preferred alternative now is to prune trees to all the same diameter

The impetus for adopting the Direct Sawlog regime roughly coincided with a decline in the profitability of pastoral farming. In response to this combination, New Zealand had also developed an agroforestry regime based on converting pasture to forest plantations, with early, heavy pruning, in the course of which grazing would provide interim revenue in lieu of harvested thinnings. Large areas of pasture were converted on this basis. Australia followed suit to a significant degree, especially in New South Wales. As with the Direct Sawlog regime, however, problems emerged that had not been fully foreseen.

6.3.2.2 Some Rethinking on the "Radical" Regimes

As it turned out, the adoption in New Zealand of the "radical" Direct Sawlog and Agroforestry regimes for radiata, with early, heavy, unharvested thinnings, was far from a total success, for several reasons (Burdon et al. 2008; Burdon 2010). The problems encountered involved both tree form and basic wood quality. It slowly became clear that how well the Direct Sawlog regime (or its variants) succeeded depended strongly the degree of exposure to wind. The "one-hit" thinning study underpinning the Direct Sawlog regime's thinning prescriptions had been on a single, very sheltered site. Results were not nearly as good with Direct-regime plantations on severely exposed sites, especially if they were fertile. There, tree form was often very bad under such regimes, with prevalent forking and multileaders (very often following top breakage), gross branching, and crooked or leaning stems.

Also, while selective breeding for "multinodal" (or "short-internode") branching had become the avenue for genetic control of branch size, the control achieved proved to be inadequate under the radical silvicultural regimes, especially when growth was further boosted by use of fertiliser or by growing the trees on pasture sites where fertility was elevated. This particularly affected the second log, which Fenton had early identified as a problem zone of the bole. The second logs were judged too high to prune profitably, but contain very important wood volume, and by age 25 would typically be large enough for cutting into valuable solid-wood products.

Furthermore, the selection for the short-internode branching habit not only had limited success in controlling branch sizes under the new thinning regimes. It also greatly reduced the average length and proportion of clearcuttings available between clusters of knots. Taking such clearcuttings can recover useful lengths of clear timber from between widely separated clusters of large knots. However, as was also known by 1984, selecting trees with a long-internode branching habit incurred a penalty in growth rate, and a severe penalty in terms of tree form with low initial stockings and heavy thinning, especially on exposed sites and/or with elevated soil fertility. On the other hand, trees with a short-internode habit could be pruned satisfactorily and would yield good lengths of clear timber in pruned butt logs.

Another problem was that while some sacrifice in wood yields per hectare through adopting such radical regimes had been accepted, the actual drop in harvested volumes proved much more than anticipated (Manley 1986). Actually, an early warning of such sacrifice had been issued by Elliott and Goulding (1976), but this was not followed by the full-length paper.

Even more disappointments with Direct Sawlog regimes came in wood properties (Burdon 2010) (Fig. 6.3). Soon after breeding of radiata began at FRI in the 1950s, enough was already known about variation in wood properties (Harris 1965) to foresee some adverse effects of the lower harvest ages that Direct regimes would permit and foster. This was largely because in radiata wood density, and other wood properties such as tracheid length and microfibril angle, exhibit large and predictable changes with number of rings from the pith. The density of innermost rings, for example, is frequently about 0.32, increasing ring by ring to values near 0.40 by about ring 15 from the pith (cf Cown and McConchie 1980; Harris and Cown 1991). The innermost 10–15 rings, called corewood, make up very high proportions of wood harvested at about 24 years, but wood harvested at about age 35 years has much higher proportions of the more desirable outerwood which has longer fibres and much better stiffness and breaking strength.

Functional relationships in radiata between growth rate and wood properties can be complex. Corewood typically has wider growth rings than outerwood,

Fig. 6.3 Sixteen-year-old radiata stand planted in pasture with progenies of 1950s New Zealand plus-tree selections. Planted at 750 trees per hectare, it was thinned after eight years to 250 stems per hectare, showing outstanding growth, but good form despite low stocking. At 18 years predominant mean height was 33.5 m, with 480 m^3 stemwood per hectare (well before culmination of mean annual increment). Wood stiffness, however, was extremely disappointing, reflecting youth of trees, fertility of site, low stocking, and some effect of selection of parents for growth rate (From Burdon 2004, permission Elsevier)

partly because a given sectional-area increment in corewood means wider rings than the same increment in outerwood. This effect can lead to an element of largely spurious correlation between growth rate and wood density. However, aside from this effect of stem geometry there are additional ways in which growth rate can be adversely associated with density and other important wood properties.

Importantly, until the 1990s, selection criteria emphasized diameter growth, and largely disregarded its effect on wood density and other important wood properties. This largely reflected a lack of clear signals from industry or the log markets on the economic worth of prospective genetic improvements or deterioration in wood properties, rather than a lack of concern among breeders and other researchers (Burdon 2010). Another deterrent for breeders, however, lay in difficulty and costs of evaluating various wood properties.

In addition to the effects on wood properties of just the trees being younger at final harvest, there emerged some more direct effects of site and tending regimes on radiata's wood properties. Crucially, these effects involved wood stiffness, a key property for structural timber, and dimensional stability, an important property for many uses. Stiffness was affected to an extent far out of proportion to any effect explainable, as previously thought, by variation in wood density. The dimensional-stability element, important both during seasoning and in service, was also badly affected by growing on fertile sites and by silvicultural regimes involving low stockings. A role of microfibril angle as a determinant of wood stiffness and stability had been noted quite early (Cave 1968; Cave and Walker 1994), but this was much subordinate to the role of density in determining mechanical properties of wood from New Zealand's "old-crop" stands planted around 1925–1935. Thus, the role of microfibril angle was seriously underestimated until harvesting began in younger stands and/or on more fertile soils.

These adverse effects on wood properties were doubly bad in combination with the disappointing control of branch size and the reduced scope for cutting clear timber between knot clusters that resulted from selecting for the "multinodal" branching pattern. Furthermore, the heavy thinning and aggressive early pruning were associated, on some sites, with blemishes such as resin pockets (largely on dry sites) and "bird's eye" figure (on fertile sites) that negated the benefits of pruning. While these various ill-effects could have been mitigated by less aggressive practices, with longer rotations, doing so would have largely negated the cost savings of growing the wood on Direct or other "radical" regimes.

Yet another factor, involving the early seedlots collected from seed orchards, was an appreciable genotypic drop in wood density, although this was later rectified by adjusting clonal composition of orchards.

The agroforestry option, growing widely-spaced trees in existing pasture, mostly failed to live up to its early promise with radiata. Admittedly, it could produce greatly enhanced volumes of wood, mostly through faster diameter growth rather than enhanced height growth. This reflected a build-up of soil fertility through intensive pastoral farming, in the form of sustained nitrogen availability that could not be matched by the pulses of nutrients provided by fertiliser applications. However, the intermediate yields of pasture were less than anticipated in quantity (Percival and Knowles 1988) and especially in food value for the livestock (Percival et al. 1988). And the effects on tree form, involving large knots and prevalent bole malformation, particularly above the pruned logs, tended to be much worse than projected by the early models. That was especially so for wider spacings (Fig. 6.4) on exposed sites. Moreover, at the very wide spacings timber

Fig. 6.4 Inside a 20-year-old agroforestry stand of radiata thinned early to 100 stems per hectare on a fertile ex-pasture site. While pasture production continued well at this stocking, branch sizes above pruning were excessive despite seed-orchard parentage. Also height growth was strongly depressed (largely through increased top breakage), and sacrifice in wood volume at the low stockings was major, nearly half. With genetically unimproved stock tree form would have been far worse

yields were disappointing, and wind breakage contributed strongly to the poor form (Maclaren and Knowles 1999). Furthermore, the effects on wood quality were often very bad, with poor stiffness and dimensional stability, and prevalence of blemishes, in addition to the increased knottiness (Fig. 6.5).

6.4 Activities and Progress in Operational Genetic Improvement

6.4.1 Orchard Developments

PROSEED was set up in New Zealand in 1987 exclusively as a seed producer *cum* seed merchant. The option of vegetative multiplication of limited amounts of control-cross seeds was therefore not made available to that company, thus motivating it to refine seed-orchard technology. Its existing open-pollinated seed orchards were by then producing large quantities of seeds of proven (if not the most advanced) radiata production populations and, with the help of export markets, easily-shipped seed-orchard seed provided PROSEED with good cash returns.

Its Amberley seed orchard, with its known excellent seed-producing performance, was soon chosen for large-scale production of control-cross seeds from top-ranked parents. This possibility was enhanced by very close planting of ramets, adapting the "meadow-orchard" system that had been developed for fruit trees (Sweet and Krugman 1978; Arnold 1990). PROSEED's ramets in the Amberley Seed Orchard were available to meet the needs of clients who included STBA,

Fig. 6.5 View of stand on same site at much higher stocking, at age 30. At 400 stems per hectare (sph) mean annual stemwood increment (mai) culminated at ca. 37 m^3 ha^{-1} yr^{-1} at ca. 27 years, at 300 sph it culminated at ca. 32 m^3 ha^{-1} yr^{-1} at ca. 32 years (from Maclaren and Knowles 1999). Better tree form and higher mai at higher stockings came with much less pasture production

various private and corporate forest nurseries and NZRPBC members for such controlled pollinations. The clients could order customised groups of pair-crosses to be done by PROSEED's Amberley staff, usually based on information available for the newly devised seed certification system (Sect. 6.6.4.1). This practice was particularly effective because of the "meadow orchard" management system and prolific seed production by Amberley's ramets (Fig. 6.6).

Fig. 6.6 Amberley Seed Orchard showing large-scale controlled pollination in closely spaced "meadow orchard" area two years after planting (From Burdon 2004, permission Elsevier)

For making crosses for advancing the breeding population, however, crossing archives maintained by FRI and the Centre for Advanced Biotechnology in the North Island continued to play key roles.

In Western Australia, a variant of the Sweet-Krugman system was adapted to produce HAPSOs (Hedged Artificially Pollinated Seed Orchards) (Butcher 1993). Begun in 1985, they did not aim at total control of pollination; rather, the configuration of ramets was adjusted to take advantage of known pollen- and seed fecundity of various clones, and to thereby achieve a high proportion of desired pollinations at limited cost.

6.4.2 Breeding Work

6.4.2.1 Australia

The fires in South Australia in early 1983 had led to a catastrophic loss of breeding-population material, including local plus-tree ortets and ramets, stored pollen and seeds from those and other trees, and established seed orchards, as well as such things as pedigree records, pollination bags and orchard equipment. That required starting almost afresh but with much better breeding theory and technology than before. Accordingly, a round of fresh plus-tree selection was embarked upon (Boomsma 1997; Wu et al. 2007), with 600 new plus trees being selected under the auspices of the STBA in the Green Triangle of South Australia and western Victoria. That was promptly followed up by the establishment of open-pollinated (OP) progeny tests, strictly limited to families from those Green Triangle land races.

In New South Wales (NSW) (Johnson et al. 2008), fresh plus-tree selection had begun in 1980 in post-war land-race plantations, with strong emphasis on vigour being sustained until near harvest age. That new search continued until 1986, producing 347 new plus trees. From these, OP progeny-tests were established during 1985–87. Some further plus trees were selected from families growing in NSW replications of Shelbourne's International Gene Pool (IGP) trials; seeds from those trees were likely to contain genes from males also within those trials, and thus both parents could have been from any of the land races and native populations included in the IGP collection, a possibly important addition to NSW's radiata breeding base. Those seeds were used to establish new and fresh "widely-OP" progeny tests during 1984–87. During 1984–89, three seed orchards were established, largely with reselected first-generation plus trees and some second-generation selections coming from their local OP progenies; one replacing the oldest seed orchard, and two *de novo*.

As noted above, the NSW breeding operations were brought within the scope of the New Zealand Radiata Pine Breeding Cooperative programme in 1990, whereas the other Australian states stayed with the STBA.

6.4.2.2 New Zealand

Immediate breeding decisions during the period of this chapter were driven by two considerations. First, with seed orchards on the point of being able to meet all New Zealand's planting needs, radiata land-race stocks, a huge but basically unexploited and uncharacterised gene resource, were subject to increasingly rapid liquidation through harvesting. Second, it was realised that the management of the existing breeding population (Fig. 5.20) had led to a steep decline in effective population size (status number), because many of the original plus trees and their descendants had effectively been culled and not archived. Accordingly, the decision was made, in 1985, to select 1000 additional New Zealand plus-trees, to both enlarge the breeding population and sample more of the remaining land-race stocks (Burdon et al. 2008). This operation was spread over three years: 1985 (468 plus trees), 1987

(243), and 1988 (263), although only in 1988 was a wide geographic spread of local land races sampled within New Zealand. A short-internode ("multinodal") branching habit was no longer required, although it still predominated among the selections. Clonal copies of those 974 additional plus trees were not routinely archived. However, OP progeny tests were established promptly, but some plantings soon failed, largely because of continuing institutional upheavals in the forestry sector. Assessing even the surviving material, analysing data and documenting results required a major effort on the part of FRI.

Also, from the mid-1980s, plus trees were selected in New Zealand from *ex-situ* plantings of the five native populations, with main emphases on the Monterey and Guadalupe populations. Those selections were made in 1986, in the Genetic Survey experiment planted during 1964–67, and again in 1994, selecting trees grown from the 1978 seed collection from the three California mainland populations.

From the Guadalupe samples growing in trials from the 1960s up till 1980 a total of 80 trees were selected and archived, and then pair-crossed to create a pure Guadalupe breeding population (cf Low and Smith 1997). Eventually, in 2000, that was planted out in three field progeny trials, and served two purposes. First, given that F_1 hybrids of Guadalupe and mainland-origin material were performing very promisingly, the pure Guadalupe breeding population created a base from which to generate even better F_1 hybrids. Such use of island genes offered a means of capturing some of the superior wind-firmness, bole straightness, wood density and stiffness, and resistance to western gall rust exhibited in various trials of the Guadalupe population. Second, it was seen as important to ensure the perpetuation of the Guadalupe population, which was then critically endangered by goats on Guadalupe Island. Not precluded, though, is the option of introgressive incorporation of some Guadalupe genes into the main breeding population.

Later, towards 1997, around 30 of the best Cedros-origin trees in New Zealand were selected and archived, so that all five native populations were included in the breeding effort.

Controlled crossing continued among select parents drawn both from the base populations and their first-generation offspring, generally using double-pair matings, with each parent crossed with two others. It was designed to balance several goals: keeping controlled pollinations to manageable numbers; retaining a large enough next-generation "effective population size"; capitalising on being able to use most clones as both pollen- or seed-parents; reducing risks of losing lineages through failure of some crosses; and reducing exposure to adverse or misleading non-additive gene effects occurring in the progeny of particular pairs of parents.

After the selection of additional plus trees from local land-race stocks, major effort was involved in initial assessment, data analysis and write-up of results. For assessing these and earlier trials, non-destructive evaluation of wood density, using a torsiometer and later a penetrometer (Pilodyn), became routine, along with the traditional assessment of stem diameter, tree-form traits and tree health.

By 1997, the propagation programme of Tasman Forestry Ltd and then Fletcher Challenge Corporation at Te Teko was embracing actual breeding work of its own. The operation was based on mass vegetative propagation of intensively select and tested pair-crosses and clones. During the 1990s the focus moved from multiplying select families to select clones, characterised for growth rate, branching ideotype, disease resistance and a range of wood properties, to be sold as multiclonal mixes ("varieties"). This took advantage of and built on many years' work in the national breeding programme to give parents with known high breeding values for growth, branch distribution, bole form and some disease resistance.

Breeders at Te Teko made pair-crosses among top-rated parents, and germinated (on average) about 100 seeds from each pair-cross, for preliminary multiplication, clonal storage, testing, evaluation and selection, prior to mass multiplication of the selected clones. Clonal screening among these offspring was the basis for a search for "correlation breakers" that maintained or even further improved on those breed traits but had above-average wood properties as well. Clones were variously maintained as hedges or cool-stored tissue cultures, for mass-propagation as cuttings or tissue-culture plantlets. They were marketed to various clients as "multiclonal varieties" which were mixes of clones that were customised for specific sites and regions.

6.4.2.3 Chile

After first-generation selections were made by Chile's companies in the breeding cooperative, mostly during 1976 to 1978, open-pollinated (OP) progeny-tests were established promptly (Balocchi et al. 1991). Each test set had little or no site replication and, because of the program independence of the different companies, generally almost no replication beyond the planting zones containing the parental plus trees of each of the cooperative's members.

Six-parent disconnected diallel crosses (crossing each plus-tree parent with each of the five others within the group) were soon made, following latest practice within the North Carolina Tree Improvement Cooperative. From these, control-pollinated progeny tests were established. These more elaborate trials had greater replication across sites, but still within each member company's planting zone(s). Because completing successful control-crosses for the diallels typically took several years, establishing these trials did not begin until about 10 years after planting the OP progeny-tests. That not all the diallel progeny sets were planted in the same year weakened or precluded many same-year or even same-site cross-comparisons among the six-parent diallel sets. Row-plot layouts continued to be used, rather than single-tree plots, sacrificing substantial precision of comparisons. These diallel trials took much time and effort to learn about very few families in this population of plus trees, but could supply some good research information on the additive and non-additive inheritance of several traits of interest.

6.4.2.4 Other Countries

Work on radiata breeding in South Africa dwindled to a very low level following institutional changes. In Spain, however, a radiata breeding programme was initiated around 1990 in the Basque Autonomous Region (BAR) with the support of the regional government in Vitoria. A modest number of plus trees were selected, mainly on the basis of stem diameter (M. Michel, pers. comm. 1992). Field tests of OP progenies were established. Work, however, was hampered by difficulties of finding a single good, level site for a nursery and a single satisfactory seed-orchard site. In 1992, another breeding programme was initiated in Galicia, in the north-west corner of Spain, also with the support of regional government. It began with the selection of 50 plus trees from which, along with six from the BAR, progeny-tests were established, and which were used to establish a clonal seed orchard (Codesido and Fernández-López 2009).

In France, pure radiata proved too vulnerable to continental winter cold (Chaperon and Fraysse 1986), while the hybrid with knobcone pine proved too vulnerable to dothistroma needle blight.

6.5 Delivery of Improved Seed

In New Zealand, seed-orchard production of improved seed caught up with national planting requirements by 1986 (Vincent 1993). From 1987, a drop in new planting, following more institutional changes—particularly Parliament's removal of some afforestation incentives—led to a surplus of seed-orchard seed being available. This in turn led to considerable exports of such seed. Most went to Australia, to serve areas where New Zealand selections had generally performed well in progeny tests. This lower local New Zealand demand also provided an opportunity to increasingly phase out first-generation orchards in favour of those with the newer and still better selections, and which were mostly on better and more convenient sites. Much of the early effort at developing new types of orchard, such as the meadow option, was in the Bay of Plenty, within easy reach of Rotorua, but conditions for them there were not very suitable. However, it became increasingly clear that parts of the eastern South Island offered far better sites for such orchards.

Australian seed orchards had performed variably. The first orchard, at Tallaganda, in New South Wales near Australian Capital Territory, had performed quite well (Johnson et al. 2008), and one near the coast in far eastern Victoria had performed well, as had one near Mt Gambier in South Australia, but other orchards had been less satisfactory. In 1983, however, the fires in South Australia destroyed most of the local orchards and created a vastly increased requirement for seed in order to replace burned plantations.

In Chile, their early seed orchards, which had been sited with the help of experience in New Zealand and Australia, generally came into good seed production within eight years of planting ramets (Balocchi et al. 1991).

South African seed orchards proved to be not well-sited, but this was not crucial because of a shortage of suitable land for expanding the estate of radiata.

In Spain, an early attempt at developing a seed orchard failed, with sites within the Basque planting zone proving very unsuitable. In response, a new seed-orchard site was chosen in a very different region, in Andalusia in the lower Guadalquivir valley (T.G. Vincent unpubl.).

Seed orchards, however, tend to be slow to deliver genetic gain, so they often came to be supplemented or even supplanted by vegetative multiplication of controlled crosses (Sect. 6.8.2). By 1997, three countries had begun using this option with radiata, but not all for the same reasons.

6.6 Demonstrations of Economic Payoffs from Breeding

Despite the acclaimed success of crop breeding and very favourable comparisons of offspring of plus trees with unselected control seedlots in progeny tests, there was scepticism in some quarters of the net benefits of tree breeding. Financial gains from tree breeding arise in three ways:

- improvement in recoverable log yields, which reflect both bole volume and form, and thus both more wood and less harvesting waste per cubic metre harvested
- reductions in production costs, which reflect better survival and fewer or no cull and low-value trees, reduced tending costs, and less expensive harvesting
- improved recovery and value of product per unit volume from harvested wood.

All these sources of financial gain are well recognized qualitatively, and to the extent that well-tested and well-characterised genetically-improved planting stock is understood and available, it will usually be eagerly sought by growers of radiata.

Less easy is quantifying the net economic benefits. For one thing, the benefits will depend on the interplay between management practice and genetic improvement, adapting one to the other, and hopefully profiting from synergisms between the two. For another, the wider benefits may depend greatly on the improved trees allowing development of better and more-reliable markets. These then support a more competitive industry, resulting in increased local employment and local infrastructure development, as well as serving such broad societal goals as good water quality and wildlife habitat.

Actually, some of the benefits of tree breeding, along with its downsides, came to figure slowly, if at all, in the computer models of crop growth that have served as decision aids for forest managers.

6.6.1 Field Performance

Trials comparing orchard seedlots and clones with unimproved controls, in order to prove and quantify their levels of genetic improvement, have given very

Fig. 6.7 Two adjoining plots in a genetic-gain trial in Kaingaroa Forest, New Zealand. On the left is a seedlot from a seed orchard based on selections from Canterbury (eastern South Island); on the right, a well-performing cross between two parent clones selected within the region. The latter was superior even when tested in Canterbury

consistent pictures of greater growth and harvestable volumes of useful wood in New Zealand (Figs. 6.7, 6.8 and 6.9; Table 6.4). Indeed, the consistency across sites, rather than statistical significance of differences at individual sites, was by the early 1990s a salient feature of the available early growth-and-form data. Early indications were very encouraging that such improvement expressed in young trees could satisfactorily predict performance to harvest age (Carson et al. 1999; see also Table 6.5), and have since been borne out by Kimberley et al. (2015). Interestingly, those early results indicated that the gains in volume production arose more from greater diameter growth rather than from taller trees. Not pursued, however, was if and how much reduced wood density may have contributed to those wood-volume gains.

The reduced bole malformation in the genetically improved stocks did undoubtedly reduce harvesting waste substantially, as well as making the harvesting easier, less expensive and safer. The improved safety, while not formally documented, is also an important societal benefit.

In addition to the setting up of genetic gain trials by FRI in New Zealand, Mike Carson was charged in the mid-1980s with researching the silvicultural performance of genetically improved seedlots (e.g. Carson 1989; Carson and Inglis 1989; Carson et al. 1991; see Burdon et al. 2008 for further details).

6.6.2 Impacts on Growing Costs

Reductions in growing costs began very early in the rotation with genetically improved planting stock. Far fewer such trees needed to be planted and tended, given confidence there would be less wastage through malformed boles, so that initial stockings then typically dropped in New Zealand by about 50%. Tending

Fig. 6.8 Two row-plots in New Zealand representing offspring of parents of contrasting ratings for growth and form

operations such as pruning also became cheaper with better-formed boles. And early costs such as spraying against dothistroma blight were reduced in material successfully bred for resistance and deployed to at-risk sites.

It was becoming appreciated that growing costs could be reduced by improving planting stock through means additional to genetic improvement and better nursery practice. Use of cuttings with some maturation had been shown to improve

Fig. 6.9 Inside 22-year-old radiata stand on a high-quality Kaingaroa Forest site, representing block-plots of controlled crosses between plus-tree clones selected in the 1950s. The better crosses gave stemwood mean annual increments of around 45 $m^3\ ha^{-1}\ yr^{-1}$ after 30 years

tree form in a California experiment (Tufuor and Libby 1973), although greater maturation meant much reduced diameter growth later in that same experiment (Bolstad and Libby 1982). Taking advantage of maturation state was also proving to reduce some growing costs substantially, although adolescent (with some maturation) cuttings were more expensive to produce than seedlings or juvenile cuttings. Neil Barr, the eminent New Zealand farm forester, had kept meticulous

Table 6.4 Percent improvement of mean age-12 stem diameter for seedlots of different Improvement Ratings (GF values – Sect. 6.6.4) at a country-wide selection of New Zealand sites, relative to performance of GF2 (Burdon et al. 2008)

GF value	Site														Mean
	1	2	3	4	5	6	7	8	9	10	11	12	13	14	
8	−4	2	6	6	1	4	4	3	0	0	−2	5	6	5	2.6
10	0	4	6	4	4	4	1	3	2	5	5	8	2	4	4.5
14	2	12	18	4	8	6	15	6	9	17	12	6	12	4	9.4
16	9	10	12	9	9	9	16	11	12	16	9	9	11	17	11.1
22	4	16	20	15	15	16	12	19	19	24	19	10	18	25	16.6

Table 6.5 Observed (in boldface) and projected percentage gains in stand stem volume production in two improved seedlots (GF14 and GF22) relative to climbing select (GF7) (From Carson et al. 1999)

Improvement rating	Age			
	15–17	20	25	30
GF14	**17**	15	14	13
GF22	**28**	26	24	22

cost figures while producing an 8-year-old agroforestry plantation of radiata at final-crop stocking, using different categories of planting stock (pers. comm. with WJL based on letter from Barr). He found that the growing costs were, even in that early stage of developing propagation technology, in inverse relationship to the initial cost of the propagules. His unimproved seedlings entailed the highest total early growing costs, his seed-orchard seedlings were intermediate, and his "adolescent" cuttings had the lowest growing costs. His cost savings were obtained by planting fewer seed-orchard seedlings and cuttings per hectare (not having to allow for culling malforms), by thus having fewer of those trees to be thinned pre-commercially, by fewer trees having to get first-lift pruning and, in the case of the adolescent cuttings, by lower per-tree pruning costs (Tufuor and Libby 1973) (note that because of their faster growth, the branches on his seed-orchard seedlings were the most expensive to prune).

Growing trees faster, thereby lowering harvest age, effectively reduces growing costs, by reducing interest charges (real or notional) against investment. Thereby genetic gains in growth rate could attractively reduce growing costs. On the other hand, the reductions in harvest age that that were being adopted with radiata, especially with the Direct regime, also reduce biomass productivity. Genetic gains in growth rate certainly could increase harvest volumes per hectare at both early and later harvest ages, thus promising better returns on outlays (Carson 1990). However, it only later became fully clear how both reduced rotation age and faster growth could have serious and negative effects on several wood properties.

6.6.3 Impacts on Financial Returns

Gains in profitability that would accrue purely from genetic gains in growth are, in principle, easily calculated. Not quite so easily calculated would be financial payoffs, direct and indirect, from improvements in tree form. Changes in value per unit of wood are more problematic, for two main reasons. First, genetic gains in wood properties tend to be interdependent with gains in growth rate: directly because of adverse genetic correlations, and indirectly, for example through growth-rate gains potentially reducing harvest ages to the detriment of wood quality. Second, markets may fail to give timely signals that reflect the true value changes resulting from genetic shifts, good or bad, in the wood properties.

Genetic gains in growth rate of radiata can reduce effective growing costs, but often with the unintended consequence of seriously reducing wood quality by adverse genetic correlations and allowing shorter rotations. Similarly, use of fertilizers to increase growth rates could apparently reduce growing costs by both increasing productivity and meeting log-size specifications at earlier ages, but again to the detriment of wood quality. Reduced stockings, at planting and/or through heavier thinning, also tend to reduce growing costs by reducing harvest age, but again at the expense of wood quality, with additional penalties in tree form in terms of larger knots and more prevalent bole malformation.

Even if genetic correlations are known, the impacts of genetic trade-offs between growth rate and wood properties can be strongly influenced by competitive interactions among trees in a stand, interactions that are still poorly known quantitatively. A case in point involves the trade-off between wood density and bole volume, a case where the pattern of genetic correlations is now reasonably known, with quite a strong adverse correlation between density and stem diameter but an appreciable positive correlation between wood density and tree height for constant diameter (Burdon and Low 1992; Burdon 2010). But the genetic correlations are only part of the story, since an unresolved issue is how individual-tree performance early in the rotation projects forward into whole-crop performance at harvest age. At present, however, the financial effects of genetic shifts are typically calculated quite early in the rotation. If based on single-tree performance, they do not account for the greater competition when most or all trees in a plantation are genetically superior for growth rate. And even if based on early whole-crop performance of advanced breeds or clones, these costs and benefits are usually calculated long before actual market demands and product prices can be known.

6.6.4 Certification of Genetically Improved Material

There has been a considerable history of seed certification for crop and pasture plants. It has traditionally included degree of purity of the seedlots and/or veracity of **cultivar-variety**^{G} identity. This approach was largely based on a system of producing and marketing distinctive and stabilised varieties, often through generations of inbreeding with or without subsequent hybridisation. As such, its protocols were not well suited to certifying genetic improvements of various forest-tree species, which have involved progressive population domestication over the few generations to date, rather than the production of either discrete, stable, sexually-produced breeds or well-characterised clonal varieties that represent crops in their own right.

From quite early on, in the 1950s, seed-certification schemes for forest trees have been based on features such as verification of provenance, selection of plus-trees, and whether the seed came from seed orchards that had been progeny-tested and then rogued. It was in the context of breeding radiata that tree-seed certification, and later clonal certification, broke out of this traditional framework for forest-tree planting stocks.

6.6.4.1 New Zealand

By 1986 the need for actively marketing genetically improved seed and planting stock of radiata was becoming increasingly apparent. New Zealand's seed orchards were on the point of satisfying the country's entire planting programmes' requirements for radiata with their genetically improved outturn (Vincent 1986; Carson et al. 1992). Also, it had become possible to use vegetative multiplication (VM) , by using nursery-produced cuttings or even tissue-cultured plantlets to produce planting stock. Such VM planting stock would be produced from top-rated pair-cross families, and thus offering higher and more consistent genetic quality than seedlings coming directly from OP seed orchards, although initially at higher cost per plant. Moreover, pressures were increasing to make plantation forestry more commercial, and opportunities were developing to export surplus improved seed profitably. Accordingly, a working group was set up, within FRI but including some industry personnel, to devise a certification scheme that would support effective marketing of genetically improved seed externally, and increase understanding and acceptance of radiata improvement within New Zealand. It first reported its recommendations in 1986 (S.D. Carson et al. 1986).

It was becoming possible to offer planting stock of various kinds and levels of genetic improvement, at various prices, presenting forest owners with much wider and clearer choices of seedlots and/or characterised batches of planting stock. To exercise such choices to full advantage, clients needed good, easily-comprehended information, and the Working Group's certification scheme was devised accordingly. It had two key features (Vincent 1987):

1. The scheme accommodated a way to express progressive genetic improvement of the different developing radiata breeds as their breeding- and production populations were advanced, rather than attempting to certify discrete, stabilized, uniform breeds. In this respect, tree-breeding and its products' certifications paralleled progressive herd improvement in animal-breeding schemes, rather than following plant-breeding traditions then common with long-bred,

short-generation plants such as grains. It also anticipated using a variation of the fruit-tree model of named clones.
2. The proposed new certification code that after 1987 characterised seedlots of radiata in New Zealand was descriptive of achieved genetic improvement, combining breed (defined by the set of traits that had been chosen for improvement) and average level of demonstrated change in those selected traits. This code then superseded those that had been based on NZ Forest Service (and various overseas) protocols, which had typically provided only year of collection, seedlot number, and a rough (and often loosely applied) categorization for "genetic quality" (Vincent and Dunstan 1989).

The new certification scheme provided seedlot codes with the following descriptors (Vincent 1987):

- breed (see Sect. 6.10), denoted by two letters: GF denoting Growth and Form (then the mainstream breed), LI the Long Internode breed, and DR the Dothistroma Resistant breed;
- improvement rating, denoted by a number. A higher number indicated a greater level of demonstrated genetic improvement for the breeding goal denoted by the breed name, rather than any absolute measure of performance; this meant that numbers assigned were to a degree arbitrary. For non-GF breeds, a supplementary GF rating could be added.
- collection year
- collection number
- for vegetatively propagated planting stocks, a supplementary alphabetic descriptor, initially either "S" for seedlings or "C" for cuttings.

For example, non-select seedlots from native populations would be assigned an Improvement Rating of 0, climbing select seedlots were designated GF7, and the then-available orchard seedlots were evaluated as GF14 to GF17 (Table 6.6). Later, GF Ratings climbed quite rapidly, to as high as 30 in 1991, but in material in which wood properties were disregarded for selection.

Improvement Ratings were assigned on the basis of two main sources of information: performance of the same or closely similar planting-stock mixes in genetic-gain trials relative to various controls, performance of the orchard parents' offspring in progeny tests, with number of seed parents contributing to the orchard seedlot mixtures modifying the rating numbers. (The Ratings were shaded downwards with low effective numbers of parents, on the basis of uncertainties concerning average merit of only a few unrelated families, or of a few large families dominating the mixture, rather than concerns over elevated biotic or abiotic risks). Anyway, reliable absolute figures for gains from breeding were initially elusive.

For the GF breed, Improvement Ratings were jointly based on changes in growth rates and bole-form, and their relative importance in this joint value was somewhat arbitrary (although favourable genetic intercorrelations among the traits concerned made this only a minor problem). In consequence, judgment

Table 6.6 Improvement ratings, for growth and form (GF), given to different categories of New Zealand seedlots (Burdon et al. 2008)

Category	Years	GF rating
Native-population collection	–	0
Unimproved "bulk" New Zealand collection	to *c.* 1962	[1–2]
Felling select	*c.* 1962–	<7?
Climbing select	1968–	7
850 seed orchard (rogued/reselected)	*c.* 1985	14*
268 seed orchard (rogued/reselected)	*c.* 1985	15*
268 seed orchard – rogued, top 16 seed parents	1985	19*
268 seed orchard (rogued/reselected)	1988	16
Top controlled crosses (for vegetative multiplication)	1988	22
"	1989	25
"	1991	26–28
"	1995	30

*Retrospective certification for rating of seed-orchard seedlots after the certification scheme came into effect in 1987.
Figures are given for "leading-edge" seedlots, with the proviso that they were based on multiple parents

calls were needed, which later led to a jibe that GF stood for "Gut Feelings" of the certifiers. But the heuristic value of the GF ratings was enormous at that time, as clients at least thereby perceived the advantages of the higher GF values and were willing to pay premium prices to obtain stocks with higher ratings within the GF breed.

In the late 1980s, during a partial lull in new planting, about 35% of the seeds from orchards that were rated in the range **GF14-to-GF17** were exported from New Zealand. However, New Zealand plantations had priority for seedlots with the better GF ratings, and New Zealand radiata planting-demand had again increased in the early 1990s. By 1993, only 5% of the GF breed's seeds rated **GF16** and above were exported, and these only to those Australian organizations that were members of the NZRPBC (Vincent 1993), an apparently substantial advantage of cooperative members.

This certification scheme was given trademark status as ***GFPlus*** in 1998 (RPBC 2002), with intellectual property rights held by the RPBC but with its secretariat still based within FRI/Scion.

Later, following the marketing practice being adopted in Australia, assignment of explicit **breeding values** to parents was subsequently pursued. (At Te Teko, breeding value of both the seed parent and pollen parent was used for predicting any pair-cross family merit.) Given breeding values of parents for various traits, and separate seed collections from individual certified parent clones in the OP orchards, together with knowledge of or variously confident assumptions concerning individual pollen-parent contributions, there then developed the basis for refining and customising the proportional representation of parents in seedlots and planting stock (Vincent 1990).

6.6.4.2 Australia

Certification of genetically improved radiata seedlots or planting stock being made available to its members and a market of non-members was developed within the STBA on a different basis. It depended on assigning trait-explicit breeding values to the performance of individual members of the breeding population(s), and averaging such values for the sets of parents of different orchard seedlots.

The individual breeding values of the parents were arrived at from the performance of offspring (OP and/or pair-cross) of the parents in progeny tests, along with any other available information on relatives. This was adapting methodology from herd- or flock-improvement schemes in animal breeding, which had been developed to a high level of computational sophistication. Assigning breeding values to particular assemblages of parents of their offspring, however, does depend heavily on adequate cross-referencing, through "**genetic connections**"[G] among progeny tests or genetic-gain trials.

The necessary computational software was progressively developed and eventually given proprietary status under the name TREEPLAN®, operated by Plantplan Genetics Pty Ltd, a subsidiary of STBA. It was not applied to radiata until 2002, and was subsequently to genetic data for several eucalypts in Australia and other species in Sweden and France.

The process of breeding-value estimation was superimposed as a refinement within the New Zealand certification scheme, which was superseded in 1998 as *GFPlus*®. Its application, however, has been complicated by limited connections between the large field trials which were historically planted in New Zealand. Those trials gave powerful cross-comparisons among large numbers of parents and offspring represented within individual trials, but only very weak cross-comparisons between material in different sets of trials.

6.7 Increased Knowledge of Genetic Architecture

Better knowledge of the attributes of the various native populations of radiata and of the inheritance of traits within populations was accumulating, and was being applied through publications, conference presentations and less formal communication among researchers and breeders.

6.7.1 Genetic Variation Among and Within the Natural Populations

Seeds from the 1978 collections in natural stands within the five native populations were, in the following few years, used to establish networks of provenance trials in both Australia and New Zealand (Eldridge 1998), with a few additional such trials in South Africa, Chile and Greece. Supplementing these trials in Australia and New Zealand were gene resource plantings, featuring large numbers of trees per population-sample within which selections could be made for performance in the exotic environments.

6.7.1.1 New Zealand

In New Zealand, these new provenance trials included only the three Californian mainland populations, although one small pedigreed planting was made from the Guadalupe Island collection. These mainland-population trials included a total of 13 "ecological subpopulation" seedlots, in each of which OP seeds from 22 to 70 seed parents were "bulked" together.

The trials were planted on 23 sites in 1980, spread throughout the country, and on one more site in 1983. Some of the sites were chosen to approach the limits of radiata's likely tolerances, hoping to give good resolution of provenance differences in survival and subsequent performance in marginal conditions. Assessment was done during ages 6–12 years from planting, generally starting on the sites with the fastest early growth. Follow-up assessments were done on seven sites at age 15, but showed little change from the earlier results.

From these trials, and from the earlier Genetic Survey results (Burdon 1992), there emerged a strong and clear picture of the adaptive profiles of the different mainland provenances (Burdon et al. 1997a, 1998; see also Table 6.7).

Important features of the adaptive profiles were:

- Año Nuevo—Generally well-adapted, except for poor growth on infertile, phosphorus-deficient clays, with somewhat better resistance than the other mainland provenances to frost and snow damage.
- Monterey—Also generally well-adapted, especially to poor soils and the general warmer conditions in the north of the country, but more vulnerable to snow damage in the south.
- Cambria—Highly vulnerable to needle-cast diseases, shoot dieback and snow damage, but tolerant of poor soils, and can show good stem straightness in absence of shoot dieback.

Differences in performance among subpopulation samples within populations, while not totally absent, were minor, despite major geochemical variation among the sites sampled within the Monterey locality. An exception was the northern (Pico Creek) subpopulation sample from Cambria, which had slightly poorer growth and markedly thicker bark. There were also surprising local variations among the Año Nuevo subpopulations in turpentine composition of twigs (Burdon et al. 1997b), which likely reflect genetic drift due to local fluctuations in population sizes resulting from episodes of stand clearances and recolonisation (cf Forde 1966, Fig. 12).

Table 6.7 Mean relative performance figures for diameter growth among mainland-origin provenances in different New Zealand site categories (From Burdon et al. 1997a, 1998)

Trait/Site category	No. of sites	Provenance			
		Año Nuevo	Monterey	Cambria	New Zealand
Stem diameter					
Infertile clays	3	92	105	102	99
Coastal dunes[a]	2	97	102	100	103
Volcanic plateau[a]	4	97	102	93	106
Central	5	98	100	96	105
Southern S. Island	6	99	100	96	104
Plot basal area					
Infertile clay	1	68	121	109	94
Coastal dunes[a]	1	96	103	99	102
Volcanic plateau[a]	2	91	101	96	116
Southern S. Island	2	126	89	57	127
Dothistroma resistance	3	108	104	71	122
Uprooting resistance[b]	5	133	91	84	100
Snow break resistance	3	106	94	86	125
Total snow resistance	3	124	90	81	104
Branch cluster frequency	18	92	100	98	114
Stem acceptability	14	99	96	96	112

[a] All in North Island
[b] Including uprooting by weight of snow
Value of 100 equals trial mean. Assessments made at 5½–11 years from planting, at roughly the same tree heights

Results for the three New Zealand "control" lots, included at each site, told an interesting story. Two of the lots, from North Island's Kaingaroa Forest and South Island's Nelson region, were evidently similar to the corresponding but independent regional land-race population samples of the Genetic Survey experiment planted in the 1960s. The third control ("Southland") came from the far south of New Zealand. Basically, the control-lot data supported or complemented the Genetic Survey results (Burdon 1992). Several generations of growing in plantation conditions appeared to have benefitted tree form and at least early vigour of the three control lots, which confirmed that they all came from developing regional land races.

Kaingaroa and Southland were evidently both founded mainly (two-thirds to three-quarters) from Año Nuevo, the only other apparent progenitor population being Monterey (Burdon et al. 1997b). The Nelson "control" seedlot, however, showed evidence of more Monterey ancestry (50% or so), with the balance from Año Nuevo (*op. cit.*).

The Nelson control lot did better than the other land-race lots in trials on phosphorus-deficient clay soils. This is consistent with Nelson's higher proportion of Monterey ancestry. However, the relative importance of origin of founding-seed origins versus several subsequent generations of natural selection on infertile Nelson soils is unclear.

Comparison of branching data from the Kaingaroa and Southland lots together with native-population samples indicated a clear genetic shift in response to natural and silvicultural selection in New Zealand. A marked increase in branching frequency (meaning a more "multinodal" habit) was particularly evident in the Kaingaroa lot (Burdon et al. 1997a, 1998).

Little additional information was obtained concerning the two island provenances, which were not included in the country-wide trial series. However, their samples imported in the 1960s both performed well in plantings on coastal dunes, and in hybrid combinations with mainland-origin radiata provenances (FRI unpublished reports).

6.7.1.2 Australia

As in New Zealand, some early differences among population samples from the 1978 collections were more marked and found earlier in provenance trials on sites characterised by specific limiting factors. The Cambria provenance was the most resistant to or at least tolerant of the important root pathogen *Phytophthora cinnamomi* (Butcher et al. 1984; Butcher and Stukely 1986), a serious limiting factor for growing radiata on soils subject to seasonal waterlogging in Western Australia. This led to a focus in Western Australia on this provenance for breeding.

A surprising early finding came from evaluations of salinity tolerances of samples from the 1978 collections in the native populations (Cromer et al. 1982). Young seedlings from the most-coastal subpopulations sampled at Año Nuevo and Monterey were the least tolerant of salinity; this difference in adaptation was tentatively attributed to the near-coastal localities perhaps being subject to the greatest dilution of salinity because of the heaviest fog drip.

A total of 67 trials of the 1978 provenance collections had been established at field sites, mostly during 1979–1981 (Eldridge 1998). For some trials, collaborators maintained seed-parent pedigree, but in others only subpopulation origin was tracked. Also, some trials included samples from both island and all three mainland native populations, while others included only subsets of the 1978 collections. Results have been reported piecemeal, largely because the trials were scattered among different state jurisdictions.

For eight New South Wales provenance trials at age 8 from planting (Johnson et al. 1997), the pattern of results showed some consistent features and some site-dependent ones. Among the mainland populations, on average, Monterey samples grew best. Año Nuevo, however, grew marginally best on the higher-quality sites, while Cambria samples grew as well as the other two or even marginally best on some of the poorer sites. Where included, the Guadalupe Island trees grew slower and Cedros Island trees consistently slowest of all.

As in New Zealand, variation among subpopulations samples within populations, while not absent, was minor. A local seed-orchard "control" lot did consistently best in the trials that were assessed.

6.7.1.3 Other Countries Trialling 1978 Native-Stand Seed Collections

In southern Greece, results from two sites at 12 years from planting (Matziris 1995) largely matched results being obtained in other countries. Among the native populations Año Nuevo and Monterey samples grew best, with Cambria slightly behind, Guadalupe further behind, and Cedros did worst with poorest survival as well as slowest growth. For tree form, Guadalupe rated best for stem straightness, and Año Nuevo the worst. Cold tolerance among the mainland populations was greatest in Año Nuevo and least in Cambria, in accordance with earlier results elsewhere (Hood and Libby 1980; Alazard and Destremau 1982). Interestingly, the Guadalupe-origin material compared well for frost tolerance, occasional freezing winds having been a factor in Greece. This does fit with the occurrence of freezing winds on the island (Sect. 2.1.3). Locally collected control seedlots grew at least as well there as any native-population material. The fastest growth, however, was observed in a seedlot collected from Guadalupe trees near Canberra (and presumably almost all hybrids with local, mainland-origin stock) and an Australian seed-orchard lot. Moreover, "Canberra Guadalupe" material was clearly the best radiata seedlot tested in Ireland for growth, tree form and resistance to *Cyclaneusma* needle cast ("yellows") (Thompson 2000). An earlier, anecdotal observation was that pure Guadalupe trees have shown best resistance to cold in Ireland (Alastair Pfeiffer pers. comm. to RDB).

In South Africa, six provenance trials from the 1978 collection, but including samples from only the mainland populations and two local control lots, had been planted in 1980 and 1981 (Falkenhagen 1991). After eight years on site, differences in growth and tree form among the three provenances were generally minor. However, Monterey grew slightly faster on some sites, while Año Nuevo was less susceptible to aphid (*Pineus pini*) attack. Differences among subpopulation samples within populations were generally very minor. Also as elsewhere, the local "control" lots, one from land races and the other from a seed orchard, tended to perform better with respect to survival, early growth, and tree form, but they did not differ between them, or excel for aphid resistance.

In Chile, seed samples from the 1978 native-populations collection, along with several kinds of control lots, were used for provenance trials on three sites spanning 4°22′ latitude (Jayawickrama and Balocchi 1993). After six years on site, results closely paralleled those from elsewhere: differences in average growth among the mainland provenances were minor but Monterey averaged marginally the best growth; pure Guadalupe was substantially slower than mainland material but had superior bole straightness; and Cedros was much the slowest. Again, the best-performing provenance seedlot was the one collected from open-pollinated Guadalupe trees growing in Australia; in it, release from inbreeding, hybrid vigour from much pollination by Australian land-race trees, and responses to local selective forces must have contributed in uncertain proportions to its superiority. All of the several kinds of seed-orchard lots, especially those from Chile, otherwise grew best, followed closely by Chilean land-race collections.

In Turkey the Año Nuevo provenance samples were inferior to those from Monterey and Cambria for diameter and stem volume on a clay site at six years from planting (Toplu et al. 1987—cited by Johnson et al. 1997).

6.7.2 *Within-Population Variation and Inheritance*

In several countries, advances in information about genetic parameters of various important traits in radiata were becoming available, as summarised by Wu et al. (2008). They mainly involved extensions and refinements of knowledge of: variances, heritabilities and coefficients of variation for individual traits, including growth and tree-form variables, disease resistances, and wood-property traits, at various ages; and genetic correlations, between traits at the same ages and different ages, between expressions of the same traits at different ages ("age-age" correlations), and between expressions of the same traits in different environments (an inverse measure of genotype-by-environment interaction). Moreover, the list of traits, especially several wood properties, was expanding. Rather than present details here, this information as of 2012 is summarised in Chap. 7. Instead, we now review developments in relation to some key issues that concerned tree breeders during 1984–1997.

6.7.2.1 Non-additive Inheritance of Individual Traits

A long-standing question had been whether non-additive gene effects are worth trying to exploit in delivering genetic gain through producing seed. Some of these effects (Sect. 4.3.2) are manifested as specific combining ability (SCA) which is basically the departure of the average of offspring from a pair-cross between two parents, from the average of the parents' breeding values or the sum of their general combining abilities (SCAs). Enticingly, non-additive gene effects have been one key to the superiority of "hybrid corn" varieties in maize. However, effective exploitation of such effects for hybrid corn has also depended on several other factors: being able to turn over generations quickly in order to develop uniform highly-inbred breeding lines; the desirability of such inbred lines for assuring crop uniformity; feasibility of maintaining numerous viable, if debilitated, inbred lines; feasibility of testing very large numbers of crosses between such inbred lines; occasional spectacular heterosis between such crosses due to SCA over and above ordinary release from inbreeding depression; and commercially most important, genetic segregation in the hybrids' offspring. That segregation means a loss of crop uniformity and vigour in farmer-saved seed, which sends farmers back to the seedsman each year. There is reason to believe that breeding maize to exploit purely additive gene effects could achieve as much or even more genetic gain, but between them the lure of crop uniformity and the seedsmens' commercial interests have prevailed.

Despite the obvious differences between forest trees and maize, the early forest tree breeding, including radiata breeding, generally entailed making multiple crosses per select parent. Reasons for this were several. They included inappropriate advice from animal breeders (Sect. 4.7.6), and less than due appreciation of the contrast with maize breeding, but also legitimate study of the comparative importance of non-additive and additive gene effects represented in SCA and general combining ability (GCA) respectively.

Results from these laborious pair-cross studies generally showed SCA effects to be less important than GCA effects for most of the traits studied. Of traits studied in radiata, bole diameter (and thence volume) generally showed the greatest importance of SCA relative to GCA, tree-form traits less, and wood density the least. Even for bole diameter, however, the contribution of SCA to its genetic variation appeared quite inadequate to justify making large numbers of pair-crosses per parent in search of immediate dramatic SCA-based responses, particularly when compared with the alternative of capturing genetic changes using additive breeding schemes and seed orchards (e.g. M.J. Carson 1986a). This conclusion was supported by Dean et al. (2006), who reported a strong decline in importance of SCA relative to GCA for bole sectional area with increasing age. Therefore, no justification was evident for involving individual members of the breeding population in numerous pair-crosses, either to achieve better estimates of additive and various kinds of non-additive genetic variances by employing such designs as complete diallels or to seek exceptional SCA in thus identified pair-crosses. Accordingly, practice in New Zealand then moved towards crossing each parent with only two others, partly to avoid losing two parents with the failure of a single cross. Using each parent in two pair-crosses rather than one would entail a cost in effective population sizes through not including lower-ranked candidate parents in ongoing breeding (Libby 1973). However, this scheme still allowed for using top-ranked parents in supplementary crosses, albeit at some additional cost. The option of practising inbreeding over multiple generations, in order to seek spectacular heterotic crossing combinations at some time in the future, was recognised as being very slow and laborious, with greater genetic gains usually achievable at much less cost through feasible schemes to achieve additive genetic gains. Moreover, the genetic uniformity that such preparatory inbreeding would promise is achievable with successful use of clonal material which also allows full capture of non-additive gene effects compared with their very partial capture in SCA by immediate pair-crossing.

6.7.2.2 Age–Age Genetic Correlations

Genetic correlations between expressions of traits at different ages can govern the age at which trees can be satisfactorily selected. These "age–age" genetic correlations depend in large part on whether different genes are affecting a trait at different stages of maturation, and/or whether the same genes are being differently regulated as the trees mature. In that sense, age-age correlations of the "same" trait are no different than genetic correlations among different traits that share pleiotropic genes.

High genetic correlations between early expression and harvest-age performance allow early selection, provided early heritability is satisfactory and all traits under current selection can be effectively selected early. Even if early selection gives markedly sub-optimal genetic gain per generation, it may be clearly optimal per unit of time because genetic changes can be accumulated over more generations.

Age–age genetic correlations for growth traits in radiata have generally been good enough to allow satisfactory selection to improve growth rate to harvest age by evaluating growth rate at age eight years or so (Cotterill and Dean 1988; King and Burdon 1991). For tree-form traits and corewood properties, such early selection is similarly satisfactory. For dothistroma resistance, extremely early selection among a much larger number of candidates appears satisfactory, because rankings of candidates in progeny trials have proved extremely consistent among ages as well as among sites (Carson 1989). Indeed, it is possible to use field trials planted at very close spacing and dedicated to rapid field screening for such resistance (although the trials could be thinned and later used to evaluate and select for other traits).

For wood properties, by the late 1980s, corewood was the main concern. Gains in outerwood properties expected from selecting on corewood properties, in conjunction with the typically high age-age correlations for wood traits, were seen as less critical. Indeed, the effect of imperfect age-age correlations could have a benefit in slightly reducing corewood–outerwood differences in the traits selected for in corewood. Similarly, genetic gains in tree form, and particularly in dothistroma resistance, are typically much more critical in young trees than in older ones.

6.7.2.3 Genotype–Environment Interactions

Genotype–environment (G × E) interactions were and are of key interest when considering whether to regionalise breeding programmes, or else whether to deploy within breeding regions specific, characterised breeds or clones by matching them either to sites that impose known demands or to different silvicultural regimes.

Such interactions represent differences among environments in comparative performances of genotypes, and can belong to one or both of two main forms. The first involves differences in *ranking* of families or clones among environments. The second involves variation in the *magnitudes*, relative or absolute, of differences among families or clones, without necessarily involving rank changes; and as such it involves different *levels of expression* of genetic differences. Both can be of major interest to the breeder and silviculturist. Patterns that have emerged in the early radiata trials evaluating several important traits have been mixed.

In New Zealand, substantial rank- and level-of-expression differences among sites had been observed for growth rate in a group of radiata clones, mainly involving a phosphate-deficient Northland clay site and elsewhere (Burdon 1971), matching earlier results in Australia (Fielding and Brown 1961). Later, substantial rank-change interactions were observed for growth in select-parent open-pollinated families between two Northland clays (varyingly phosphorus-deficient) and the pumicelands of central North Island (Johnson and Burdon 1990). Another study, involving pair-cross families on 11 widely scattered New Zealand sites, showed generally minor rank changes in growth rate (Carson 1991). This pattern agreed with some earlier, less-extensive New Zealand findings (Shelbourne and Low 1980). In Australia, however, no coherent pattern of rank-change interaction for growth was discernible in a set of pair-cross families trialled on 11 sites (Matheson and Raymond 1984).

Among early-evaluated traits in New Zealand and Australia, bole diameter tended to show more rank-change interaction than did tree-form traits or wood density (e.g. Shelbourne and Low 1980; Pederick 1990). However, tree-form traits such as stem straightness, malformation rate, or size and angle of branches, while showing little rank-change interaction, could show practically important level-of-expression variation within quite stable rankings among sites. Such interactions, however, were not clearly expressed in data from visual scoring aimed at resolving differences among clones or families within each site. For disease resistance, such G×E interactions were often huge, since susceptibility/resistance can only be expressed where the disease occurs.

6.7.2.4 Additional Wood Properties Beyond Density

By the 1990s, there was a growing concern in New Zealand over deterioration in wood properties of radiata. This was largely associated with new silvicultural regimes (Sect. 6.3.2.2) that facilitated early harvesting which led to higher percentages of corewood, and was exacerbated by adverse genetic correlations that could affect the faster-growing breeds. The concern led to a broader focus by radiata breeders on wood properties, with several properties other than density becoming identified as important. The properties that were proving troublesome included poorer stiffness, affecting structural sawn timbers and plywood; increased dimensional instability and consequent warping, affecting both structural and appearance-grade products; increases in prevalence of blemishes such as resin pockets, affecting appearance and such specialty uses as bar stools; and increases in the frequency and severity of internal checking during wood drying, all affecting recovery or values of the potentially lucrative solid-wood products.

Three complementary lines of study were then initiated or intensified:

- inheritance of basic physical, anatomical and chemical wood properties
- relationships between those basic wood-properties and product performance, the latter including frequency and severity of timber defects, mechanical properties, dimensional stability, and properties of paper and reconstituted-wood products
- measures, where achievable, of the inheritance of "product-performance" properties.

Much of this work was summarised by Shelbourne (1997) and Matheson et al. (1997). As well as the basic wood properties, the work included processing properties in the various kinds of mill, pulp and paper yields and properties, as well as defects, grades and values of sawn-timber outturns.

6.7.2.5 Inheritance of Basic Properties

Basic wood properties, while generally involving cellular or subcellular features, can include bulk properties such as density (specific gravity). They govern the product-performance properties, which include stiffness, load-bearing strength, dimensional stability and, for paper products, a paper sheet's tearing strength and brightness, and for tissue its fluffiness. So, by modifying the basic properties of cells, breeders could and did change product performances.

Research continued to show that wood properties of radiata are typically highly heritable (Matheson et al. 1997; Shelbourne 1997) and subject to relatively small non-additive gene effects and G × E interactions. Observed heritabilities are particularly high for wood density, fibre dimensions and some chemical properties. For some of these basic wood properties, the limiting factor appeared to be not biological heritabilities (generally relatively high) but the cost and difficulty of obtaining satisfactory trait evaluations on selection candidates, which would reduce selection intensities or observational heritabilities, or both (Fig. 6.10). In the early 1990s this was particularly true of microfibril angle within tracheids, which had recently been found to largely determine stiffness and dimensional stability of sawn timbers. An instrument, SilviScan, was then developed for automated determination of this and

Fig. 6.10 Field operation of penetrometer used to measure wood density on standing radiata pine. This gave less accurate determinations than using increment cores, but that was offset by allowing determinations on more trees for given time and budget

Fig. 6.11 Collection and preliminary field evaluation of increment core which can be used for determination of several wood properties in addition to density

several other features of fibre anatomy (Evans 1994; Evans et al. 1995; Matheson et al. 1997) (Fig. 6.11). Using a combination of technologies based on conventional optics, X-ray transmission and diffraction, and spectrography, SilviScan vastly reduced the human work needed for detailed determinations, although it was still costly to operate.

Not only are heritabilities of corewood properties often high, but some wood properties were also showing high coefficients of variation (Shelbourne 1997). This was and is important, because the potential for genetically changing any one trait is proportional to its heritability times its coefficient of variation (standard deviation divided by the mean)—provided selection candidates can be assayed at acceptable costs and the trait is not in strong adverse genetic correlations with one or more other important traits.

Genetic correlations between important wood properties and wood-volume production are proving inherently problematic. They were and are hard to estimate with adequate precision. More importantly, when adverse as they frequently are, genetic correlations can seriously limit the genetic gains simultaneously achievable in the traits concerned. This in turn particularly imposed a need for good knowledge of the comparative economic worth of the adversely-correlated traits, which can even be crucial to assuring positive economic benefits from breeding. By the early 1990s the breeders were becoming increasingly aware that important wood properties of radiata tend to be adversely correlated with stem diameter and thence volume growth which has undoubted economic value. However, technical and institutional factors were continuing to prevent clear market signals on the value of basic wood properties. Exactly how adversely the genetic trade-off between wood density and bole volume operates at harvest age, at the whole-crop level, is a key question (Burdon 2010), but answering it poses a huge challenge.

6.7.2.6 Relationships Between Basic Wood Properties and Processing- and Product-Performance Properties

The traits of economic interest influence not only efficiencies of *growing* radiata, but also processing- and/or product-performance properties following harvest. These end-of-rotation properties are often expressed late and can be prohibitively slow and expensive to evaluate routinely on selection candidates. Moreover, product performance may only become clear after years in service.

Many basic wood properties, be they chemical, anatomical or gross physical, might be assayed. The challenge, though, has been to identify with confidence which ones both govern key economic outcomes and could be assayed cheaply and reliably, and on young trees.

Prior to the mid-1970s, wood density remained the easiest wood property to determine on candidate ortets. This trait could be measured directly by weighing dried and water-saturated increment cores of their wood at some key site(s) within the tree. But even this was too slow and laborious for assaying large numbers of individual candidate trees, so torsiometers or penetrometers came into use by the late 1970s (Cown 1978). While giving determinations that were less precise and confined to outerwood, this was outweighed by the greater speed and cheapness of determinations. But, while wood density could be adequately assayed on quite young trees in progeny tests and clonal trials, it was becoming clear by the 1990s that other basic properties were too important to be ignored. In particular, microfibril angle (MfA) in radiata's fibres was found to be a main determinant, along with density, of corewood stiffness (Cave and Walker 1994; Walker and Butterfield 1996; Butterfield 1997) and a key determinant of dimensional stability. While MfA's heritability was promising (Donaldson and Burdon 1995), the direct microscopic measurement of cellular microfibrils was very slow and costly, while even automated SilviScan determinations were still costly. As a result, more direct measurements of stiffness were adopted from the end of the 1990s.

6.7.2.7 Direct Inheritance of Processing- and Product-Performance Properties

Processing- and product-performance properties could actually show quite high heritabilities, despite being very complex traits. The pulping properties include energy requirements (Corson et al. 1989) and various actual pulp properties (Shelbourne 1997), while solid-wood product properties include sawn-timber grades and actual product value (*op. cit.*). For such properties, however, direct evaluation of selection candidates would be prohibitively slow and expensive, and far too late. Nevertheless, demonstrating such properties' genetic variation and heritability for such properties was useful, both in itself and for the reassurance that screening and selecting for quickly and easily evaluated component or indicator traits is worthwhile. However, these forward-looking analyses could not shed much light on the strength of the adverse genetic correlations between the post-harvest properties and wood volume production or on how best to cope with them.

6.8 Propagation Options Increase

6.8.1 The Evolution and Decline of Seed Orchards

As traced in the previous chapter, the development of clonal seed orchards exposed some major biological and operational problems. Given problems of pollen contamination in wind-pollinated orchards and the inconvenient height to which trees in radiata orchards tended to grow, Geoff Sweet at the New Zealand FRI had embarked on developing new types of orchard (Sweet and Krugman 1978). Orchard ramets were pruned much like fruit trees, to keep them low and facilitate controlled pollination. In some cases, separate "male" and "female" orchards were created and then managed to produce either pollen or seed-cones respectively (Fig. 5.22.). (In conjunction, techniques such as bagless "wet pollination" were tested for achieving cheaper semi-controlled pollination.) In the region around Rotorua, however, this new type of orchard system was not a practical success because of difficulty in containing the ramets to a convenient height while still producing enough seed per pruned tree. However, they were useful there for producing limited numbers of bagged controlled crosses for breeding and vegetative multiplication (VM) purposes (see Sect. 6.8.2).

Meanwhile, VM of valuable seedlings, by means of cuttings set directly in nurseries, had developed as an option for rapid and large-scale capture of genetic gain. At first, this was typically accomplished by producing juvenile or early-adolescent cuttings derived from seedling ortets. The cuttings were taken from the most promising full-sib families, initially based on the (mostly GF) breeding values of their parents (Arnold 1990). By doing so, the high costs of control-pollinated seeds from the latest top-rated parents could be greatly diluted (*op. cit.*; Menzies et al. 1992). Those early 1990s cuttings, initially representing multiplication factors of around 10, were still about twice the cost of unpedigreed seedlings. Importantly, however, they quickly proved to be much less expensive than seedlings produced from super-expensive pedigreed seeds (Sect. 6.11). Such nursery-cuttings technology greatly reduced the requisite quantities of pedigreed seeds, often produced in variants of Sweet's crossing archives rather than in orchards meant to supply planting demands using seedlings.

The option of being able to mass-produce planting stock from controlled crosses also had a major implication for population management, which we will address later (see Sect. 6.12.1).

Despite the strong move to VM from the mid- to late 1980s, large clonal seed orchards were not fully replaced by crossing archives. For one thing, not all nursery climates were well-suited for mass-scale VM, and these nurseries thus continued to rely on seedling production. For another, there was the nature and designated mission of the PROSEED operation in New Zealand which did not include nursery work (Sect. 6.2.2.2). The PROSEED operation was based mainly on large seed orchards, which could continue to supply genetically improved seed to nurseries unable or unwilling to use VM.

The chief of PROSEED, Peter Bolton, embarked on an ambitious plan of producing less-expensive control-cross seeds on a commercial scale. For this, PROSEED'S Amberley seed orchard, located coastally near Christchurch (Sect. 4.7.5), was ideal. There, grafts of orchard clones were little affected by graft incompatibility, and could produce abundant seed-filled cones close to the ground, while a combination of soil, climate and management prevented rampant height growth. Bolton adapted the concept of' meadow orchards' from fruit-growing (cf Sweet and Krugman 1978), in which grafts with dwarfing rootstocks, planted at very close spacings, allowed precocious and abundant low fruiting. Under such a system radiata grafts at Amberley came into full pollen and cone production very quickly, without dwarfing rootstocks but with some help from gibberellins treatment (Sect. 5.5.5). Even so, the use of such orchards to produce enough seed to allow large-scale planting of genetically improved seedlings only remained a competing option, as we explain below.

6.8.2 Vegetative Propagation and its Objectives

During and even well before the 1970s, much work had been done in China, Japan, Italy, Brazil, Canada, Germany, Scandinavia, the US, France, Australia, New Zealand and elsewhere to make possible large-scale production of forest-tree planting stock by vegetative propagation (Ahuja and Libby 1993, Volumes 1 and 2). Centuries of horticultural experience of the benefits of clonally propagating fruit and nut trees, and some existing successes with forest trees, had surely provided a stimulus.

Ill-fated early attempts had been made in the 1960s to select and test radiata clones for field performance in both Australia and New Zealand (e.g. Wilcox et al. 1976). The reasonable quest to use and then test propagules (both cuttings and grafts) from harvest-age trees did not anticipate that it then proved impracticable to mass-propagate as clones their mature selections after field testing (Menzies and Aimers-Halliday 2005). As the relationship of successful rooting of radiata cuttings to the cuttings' maturation state was increasingly understood, the focus was shifted to propagating cuttings that were fully juvenile, or nearly so.

6.8.2.1 Vegetative Multiplication of Young Seedlings

Vegetatively multiplying young radiata seedlings by harvesting and rooting cuttings from them was refined in the 1970s, largely through "hedging" the ortet seedlings, (and for greater multiplication, their derived juvenile cuttings) close to ground level (cf. Libby et al. 1972). Such hedging evidently halted or at least delayed maturation of these cutting donors (stools), so that for several years they could each continue to produce many late-juvenile or early-adolescent regrowth shoots that could be harvested, effectively rooted, and easily and reliably raised in a nursery as cuttings for field deployment. Various alternative systems of producing cuttings from juvenile material were explored during the 1980s and later (Menzies et al. 1988). Remarkably, under extremely favourable conditions cuttings could be set directly in the field, without the expense of raising them in the nursery, and with the bonus of assuring stable root systems that would avoid toppling in the field. However, without good assurance of such favourable conditions this is a risky practice.

Meanwhile, despite the increasing feasibility of producing adequately juvenile cuttings on a large scale, concerns existed over the rate at which multiplication could be achieved cost-effectively. Vegetative multiplication (VM) rates could be boosted by an order of magnitude by learning to convert juvenile radiata's short shoots (needle fascicles) to long-shoot growth and use them as cuttings, but this proved difficult to manage operationally. However, VM was achieved effectively by producing cuttings from hedges of seedlings, and with "serial propagation" of "secondary" or even "tertiary" cuttings from hedged juvenile cuttings. Thus mass-propagation of juvenile cuttings became operational practice, producing an estimated 14% or so of commercial radiata planting stock in New Zealand by 1995 (J. Novis pers. comm. 2013; see also Table 6.8). In New Zealand, this practice, with some variations, was led by large corporates operating in central North Island. Similar success was achieved in eastern Victoria, Australia. There, APM Forests, under the leadership of John Cameron, were by around 1990 planning to replace seedling planting stock entirely with nursery cuttings.

On the other hand, use of VM in South Australia to multiply severely depleted seed stocks to meet the huge demand for planting stock after the 1983 fires was less successful, presumably because of the harsher climate combined with the hardness of the water supply and/or the lack of lead time for adapting the technology to local conditions.

As we have seen in Sect. 6.8.1, the development of VM had the soon-realised potential to reduce the scale and role of seed orchards, and instead to refocus on crossing archives coupled with higher-tech VM of pedigreed radiata pines.

Given the technical problems encountered with the early clonal seed orchards, use of VM by producing nursery cuttings using pedigreed seeds from crossing archives provided a compelling alternative. This paradigm shift could have delivered current genetically improved planting stock for an important part of New Zealand much faster than the seed orchards did. However, early adoption, even if it had been urged by the tree breeders, would almost certainly have been blocked by institutional resistance. Yet, with increasing demonstration of technical feasibility of such VM, and the enlightened foresight of a few leaders in the forestry sector, institutional resistance was overcome. As it turned out, such institutional inertia only slowed the shift to VM.

VM using nursery cuttings became embraced in parts of Australia, New Zealand and Chile. In Australia it was partly spurred by the loss of seed orchards in the fires

Table 6.8 Recorded sales of radiata pine planting stock (millions) in New Zealand, by years, propagule type and improvement rating or status (*GFPlus*™) (J. Novis pers. comm. 2013)

Year	Seedling rating					Cuttings[a]	Total
	<14	14–19	>19	GFPlus[b]	Stand select[c]		
1992	1.5	54.4	0.9	–	–	6.1	63.0
1993	1.3	69.3	2.3	–	–	8.1	81.1
1994	2.2	105.2	4.2	–	–	11.2	122.9
1995	0.4	74.9	7.8	–	–	13.6	96.7
1996	0.6	65.1	9.8	–	–	24.0	99.5
1997	0.1	45.4	8.4	–	–	22.9	76.8
1998	0.1	30.3	14.5	–	–	19.8	64.7
1999	0.2	22.5	7.1	0.2	–	23.0	53.1
2000	0.2	26.1	8.5	1.1	–	20.6	56.5
2001	0.0	21.2	8.3	6.3	–	19.9	55.8
2002	0.0	18.0	8.9	7.2	–	16.0	50.2
2003	0.1	14.6	7.6	11.2	–	14.0	47.4
2004	No reliable breakdown						43.5
2005	"						32.6
2006	0.0	7.6	11.3	–		9.9	28.8
2007	0.1	11.2	17.2	–	–	8.1	36.7
2008	0.2	11.7	15.4	–	–	8.8	36.2
2009	0.2	10.9	17.0	–	–	9.6	37.7
2010	1.0	17.6	13.2	–	1.9	12.6	46.4
2011	0.4	16.1	16.8	–	13.4	12.2	58.9
2012	0.7	15.1	20.3	–	16.6	11.9	64.6

[a]Predominantly of the higher ratings
[b]Generally ratings of GF19 or better. Some sales are believed to have been included as "cuttings". Category abandoned after 2003
[c]Collected from stands of seed-orchard source (GF14–16?), but not formally rated for improvement

of 1983. In New Zealand the impetus came more predominantly from the desire for faster capture of genetic gain, as was the case in Chile (Balocchi 1997).

6.8.2.2 In-vitro Culture Systems

Concerns over the costs and multiplication rates achievable through nursery production of cuttings heightened interest in developing pure tissue-culture systems for mass-multiplication of radiata as pedigreed plantlets. A hope was that the benefits of higher multiplication rates could offset the additional costs of *in-vitro* culture and then acclimating the propagules to nursery conditions. A few forward-thinking leaders, however, saw tissue culture as a preliminary, rapid-multiplication phase that could be followed by more conventional production of pedigreed cuttings.

Fig. 6.12 Axillary shoots produced by *in vitro* culture, awaiting separation for rooting or further multiplication (Photo Keiko Gough)

Accordingly, *in-vitro* tissue-culture systems, including cultures of callus masses and of whole organs (small shoots) began to be developed for radiata at FRI during the late 1970s, becoming reliable by 1981 (Horgan and Aitken-Christie 1981) (Fig. 6.12). Serious attempts at operational applications began around 1984. Despite starting with the "zero" maturation state of seedling embryos, achieving high multiplication rates in such systems additionally depended on propagating adventitious shoots. Rather than being grown out from natural growing points in organ culture, these adventitious shoots were often initiated in cotyledon-origin callus masses. However, such adventitious plantlets became suspect when they were found to show elevated rates of **somaclonal variation**[G]. As part of that variation, they exhibit variably accelerated maturation (Horgan et al. 1997), which subsequently affects the vigour of the derived plantlets, and their increased somaclonal variation would create unwanted variability in the plantation crops. However, some hoped that the worrisome problem of radiata's slow but continued clonal maturation in tissue culture could be solved by physical or chemical manipulation of the tissue-cultures, and/or by cool-storage and later recovery of tissue-culture samples of the interesting clones.

One New Zealand company, Tasman Forestry Ltd, prepared to commit to a large-scale tissue-culture facility under the skilful and persuasive leadership of John Gleed. This venture was embarked upon after 1984, despite the company having operated a successful seed orchard for several years. FRI, on scientific advice that there was no prospect of producing plantlets cheaply enough, declined an invitation to join in the venture. Nevertheless, Gleed and Tasman Forestry went ahead, siting their new Centre for Advanced Forest Biotechnology at nearby Te Teko. In fact, tissue-culture plantlets did remain expensive to produce. Also, they were as prone to unwanted maturation as cutting-donor hedges, and could not be produced with every clone, necessitating quite rapid phasing-out and replacement of individual clones for multiplication of planting stock. Nevertheless, the Te Teko venture worked much to the company's political advantage. Its reputation was enhanced as both a forward-looking innovative forestry company and one willing to invest heavily in the future of New Zealand's economy through its increasingly domesticated forests.

At Te Teko, and for a time ignoring or winning-over Tasman's accountants, Gleed had recruited Bev Nairn from Vancouver, Canada, where she had been running a human tissue-culture lab. She designed, outfitted and operated the Te Teko lab to human-research standards, which led to some of its early successes with radiata and several other conifer and eucalypt species culture there. Thus Tasman became radiata's pioneer in widespread deployment of partially characterised promising clones in its plantations and in plantations of its nursery's clients in New Zealand.

Meanwhile, Tasman Forestry had planted areas with OP families as blocks of seedlings from different cloned parents in its seed orchard, and thus demonstrated to company executives and foresters big differences in growth performance among their seed-orchard families. Their progenies of the celebrated Clone NZ55 performed outstandingly in early growth and form. These results strengthened the company's commitment to genetic improvement of its future forest assets. Indeed, the company even temporarily adopted, as a back-up option, the harvest of cuttings from above-average young trees in the field for nursery production and multiplication as further cuttings (Arnold and Gleed 1985). Subsequently, that back-up morphed into a nursery operation producing millions of cuttings annually (Gleed 1993), later relying on cuttings from hedges originating from promising pedigreed seedlings, or from plantlets of clones in their repeatedly updated and analysed clonal tests. But continuing slow maturation of those donor hedges remained a daunting problem.

By the early 1990s, adventitious callus-origin plantlets had been replaced at Te Teko by organ-culture plantlets, derived from shoots originating from normal bud-meristems in the axils of cultured shoots. That change traded the very high multiplication rates obtained with adventitious shoots for the lower rates using axillary shoots, which were less prone to problems of increased maturation, somaclonal variation, and other epigenetic consequences. Slow maturation, however, persisted in the axillary-shoot organ cultures.

Tissue culture with radiata had developed at FRI by adapting previous organogenesis work with other species. That typically entailed initiation of shoots on one culture medium, and then detaching and transferring them to another medium with different nutrients and plant-growth regulators on which they could thus be induced to develop roots and eventually become free-living plantlets (Menzies and Aimers-Halliday 2005 and references therein). Responding in part to concerns over the cost of such tissue-culture plantlets, and in part to the prospects of achieving higher multiplication with greater automation, interest then developed in embryogenesis as an option for mass-multiplication of radiata.

In some plant species, e.g. carrot, it was possible to create a suspension of individual cells that could be induced to become truly somatic embryos genetically identical to the donor carrot, and then germinated and grown as normal plants. Attempts to achieve this with radiata have, as of 2015, still not succeeded. However, it has proved possible, using proper culture media and conditions, to induce cultured tissue from excised seed embryos to produce multiple so-called somatic embryos that can be reared into field-ready plants called **emblings**[G] (Hargreaves et al. 1997). Even more ambitious is the hope of embedding such induced embryos in artificial seeds encapsulated with nutrients and growth regulators in artificial coats, for later germination. The focus broadened to use of embryogenic tissues for clonal storage and as targets for genetic engineering (Sects. 6.13.3 and 7.11.2).

The clonal storage option became envisaged in combination with cryopreservation (Menzies and Aimers-Halliday 2005), storing embryogenic tissue in liquid nitrogen so that samples of it can be revived. This revived tissue can then initiate new cultures of somatic embryos at its embryonic state, after testing of the clone has already proceeded to its deployment and characterisation for an increasing list of important traits. This can apparently continue indefinitely (Sect. 7.10), a notable breakthrough supporting clonal forestry as an important part of the domestication of radiata.

While embryogenesis was originally seen as a basis for mass-multiplication, and still is, it has also come to be seen as a platform for genetic engineering, producing cells that are amenable to inserting transgenes (Walter et al. 1995; our Sects. 6.13.3 and 7.11.2).

6.9 Clonal Forestry Beckons Again

The early attempts to achieve full clonal forestry with radiata, namely selection of clones that could be mass-propagated commercially, testing the clones to characterise them accurately for important traits, accurately matching clones to sites and grower objectives, and being able to indefinitely produce those clones in a state that reliably repeats its characterised performance, had thus far failed with radiata. However, interest in that goal (Box 6.1) was far from lost (M.J. Carson 1986b; Miller 1991).

Box 6.1 Clonal Forestry (CF)

This entails mass propagation for plantation establishment of known, proven clones. Such clones are often referred to misleadingly in forestry as varieties. CF offers advantages in terms of: greater genetic gain; elimination of unwanted genetic variation; and much more precise targeting of clones to planting sites, tending regimes, wood processing, and end-uses. The risks associated with genetic uniformity within clones can be countered by

(continued)

Box 6.1 Clonal Forestry (CF) (continued)
deploying risk-spread clonal portfolios, which can be deployed either in monoclonal blocks or in intimate **mixtures**. CF, however, is a deployment system, not an actual breeding method.

In radiata, implementation of CF has been dogged by imperfect control of maturation. While maturation can now be controlled by cryopreservation of tissue, the ideal would be to achieve true rejuvenation of clones without incurring the genetic recombination associated with seed production.

For one thing, advances in propagation technology and clonal storage were giving promise of both the biological and financial feasibility of full clonal forestry with radiata. For another thing, the potential advantages were considerable: among a longer list (Libby and Rauter 1984), being able to capture favourable non-additive gene effects that get lost in producing seedlings, avoiding the genetic segregation that creates unwanted variability among trees in plantations, and exploiting the uniformity and general repeatability of well-known clones.

In a number of tree species, greatly decreasing the heritable tree-to-tree variation by deploying characterised clones in medium-size monoclonal stands within plantations has proven to be a considerable advantage in managing growth and silviculture, harvesting, processing and utilisation of the wood. Moreover, in radiata, there is potentially a special advantage of clonal forestry involving branching habit. A less-than-extreme long-internode habit can be the best internode-length compromise for giving useful clear lengths of sawn timbers that can be cut out between knot clusters. On the one hand, these clear lengths can either be used directly or allow remanufactured products such as finger-jointed knot-free boards and beams. On the other hand, such a branch habit should not incur too much malformation or loss of growth rate in plantations. Such an intermediate branching habit is not likely to "breed true" with seed production, but should be effectively secured by clonal propagation (Burdon 1989).

In response to this advancing knowledge, active research on radiata clonal forestry resumed in New Zealand in the mid-1980s. Tasman Forestry Ltd developed its propagation facility with the express intent of practising full clonal forestry, whereas FRI lacked a mandate to commercially exploit clonal forestry. The Forestry Corporation of NZ, newly-created from parts of the dissolved Forest Service, began to follow suit by developing the VM technique of producing easily-harvesting long shoots stimulated to grow from numerous fascicle shoots, as a means of greatly increasing the multiplication rate as nursery cuttings. Meanwhile, in New Zealand and overseas, the physical research was complemented by much thinking and writing on the topic (Miller 1991 and papers therein). Libby's long-standing interest in clonal forestry found full expression in a two-volume book on clonal forestry (Ahuja and Libby 1993). To emulate the successes that China and Japan had achieved historically with clonal forestry in Chinese-fir (Li and Ritchie 1999) and sugi (Ohba 1993) was indeed tantalising.

Despite early disappointments with adapting clonal forestry to radiata, it emerged that expression of genetic variation in its seedlings is closely matched in vegetative propagules, provided they begin in a sufficiently juvenile state (Miller 1991). Furthermore, the benefits of a modest initial level of maturation for tree form (especially on fertile sites) are considerable (Menzies and Klomp 1988). They include some control of branch diameter, and straighter and less-tapered boles. And they also include resistance to some important juvenile-phase diseases of radiata, notably the potentially devastating western gall rust (Old et al. 1986). Against the advantages of some maturation, rooting of cuttings becomes progressively more difficult and thus more expensive as maturation progresses. Even when more-mature cuttings can be satisfactorily rooted, their increased maturation incurs reduced vigour, at least in the early stages of growth (e.g. Sweet 1973; Brown 1974). This can be accompanied by exaggerated within- and among-clones variation in growth rates, through increasing variation in the state and effects of maturation on cuttings' vigour. More clearly, the correspondence of clonal growth performance compared with seedling ortet's growth performance can decline sharply with increasing maturation (e.g. McGranahan et al. 1997). However, as already mentioned, general tree form is improved by increasing maturation (Figs. 4.18 and 5.26). Also, and in contrast to the case with growth rate, clonal variation in bole straightness and regularity of branching often becomes far less evident with increased maturation state of the young propagules. However, despite such effects of maturation on variability, genetic rankings for tree form have been found to remain essentially the same between adolescent cuttings and seedling trees (McGranahan et al. 1997).

Attention thus shifted from maintaining and using full juvenility in clonal planting stocks to identifying and then managing for a "Goldilocks" level of propagule maturation that is "just right." That would entail accepting some loss in ease of propagation and early growth vigour; in return would be greater capture of genetic variation, avoidance of unwanted genetic variation, and maturation-related improvement of tree form, resistance to juvenile-phase diseases, and some wood properties. As of 2015, that desired propagule maturation state for use in radiata plantations seems around "late-juvenile" to "early-adolescent." Indications are, however, that the reduced vigour of propagules associated with greater maturation may not adversely affect harvest-age financial and product yields (Eldridge and Spencer 1988), largely due to the better log form favouring higher recovery of more valuable grades of sawn timber per harvested hectare.

Nevertheless, despite technical advances, operational challenges have remained in being able to store clones in a juvenile state during and after field evaluations. Cool-storage of tissue cultures in refrigerators is possible over several years, but does not fully halt maturation; moreover, such cool-stored cultures are prone to attrition due to mistakes and institutional changes, and are vulnerable to equipment failure. Cryopreservation of clones, storing embryogenic tissue in liquid nitrogen, apparently halts maturation, but being able to revive the clones after such storage requires careful prior preparation of the tissue and leaves little margin for error. Nevertheless, the ideal remains being able to rejuvenate at will any tree at any maturation state, in a manner similar to the production of a seed, but without unwanted genetic recombination.

Despite successes already achieved, refinements of technology for storage and mass-propagation of clones remain ongoing pursuits as of 2015.

Box 6.2, concerning the efforts of a private entrepreneur to operate a clonal forestry propagation business, illustrates a number of technical points, the influence of institutional changes, and commercial and ethical questions surrounding increasing pressures to appropriate hitherto freely-available intellectual property.

Box 6.2 An Unhappy Tale of Commercialising Mass Vegetative Propagation

One venture, which illustrates several points, began before and ended during this 1984–1997 period. In 1973, a cashed-up Australian entrepreneur and tree enthusiast, Francis Clarke, with a large property "Pinebank" near Tarago, New South Wales, began planting his version of New Zealand's Agroforestry Direct regime with wide-spaced radiata seedlings.

In 1977, with his trees growing well, he sought guidance on tending from CSIRO's Division of Forest Research, whose silviculture section was planning a large agroforestry trial, based on the Tikitere trial near Rotorua. Their trial was established in 1978 within Clarke's larger commercial agroforestry planting.

Clarke's 1973 plantings of radiata seedlings were by 1978 entering their ugly adolescent phase, with very poor tree form, and the potentially valuable butt logs were in a daunting need of pruning. Reaction wood was prevalent in such trees' lower boles. The CSIRO silviculturists suggested that planting cuttings from adolescent trees might largely solve those problems.

In June of 1981, Clarke made a voyage of discovery. In California he saw the virtues of clones from selected high-performance control-pollinated families. In France he learned how to obtain high multiplication rates (×100/yr) by producing fascicle cuttings. In Sweden he learned "fertigation" techniques (a combination of fertilisers and irrigation schedules) that grew fascicle-donor plants very rapidly. Then, in New Zealand, he learned some nursery practices to produce sturdy cuttings for field planting. Neil Barr, from his own observations of agroforestry trials, supported the idea of using adolescent cuttings to improve tree form and pruning efficiency (see also Tufuor and Libby 1973).

Clarke then presented his concepts to CSIRO silviculturists, who then arranged with FRI New Zealand to supply top-rating control-pollinated seeds, and they added similar seeds from nearby Australian breeding programmes. CSIRO's radiata breeders were then under increasing budgetary pressures, and being redirected to work on eucalypts and other species. But with some available Australian select families and some contributions from

(continued)

Box 6.2 An Unhappy Tale of Commercialising Mass Vegetative Propagation (continued)

FRI breeders included, the CSIRO silviculturists launched a series of trials begun with seedlings from the 40 top-performing full-sib families that they had acquired. They soon chose the 20 "then-best" families from the nursery and began rooting cuttings from the (hedged) five best-appearing seedlings per family. Enough cuttings were thus produced for tests of those 100 clones to be organized by CSIRO and planted in 1984 on seven eastern Australian sites. Besides characterizing those 100 clones, the trials would compare clones from the best of the New Zealand and south-eastern Australian breeding programmes in Australian environments. Clarke was then permitted to take fascicle shoots from those 40 families, and the cuttings became donor plants for 360 clones marketed from his nursery.

Clarke had assisted in planting one of the seven 100-clone trials at Pinebank and on another private site. He personally measured those two trials for four years after planting, but the other five trials were not measured. His data were sent to CSIRO on disc; resources were not available to analyse them.

In 1986, Clarke established three large demonstration trials, which included 49-tree monoclonal block plantings using good Australian and New Zealand clones. Four years after planting, the best clones were showing with excellent vigour and near-perfect tree form. Meanwhile, Clarke had joined the "Management of Improved Breeds Cooperative" in New Zealand, and invited co-op members to visit the demonstration trials. Clarke's in-kind contributions to the co-op included a study of clonal maturation, where cuttings from the original hedged donor plants were compared with ones from five-year-old tree-form cuttings and with "fifth-serial" annually repropagated cuttings, all of the same clones. Cuttings from one- and two-year-old seedlings were used as controls. Observations of their subsequent performances were to be freely shared.

In November 1991, co-op members (including two New Zealand corporate members) accepted Clarke's invitation and saw the excellent performance of the New Zealand clones in Australia. Within three weeks of their return home, a total ban on sharing and exporting of advanced New Zealand radiata breeding material was imposed, and the Management of Improved Breeds co-op was transferred to the NZRPBC. Cut off from fresh and better New Zealand clones, Clarke tried to upgrade his maturing stable of clones with the latest and best local Australian clones. Intellectual property had become important, with ugly responses and murky ethical questions. A standoff resulted, and Clarke's clonal business struggled on for a further 10 years, until his donor hedges had matured too much for reliable rooting. Thus radiata clonal forestry again died in NSW and in other parts of Australia.

6.10 Evolution of Breeding Goals

6.10.1 Shifts in Emphasis on Various Traits

Assessment and inheritance of wood properties had been pursued, mainly in research, from the early days of radiata breeding in both Australia and New Zealand. However, commitments to active breeding for wood properties did not come readily or quickly (Burdon 2010). Industry participants were for long reluctant to call for genetic improvement of wood properties in the face of any trade-off with growth rate. Even without adverse genetic correlations a "soft" trade-off between increased wood volume production and wood quality would be expected through dilution of genetic gain in individual traits, a well-known consequence of simultaneous selecting for additional traits. And a stronger ("hard") trade-off would be expected with the presumed negative genetic correlation between wood density and growth rate in terms of wood volume. Moreover, the markets were often failing to reward the forest grower for better wood quality, rather than basically for the size, straightness and knot characteristics of logs.

Dramatic genetic changes in tree form and substantial changes in growth rates were quickly and easily obtained in most programmes in all countries doing radiata breeding. On the other hand, the difficulties and costs of assaying candidate trees for wood properties, together with adverse genetic correlations of wood properties with growth rate, made committing to and achieving desired genetic improvement of wood properties much more challenging.

Nevertheless, by the 1990s, continuing research and operational experience cumulatively revealed an undeniable deterioration of important wood properties in much of the radiata coming due for milling into sawn products and plywood. Contributing factors also included shorter rotations, heavier thinnings and elevated soil fertility (Sect. 6.3.2.2), but the adverse genetically-correlated response to selection for growth rate (cf Sects 6.10.1, 7.6.2.2) was a particular problem for the breeders and for the credibility of radiata breeding programmes. That unwelcome realisation eventually generated a call for active genetic *improvement* of important wood properties, or in some programmes a strong call for at least using breeding to hold the line against such degradation in wood properties by genetically countering the effects of shorter rotations, heavy thinnings, fertile sites and adverse genetic correlation. Given the accountants' strong preference for shortening rotations, it was much easier said than done. Nevertheless, that call gave impetus to developing improved assay technology for wood properties.

Wood properties affecting pulping efficiencies and paper properties, although of substantial economic importance to the pulp and paper mills while the properties were being researched for radiata, did not involve the parts of the crop that dominated financial returns to the grower. The cellular delamination property of NZ Clone 55's progeny (Corson et al. 1989), so financially and environmentally attractive in mechanical pulping, was discovered fortuitously. However, by the 1990s in New Zealand, the valuable radiata logs were being exported or sent to sawmills, while the pulp mills got and used the small proportion of lower-value logs that remained. In Australia the situation was generally similar. Thus the pulp and paper mills were generally in no position to make effective input to radiata breeding. This situation was very different from the pioneering pine breeding programmes in south-eastern USA. There the pulp and paper industry supported and partly funded the tree-breeding programmes, and strongly influenced their priorities.

Despite the awareness by 1997 of the need for more positive genetic improvement of wood properties, rather than just defensive maintenance, and despite the availability of some better assay methods, the breeding programmes were still not quite ready for intensive commitment to select for wood properties. While grain spirality was also recognised as a key trait governing dimensional stability (Sorensson et al. 1997), and showed early indications of good heritability, it has until very recently proved difficult to assay and select for effectively. Although the SilviScan instrument (Evans et al. 1995; other references in Matheson et al. 1997) had recently provided a major technological advance for assaying detailed wood anatomy there was not yet sufficient impetus for routine large-scale assays of candidate trees. Assay costs per candidate were still high, while there was still insufficient understanding of how which basic wood properties governed what processing- and product-performance properties, and of their inheritance patterns. Chap. 7 reviews the advances made in the next stage of radiata domestication.

The shift in emphasis among breeding-goal traits was associated with a shift in the exact role for continuing breeding in this new phase of domesticating radiata. To a large extent, this changed role amounted, implicitly rather than explicitly, to the breeders being called upon to offset the adverse effects on crop quality of silvicultural practices that served to reduce the effective growing costs by boosting productivity and shortening rotations. Those practices included wider initial spacings, use of fertilisers and heavier thinning. Their resultant adverse effects involved both poorer tree form and poorer-quality wood. Tree form was thereby degraded by mainly by larger branches and more forking, and greater sweep and consequent asymmetry of stem cross-section. Some tree-form effects were site-related, especially on fertile and windy sites, which resulted in much larger knots in sawn timbers. Degrading of wood properties was particularly troublesome, and those effects included reduced density, much-reduced stiffness, and increased dimensional instability which could be manifested in service as well as during drying. Tree-to-tree variation in wood properties was particularly awkward, because even occasional pieces that distort in service are very costly to rectify, favouring consumer shifts to less environment-friendly alternatives such as aluminium.

6.10.2 Breed- and Multiclonal Variety Specialisation

By 1970 in New Zealand, the concept of differentiated breeds had taken hold, addressing different breeding goals defined by different selection weightings or even contrasting expressions of different traits (Sect. 5.5.2). At FRI, a "side programme' selecting for "long-internode" trees was begun in 1970 to allow remanufactured sawn timbers without knots and without having to prune. While this "breed" had initially met with essentially no market demand, the concept of differentiated breeds was not abandoned. The Long Internode (LI) breed was retained, as a specialised breeding population, and the Dothistroma Resistant (DR) breed became instituted in 1983, as a mere-subset of the mainstream breeding population embodied in the Growth and Form (GF) breed (Vincent 1987; Jayawickrama and Carson 2000). As such, the emphasis on resistance to dothistroma entailed little sacrifice in the genetic gain already achieved in the GF breed at FRI.

The Te Teko clonal programme by 1994 found it worthwhile to deploy highly dothistroma-resistant clones as regional **multiclonal varieties**[G] to sites with high disease hazard, deploying clones top-ranked for other desired qualities to sites with low dothistroma hazard. In similar manner, LI clones were deployed to low-fertility sites where tree form would not be a problem.

In Australia the LI breed option had been seriously considered in the 1960s, but was not really pursued operationally. Some breed differentiation was pursued by the STBA in the 1980s, but without strong and sustained commitment.

Several factors governed the setting up of differentiated breeds in New Zealand:

- The fact that the two contrasting branching patterns each had its own attractions, whereas intermediate branching patterns would often have the advantages of neither. Selecting for the two extremes was hoped to minimise unwanted variation by effectively "squeezing the GF and LI breeds (and particularly their subsequently selected clones and multiclonal mixes) towards the end walls of biological feasibility." Consistently achieving a highly desirable intermediate branching pattern in seedling offspring is liable to be frustrated by genetic segregation, but it should to be possible through mass-producing clones that represented relatively rare genetic segregants, perhaps by crossing parents of the contrasting breeds.
- The range of processes and end-uses for which radiata was being domesticated, including both structural- and appearance-grade solid-wood products. Thus far, however, this has proved less of a driver of breed differentiation than was earlier envisaged.
- The fact that dothistroma hazard is very site-dependent, leading to the organisation of a special DR breed and multiclonal varieties to serve only a characterised subset of planting sites. That principle may hold as later domestication takes in resistance to other important pests and diseases of radiata.
- Radiata's notable environmental plasticity, whereby some wood properties and tree form can be dramatically influenced by site characteristics. For some sites (e.g. nutrient-rich volcanic plateau or ex-pasture), poor tree form and some wood-properties can be a serious concern, those problems sometimes completely overshadowing the importance of potential gains in growth rate. On some other sites (e.g. infertile coastal sand dunes), basic wood properties and tree form tend to be consistently very acceptable, despite heritable variation, leading genetic gains in growth rate being of paramount interest to the forest growers and managers.

With time the New Zealand's radiata breed portfolio was broadened and changed. Its High Wood Density breed was designated in 1986, and it morphed later into a

Structural Timber breed as it became clear that density was not the only important determinant of stiffness. At essentially the same time, the LI breed was redesignated Clear Cuttings [between knot clusters]. With these changes came some realignment of selection criteria.

Moreover, with the option of combining controlled crossing and vegetative propagation, it had become possible to customise seedlots and batches of nursery stock, by specific choice of parents and VM, without even depending on discrete breeds. This led to "family forestry," where the unit of deployment was neither "breed" nor "multiclonal variety" but rather a specific, vegetatively multiplied (VM) full-sib family or pair-cross. As such it represented an intermediate step towards clonal forestry which would require characterisation and storage of individual offspring clones.

6.11 Deployment Options Multiply

As genetic changes were achieved, characterised and demonstrated, nursery managers and forest growers had increasing choices of what kinds of genetically improved propagules to use, where to plant which breeds, VM families or multiclonal varieties, the latter at which maturation state, and how to grow them. One element of choice came directly from the fact that both trait-specific improved planting stock and current land-race seed stocks were available, the former in alternative breeds with an increasing range of improved traits and improvement levels, breed alternatives, VM pedigreed families, multiclonal portfolios, and even individual well-characterised cutting or plantlet clones, perhaps at different maturation states.

Level of improvement and propagule type, however, were partly interdependent. Seeds of the putatively very highest currently-available genetic quality, coming from the very best proven parents for various traits, were initially available in only very small quantities. That was not only because of the lead time in obtaining and characterising those parents, the latter usually based on both ortet- and progeny-test data, but the additional time needed to clonally multiply them in various kinds of seed orchards to produce abundant, easily-harvested seeds.

More-complete pedigree of seeds incurred the costs of controlled pollination. That allowed more precise characterization of full-sib seedlots by the breeding values of both parents, compared to seeds collected from an open-pollinated seed orchard where the male parents could not be identified even using current DNA analyses. Such full-sib seedlots turned out to have a "scarcity value" plus a perceived final-crop value well above their higher cost of production.

The "scarcity cost," however, could be greatly diluted by VM, the more so the higher its multiplication factor, such that even very expensive seeds contributed only a very small part of the total cost of planting stock. The vegetative propagation system dominated the cost of VM planting stock (Arnold 1990; Menzies et al. 1992). With continuing refinements of hedging, collecting, setting and rooting the

Table 6.9 Estimated comparative planting stock costs (1995 NZ$/1000) for various seed sources and stock types (From Menzies and Aimers-Halliday 2005)

Stock type (seed source)	Cost/1000 plants
Seedlings (open-pollinated seed)	110
Seedlings (control-pollinated seed)	460
Cuttings from stool beds (control-pollinated seed)	200
Cuttings from field collections	370
Micropropagated plantlets through organogenesis	~700
Cuttings from such plantlets	220

cuttings, and subsequent nursery operations, cuttings could by 1990 be as little as twice as expensive to produce as seedlings from open-pollinated seed orchards (Table 6.9). Field-ready tissue-cultured plantlets, however, remained more expensive, at 3–5 times the cost of seed-orchard seedlings, and, like cuttings they posed some additional unsolved problems in controlling maturation.

6.12 Developments in Breeding Strategy and Population Management

During this period, some major developments occurred in breeding strategy and in the population management that those developments allowed. Of great significance was one based on the realisation that, by combining production of control-pollinated commercial seeds with VM, the implementation of **sublining** became vastly easier.

6.12.1 *Implementation of Sublining*

The concept of sublining (Burdon and Namkoong 1983), whereby a breeding population is divided into strictly unrelated sublines *within* which inbreeding could be tolerated, had been developed to assure outcrossed commercial planting-stock at any future date. That desired outcome is achievable by controlled crossing *between* the sublines when producing seeds for commercial stocks. However, rapid and/or unven accumulation of inbreeding within sublines could cause difficulties in maintaining the sublines.

For an open-pollinated clonal seed orchard, trying to avoid all related matings would require as many sublines as clones in the orchard, and yet would not preclude both self-fertilisation and pollination by unwanted external unknown parents. If, however, pollination can be fully controlled, as had long been possible on an operational scale in South Korea hybridizing pitch and loblolly pines, then only two unrelated sublines are needed to guarantee complete outcrossing (Burdon 1986), and those sublines would be easier to manage.

By 1986, there was so much unknown and known inter-relatedness within the New Zealand radiata breeding population that constructing as many more-or-less unrelated sublines as clones wanted in a seed orchard (12 at the very least) was posing a major problem. To ensure their unrelatedness would entail discarding an unacceptably large number of families, but constructing only two sublines from the main GF breeding population could be done with minimal sacrifice of families. Moreover, within each such subline there can be differentiated breeds, which have different breeding goals, but members of the different breeds within a subline need not be strictly unrelated to each other; in fact, a single parent clone could belong to more than one breed. Indeed, the two-sublines scheme has also been adopted in some other parts of the world, and applied in other forest-tree species as well as in radiata in Southern Tree Breeding Association in Australia (Boomsma 1997; White et al. 1999).

6.12.2 Developments in Breed Differentiation and Stratifying Breeding Populations

By 1986 in New Zealand, the basis for differentiation of the original and then enlarged breeding population into "regional breeds" had changed radically. Originally it had entailed a three-breed regional differentiation: the main group of predominantly central North Island selections, which were designated to also serve Nelson (northern South Island), Canterbury (eastern South Island) selections and Southland-Otago (far south) selections. Thus, New Zealand's provisional radiata breeds had initially been designated according to their regional groups of field selections and their descendants. This regional differentiation, however, was dropped when it was found that the South Island plus-tree progenies (and seed-orchard seedlots) were generally inferior in both growth rate and general tree form, even when planted in the regions that their seed orchards were designed to serve.

Later, in the early to mid-1980s, breeds were reorganised according to breeding goals (Burdon 2008), as we have already traced in Sect. 6.6.4.1. The then-called "uninodal" selections, first made in 1970, were the initial basis of the Long Internode (LI) breed. The main component of the then New Zealand-wide breeding population, which had become defined predominantly by its members' "multinodal" (or "short-internode") branching pattern, was then designated the Growth and Form (GF) breed. Additional breeds were mostly offshoots of the GF, their members mostly drawn from subsets of it.

The resulting breed structure has been portrayed in various ways. However, a common heuristic device has been an analogy to an amoeba (Shelbourne et al. 1989), with the main (GF) component of New Zealand's breeding population likened to the cell body and the offshoot breeds to the pseudopodia. The original LI breed, however, belonged as a separate entity.

Developments in breeding strategy within the STBA in South Australia have been traced by Boomsma (1997). Initially, Paul Cotterill of CSIRO drew up a proposed breeding plan (Cotterill 1984). The starting foundation population was 700 open-pollinated (OP) families from plus trees selected in the South Australian radiata land races. After a few years, one promising tree would be selected from each of 400 of those OP families planted in "existing progeny tests," and then mated to produce 200 pair-cross families, and their seedling offspring planted on one large progeny-trial replicated on two sites. Within each of those 200 families, four individuals would be selected after eight years on site for tests of their open-pollinated progeny, and scions of each would be grafted and archived. Then, after four years of testing their progeny, these 800 candidates would be culled to two per cross the give 400 parents to produce 200 pair-crosses. STBA would produce the next generation repeatedly by crossing in pairs. Cotterill's proposed strategy was fundamentally the "single-pair-cross" mating scheme earlier proposed by Libby (1973).

Cotterill and colleagues revisited the breeding strategy (Cotterill et al. 1989), picking up on earlier work by his past doctoral supervisor, John James. They introduced the concept of "nucleus breeding" to STBA, which stratified its breeding population into two levels, creating a population hierarchy: the "elite" or "nucleus," produced by matings between 40 top-rated parents, and the "main" which was based on 300 OP families but contained additional genetic diversity and could later feed additional parents into the elite.

This was a shift from the single-pair scheme of Libby (1973), concentrating on the top-ranked parents to produce the next generation of plantations with the help of VM and increasing their representation in the breeding population, yet allowing parents in the elite to be demoted into the main, instead of being culled and lost. This continued a traditionally closer relationship of forest-tree breeding to animal breeding than to most plant breeding. A key consideration for Cotterill was operational simplicity, given past experience in Australia with elaborate mating schemes that looked good on paper but took an inordinate time to complete.

Following additional input from other breeders, notably Tim White of University of Florida, the following features were adopted for the implemented STBA breeding strategy (Boomsma 1997):

- stratification of the breeding population (Fig. 5.21) into two levels – small, elite Nucleus, and larger, less-select Main
- pair-crossing and **polymix**^{G} crosses within Nucleus, but only polymix crosses within Main, polymixes differing from open pollination in ensuring that all pollen parents are select
- designating three breeds: Growth + Stem quality + Branch quality, Wood density + Growth, and Growth + Resistance to *P. cinnamomi*
- inclusion of the three breeds being equal and overlapping (not mutually exclusive) in the Elite, but 50% Growth + Stem Quality + Branch Quality without overlap of breeds in the Main
- superimposed upon the 2 strata × 3 breeds structure, two unrelated sublines
- planting progeny tests (for both forwards and backwards selection) on multiple sites
- clonal replication as cuttings of at least some individual offspring, meaning that forwards selection could be made on performance across multiple sites.

Soon afterwards, however, several variations of the plan were placed on offer to the STBA (White et al. 1999), but the course taken by STBA is presented in the next chapter.

The classification of the breeding population into Elite and Main components was subsequently applied within the New Zealand breeding population (Jayawickrama et al. 1997; Jayawickrama and Carson 2000). Members of existing specialised breeds were assigned to the Elite stratum, along with top-ranked members of the existing mainstream GF (Growth and Form) breed. As at 1997 the existing specialised breeds were Dothistroma Resistant, Long Internode and High Wood Density, although some revision of this "breed portfolio" was envisaged. A separate entity, with an additional, conservation-related objective, was a pure Guadalupe Island breed (Low and Smith 1997). Actually, designating specialised breeds derived from the original GF into elites represented no substantial change, but stratifying the GF component into Elite and Main was a strategic change.

6.12.3 Coping With Increased Population Complexity

As breeding programmes developed, the breeding-population structures became more complex. The generations of breeding to produce the next selections became no longer discrete, and were thus subject to various overlaps. Field trials proliferated, in which new progeny served both as selection candidates and for evaluating their parents' breeding values. Comparisons between trees of different generations and/or in different field trials became more and more challenging because of the difficulties of cross-comparisons between different subsets of breeding populations. This problem had been deferred, but only deferred, in New Zealand, by the breeders having mounted very large field trials containing enormous numbers of trees. The increasingly complex breeding-population structures produced varied responses.

Mike Wilcox, at FRI in New Zealand, aimed to combat this proliferation within FRI's programme in the early to mid-1980s by consolidating crossing programmes, and field trials into the so-called "Big Bang," which would allow better cross-reference between different segments of the still-single NZ breeding-population. However, this endeavour fell victim to institutional changes affecting FRI and the broader forestry sector, together with logistical difficulties.

Elsewhere, with slash pine breeding in Florida, USA, and thence with radiata breeding in the STBA in Australia, different approaches were adopted, in both quantitative selection methodology and physical logistics. These reflected combinations of circumstances, among which was a shortage of sufficiently large progeny trials within which large numbers of selection candidates could be compared with adequate and similar precision. Thus, in the early decades of radiata's intensive breeding, many of the advances were obtained in New Zealand, but with local circumstances meaning that some leading-edge developments then came in Australia, particularly at STBA.

As the breeding-population structure became more complex, selection methodology was increasingly adapted from animal-breeding methods. Among other things, this helped efficient cross-reference between candidates selected in different kinds of field trials and/or different breeding generations, and about which there were varying amounts of pedigreed data on traits of interest. The last-mentioned feature meant that estimated heritabilities for candidates could vary widely according to the amount of pedigreed data that was available from ancestors, current sibs and cousins, and descendants. This methodology, called Best Linear Unbiased Prediction (BLUP), was far more flexible than a simple selection index, which depended on similar levels of data on all selection candidates. However, it had one important failing to be rectified for forestry applications, in that it was not set up to cope with cloning of offspring.

Adoption of BLUP (White and Hodge 1989; White et al. 2007) began in slash pine by the Florida Cooperative Tree Improvement Program, in which data from relatives was being derived from large numbers of small, variously-related progeny trials. BLUP was first adopted for breeding radiata by the STBA, based in South Australia, using the TREEPLAN suite of software. The flexibility of using BLUP increased the attractiveness of breeding on a "rolling front," in which discrete generations are unimportant and which has the logistical attraction of smoothing the breeding workload over successive years. That "smoothing" is another element of the parallels of forest-tree breeding with animal breeding, rather than with most other plant breeding.

While BLUP selection methodology assures a rigorous process, personal experience (WJL) is that, for an experienced breeder, comparing trees of clonal selection candidates with their immediate neighbours, comparatively ad-hoc selection algorithms can actually deliver closely equivalent results.

6.12.4 Cloned Breeding Populations?

In advancing a breeding-population, "forwards selection" among the offspring representing the new generation is essential. In contrast to "backwards selection," which re-evaluates relatively few parents, belonging to the preceding generation, using their offspring's average performance in progeny tests, forwards selection can be far more intensive because breeders can have far more candidates for selection. However, as in the Cotterill scheme for STBA, it may be useful to limit the number of selected candidates to a few per family, in the interests of being able to handle a broader genetic base.

Effective heritabilities are far lower for individual seedlings in progeny trials than in most progeny tests that serve backwards selection (reslection of parents on the basis of progeny performance). This is because, unlike in a progeny test, there is no statistical damping-down of the environmental variation influencing the candidate seedlings. Thus, while expected genetic gains from forwards selection may often be greater than such gains from backwards-selection, limited confidence can be placed in the genetic merit of any one selected candidate in forwards selection—a matter that can cause much consternation among tree breeders' clients.

That problem has sometimes been overcome by cloning all candidate offspring, thereby raising effective heritabilities for all traits, but at some loss of selection-intensity. John Russell exhaustively explored the trade-offs between selection intensity and effective heritability (Russell and Libby 1986). With high-heritability traits, such clonal replication provides no advantage for forwards selection. With medium heritability traits, two ramets per clone seems optimal. With low-heritability traits, three or more ramets per clone can be justified.

However, one doesn't clone traits, but the entire seedling with all of its traits must be cloned. Thus, the breeders must decide which traits are most important and justify cloning or not on that judgment. If important traits exhibit appreciable genotype-environment interactions, such clonal replication with trials in different environments can also, and importantly, lead to enhanced genetic gains from forwards selection, and will better characterise the individual selections for later deployments. By 1992, cloning within progeny trials was being vigorously promoted in New Zealand by FRI's Shelbourne (1992) for radiata and some other species.

6.13 The Molecular Revolution Develops Momentum

The practice of tree breeding had been based heavily on the premises that many individual genes control variation in traits of interest, that selection had to be based predominantly on field performance, and that new genetic combinations would be created entirely by pollination to produce seeds. Progress towards removing these limitations began during 1984–1997, but few operational solutions could be delivered very quickly. But because R&D funding was static or even declining in real terms, the progress tended to be achieved at the expense of resources for classical tree breeding. Tensions over research funding were reinforced by a "cultural divide" between operational tree breeders and researchers in DNA technology.

6.13.1 Background

Mendel's work with garden peas proved that inheritance could be governed by discrete factors that maintained their integrity and identity over generations. To achieve this, he studied traits that showed obviously discontinuous variation, such as a tall versus dwarf habit, pink vs white flower colour, or smooth vs wrinkled seeds. Such discrete variation, however, is the exception rather than the rule for most "important" traits in most organisms, including forest trees, which posed a conundrum for the early geneticists. This was resolved in the neo-Darwinian synthesis spearheaded by R.A. Fisher and J.B.S. Haldane in England, Sewall Wright in the US, and belatedly recognised but most completely resolved by S.S. Chetverikov in Russia (Lerner and Libby 1976).

Multiple genes, individually exerting relatively small or even some medium-to-large effects on a trait, but each independently observing Mendelian inheritance, could account nicely for continuous variation in such traits as diameter growth in radiata. That theory held up and formed the basis for a model of inheritance that both accounted for observations and could predict responses to selection for such traits. So came into being classical quantitative genetics, based on the so-called "polygene" model. Indeed, that conceptual model has proved extremely useful, being both powerful and robust, but it effectively abdicates from the task of assigning expressed trait variation to individual genes. Getting past that limitation with species like radiata has been a slow and difficult task, which is still in its early stages.

A central problem with classical plant- and animal breeding is that whenever two parents are mated, their offspring all get different halves of their genes from each parent. Thus, no two such full-sibs are genetically identical to each other, or to either parent. Furthermore, if one of the parents is used to confer on its many offspring a particular allele that affects a trait in a desired way, some alleles of its other genes may be downright unwanted. Worse, the parent confers different sets of these unwanted alleles to each offspring, and these unidentified alleles could only be eliminated slowly and imperfectly by selection against their effects over many subsequent generations. Worse still, desired alleles may not be available in any parents that can be successfully crossed with the species one is using. All three of these situations could, by the 1970s in principle, be rectified by **transgenic genetic engineering** (Lerner and Libby 1976). Technology allowing short and highly specific sequences of DNA ("**transgenes**") to be inserted into recipient cells of radiata would avoid the unwanted "genetic baggage" that comes with the sexual reproduction that produces seeds. The technology's continuing development is outlined in Sect. 6.13.3 and the next chapter.

Unfortunately, rapid developments in biotechnology brought institutional strains. Those changes accentuated differences among member companies in breeding cooperatives in the levels of commitment to state-of-the-art technology, which in turn tempted some to develop and then appropriate new technologies for themselves—quite contrary to the original basis of collaboration in breeding cooperatives.

The advent of gene technologies also created a divide within the scientific community that had been engaged in the common-cause goals of radiata breeding. On the one hand, an imperative was seen to engage in what appeared to be the technology of the future, and support was diverted to that goal. Not surprisingly, those new molecular technologies delivered few immediate payoffs. Meanwhile, support for breeding, selection and field-testing was reduced, incurring increasing opportunity costs in continuing the accumulation and deployment of genetic gain (Libby 1991).

6.13.2 The Search for Individual Genes and their Desired Alleles

Occasionally, evidently mutant seedlings appear that are obvious major variants from the normal state. They typically reflect deleterious recessive genes that are seldom expressed until inbreeding has occurred, although they may much less commonly be the expressions of new, dominant mutations. While these mutant alleles can serve as genetic markers, their practical value is limited. Moreover, the chromosomes of pines are relatively featureless under conventional light microscopes (Pederick 1987), further frustrating efforts of radiata breeders to relate trait variations to chromosome features and thus to the possible locations of genes that affect the traits.

An alternative approach to locating genes of interest in radiata and other pines was studying variants of secondary metabolites, such as components of turpentine. Those compounds exhibit more direct expressions of genes than most traits that are observable in the field. In radiata, the average proportions of various terpenes differ substantially among the five native populations (Burdon et al. 1997b and references therein). The terpenes have thus given useful information on the origins of land races that have provided the early breeding populations for intensive domestication. But, while the data contain suggestions of some terpene genes with large effect, nothing more was thereby revealed about the genes themselves. Unlike the early situation with slash pine in Florida, in which resin products were important, none of these secondary metabolites has merited a place in breeding goals for radiata.

A closer approach to the nature of individual genes is expressed by variants of a suite of enzyme proteins that catalyse biochemical reactions. The sequence of amino acids in those enzymes is read directly from the sequence of bases in the RNA that delivers information from the DNA sequences of the alleles to the sites of protein synthesis. Reading backwards decoded the allele, but did not yet locate it. Called either isozymes (because they are involved in the same reactions) or allozymes (because they are variants of the enzyme), these proteins typically occur as a few discrete variants that can be readily discriminated in the laboratory. They were, from the late 1960s, the default tool for studying allele frequencies of individual genes and mating patterns in many organisms, including forest trees (e.g. Plessas and Strauss 1986; Moran and Bell 1987 for radiata), until the early 1990s.

Allozymes have been especially valuable for studying the breeding system of radiata. They were good at indicating the relative amounts of outcrossing and related matings that occurred in natural populations (Moran and Bell 1987), plantations and seed orchards (Moran et al. 1988), and in very large studies could even indicate the types of inbreeding that occurred (selfing or other kinds of related matings). That, in turn, was valuable for interpreting comparisons of growth-performance of samples of natural stands and plantations in provenance trials, the natural stands usually having some family population structure and thus being subject to "neighbourhood inbreeding" effects. They also revealed a few large differences in allele frequencies among natural populations of radiata (Moran and Bell 1987). When used in seedlings from seed orchards or from "controlled" crosses to verify parentage, they were the first class of markers to reveal embarrassing amounts of pollen contamination in OP seed orchards, and even in so-called "controlled" crosses, calling into question the quality the crossing protocols and allowing correction of pedigree records.

The limitations of allozymes as an investigative tool in radiata, however, were major. Out of 20-odd such enzyme systems, not all varied enough (technically **polymorphic**[G]) to be useful. Among those that qualified, if one allele was present at a very high frequency that gene added little information. This meant that, of the many thousands of genes present in a radiata, worthwhile allozyme-based information was typically available for only 2–4 of them. Furthermore, it became clear that isozyme variants in pines can be far from selectively neutral (e.g. Strauss and Libby 1987), and were thus less useful than neutral DNA sequence variants in establishing long-term or even recent population histories. Non-neutrality, however, points to a potential role in tracking adaptive genetic shifts in populations in new or changing environments.

More specific information on the DNA of individuals, families or populations could be obtained from variants in sequences of DNA's four bases, the "letters" of the genetic code. Such variants could be revealed by **RFLPs** (**R**estriction **F**ragment **L**ength **P**olymorphisms), or **AFLPs** (**A**mplified **F**ragment **L**ength **P**olymorphisms), which depend on specific enzymes cutting DNA strands at specific sequences, or by **RAPDs** (**R**andom **A**mplified **P**olymorphic **D**NA sequences) which depend on whether short, arbitrarily chosen DNA sequences will amplify at chromosome sites where they are present.

It was RAPDs that opened the door to achieving chromosome maps with conifers. In a conifer seed, its nutrient-storage organ (the "endosperm") distinctively remains **haploid**[G] after the egg is fertilised. Thus, this relatively large tissue, having only a single set of chromosomes of maternal origin, can have its DNA compared to its contained embryo or germinated seedling, which is **diploid**[G] with both the maternal and paternal sets of chromosomes and their DNA. This feature was exploited with loblolly pine in 1991 (Grattapaglia et al. 1992; O'Malley et al. 1996), and what worked with loblolly duly worked with radiata (Devey et al. 1996). Although the RAPDs are dominant rather than recessive, which for several purposes is a disadvantage, this did not matter in haploid tissue. Thanks to the development of **Polymerase Chain Reaction (PCR)**[G] technology, such polymorphisms were soon easily and cost-effectively identified using minimal tissue of conifer seeds.

Another type of DNA marker is **microsatellites**[G] or **Simple Sequence Repeats (SSRs)**[G], for which variation exists in the number of adjoining repeats of short, simple DNA-base sequences. These have the two advantages of being codominant, allowing alleles on both homolog chromosomes of an individual to be recognised, and of typically having several common variants (also called alleles) at the same small DNA-sector. These became the typical DNA marker type used for "bar-coding" individuals in crime forensics.

Those new markers formed the basis for mapping chromosomes and their resident DNA, in which huge groups of genes are linked in a linear order. Once identified, these linked groups formed a basis for relating observed trait-variants to particular segments of DNA, thus approximately locating desired genes if their effects on the trait are fairly large. For this, it is not necessary to identify actual genes, only to identify specific DNA variants (or polymorphisms) that were linked to the genes influencing important or interesting traits.

While the old pre-DNA polygene model was a great predictive tool for selecting desired variants for polygenic traits, its predictive value did not preclude the possibility that some genes exert quite large effects in continuous-variation traits and thus could be of unusual importance in the genetic variation in those traits. This initiated a search for those unusual major-effect genes (or groupings of genes), termed **Quantitative Trait Loci (QTL)**^{G}, which could be selected for on the basis of the markers, to capture important desired genetic variation.

Such selection could use marker information alone or, much better, use marker information in conjunction with trait-performance information, in what is collectively termed **Marker-Aided** (or **Assisted**) **Selection (MAS)**^{G}. This has worked well with highly-inbred domesticated annual crop plants, and with domesticated apple which came from a fusion of different species, but not nearly so well with species like radiata.

Locating such relatively important DNA sequences using markers, however, depended on **Linkage Disequilibrium (LD)**^{G}, namely consistent co-occurrence of marker variants and the important genes of interest. Being an outbreeder and largely undomesticated, radiata has very limited LD, except within individual families. Not surprisingly, few large-effect QTLs have been detected and verified in radiata, false positives being all too prevalent.

The quest for QTLs, while frustrating in radiata, had an unexpected payoff. It verified the earlier allozyme-based detection of a disturbingly high incidence of mistaken pedigree, which is especially easy to detect with SSRs. That has led to much better quality control in breeding. Identity of ramets used in controlled crossing and in seed orchards could then be easily verified, as could the parentage of the next generation of the breeding population. This will become ever more valuable as a breeding programme advances and delivers cumulative genetic gain; with increasing levels of gains secured by correct identity of pedigree.

Some of the early genomic research with radiata involved not the nuclear genome, which is dauntingly large, but the organelle genome of the chloroplasts. Such work, however good, has not led to major developments, so is not cited here. Regarding the nuclear genome, it was starting to become clear by 1997 that much the same genes exist in very different plant species, a phenomenon called **orthology**. For instance, similar genes associated with flowering have been shown to exist between radiata and the tiny thale cress, *Arabidopsis*, despite radically different developmental architecture of reproductive structures. Orthologies can be used to locate genes of specific functions in a species like radiata, although practical applications with flowering have still to come. Also helpful is that among pines, and conifers in general, the same genes may be found on corresponding chromosomes (**synteny**) and in much the same order (**colinearity**). The import of these developments is outlined in the next chapter.

6.13.3 The Pursuit of Molecular-Based Genetic Engineering

"Genetic engineering" (GE) , as we consider it, is using DNA technology to confer specific attributes without the natural processes of creating new genetic combinations that arise in the process of seed production. It thus avoids introducing unwanted "baggage" genes coming from a second parent whenever a seed is produced (Sect. 6.13). Use of GE for genetic improvement effectively raises both the efficiency and the level of domestication obtainable, times and costs being equal compared with classical breeding.

Compared with domesticates such as cotton or soybean, some interrelated features of radiata (and other forest trees) created some special attractions for breeders to adapt recently developed genetic engineering, with its prospective time savings. These related to radiata's relatively long generation time and delayed expression of some key traits, and consequent and additional problems in identifying and locating QTLs. Those features, along with success with GE of some agricultural crop plants, prompted exploratory work on GE's feasibility with radiata (Walter et al. 1997).

An immediate problem was finding a preferred method of inserting DNA fragments into radiata pines. After some trials with alternatives, a biolistic technique was chosen (Walter et al. 1995), which worked well with radiata despite its crudeness. It involved bombarding tissue with tiny metal pellets coated with short "transgene DNA" sequences. By creating tightly-linked "constructs" of a desired transgene and an accompanying "marker" transgene, stable incorporation into callus tissue was initially verified by observable expression of the accompanying marker transgene.

Meanwhile, advances in *in-vitro* propagation technology had provided promising tissues whose cells could be transformed, and then grown into plant cultures that could be clonally propagated as complete plantlets. Choice of which transgenes to use depended first on what transgenes could show demonstrable activity, so knowing pathways of the transgenes' expression was by 1997 a topic of considerable research (Walter et al. 1997). An immediate, proof-of-concept target of GE was the incorporation of herbicide resistance in several crop plants, although that modification is less important for growing radiata than for many other species. Another other early research target with pines was modified lignin production, prompted by the fact that, in chemical pulping, conifer lignin is not as readily and cheaply dissolved as hardwood lignin. This, however, led to worries that such transformed pines would be less robust in field environments, a worry now supported by recent field trials of such lignin-modified pines and other tree species (Wagner et al. 2009; Voelker et al. 2011—and some references therein).

6.14 Summary of Domestication Progress

Active afforestation continued in the main grower countries, some of it being conversion from marginally profitable pastoral farming. Afforestation rates, however, were fluctuating, in responses to changes in government policies and in export markets.

Large genetic gains from radiata breeding were demonstrated for quality of radiata's tree form and for its growth rate. However, the standard of branch size and wood properties was often markedly poorer in a new generation of commercial stands of breeds selected for faster growth and better form. While effects of adverse genetic correlations between growth rate and wood properties contributed, this situation arose mainly through a combination of direct effects of the aggressive Direct Sawlog silvicultural regime and the indirect effects of unforeseen reductions in harvest ages that resulted partly from adopting that regime.

The situation prompted a major shift in interest towards genetic improvement of wood properties, a pursuit that had long been hampered by both a lack of direction from industry and a shortage of cheap and efficient assay technologies. There followed much increased research on the importance and inheritance of several wood properties, and improved assay technologies were adopted. This refocus of research, during the mid- to late-1990s, delivered improved assay technologies and a better knowledge of variation and inheritance of performance-related wood properties, especially stiffness and dimensional stability, allowing a serious commitment to intensive genetic improvement of those properties. This development represented a shift in the appropriate focus of genetic improvement in the direction of counteracting drops in quality that were resulting from management practices designed to boost productivity and otherwise reduce effective costs of growing radiata wood.

Seed-orchard outturns reached the stage at which genetically improved stock of radiata could cover planting needs for radiata in New Zealand, parts of Australia, South Africa, and then Chile. This helped prompt in New Zealand a new system of seed certification that served as an effective educational and marketing aid for genetically improved seed and nursery stock. The self-sufficiency for seed-orchard seed also meant that seed was no longer wanted from the passively improved land-race stocks, which then became subject to progressive liquidation by replacement with seed-orchard stock. In response, some major plus-tree selection operations began afresh in some previously untapped land-race stocks before they were harvested.

New seed-orchard systems were developed, based on controlled pollination, which allowed much more precise targeting and better capture of genetic gains. At much the same time, vegetative multiplication of select radiata families and clones became widely adopted, first for supplementing seed-orchard production and then for otherwise accelerating the capture of genetic gain. To achieve this, one commercial operation was mounted, based on initial in-vitro organ culture. In other cases, and less expensively, direct use of nursery cuttings was adopted, helped by practices to slow the maturation of cutting-donor plants. Continued propagation of outstanding clones, however, was limited by the fact that maturation could only be delayed by then-current techniques, not reversed or even halted.

Use of control-pollinated seed production systems also allowed important changes in the management of the breeding populations, to better generate cumulative genetic gain over generations.

Far more became known about the desirable features—and the drawbacks—of the five natural populations, and thence of how their genes might best be used in future breeding work. From *ex-situ* plantings of them in New Zealand modest numbers of plus trees were selected, to augment the main breeding programme.

Much research and development work was done towards achieving feasibility of large-scale clonal forestry in radiata, including refinements of in-vitro culture systems. This pursuit, however, remained hampered by limited capability to check progressive maturation of clones. Towards overcoming this, radiata scientists embarked upon adapting and emulating embryogenesis technology that was succeeding with spruces.

Pilot research in radiata began on technologies for genetic engineering (transformation), use of DNA markers for selection, and cryogenic storage of embryogenic tissue.

References

Ahuja MR, Libby WJ (eds) (1993) Clonal forestry. Vol. 1. Genetics and biotechnology (15 chapters, p ix + 277). Vol. 2. Conservation and application (13 chapters, p 240). Springer-Verlag, Heidelberg

Alazard P, Destremau D-X (1982) De l'expérimentation en France de *Pinus radiata* – résultats préliminaires. 1981 Annales de Recherches Sylvicoles. AFOCEL, Paris, pp 5–33

Arnold RJ (1990) Control-pollinated radiata pine seed—a comparison of seedling and cutting options for large-scale deployment. N Z [J] For 35(3):12–17

Arnold RJ, Gleed JA (1985) Raising and managing radiata pine vegetative cuttings for production forests. Aust For 48:199–206

Balocchi CE (1997) Realised versus operational gain: the role of propagation strategies. In: Burdon and Moore, pp 253–255.

Balocchi C, Jayawickrama K, Perez E (1991) Octavo informe annual (1989–1991). Cooperativo de mejoramiento genetico – UACH – CONAF – Empresas forestales. Universidad Austral de Chile, Valdivia, Facultad de Ciencias Forestales

Birk EM (1990) Poor tree form of *Pinus radiata* D.Don on former pasture sites in New South Wales. Aust For 53:104–112

Birk EM, Bowman VJ, Fulton JA et al (1993) Merchantability of *Pinus radiata* [radiata pine] in relation to previous land use. Aust. For 56:157–164

Bolstad PV, Libby WJ (1982) Comparisons of radiata pine cuttings of hedge and tree-form origin after seven growing seasons. Silvae Genet 31:9–13

Boomsma DB (1997) Development of the Southern Tree Breeding Association breeding population. In: Burdon & Moore, p 211–216

Brown AG (1974) Comparison of early growth in radiata pines raised by cuttings from parents of different ages with that of seedling trees. Aust For Res 6:43–47

Burdon RD (1971) Clonal repeatabilities and clone-site interactions in *Pinus radiata*. Silvae Genet 20:33–39

Burdon RD (1986) Clonal forestry and breeding strategies—a perspective. In: Proceedings, IUFRO genetics group meeting. Williamsburg, VA, USA, October 1986, pp 645–659

Burdon RD (1989) When is it appropriate to clone on an operational scale? Pp 9–27 In: Matheson AC, Gibson GL (eds) Breeding tropical trees: population structure and genetic improvement strategies in clonal and seedling forestry. Proceedings, IUFRO conference, Pattaya, Thailand, November 1998. Oxford Forestry Institute, Oxford, United Kingdom and Winrock International, Arlington

Burdon RD (1992) Genetic survey of *Pinus radiata*. 9. General discussion and implications for genetic management. N Z J For Sci 22:274–298

Burdon RD (2004) Genetics of *Pinus radiata*. In: Burley J, Evans J, Youngquist JA (eds) Encyclopedia of forest sciences vol 3. Elsevier Academic Press, Oxford and San Diego, pp 1507–1516

Burdon RD (2008) Branching habit in radiata pine – breeding goals revisited. N Z J For 53(1):20–23

Burdon RD (2010) Wood properties and genetic improvement of radiata pine. N Z J For 55 (2):22–27

Burdon RD, Low CB (1992) Genetic survey of *Pinus radiata*. 6. Wood properties: variation, heritabilities and interrelationships with other traits. N Z J For Sci 22:228–245

Burdon RD, Moore JM (eds) (1997) IUFRO '97 Genetics of radiata pine. In: Proceedings of NZ FRI-IUFRO conference 1–4 December and workshop 5 December, Rotorua, New Zealand. FRI Bulletin No. 203. p 354

Burdon RD, Namkoong G (1983) Multiple populations and sublines. Silvae Genet 32:221–222

Burdon RD, Firth A, Low CB et al (1997a) Native provenances of *Pinus radiata* in New Zealand: performance and potential. N Z [J] For 41(4):32–36

Burdon RD, Broekhuizen P, Zabkiewicz JA (1997b) Comparison of native-population and New Zealand land-race samples of *Pinus radiata* using cortical oleoresin monoterpenes. In: Burdon & Moore, pp 50–56

Burdon RD, Firth A, Low CB et al (1998) Multi-site provenance trials of *Pinus radiata* in New Zealand. In: Forest genetic resources No. 26: 3–8. FAO, Rome

Burdon RD, Carson MJ, Shelbourne CJA (2008) Achievements in forest tree genetic improvement in Australia and New Zealand 10: *Pinus radiata* in New Zealand. Aust For 71:263–279

Butcher TB (1993) HAPSO development in Western Australia for mass production of full-sibling seed. Pp 39–40 In: Slee M (ed) Proceedings 12th meeting, working group #1 (forest genetics) of the Australian Forestry Council. Department of Forestry, Australian National University, Canberra

Butcher TB, Stukely MJC (1986) Update of *Pinus radiata/Phytophthora cinnamomi* studies in Western Australia. In: Griffin R , Volker P (eds) Proceedings of the ninth meeting of representatives held at Burnie, Tasmania. Research working group no. 1 of the Australian Forestry Council. Forest genetics. Australian Forestry Council, pp 36–39

Butcher TB, Stukely MJC, Chester GN (1984) Genetic variation in resistance of *Pinus radiata* to *Phytophthora cinnamomi*. For Ecol Manage 8:197–220

Butterfield BG (ed) (1997) Microfibril angle in wood. In: Proceedings, IAWA/IFRO workshop on the significance of microfibril angle to wood quality, Westport, New Zealand, November 1977, p 401

Carson SD (Chair), Burdon RD (ed.), Buddle JN, Field DA et al. (1986) Report of Working Group on certification of forest tree seed and planting stock. NZ For Serv, Forest Health & Management Div, For Res Inst and For Management Div, Head Office, Wellington (unpubl.)

Carson MJ (1986a) Control-pollinated seed orchards of best general combiners. In: Williams GA, Wratt GS (eds) Agronomy Society of New Zealand special publication No. 5, pp 144–149

Carson MJ (1986b) Advantages of clonal forestry for *Pinus radiata*—real or imagined? N Z J For Sci 16:403–415

Carson MJ (1990) Comparison of improved radiata pine breeds using STANDPAK. In: James & Tarlton, pp 184–194

Carson MJ, Inglis CS (1989) Genotype and location effects on internode length of *Pinus radiata* in New Zealand. N Z J For Sci 19:267–279

Carson MJ, Tombleson JD, Wilcox PL (1991) Research for stand management of clones. In: Miller, pp 147–149

Carson MJ, Vincent TG, Firth A (1992) Control-pollinated and meadow seed orchards of radiata pine. Pp 13–20 In: Bey M-N, Leroy M, Verité S (eds) Mass production technology for genetically improved fast growing forest tree species. Tome II. AFOCEL, Nangis, France

Carson SD (1989) Selecting *Pinus radiata* for resistance to *Dothistroma* needle blight. N Z J For Sci 19:3–21

Carson SD (1991) Genotype × environment interaction and optimal number of progeny test sites for improving *Pinus radiata* in New Zealand. N Z J For Sci 21:32–49

Carson SD, García OP, Hayes JD (1999) Realized gain and prediction of yield with genetically improved *Pinus radiata* in New Zealand. For Sci 45:186–200

Cave ID (1968) The anisotropic elasticity of the plant cell wall. Wood Sci Technol 2:268–278
Cave ID, Walker JCF (1994) Stiffness of wood in fast-grown plantation softwoods: the influence of microfibril angle. For Prod J 44(5):43–48
Chaperon H, Fraysse JY (1986) Frost resistance of the principal pine species introduced in southwestern France. In: 1985 Annales de Recherches Sylvicoles. AFOCEL, Paris, pp 219–259
Codesido V, Fenández-López J (2008) Juvenile genetic parameter estimates for vigour, stem form, branching habit and survival in three radiata pine (*Pinus radiata* D.Don) progeny tests in Galicia, NW Spain. Eur For Res 127:315–325
Codesido V, Fenández-López J (2009) Implications of genotype × site interaction on *Pinus radiata* breeding in Galicia. New For 17:17–34
Corson SR, Foster RS, Richardson JD (1989) New Zealand-grown spruce and radiata pine can have similar TMP properties. Appita 42:345–349
Cotterill PP (1984) A plan for breeding radiata pine. Silvae Genet 33:84–90
Cotterill PP, Dean CA (1988) Changes of genetic control of growth of radiata pine to 16 years and efficiencies of early selection. Silvae Genet 37:138–146
Cotterill PP, Dean C, Cameron J et al. (1989) Nucleus breeding: a new strategy for rapid improvement under clonal forestry. In: Gibson GL, Griffin AR, Matheson AC (eds) Breeding tropical trees. Proceedings IUFRO conference, Pattaya, Thailand, November 1988. Oxford Forestry Institute, pp 39–51
Cown DJ (1978) Comparison of the Pilodyn and torsiometer for the rapid assessment of wood density in living trees. N Z J For Sci 8:384–391
Cown DJ, McConchie DL (1980) Radiata pine: wood age and wood property concepts. N Z J For Sci 10:504–507
Cromer RN, Eldridge KG, Tomkins D et al. (1982) Intraspecific variation in the response of *Pinus radiata* to saline and waste water. Aust For Res 12:203–215
de Ronde C, Donald DGM (1993) Working towards a P-response model for *Pinus radiata* in the Tsitsikamma. S Afr For J 166:27–35
Dean CA, Cotterill PP, Burdon RD (2006) Early selection of radiata pine. I. Trends over time of additive and dominance genetic variances and covariances of growth traits. Silvae Genet 55:182–191
Devey ME, Bell JC, Smith DN et al. (1996) A genetic linkage map for *Pinus radiata* based on RAPD, RFLP and microsatellite markers. Theor Appl Genet 92:673–679
Donald DGM, Glen LM (1974) The response of *Pinus radiata* and *Pinus pinaster* to N, P and K fertilizers applied at planting. S Afr For J 91(1):19–28
Donald DGM, Lange PW, Schutz CJ et al. (1987) The application of fertilisers to pines in Southern Africa. S Afr For J 141(1):53–62
Donaldson LA, Burdon RD (1995) Clonal variation and repeatability of microfibril angle in *Pinus radiata*. N Z J For Sci 23:164–174
Eldridge KG (1998) Australian radiata provenance trials: an annotated catalogue of Australian provenance trials and genetic conservation plantings of *Pinus radiata* of Californian origin, up to May 1977. Client Report No. 416. CSIRO FFP, Canberra. p 25
Eldridge KG, Spencer DJ (1988) Field performance of cuttings in Australia. In: Menzies et al, p 42–55
Elliott DA, Goulding CJ (1976) The Kaingaroa growth model for radiata pine and its implications for maximum volume production (Abstract). N Z J For Sci 6:187
Evans R (1994) Rapid measurement of variation of tracheids transverse dimensions in radial wood sections from *Pinus radiata*. Holzforschung 48:168–172
Evans R, Downes G, Menz D et al. (1995) Rapid measurement of variation in tracheid dimensions in a radiata pine tree. Appita 48:134–138
Falkenhagen ER (1991) Provenance variation in *Pinus radiata* at six sites in South Africa. Silvae Genet 40:41–50
Fielding JM, Brown AG (1961) Tree-to-tree variations in the health and some effect of superphosphate on the growth and development of Monterey pine on a low quality site. For & Timber Bur, Canberra, Leaflet No 79

Forde MB (1966) *Pinus radiata* in California. N Z J For 11:20–40

Gleed JA (1993) Development of plantlings and stecklings of radiata pine. Pp 149–157 in Ahuja MR, Libby WJ (eds) Clonal forestry II. Conservation and application. Springer-Verlag, Heidelberg

Grattapaglia D, Chaparro JXC, Wilcox PL et al. (1992) Mapping woody plants with RAPD markers: application to breeding in forestry sand horticulture. In: Applications of RAPD technology to plant breeding. Joint Plant Breeding Symposia Series, November 1, 1992, Minneapolis

Hargreaves CL, Smith DR, Foggo MN et al. (1997) Cryopreservation of zygotic embryos of *Pinus radiata* and subsequent plant regeneration. In: Burdon & Moore, pp 281–284

Harris JM, Cown DJ (1991) Basic wood properties. Chapter 6. In: Kininmonth JA, Whitehouse LJ (eds) Properties and uses of New Zealand radiata pine. Volume 1. Wood properties. New Zealand Forest Research Institute, Rotorua, pp 6–1 to 6–28

Hood JV, Libby WJ (1980) A clonal study of intraspecific variability in radiata pine. I. Cold and animal damage. Aust For Res 10:9–20

Horgan K, Aitken-Christie J (1981) Reliable plantlet formation from embryos and seedling shoot tips of radiata pine. Physiol Plant 53:170–175

Horgan K, Skudder D, Holden G (1997) Clonal storage and rejuvenation. In: Burdon & Moore, pp 273–279

Instituto Forestal (2011) Recurso forestal. Cuadro4. <www.infor.cl> Santiago de Chile

James RN, Tarlton GL (eds) (1990) New approaches to spacing and thinning in plantation forestry. In: Proceedings, IUFRO symposium, 10–14 April 1989, Rotorua, New Zealand. Ministry of Forestry, FRI Bulletin 156. pp 360

Jayawickrama KJS, Balocchi C (1993) Growth and form of provenances of *Pinus radiata* (radiata pine) in Chile. Aust For 56:172–178

Jayawickrama KJS, Carson MJ (2000) A breeding strategy for the New Zealand Radiata Pine Breeding Cooperative. Silvae Genet 49:82–90

Jayawickrama KJS, Carson MJ, Jefferson PA et al. (1997) Development of the New Zealand radiata pine breeding population. In: Burdon & Moore, pp 217–225

Johnson GR, Burdon RD (1990) Family–site interaction on *Pinus radiata*: implications for progeny testing strategy and regionalised breeding in New Zealand. Silvae Genet 39:55–62

Johnson IG, Ades PK, Eldridge KG (1997) Growth of natural Californian provenances of *Pinus radiata* in New South Wales, Australia. N Z J For Sci 27:23–38

Johnson IG, Cotterill IM, Raymond CA et al. (2008) Half a century of *Pinus radiata* tree improvement in New South Wales. N Z J For 52(4):7–13

Keeves A, Douglas DR (1983) Forest fires in South Australia on 16 February 1983 and consequent future forest fire management aims. Aust For 46:148–162

Kimberley MO, Moore JR, Dungey HS (2015) Quantification of realised genetic gain in radiata pine and its incorporation into growth and yield modelling systems. Can J For Res 45:1676–1687

King JN, Burdon RD (1991) Time trends in inheritance and projected efficiencies of early selection in a large 17-year-old progeny test of *Pinus radiata*. Can J For Res 21:1200–1207

Koehler AR (1984) Variable lift pruning. New Zealand Forest Service, Forest Research Institute, Rotorua, FRI Bulletin No. 78. p 38

Lanfranco DM, Bűchner JF, Aguilar AM et al. (1991) Parasitoidos nativos en el control de la polilla del brote del pino (*Rhyacionia buoliana*): avances en la identifcacíon del complejo y de sus estrategias de desarollo. Bosque 12(1):69–74

Lavery PB (1990) Radiata plantation statistics NEFD and notes on Chilean, Australian and other resource descriptions. N Z [J] For 35(3):24–29

Lerner IM, Libby WJ (1976) Heredity, evolution and society, 2nd edn. WH Freeman & Co, San Francisco, p 431

Li M, Ritchie GA (1999) Eight hundred years of clonal forestry in China. I. Traditional afforestation with Chinese-fir. II. Mass production of rooted cuttings of Chinese-fir. New For 18:131–159

Libby WJ (1991) The problem of biotechnological constipation. Pp 323–328 In: Ahuja MR (ed) Woody plant biotechnology. Plenum Press, New York.

Libby WJ (1973) Domestication strategies for forest trees. Can J For Res 3:265–276
Libby WJ, Rauter RM (1984) Advantages of clonal forestry. For Chron 60(3):145–149
Libby WJ, Brown AG, Fielding JM (1972) Effects of hedging radiata pine on production, rooting and early growth of cuttings. N Z J For Sci 2:263–283
Liebenberg F, Bientema NM, Kirsten JF (2004) South Africa. Agricultural science and technology indicators, ASTI Country Brief No. 14, p 10 <http://www.asti.cgiar.org>
Lisboa HB (1993) The Chilean radiata pine sector. In: Lewis & Ferguson, pp 365–379
Low C, Smith T (1997) Use of the Guadalupe provenance in *Pinus radiata* improvement in New Zealand. In: Burdon & Moore, pp 57–61
Maclaren JP (1993) Radiata pine growers' manual. New Zealand Forest Research Institute, Rotorua, Bulletin No. 184. p 140
Maclaren JP, Knowles RL (1999) Economics of final crop stocking at the Tikitere field trial. Part 1: volume and quality comparisons. N Z J For Sci 29:165–174
Manley B (1986) Performance of the Kaingaroa growth model at low-stocking regimes. New Zealand Forest Service, Forest Research Institute, Rotorua, FRI Bulletin 113. p 21
Matheson AC, Raymond CA (1984) The impact of genotype × environment interactions on Australian *Pinus radiata* breeding programs. Aust For Res 14:11–35
Matheson AC, Spencer DJ, Nyakuengama JG et al. (1997) Breeding for wood properties in radiata pine. In: Burdon & Moore, pp 169–179
Matziris DI (1995) Provenance variation of *Pinus radiata* grown in Greece. Silvae Genet 44:88–96
McGranahan MF, Borralho NMG, Greaves BL et al. (1997) Estimates of variances and heritabilities, and genetic correlations between propagule types, for growth and form between seedlings and cuttings of *Pinus radiata*. In: Burdon & Moore, pp 125–128
Menzies MI, Aimers-Halliday J (2005) Propagation options for clonal forestry with conifers. Pp 255–274 In: Walter C, Carson MJ (eds) Plantation forest biotechnology for the 21st century. Research Signpost, Trivandrum, Kerala, India
Menzies MI, Klomp BK (1988) Effect of parent age on growth and form of cuttings, and comparison with seedlings. In: Menzies et al., pp 18–41
Menzies MI, Aimers JP, Whitehouse LJ (eds) (1988) Proceedings, workshop on growing radiata pine from cuttings, 5–7 May 1986, Rotorua, New Zealand. Ministry of Forestry, FRI Bulletin No. 135. p 197
Menzies MI, Faulds T, Dibley MJ (1992) Production of radiata pine cuttings for plantation forestry in New Zealand. Acta Hortic 319:359–364
Michel M (2006) El pino radiata en la historia forestal vasca. Sustral: Rev Agropesquera 2006:56–59
Miller JT (compil. & ed.) (1991) Proceedings, NZFRI/NZFP Forests Ltd clonal forestry workshop, 1–2 May 199, Rotorua, New Zealand. FRI Bulletin, Rotorua No. 160. p 200
Mitchell RG, Steenkamp ET, Coutinho TA et al. (2011) The pitch canker fungus, *Fusarium circinatum*: implications for South African forestry. South For 73(1):1–13
Moran GF, Bell JC (1987) The origin and genetic diversity of *Pinus radiata* in Australia. Theor Appl Genet 73:616–622
Moran GF, Bell JC, Matheson AC (1988) The genetic structure and levels of inbreeding in a *Pinus radiata* D.Don seed orchard. Silvae Genet 29:190–193
Old KM, Libby WJ, Russell JH, Eldridge KG (1986) Genetic variability in susceptibility of *Pinus radiata* to western gall rust. Silvae Genet 35:145–149
O'Malley D, Grattapaglia D, Chaparro JX et al. (1996) Molecular markers, forest genetics and tree breeding. In: Gustafson JP, Flavell RB (eds) Genomes of plants and animals: 21st Stadler genetics symposium. Plenum Press, New York
Ohba K (1993) Clonal forestry with sugi (*Cryptomeria japonica*). Pp 66–90 in Ahuja MR, Libby WJ (eds). Clonal forestry II: Conservation and application. Springer-Verlag, Heidelberg
Payn T, de Ronde C (1988) Phosphate fertilisation of mature *Pinus radiata*. S Afr For J 147:26–31
Pederick LA (1987) The structure and identification of the chromosomes of *Pinus radiata* D.Don. Silvae Genet 16:69–77

Pederick LA (1990) Family × site interactions in *Pinus radiata* in Victoria, Australia, and implications for breeding strategy. Silvae Genet 29:134–140

Percival NS, Knowles RL (1988) Relationship between radiata pine and understorey pasture production. In: Maclaren, pp 152–164

Percival NS, Hawke MF, Jagusch KT et al. (1988) Review of factors affecting animal performance in pine agroforestry. In: Maclaren, pp 165–174

Plessas ME, Strauss SH (1986) Allozyme differentiation among populations, stands, and cohorts in Monterey pine. Can J For Res 16:1155–1164

RPBC (2002) Rating the genetic quality of radiata pine. RPBC Information Bulletin, Rotorua, New Zealand, No. pp 1. p 4

Russell JH, Libby WJ (1986) Clonal testing efficiency: the trade-offs between clones tested and ramets per clone. Can J For Res 16:925–930

Shelbourne CJA (1992) Genetic gains from different kinds of breeding population and seed or plant production population. S Afr For J 160:49–65

Shelbourne CJA (1997) Genetics of adding value to the end-products of radiata pine. In: Burdon & Moore, pp 129–141

Shelbourne CJA, Low CB (1980) Multi-trait index selection and associated genetic gains of *Pinus radiata* progenies at five sites. N Z J For Sci 10:307–324

Shelbourne CJA, Burdon RD, Carson SD et al. (1986) Development plan for radiata pine breeding. New Zealand Forest Service, Forest Research Institute, Rotorua. p 142

Shelbourne CJA, Carson MJ, Wilcox MD (1989) New techniques in the genetic improvement of radiata pine. Commonw For Rev 68:191–201

Sorensson CT, Burdon RD, Cown DJ et al. (1997) Incorporating spiral grain into the New Zealand radiata pine breeding programme. In: Burdon & Moore, pp 180–191

Strauss SH, Libby WJ (1987) Allozyme heterosis in radiata pine is poorly explained by overdominance. Am Nat 130:879–890

Sweet GB (1973) The effect of maturation on the growth and form of vegetative propagules of radiata pine. N Z J For Sci 3:191–210

Sweet GB, Krugman SL (1978) Flowering and seed production problems—a new concept of seed orchards. Paper 4/2. In: Brown AG, Palmberg CM (compil.) Proceedings of the 3rd world consultation on forest tree breeding, 21–26 March 1977, Canberra, Australia. CSIRO, Canberra, pp 749–759

Thompson D (2000) Monterey pine – the New Zealand wonder tree – does it have a role in Irish forestry? Coillte (Ireland) Research and Development Information Note No. 20, p 4

Toplu et al. (1987) (Cited by Johnson et al. 1997).

Toro J, Gessel S (1999) Radiata pine plantations in Chile. New For 18:33–34

Tufuor K, Libby WJ (1973) First-lift pruning times of radiata pine seedlings and rooted cuttings in a small California experiment. N Z J For 18:124–132

Viljoen A, Wingfield MJ (1994) First report of *Fusarium subglutinans* f. sp. *pini* in South Africa. Plant Dis 76:309–312

Vincent TG (1986) Seed orchards. In: Shelbourne et al., pp 62–97

Vincent TG (1987) Certification scheme for forest tree seed and planting stock. Ministry of Forestry, Rotorua, New zealand, FRI Bulletin No. 134, p 17

Vincent TG (1990) You choose the parents. What's new no. 182. Ministry of Forestry, Forest Research Institute, Rotorua, New Zealand, p 4

Vincent TG (1993) Supply of improved radiata pine seed. N Z [J] For 38(3):45–48

Vincent TG, Dunstan JS (1989) Register of commercial seedlots issued by the New Zealand forest service. Ministry of Forestry, New Zealand Forest Research Institute, Rotorua, FRI Bulletin 144, p 155

Voelker SL, Lachenbruch B, Meinzer FC, Strauss SH (2011) Reduced wood stiffness and strength, and altered stem form, in young antisense 4CL poplars with reduced lignin content. New Phytol 189:1096–1109

Wagner A, Donaldson L, Kim H et al. (2009) Suppression of 4-coumarate-CoA ligase in the coniferous genome *Pinus radiata*. Plant Physiol 149:370–383

Walker JCF, Butterfield BG (1996) The importance of microfibril angle for the processing industries. N Z [J] For 40(4):34–40

Walter C, Smith DR, Grace LJ et al. (1995) Plant regeneration and transformation studies with *Pinus radiata* embryonic tissue. In: Technical paper, International Wood Biotechnology Conference, Tokyo, Japan 1994

Walter C, Carson M, Charity J et al. (1997) Transformation and gene expression in *Pinus radiata*. In: Burdon & Moore, pp 319–332

White TL, Hodge GR (1989) Predicting breeding values with applications in forest tree improvement. Kluwer, Dordrecht, The Netherlands, pp 367

White TL, Matheson AC, Cotterill PP et al. (1999) A nucleus breeding plan for radiata pine in Australia. Silvae Genet 48:122–133

White TL, Adams WT, Neale DB (2007) Forest genetics. CABI Publishing, Wallingford and Cambridge, p x + 682

Wilcox MD, Thulin IJ, Vincent TG (1976) Selection of *Pinus radiata* clones for planting from cuttings. N Z J For 21:239–247

Wingfield BD, Coutinho TA, Viljoen A et al. (1999). Pitch canker: a South African perspective. In: Devey ME, Matheson C, Gordon T. (eds) (1999) Proceedings of the IMPACT Monterey Pine Workshop, Monterey, CA, 30 Nov to 3 Dec 1988. CSIRO Forestry & Forest Products, Australia Tech Rep No 111. pp 62–69

Wu HX, Eldridge KG, Matheson AC et al. (2007) Achievements on forest tree improvement in Australia and New Zealand 8: successful introduction and breeding of radiata pine in Australia. Aust For 70:215–225

Wu HX, Ivković M, Gapare WJ et al. (2008) Breeding for wood quality and profit in *Pinus radiata*: a review of genetic parameter estimates and implications for breeding and deployment. N Z J For Sci 38:56–87

Zobel BJ, Wakeley PC, Sprague JR (1993) A forestry revolution. A history of tree improvement in the southern United States. Carolina Academic Press, Durham, NC p 161

Chapter 7
A Wild Ride, 1998 Onwards

7.1 Introduction

The last chapter covered major developments in radiata's domestication, namely:

- propagation technology, which offered speedier and better targeted capture of genetic gain
- monitoring of genetic gain and systems for guaranteeing genetic quality
- much more focus on genetic improvement of wood properties
- institutional changes, in ownership of forests and in R&D organisations
- mounting of molecular genetics programmes with radiata
- continuing development of decision-aid models for the growers.

This chapter continues with numerous threads to the story. Major threads, however, are

- ongoing institutional changes, in both forest ownership and R&D organisations
- continued and intensified focus on the wood properties, in both research and operational breeding
- periodic biotic alarms
- refinements of predictive and decision-aid models for growth and wood properties in conjunction with advances in remote-sensing technology
- emphasis on molecular genetic research.

The 1997 IUFRO Conference *Genetics of Radiata Pine* (Burdon and Moore 1997), drew together much of the existing literature on the genetics. That meeting and collected proceedings having not been repeated; such literature now tends to be more dispersed. Another factor, however, is the increasingly proprietary nature of genetic information, which tends to delay or even prevent publication of some research results. However, literature reviews of genetic parameter information for wood traits in radiata have become available (Wu et al. 2008; Gapare et al. 2009), as have three historical accounts of breeding radiata in Australia (Wu et al. 2007a;

R. Burdon et al., *Domestication of Radiata Pine*, Forestry Sciences,
DOI 10.1007/978-3-319-65018-0_7

Johnson et al. 2008) and New Zealand (Burdon et al. 2008). Moreover, Mead (2013) recently completed a book *Sustainable management of* Pinus radiata *plantations*.

7.2 Institutional Developments

Institutional changes have continued rapidly, and some have been dramatic (Box 7.1). Not only that, some are still in progress, with their full impacts yet to be felt. Often corporate changes have involved changes not only in forest ownership, but also in ownership of processing plants and in their technology arms. The following paragraphs are not an exhaustive coverage of name and ownership changes and lesser reorganisations within the various institutions involving radiata worldwide, but they are meant to illustrate the trends.

Box 7.1 Institutional Turbulence

The last 20 years or so have been turbulent for many of the institutions involved in the various aspects of growing radiata. This, however, has often been true for corresponding institutions involved in various other major plantation forest species. Ownership of and management of forest have seen major changes, which have often been disruptive. R&D organisations have tended to come under pressure, and one key organisation, CSIRO Forests and Forest Products in Australia, has been fully disbanded. Commitment to conservation and management of gene resources—the back-up genetic variability—has often been manifested more in words than in provision of funding and other resources on the part of industry.

In Australia (Productivity Commission 2008) and New Zealand, there was a general trend for state or national governments to restructure their forest services. The commercial forest estates and their management were segregated into corporatized trading organisations, with management of ecological conservation forests coming under separate agencies. This had already happened in 1987 in New Zealand when the exotic forest plantations, which were very predominantly radiata, were placed within the newly created New Zealand Forestry Corporation (NZIER 2000). In 1993, in Australia, the Victorian Plantations Corporation was similarly created from within Forests Commission, Victoria. In 2001, the South Australian Forestry Corporation was established, with ForestrySA as its trading arm, and it opened a new head office in Mt Gambier in the Green Triangle. In NSW, Australia, the Forestry Commission of New South Wales had changed name to State Forests NSW, with a separate commercial trading arm, but full corporatisation took until 1 January 2013. Such government corporatisation has generally been followed by privatisation: in New Zealand during 1992–1996 (NZIER 2000), in Victoria in 1998 (Anon 1998) and in the Green Triangle region of South

Australia in 2012 (Smith 2012). Similar developments have occurred in Queensland, but have involved only very minor areas of radiata.

The pattern for actual privatisation has been for the governments concerned to retain land title, the sales involving the remaining assets, such as the growing trees, the harvesting rights and some processing plants. In some cases, however, the transfer of assets was subject to covenants relating to land use, existing research plantings including genetic trials and tree-breeding material, or wood-processing activities. In New Zealand, the land titles concerned have generally devolved from government ownership into Maori tribal ownership under the terms of Waitangi Treaty settlements.

7.2.1 *Forest Ownership and Activity*

Forest and associated land ownership continued in a state of flux (cf Burdon and Carson 1999). Indeed, major developments occurred in Australia, New Zealand and South Africa, largely involving new types of forest owner. Changes in controlling forest ownership have brought changes in management, which have often been followed by further changes occasioned by internal corporate reorganisations.

In Australia and New Zealand major buyers of forest estate included **Timber Investment Management Organisations (TIMOs)**, the larger transactions being listed in Table 7.1. Of these TIMOs, Hancock (variously Hancock Forest Management or Hancock Timber Resources Pty Ltd) was largely managing pension funds, while Harvard (as Harvard Management Company) runs a research endowment fund. Pru Timber was basically Canadian, being mainly owned by the Ontario Teachers Pension Fund, but had actually been taken over by Hancock.

Note that these figures do not include short-term double transactions or some subsequent changes of ownership. The TIMOs concerned operate strictly as forestry companies, not as processors, maintaining a widespread trend away from vertical integration of forest growing and wood processing. They have had a two-fold advantage over traditional corporates for running forestry as a business: some tax advantages, and the fact that the investment money comes in as retirement or

Table 7.1 Some TIMO purchases of radiata plantation forest (as cutting rights) (After Horgan 2004)

Year	Where	Seller	Buyer	Radiata (ha)
1998	Australia (Victoria)	Forests Commission	Hancock	170,000
2006	New Zealand (Kinleith Forest)	Carter Holt Harvey	Hancock	64,000
2004	New Zealand (Tarawera Forest)	Fletcher Challenge*	Hancock	36,200
2006	New Zealand (Kaingaroa Forest, etc.)	Fletcher Challenge*	Harvard	166,000
2006	New Zealand (misc. regions)	Fletcher Challenge*	Pru Timber	65,000

*Either Fletcher Challenge Forests (a publicly listed company) or partly owned subsidiary

endowment contributions without the added costs of borrowing fees from banks. Less favourably for forest grower countries, the TIMOs' commitment could end very quickly with onselling (Neilson and Evans 2009). That can occur when the forest asset values have appreciated to the point when onselling becomes too attractive to resist, or it becomes necessary to cover needed payouts, or it might be premeditated from the outset. Whatever happens, the situation militates against long-term commitments to R&D activities. Moreover, the TIMOs' forestry operations, especially within individual countries, typically belonged within very broad risk-spread portfolios.

A strong driver of forest ownership changes in New Zealand was the effective collapse of the Fletcher Challenge corporate empire around 2000. As a result Fletcher Challenge Forests (FCF) went from owning or managing 275,000 of forest (almost all radiata) in the key Central North Island Region around the end 2002 to none by early 2004. Prominent among the several causes of the collapse was the disastrous acquisition of UK Paper (Wallace 2001), but there were other factors. In acquiring state-owned forests in the region in 1996 FCF hoped to secure a strong revenue stream, but a financial crisis in East Asia soon reduced that sharply. In trying to maintain net cash flow, FCF over-cut their forests, harvesting stands while they were still too young, and further compromised future value by cost-cutting measures. This was part of a pattern in New Zealand at the time—failing to achieve makeable returns on radiata plantations by pursuing what proved to be unmakeably high rates of return.

In New Zealand, there were two notable exceptions, both in Asian ownership, to the trend against vertical integration. Ernslaw One, which is Malaysian-owned, holds around 70,000 of radiata plantation. Juken Nissho New Zealand, a Japanese-owned company holding around 50,000 ha of radiata, was the other, although it has recently done partial de-integration. Juken Nissho was also notable for not following the widespread trend towards shortening rotations for radiata.

In South Australia, privatisation has involved a major sawmill at Nangwarry which was specifically designed to cut radiata timbers. The New South Wales state forests, which contain the largest resource of radiata plantations in Australia, are now vested in Forests NSW, which is a state trading enterprise within the NSW Department of Primary Industry. The nature of Forests NSW as a trading enterprise is drawing speculation concerning its possible future privatisation.

In New Zealand there is also the complication of much forest land being vested in Maori ownership, with forest ownership taking the form of finite-period management and cutting rights. The split between land- and forest ownership does not of itself favour commitment on the part of forest owners to supporting long-term R&D.

In New Zealand the privatisation of state-owned forest plantations was essentially complete by 1996. Forest ownership, however, was not thereby stabilised. Indeed, the passing of Kaingaroa Forest and nearby areas of other former state-owned forest plantation into the hands of Harvard, to become the entity Kaingaroa Timberlands Ltd, took place in several steps. Along the way, the controlling owners made part-sales of the forest assets in order to dilute their business exposure to the forest holding. Moreover, these forest assets have since changed hands again, into a consortium of a New Zealand pension fund, a Canadian one, and a local Maori tribal incorporation.

Following the dissolution of the Forest Service in 1987, its suite of plantations was subject to land claims in connection with New Zealand's founding Waitangi Treaty of 1840, which are still being settled. This often leaves the forest owner effectively operating on a forest management licence with the land title vested in Maori interests. Encouragingly, however, Maori ownership of New Zealand plantation forests, in addition to just the land under them, has increased in recent years, and may well increase further. This promises greater future commitment to supporting radiata R&D, and some has started to materialise.

In New Zealand the economic context has changed markedly in the last few years. The dairying sector has experienced a strong boom. At the same time, the Global Financial Crisis, which began in the US in 2008, has severely affected the forestry sector. The resulting acute recession in the US housing industry, and associated weakening in the US dollar against the New Zealand dollar, severely depressed that key market for high-quality appearance-grade timber at a time when very large volumes of pruned logs were coming on stream. Even for other export markets, a strong New Zealand dollar, driven by the dairying boom, much reduced the profitability for forestry and forest products. Between these factors, and fears of carbon-market penalties arising from government management of that market, around 58,000 ha of plantation forest land was converted to dairying between 2008 and 2013 (Fallow 2013), despite major concerns about its effects on water quality in streams and lakes. These developments were surely having adverse flow-on effects for R&D expenditure in the forestry and wood processing. The adverse effects on wood exports, however, became mitigated by increased demand for wood in China, albeit primarily for lower-quality logs. Also reassuring is a very recent upsurge in the US housing market. Moreover, since mid-1994 international dairy prices to the farmers have fallen by over 50%. Furthermore, recent economic analyses (Monge et al. 2016; cf Yao et al. 2017) have explored the significance of likely dollar values of environmental externalities associated with intensive dairy farming compared with plantation forestry, showing forestry to be a much more economically competitive land use than was widely assumed. In fact, one major operation converting land from forestry to dairying has been terminated, albeit shortly before its completion.

A recent overview of the role of forestry—and by implication radiata—in New Zealand's economy is given by NZIER (2017).

In **Chile**, the forestry sector has tended to consolidate, with the two main corporate groups, Arauco and Mininco which are both controlled by Chilean families, becoming increasingly dominant. This has been especially in respect of new planting and industrial ventures, although a number of significant parcels of radiata plantations have changed hands, with the effect of medium-sized forest owner companies tending to disappear. Pulping capacity has continued to increase, by around 800,000 tonnes to nearly 3 million tonnes of radiata pulp (almost all chemical) per year, plus over 2 million tonnes of eucalypt pulp (J.C. Carmona pers. comm. 2012). Vertical integration remains a feature of the forestry sector there, unlike in other grower countries.

An important development in the context for forestry in Chile has been increasing acceptance of wood as a building material. This was long overdue in the light of how well wooden buildings withstand the earthquakes that occur there.

In **South Africa**, ownership changes of forests continued (John Mather pers. comm. 2011). The commercialisation of state forest plantations into the state-owned company SAFCOL was completed by 2000. This process, however, entailed the transfer of a high proportion of the radiata plantation estate to state conservation areas, which also entailed effective loss of a high proportion of the experimental trials. The remainder of the state-owned radiata plantations was privatised to Mountain to Ocean Forestry Company (MTO), which was then effectively cut off from all expert advice and assistance from SAFCOL. In turn, MTO was sold to Cape Pine Investment Holdings, which has since been sold to Global Environment Fund in which it remains a separate business arm. Website figures (http://www.capepine.co.za 2012) indicate that current holdings of radiata plantation are about 24,000 ha. Most radiata plantings in South Africa, however, have become subject to heavy post-planting losses from pitch canker infection that had originated in the nurseries (Sects 7.3.1, 7.3.2, 7.3.5).

The widespread abandonment of vertical integration of forestry, wood processing and marketing generally continued, except in Chile. In addition to the sale of the state-owned sawmill in South Australia, the two state-owned sawmills in New Zealand were sold separately from forest holdings.

7.2.2 R&D Organisations

For various reasons, the major publicly funded R&D organisations serving the forestry sectors tended to come under increasing pressure. Governments were under a two-way squeeze, coming under political pressure to reduce tax levels and expenditure, and yet facing increasing welfare budgets that were largely driven by a combination of ageing populations and ever-increasing costs of advancing medical technology.

Government support for R&D tended to become contingent on support from industry. Meanwhile, foreign ownership of commercial forests was increasing greatly in New Zealand and Australia. Foreign owners of both forests and industrial plant have typically represented large, multinational interests. As such, they typically had their own technology units which they might turn to, particularly for wood processing. Also, some of the risks associated with huge local dependence on radiata were often easier for such owners to address by global risk spread than by supporting R&D to develop active but expensive countermeasures to those risks. Smaller forest growers, while collectively a large element in New Zealand, were typically not funders of research. Despite being a very diverse group, they shared another feature, in their forestry interests very often being parts of risk-spread portfolios. Thus there were forces militating against industry support of government research organisations, and thence against sustained government funding of them.

We trace here the fortunes of several key R&D institutions, but do not try to cover all the advisory committees that were created and made input. With increasing privatisation of radiata-associated R&D institutions, as well as of radiata plantation, interrelationships have become more complicated. New institutions have resulted in both new alliances and new competitive relationships, with scope for conflicts. Notably among the changes, the traditional territoriality that existed among the state forest services within Australia has been much reduced.

In Australia, CRCs (Cooperative Research Centres) were created, linking government research organisations, universities and industry organisations. Some forestry research involving radiata proceeded under the CRC for Sustainable Production Forestry, based in Hobart and involving four universities, which was established in 1997 and gave way to the CRC for Forestry in June 2005 which itself was terminated in June 2012.

A more specialised organisation, PLANTPLAN, continues some research in conjunction with servicing the Southern Tree Breeding Association and other plant breeding programmes.

In 2001, the Forest and Wood Products Research and Development Corporation (FWPRDC) was set up in Australia to gather and disburse research funding. It was converted in 2007 into FWPA (Forests and Wood Products Australia), a non-profit industry-owned organisation with a strong focus on radiata. Among its activities was the Juvenile Wood Initiative (JWI) to research corewood properties of radiata, largely with a view to improving them through genetic selection. FWPA soon became included in New Zealand's Wood Quality Initiative, as an "additional stakeholder." Among other things, the JWI corewood research entailed considerable emphasis on attaching economic weights to genetic improvement of various wood properties relative to each other and to gains in volume of wood produced, in collaboration with both STBA and CSIRO. This endeavour was much simplified by the pre-eminence of light structural timber in Australia's end-product mix for radiata. However, a major imperfection in the log market in Australia has been exposed by finding very different sets of economic weights among traits for forest growers and sawmillers respectively (Ivković et al. 2006).

Despite such R&D activities and support for them, the general level of industry's political support for CSIRO [Division of] Forestry and Forest Products (FFP) dwindled to the point that the Australian government decided in 2008 to disband it. From that institution, which had a convoluted history (CSIROpedia 2013), a few researchers involved in radiata were then assigned elsewhere in CSIRO, relying on STBA and FWPA for funding support. Some CSIRO scientists had by then embarked on genomic research with radiata, with a major emphasis on discovery of genes controlling wood properties (e.g. Dillon et al. 2010), and such research is continuing within CSIRO's Division of Plant Industry.

The disbanding of CSIRO FFP was seen by Kile et al. (2014) as largely a "downstream" consequence of privatisation of plantation forests combined with abandonment of vertical integration, but Ferguson (2015) saw it more in terms of a loss of "strong linkages and networking between research organisations and research users."

An operational merger between CSIRO FFP and the New Zealand Forest Research Institute Ltd (branded as Scion from 2005) was attempted in 2004, in the form of an Unincorporated Joint Venture, Ensis. This, however, did not deliver the hoped-for benefits and proved essentially unworkable. Among other things, the links of Ensis with both STBA and RPBC, which were competing organisations, were causing much unease. The merger was formally reversed on 31 December 2007.

In Australia, the STBA developed a comprehensive data management system that was running in 2000. From that, through the Animal Breeding and Genetics Unit (AGBU) of University of New England, Armidale, NSW, scientists developed TREEPLAN®, a software package to make full use of all available pedigree information to evaluate breeding values of selection candidates. Operational use of TREEPLAN began with radiata, in 2002 (Wu et al. 2007a), and has since been extended to other tree species, in Australia and overseas. In addition, the STBA became heavily involved in the breeding of shining gum, *Eucalyptus nitens*, which had come to be planted on a large scale for pulpwood. In 2000 Western Australia's government agency exited from STBA, and soon afterwards Queensland's withdrew from breeding radiata in favour of buying its seed on the open market.

Within **New Zealand** the RPBC (Radiata Pine Breeding Cooperative) became a limited liability company (LLC) in 2001, renaming itself Radiata Pine Breeding Company, but with little immediate change in how it operated apart from greater autonomy in its financial management, and no change in the RPBC brand. Nominally at least, RPBC and Scion were designated as a consortium in 2005 for negotiating government research funding which matches industry contributions. Funding from industry, however, for some years tended to remain effectively static, despite the inducement of matching contributions from government sources. Moreover, radiata breeding research is funded on the basis of individual projects, for which contracts are awarded to a variety of providers. Actually, RPBC, like STBA in Australia, has come to represent a separation between operational breeding and associated R&D instead of tree breeding being done within R&D organisations.

In 2013, the RPBC, after Antoine Kremer of France joined Andrew Granger and Christine Dean on its advisory board, embraced research on **genomic selection**$^{\mathbf{G}}$ (Sect. 7.11.1.4), with emphasis on selecting for both disease-resistance and wood-properties. This move was accompanied by much-increased financial contributions from PRBC and increased co-funding from government.

In 2003, another government-sanctioned consortium between industry and the government's research funding agency and industry came into being, with FRI heavily involved in the actual research. Originally called WQI (Wood Quality Initiative), it was superseded in 2009 by SWI (Solid Wood Innovation). WQI and then SWI addressed a broad range of research on radiata's wood quality, including prospects for genetic improvement. SWI has maintained its link with Australia through FWPA upon expiry of its Juvenile Wood Initiative (JWI) programme.

In 2007 another such consortium, Future Forests Research (FFR) was set up, involving Scion, industry and Government. It includes almost all areas of commercial forestry research for New Zealand that were not covered by existing consortia. With NZRPBC thereby excluded, FFR has operated with a major emphasis on intensive silviculture of radiata. In 2017, FFR was renamed Forest Growers Research Ltd (FGR), with some changes to its mandate.

Changes also occurred in the Centre for Advanced Biotechnology Te Teko which was started by Tasman Forestry Ltd and had gone through several ownership and name changes; for a while the Centre was under the brand of Fletcher Challenge, and then Horizon 2 after the Fletcher Challenge corporate empire disintegrated. In the Tasman and Fletcher Challenge days, during 1994–1999, co-author WJL was engaged there for part of each year 1992–1995. In 2007 the Centre was taken over by the biotechnology company ArborGen Inc., which itself was the outcome of similar institutional and name changes in south-eastern USA and took over remaining assets from CellFor (another biotechnology company) in 2012.

A separate company, Forest Genetics (FG), was started in 2002 by Mike and Sue Carson, both formerly of NZFRI, to develop and market for commercial forestry their own radiata clones, descended from the New Zealand radiata breeding programme. They were joined by Christine Te Riini, formerly in charge of genetic improvement of radiata at Te Teko with 10 years' experience with clonal research and deployment there. FG's operation grew to include a major programme of testing a set of clones on multiple sites in New Zealand and Australia, and contracting out the propagation of their selected clones to commercial nurseries. FG was a member company within the RPBC. Since its inception, FG has taken on Kaingaroa Timberlands as a shareholding partner.

Alliances involving tree-breeding and research agencies and biotechnology companies have formed and evolved. During this period there were some takeovers or mergers involving the biotechnology companies, with changes in some of the competitive interrelationships.

In **Chile**, the breeding Cooperativa has remained in place, with more integrated management of breeding material among member companies, in contrast to the earlier model of persisting with essentially individual breeding programmes. Thus the role of the Cooperativa has continued to centre on setting and communicating protocols for operational breeding practices, reporting on activities and carrying out some data analysis. Two changes occurred around the start of the chapter period. First, Roberto Ipinza left his post of Director in 1997, with Heidi Dungey taking over as acting Director under the tutelage of the founding Director, Roberto Delmastro. Then, in 1998, Fernando Droppelmann returned as Director and still holds the post.

However, in 1997 a major change in Cooperativa membership came; one of the two main corporates, Mininco, opted out of the Cooperativa, to run its own breeding programme, with Tim White of University of Florida as external consultant. The Mininco group currently holds around 335,000 ha of radiata in Chile. Among the remaining membership, the Arauco group with ca 615,000 ha of radiata (Arauco 2012) is, through its subsidiary Bioforest which is led by Claudio Balocchi, now greatly pre-eminent within the cooperative. Bioforest is a large organisation in itself, with some 130 employees, and is involved in a wide range of forest research including eucalypts as well radiata.

A high proportion of Chile's research relating to radiata has continued to be done within various universities, commissioned with varying levels of contestability by a mix of individual companies, industry advisory bodies, and several government agencies with research funds to disburse. Among the participating universities, the University of Concepción has been playing an increasingly prominent role, as befits the proximity of Concepción to much of Chile's forest plantations and wood-processing plants. Among the governmental commissioning agencies has been CORFO (Corporación de Fomento de la Producción). Importantly, CORFO has facilitated setting up the Biotechnology Centre in the University of Concepción. This centre has a broad portfolio of projects involving radiata, ranging from pulping technology to genomic studies as a prospective tool for selective breeding. Major companies with their own research facilities are the Arauco group, in its subsidiary Bioforest SA, and the Mininco group. Bioforest has focused heavily on clonal propagation with large numbers of clones cryopreserved, and is pursuing other biotechnology including genomic selection. Within Mininco a radiata breeding programme now operates separately from the Cooperativa. Another national agency, Fundación Chile, was involved in developing technology companies to promote clonal forestry, but with only two major prospective clients could not strike a deal. In fact, the forestry department of Fundación Chile has now been abolished.

As elsewhere, intellectual property associated with breeding radiata in Chile has often been sequestered within by individual corporates rather than shared, despite the presence of the Cooperativa.

In **Spain** active research on quantitative genetics and breeding of radiata is proceeding in Pontevedra in Galicia, while the research effort on radiata breeding in the Basque Autonomous Region has been dropped in favour of a focus on propagational and genomic biotechnology.

7.2.3 Non-governmental Organisations (NGOs)

Among NGOs, historical objections to plantation forestry have increasingly given way to at least some acceptance. This acceptance is for both the potential to relieve exploitation pressures on natural forests and an appreciation of the more direct environmental services that plantations can provide, such as checking soil erosion, preserving water quality and sustaining populations of birds and even bats.

For growing radiata, amongst forest plantations in general, the Forest Stewardship Council (FSC) has become increasingly influential. It promotes and operates the certification of forest products as having been sustainably produced, placing pressure on retail outlets, especially in North America and Europe, to sell only compliant products. It has also generated business for certifying agents who scrutinise the processes of growing and harvesting.

The basic objectives of third-party certification are entirely worthy, but some proscriptions of acceptable growing practices are a concern, involving some fundamentalist positions and a poor appreciation of some local biological realities. For example, there is an absolute refusal by some such NGOs to contemplate participant growers being involved in field research on genetic engineering (GE), let alone adopting it operationally. Indeed, certain positions taken by certifying NGOs could even be counterproductive, given that they can both impede or discourage plantation forestry in general and block use of some biotechnologies that may be environmentally beneficial.

7.3 Forestry Activity and Problems

7.3.1 New Planting and Retirements

From 1998 onwards new plantings of radiata have been relatively limited, for a variety of reasons, including economic and biotic factors. Indeed, some significant areas of radiata plantation have been converted to other land uses, including cases of very young crops. Nevertheless, there is a portent, from the 2015 UNFCCC conference, that this trend could be reversed.

In **Australia**, new planting of radiata has been largely replaced by a boom in planting eucalypts, mostly Tasmanian blue gum (*Eucalyptus globulus*) (Ferguson 2014). With the promise of high returns from short, pulpwood rotations, and some tax advantages, various private forestry ("prospectus") companies, termed Management Investment Schemes, have been formed to plant eucalypts. Many of these eucalypt plantings, however, have under-performed, largely through planting on inappropriate sites, while key tax advantages had been withdrawn in South Australia in 2000, so this boom is over. Meanwhile, companies owning radiata plantations have tended to consolidate their holdings, often following or pending ownership changes (Table 7.2).

In **New Zealand**, the rate of new radiata plantings (MPI 2013) fell steeply (Figure 3.1) after 1996 following a recession in south-east Asia, and almost ceased around 2006 although harvested areas were mostly restocked. Since then, however, there was been a slight upturn with around 12,000 ha of new planting in 2011, compared with around 6,000 ha in 2010, although not all would be with

Table 7.2 Areas of different conifer species ('000 ha), by States, in Australia as at 2012 (ABARES)

State	Radiata pine	Maritime pine	Southern pine	Hoop pine, etc.	Total
New South Wales	277	0	12	7	296
Victoria	226	0	0	0	226
South Australia	129	0	0	0	129
Tasmania	75	0	0	0	75
Western Australia	56	42	0	0	98
Capital Territory	8	0	0	0	8
Queensland (&NT)	1	0	139	50	189
Total	772	42	151	49	1,024
Percentages	75.4	4.1	14.7	5.7	100

radiata. In recent years there has tended to be consolidation of forestry operations, following ownership changes. From 2003 (NEFD 2013) to 2012 New Zealand's total forest plantation area actually declined by around 5% (*op. cit.*), due mainly to a dairy farming boom, and the conversion of plantation land to dairy farms would have been almost all from radiata. Harvesting and restocking with radiata had been running at around 40,000 ha per annum (NEFD 2013, Figure 2.1), while average recovered yields per hectare rose from around 460 to around 525 m^3 per ha. In younger radiata plantations pruned area declined from 67 to 58% and areas commercially thinned declined from 21 to 17% (Table 7.3).

At one stage in very recent years, a strong market for carbon sequestration credits was starting to boost new plantings. Then, however, a collapse in carbon-credit prices was allowed to spread from eastern Europe where carbon credits had been granted to encourage compliance that did not eventuate; that created an incentive to liquidate forest plantations and many in New Zealand acted accordingly. Fluctuating planting rates are a worry for the future of the forestry sector, in which capital-intensive processing plant will depend heavily on continuity of supply of raw material.

Investment in major processing plant has lagged in New Zealand, with even some retrenchment in pulping and papermaking. By contrast, exports of unprocessed logs have boomed, making New Zealand the biggest such exporter globally. That is preventing the local supply of the sawmilling residues needed to justify building new and profitable pulping plant (Hall 2016).

In **Chile**, new planting of radiata has slackened significantly, averaging less than 20,000 ha per year from 1998. This probably reflected several factors: considerable planting of eucalypts mainly for short-fibred pulpwood; some concerns over European pine shoot moth deforming stems; and concerns over arson blamed ondisgruntled native Americans. The forestry encouragement scheme, under D.L.

Table 7.3 Geographic distribution of radiata plantation ('000 ha) in New Zealand as at April 2012 (NEFS)

Island	Wood supply region (Fig. 7.1)	Latitude range (°S)*	Seaboard(s)	Area
North	Northland	34½–37	W–E	197
	Central North Island	37–39	W–NE	519
	East Coast	38–39	E	149
	Hawke's Bay	39–39½	E	127
	Southern North Island	39–41½	W–E	160
	Total North Island	34½–41½	–	1,152
South	Nelson/Marlborough	41–42	N–NE	155
	West Coast	42–43	W	23
	Canterbury	42½–44½	E	83
	Otago/Southland	45–46½	S–E	130
	Total South Island	41–46½	–	391
National total		34½–46½	–	1,543

*Approximate range of concentrations

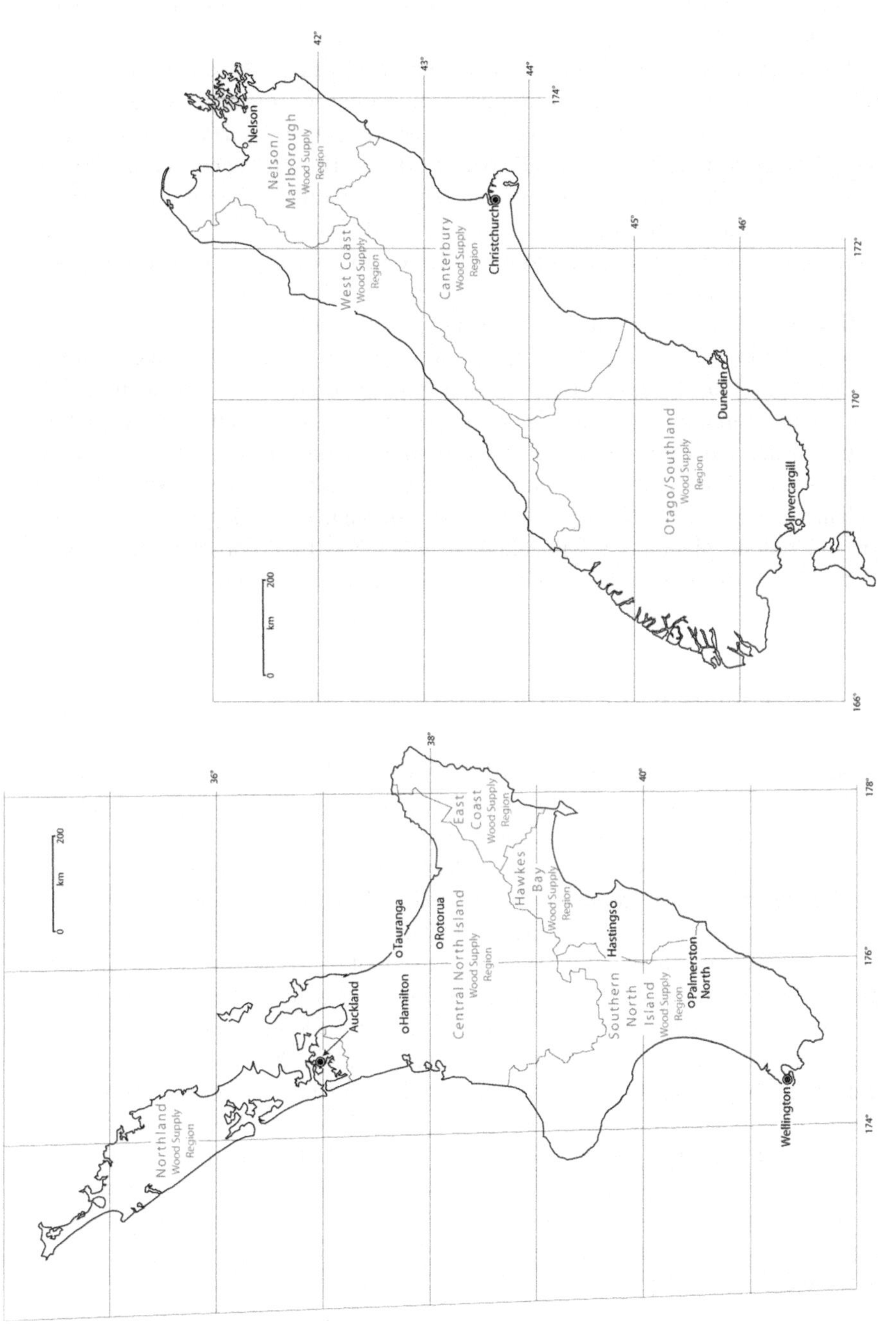

Fig. 7.1 Maps showing wood supply regions in New Zealand, right, South Island, left, North Island (After NEFD 2016, permission Ministry of Primary Industries)

701, which had effectively run on for many years after its original expiry date, was renewed as at 1 January 2013, but with focus on both encouraging small land-owners and rehabilitation of degraded soils (Anon 2013).

In **Spain**, however, new planting of radiata accelerated, especially in the province of Galicia, largely through European Community encouragement of afforestation of marginal agricultural land (Zas and Serrada 2003). By 2001, the recorded radiata plantation area had increased to 291,000 ha (R. Zas pers. comm. 2012). However, the annual planting figures have been volatile, reflecting some biotic alarms over radiata, some surges in prices for hardwood pulp logs and a regional slump in softwood log prices caused by massive salvage logging of maritime pine following an extreme storm in south-west France in 2009.

In **South Africa** new plantings of radiata virtually halted, what with changes in forest ownership (Sect. 7.2.1) and increasing concerns over pitch canker (Morris 2010, Mitchell et al. 2011). Indeed, the area of commercial radiata plantations has shrunk, with retirement of some land to conservation estate and not replanting with the species on account of problems with pitch canker. It was reported (DAFF 2010) that, as at 2008/2009, of a total listed area of 58,000 ha previously occupied by radiata, nearly 18,000 ha were "temporarily unplanted," and the area of 1- and 2-year-old radiata plantations had fallen to under 500 ha compared with around 1000 ha in each of the remaining annual age classes up to around 10 years.

7.3.2 Developments in Forest-Growing Practices

In **Australia**, the developments have been modest. Because of several factors, including the pre-eminence of the market for light structural timbers, and the sometimes profitability of commercial thinning, changes to silvicultural regimes for radiata have been less radical than in New Zealand. That meant less readjustment when problems with radical regimes came to light. Even so, stockings tended to be reduced and rotations shortened in comparison with the past.

In New Zealand especially, TIMOs were making changes in silvicultural regimes. They were effectively working to lower discount rates which reflected lower costs of investment money, and which in turn made more attractive the economics of longer rotations. One overseas-based company, however, had embarked on a dramatic but ill-fated experiment on cost-cutting, planting radiata at near-final stockings and doing almost zero tending (Dyck and Thomson 1999; Sutton 1999); subsequently, the resulting stands got largely converted by a new owner to dairy farming. Anyway, rotation ages have typically extended from 24 to 27 years to 30 to 34 years, although this has brought concerns over more heartwood which is unwanted because its appearance is disliked by many end-users and it is neither reliably durable nor reliably treatable with preservative. Also, final-crop stockings have tended to rise from around 200 to around 300 stems per ha. For the decline in the percentage of stands being pruned there are several reasons. In some cases there were the cost savings driven by immediate financial liquidity problems. Labour availability was often a problem. Moreover, a decline in the USA market for high-quality clear timber, and a boom in the Chinese market for logs (Katz 2013) with limited price premiums for quality, have doubtless influenced perceptions of the profitability of pruning which is done at some sacrifice of wood volume. Yet, if countries like China and India eventually become premium markets for radiata clearwood, failure to prune might prove costly, especially with the widespread problem of achieving high wood stiffness with radiata in New Zealand.

While New Zealand has long used almost exclusively bare-root planting stock for establishing radiata, several major nurseries have begun large-scale production of container stock. That extends the planting season, can reduce post-planting toppling and subsequent butt sweep, and promises more efficient use of scarce seedlots and cuttings material of top genetic quality. Using container stock may also compensate for some loss of field-planting skills arising from continuing rapid changes of personnel.

Reported percentages of cuttings in total nursery outturns of radiata (J. Novis, pers. comm. 2013) have fluctuated somewhat erratically (Table 6.8). Figures are by no means complete, but there was an increase from 30% in 1997 to 43% in 1999, falling back to 30% in 2003. Figures are available again from 2006–2010, varying from 34% in 2006, 23% in 2007 and 27% in 2010.

In **Chile**, no radical silvicultural changes occurred for radiata, there remaining a broad mix of regimes according to markets, terrain and site quality. By 1997, however, a substantial shift to using cuttings rather than seedlings had already been made in order achieve faster capture of the latest genetic gains from breeding (Balocchi 1997), and Arauco has been using nearly 100% cuttings for radiata planting from 2001 (Balocchi pers. comm. 2014). Arauco also made a massive shift to using container rather than bare-rooted stock; while more labour-intensive, it has served to give more efficient capture of genetic gain, and is evidently preferred for the relatively sticky soils on which much of the plantation estate grows. The other main corporate, Mininco, has not made the same wholesale conversion to container stock, having large areas of sandy or scoriaceous soils. On these poorer soils, radiata is often grown as dedicated pulpwood crops and post-planting toppling of trees is not the same issue.

In **South Africa**, the challenge has been to accommodate early mortality of radiata associated with pitch canker infection acquired in nurseries, with the complication that pitch canker has recently been found affecting older trees (Mitchell et al. 2011). Accordingly, screening for genetic resistance to canker is being intensively practised, and vigorous efforts are being made to hybridise radiata with more resistant species (A. van der Hoef, pers. comm. 2014). The eventual success is not yet known, but it will probably determine the future usefulness of radiata in South Africa.

In **Spain**, pulpwood and roundwood regimes for radiata have been mostly abandoned in favour of producing sawlogs (Rodriguez et al. 2002), with or without commercial thinning. Regimes envisaged range from 325 to around 500 stems per hectare at harvest, with rotations from 25 to 35 years, depending on site quality and ownership. Most of the Basque Country plantations that are communal forest have been managed by the regional Forest Service. For the Galicia Region much seed, from seed orchards, has been imported from New Zealand.

7.3.3 New Decision-Aid Software and Remote-Sensing Technology

Underpinning leading-edge plantation-forestry practices have been major advances in the continuing development of prediction tools for wood yields and log- and wood properties, which serve as decision aids, and are becoming increasingly linked to the availability of new remote-sensing technology. Radiata, because of its plantation-forestry status, has been heavily involved in these developments.

In New Zealand, a key resource for the growth modelling is the national Permanent Sample Plot (PSP) database, which as at 2007 comprised records of regular assessments of ca 27,000 PSPs (Hayes and Andersen 2007). These plot data have come largely from PSPs in radiata stands, nearly half of them being still current. Continuing software development work for modelling radiata crop development has been undertaken by both ATLAS Technology, a subsidiary of Scion (New Zealand Forest Research Institute Ltd), and independent operators. Harnessing databases and models concerning variation in wood properties according to site, silviculture, genetic status and tree age (e.g. Watt et al. 2011, 2013), prediction tools and decision aids for forest managers that involve all stages of the value chain have come available for radiata. The software, however, is mostly proprietary, needing to be purchased by users. Details are not only beyond the scope of this book but are also subject to continuing updating and refinement.

Aerial photography has long been available as a remote-sensing technology for forest management, but some of its role can now be supplanted by satellite imagery which has become much cheaper to access in recent years (Watt and Watt 2011). However, a major recent development in remote-sensing technology, is LiDAR (Light Detection And Ranging), which typically uses near-infrared light, in the manner of using radio waves for radar (Adams et al. 2013). It can be used to study numerous stand variables, and progress is being made in using it for inferring wood properties (Watt et al. 2013). With the availability of small drones (Unmanned Aerial Vehicles) remarkably detailed information on stand structure can be obtained and analysed.

These developments in use of remote-sensing technology and modelling of stand- and processing outturns serve to both reduce the physical tasks of obtaining information and allow managers to make better-informed decisions. Indeed, they make it possible to address questions that could not be effectively addressed in the past. All these developments have depended heavily on the continuing increases in computing power, but they still have their limitations. The genetic populations of radiata will remain a continuously changing variable. Moreover, there remain basic biological information gaps, as in how the trade-off between wood density and stem volume operates at the whole-crop level at harvest age.

7.3.4 Harvesting and Log Segregation

At harvesting, logs are cut into certain lengths and are then segregated into categories for processing, all with the aim of optimising net product value. Traditionally, logs have been segregated into categories according to size, taper, knottiness and straightness. For cutting to optimum lengths, a decision-aid programme that addressed changes in diameter up the stem had been developed and updated. But other bases for optimal cutting to length and segregation are also needed, including automatic adjustments for changes in market specifications. Pruned logs are an obvious category for special processing.

Individual trees of radiata can vary importantly in wood stiffness, which is often the main limiting factor for obtaining structural timber grades from unpruned logs. To deal with this, acoustic tools have been developed to grade individual logs for wood stiffness. These tools, which include the proprietary HITMAN and SWAT, use sound pulses into the ends of the logs. The pulses are reflected back from the opposite ends, and measure the rate of sound travel which depends on wood stiffness, thereby allowing a useful measure of stiffness. Those logs with the stiffer wood are sent to the sawmill for cutting into appropriate dimensions for structural timber, while those with less stiff wood are dispatched for cutting to dimensions for non-structural uses or else for pulping. The economic benefits of such segregation are substantial (Tsehaye et al. 1997).

7.3.5 Biotic Alarms

Several alarming outbreaks of diseases or pests have affected radiata in recent years, largely differing among countries.

Pitch canker is a fungal disease of pines, which is caused by a *Fusarium* species. It leads to shoot dieback and, when cankers affect the main stem and girdle it, can kill trees. Stem cankers typically develop slowly but steadily, and are distinctive on account of profuse resin bleeding. Hodge and Dvorak (2000), in an inoculation study, found radiata to be among the most susceptible host species. Pitch canker first became evident in and around native stands of radiata in California in 1976. There its incidence became alarming and, while it has since largely subsided, it has discouraged planting radiata.

The disease appeared in South Africa in 1990 (Wingfield et al. 1998), where it has since become troublesome. The pathogen has spread widely among forest nurseries, to cause considerable post-planting mortality. The first outbreak in a plantation of radiata there was in 2005, and was subsequently confirmed as pitch canker (Coutinho et al. 2007). Pitch canker has tended to discourage replanting of radiata (see DAFF 2010), but is being addressed by routine screening of intensively select parents for resistance.

Pitch canker has also appeared in both Chile and Spain. In Chile, it has not been reported as having caused the same trouble in plantations as in South Africa, perhaps because of a shortage of insect vectors compared with in South Africa and what appears to be relatively low climatic hazard (Ganley et al. 2009). In Spain, however, it is causing concern in the Basque Autonomous Region, leading to very little new planting and even a drop in replanting.

There the climatic hazard rates as high (*op. cit.*), while there are insect vectors that reportedly proliferate in log yards of processing plants leading to concentrations of disease nearby.

New Zealand and Australian radiata breeders have attempted proactive screening of select progenies for resistance by planting in a disease-hazard area in California, under the IMPACT project (Devey et al. 1999a). Although genetic differences in susceptibility within radiata had been observed in an inoculation trial (Matheson et al. 2006), there has been too little pitch canker in the field trial to resolve genetic differences in susceptibility. While there has been some shoot dieback there, it has since proved attributable to another, less alarming pathogen, *Diplodia*.

The presence of pitch canker in all three natural stands in mainland California now creates a major biosecurity barrier to tapping those stands for further refreshing the genetic diversity of domesticated stocks.

The **Monterey pine aphid**, *Essigella californica*, was discovered in 1998 in both Australia and New Zealand, and spread rapidly in both countries (Wharton and Kriticos 2004; Watson et al. 2008). Capable of debilitating trees of radiata under drought stress, it has caused alarm in Australia. However, despite fluctuations in abundance, it has not had disastrous effects there, let alone in New Zealand. It is suspected that some natural biocontrol has occurred.

Nectria, the fungus *Nectria fuckeliana*, was identified in the 1990s as a pathogen causing bad stem cankers following pruning of radiata in the south of New Zealand (Crane et al. 2009; Hopkins et al. 2012), despite it having no prior record as a damaging forest pathogen. It has since been identified as causing similar damage in radiata plantations in southern Chile (Morales 2009). After causing initial alarm, it appears to be controlled by avoiding winter pruning. This solution contrasts with practice in the north of New Zealand where summer pruning is often avoided because of the risk of stem cankers following infection by *Diplodia* (M.A. Dick pers. comm. 2012).

In Chile, in the Arauco Peninsula, a combination of very severe needle cast and considerable shoot dieback, which often killed very young radiata trees, was first observed in 2004 over 70 ha, the immediate cause being the fungus-like oomycete *Phytophthora pinifolia* (Durán et al. 2010). By 2006 the affected area had expanded to an alarming 60,000 ha, but mysteriously reduced to under 500 ha by 2008. Durán et al. (2008) reported the identification of the pathogen.

Similarly, the threat posed to radiata stem form in Chile by the European pine shoot moth appears to have receded, in this case largely through natural parasitoid biocontrol (Ramos and Lanfranco 2010: Mead 2013, Box 4.1). However, sirex (*Sirex noctilio*) was eventually reported in Chile in 2001 (Beèche et al. 2012). Its status there as a threat was accentuated by the importance of pulpwood production leading to maintaining relatively high tree stockings, but the threat has been mitigated by the successful introduction of a nematode and two wasp parasitoids (*op. cit.*).

However, another insect pest, the European gypsy moth (*Lymantria dispar ddispar*), has unexpectedly proved capable of causing complete defoliation and severe mortality in a pure radiata plantation in north-western Spain (Castedo-Dorado et al. 2016).

In New Zealand, a novel form of needle cast, "**Red needle cast**" (RNC), was observed in 2006, and recognised in 2008 (Dick et al. 2014), in some northern areas where needle casts had not hitherto been prevalent. It has caused severe defoliation, mainly in relatively old trees, but was evidently different from PNB (physiological needle blight), a needle cast of radiata that had long occurred sporadically following wet winters in the north of the country but could not be related to any actual pathogen. RNC has since been associated mainly with a newly discovered oomycete, *Phytophthora pluvialis*. It has spread over a wide geographic area within the North Island, being associated with mists on ridges or general humidity on lower ground. Its final extent and significance are still unclear, but it is spurring considerable research.

Thus, since around 1997, three diseases appeared in radiata, which exhibited alarming symptoms and threatened to spread widely. While the initial threats have generally not materialised into serious and widespread losses, a worrying common feature is that they have all been caused by fungi or oomycetes with no prior record of being significant pathogens. Meanwhile, some foliage diseases that had become "traditional" in radiata, notably cyclaneusma needle cast and dothistroma needle blight, remained as greater or lesser nuisances, according to local climates. In response, there has been a very recent move within New Zealand to escalate research on the development of genetic resistance in radiata to both the new and traditional diseases. One avenue of research is the search for cross-resistance whereby the same individuals can be selected effectively for resistance to more than one pathogen (S. Kennedy et al. in prep.). For these endeavours, a broad range of technologies is being enlisted.

7.3.6 Weeds and Related Issues

In addition to pests and diseases, weeds have tended to become an increasing problem over time, involving plantation forests in general and not just radiata. Typically, weeds are easily dispersed and far more easily established than eliminated, leading to progressively increasing suites of weed species that depress tree growth by competition and/or inflate plantation-growing costs. Use of fire to remove logging residues and help control weed growth has been widely discontinued, on various environmental grounds. While some traditional herbicides are becoming outlawed, new and more benign and selective herbicides have been developed.

An example of a weed acquiring relatively recent prominence in forest plantations is *Buddleia davidii* (Fig. 7.2) in New Zealand, which has spread very rapidly. Indeed, *Buddleia* has become so troublesome as to be targeted for biological control; for that, a parasitoid *Cleopus japonicus*, the buddleia leaf weevil, has been successfully introduced and promises significant control (Watson and Withers 2012) (Fig. 7.3).

Fig. 7.2 Buddleia plant in flower, Rotorua, New Zealand. The species is an imported ornamental shrub that has become invasive in forest plantations

Fig. 7.3 Buddleia plants showing damage from the *Cleopus* weevil

7.3.7 The Advent of "Carbon Forestry"?

Concern over the build-up of atmospheric carbon dioxide (CO_2) and consequent global warming has created a market for growing crops that sequester large quantities of CO_2 (Manley and Maclaren 2009). This has stemmed from the Kyoto Protocol of 1997 in which various countries committed themselves to certain net reductions in CO_2 emissions. Despite obvious deficiencies in the agreement, stumbling progress, and much current uncertainty, a market developed for "carbon credits". Radiata, having quite wide site tolerances and being generally grown on the comparatively long rotations needed for producing solid-wood products, is very suitable for carbon sequestration, with the bonus that such products have service lifetimes that mean sequestration that continues long after crop harvest.

An established, reliably high-value market for carbon credits has potentially very important implications for growing radiata. It creates an obvious encouragement for new planting, raising the profitability of plantations. This is especially so where there is pastoral land of marginal or negative profitability, which is the present situation in New Zealand. But carbon credits also lengthen the economically optimum rotation (Manley and Maclaren 2009) and favour higher stockings (cf. Kimberley et al. 2005). In turn, the longer rotations and higher stockings will mitigate the problems with radiata's wood quality, making less urgent the need for genetic improvement in this respect.

However, the impacts of "carbon forestry," while potentially great if certain shortcomings of the Kyoto Protocol are rectified, have recently become problematic. Since the global financial crisis that began in 2008, eastern European countries have been selling 'carbon liabilities' at progressively decreasing prices. In New Zealand, where the management of liability regulations has led to some perverse features and the government has not imposed a minimum carbon liability price, the cheapness of such liabilities has created a scramble to liquidate large areas of forest. This resulted in a reversal from a brief carbon-driven upsurge in planting, and conflicts with some pressing needs for soil conservation and catchment protection.

7.4 Operational Breeding and Deployment Activities

The five main grower countries all had up-and-running breeding programmes, which were set for continued advancement, maintaining cycles of evaluation, selection, intermating, testing, evaluation, and so on. However, the emphasis in breeding goals has changed appreciably in New Zealand and Australia, with an increased focus on improving several wood properties, in some trade-off with the emphasis on past breeding objectives. Indeed, this shift has driven much of the research and breeding activity since the mid-1990s. With the shift, and problems with maintaining and evaluating some progeny trials resulting from institutional turbulences, emphasis on fast turnover of generations in New Zealand and Australia has reduced, despite active research on improving early selection on young progeny.

Also, since 1997 there have been revisions of the underlying breeding strategies for radiata in both New Zealand and Australia (Sect. 7.8.2).

7.4.1 Australia

Australia's largest forest owner, State Forests NSW, has remained within the New Zealand RPBC. As a full member, it has exercised substantial influence in shaping RPBC's research programme, especially in the area of genotype-site interactions which are more important in many parts of Australia than in New Zealand.

The STBA has continued to service radiata breeding in the rest of south-eastern Australia, along with breeding of temperate eucalypts, principally shining gum, *E. nitens*.

7.4.2 New Zealand

In New Zealand, radiata breeding was affected by indifferent success of progeny trial plantings from the 1985–1988 round of fresh plus-tree selection, which resulted largely from disruptive management changes in some forest-owner companies. A further blow came when a promising progeny trial was virtually lost through defoliation by caterpillars of *Helicoverpa*, the tomato fruitworm. This was an unforseen consequence of establishing *Lotus preunculata*, which was intended to provide fodder for grazing cattle in the forest, but was attacked by the moth whose caterpillars turned to the pine foliage after running out of the *Lotus*. These problems have slowed the turnover of generations.

Breeding operations, namely assessments, selection, intercrossing and establishment of progeny trials have continued, albeit under revised selection criteria (Sect. 7.7) and an evolving strategy (Sect. 7.8.2.2).

Meanwhile, establishment of new genetic gain trials, especially ones containing the large plots needed give a measure of whole-crop performance, had been restricted by budget levels for breeding. Given the need for continued economic justification of further genetic improvement, and increasingly evident problems of assigning relative weights to growth rate, tree form and wood properties, this restriction has led to some troublesome information gaps. Since adverse genetic correlations among traits are involved here, weighing up the economic worth of the various traits is extremely important, but it is actually very difficult.

In deployment of genetically improved planting stock there were fluctuations (Table 6.8). Two contributing factors were fluctuations in both planting programmes and seed supplies, the latter being influenced by some seed orchards being phased out and others coming into production. The percentage of deployed planting stock that rated GF19 or higher rose from 11% in 1997 to 16% in 2003, ranging from 39% to 47% during 2006–2009, but dropping to 28–31% in 2010–2012. In 2008, there was a shortfall in open-pollinated seed orchard seed, leading to "Stand select" seed collections made within stands of seed-orchard source: such seedlots are of unproven genetic merit and uncertain breadth of genetic base, and therefore entail some elevated genetic risk. Apart from this brief episode, the pattern has been a progression towards greater use of vegetative multiplication in order to accelerate capture of genetic gain, followed by increasing large-scale deployment of specific and progressively better characterised clones.

7.4.3 Chile

Since 1997, there have effectively been two radiata breeding programmes running, under the Arauco and Mininco groups. Within each of these programmes, however, there has been amalgamation of breeding programmes, instead of the original disconnection among the programmes of the original individual member companies. Turnover of generations has been accelerated, pressing on to producing some third-generation breeding-population material by the early 2000s. This has been helped by switching to simpler mating designs within breeding populations. As elsewhere, there has been an increasing focus on wood properties. The arrival of the pine shoot moth in Chile did occasion a substantial research effort into both selective breeding for resistance and genetic engineering to confer resistance, but the level of shoot moth damage has naturally abated to tolerable levels (Lanfranco et al. 1991).

Mininco radiata breeders have put strong emphasis on managing the breeding population, and have been relying on both sexually and vegetatively propagated stock for operational planting. Related research, by Sergio Espinoza and colleagues (Espinoza et al. 2014, 2016) has been directed at identifying the extent and detailed basis of adaptation of intensively cultivated radiata stocks to contrasting local environments. Results to date have come from nursery-based studies, with more definitive results from field-based research still pending.

Arauco breeders and silviculturists have been more aggressively pursuing clonal deployment (C. Balocchi, pers. comm. 2014), testing large numbers of clones and deploying around 100, largely in monoclonal blocks of typically 20–40 ha. Despite work on embryogenesis, individual clones have been mass-propagated commercially for only 4–5 years, largely on account of continued maturation which costs a bit over 20% of potential gain through failure to mass-propagate all clones successfully. Because of this, and the opportunities for progressive genetic improvement over generations, a new generation of clones is now being used. Deployment is regionalised, with eight "breeding regions," given the spread of latitude (36–41°S) and rainfall in their forest estate, with very imperfect cross-correlation of individual clonal performance between some of these growing regions. Up till 2014, a total of 534 individual clonal trials had been established. The deployment of identified clones has risen from 6% in 2008 to 27% in 2009 and nearly 90% in 3013.

7.4.4 South Africa

There the radiata breeding programme was overhauled and re-sized (John Mather pers. comm. 2011), amid numerous institutional changes involving R&D

(Liebenberg et al. 2004). Camcore, a North Carolina-based forest genetics and breeding organisation, was engaged to revise the breeding and deployment strategies. As at October 2011, 240 3rd-generation selections have been made, of which the top 30 clones have been established in a seed orchard. All of those orchard clones are subject to screening for pitch canker resistance by the Tree Protection Cooperative Programme. Orchard seed production is supplemented by some vegetative multiplication as nursery cuttings (http://www.capepine.co.za 2012).

7.4.5 Spain

The operational breeding programme in the Basque Autonomous Region had been faltering since around 2006, and effectively halted around 2010, for want of funding and loss of personnel. With the extremely fragmented ownership of radiata plantation, and the plantation holdings often a secondary commercial activity for owners, the individual owners have not been ready to pay any extra for genetically superior planting stock, and none were willing to provide suitable land for field genetic trials. Moreover, with pitch canker new planting largely ceased. So political support for the breeding programme has lapsed. In that region, however, some biotechnological research on radiata has proceeded, mainly led by Paloma Moncaleán. That research is partly based on adapting propagational technology from New Zealand (e.g. Montalbán et al. 2012), but also includes basic physiological and genomic research on radiata and some of its hybrids (e.g. De Diego et al. 2013). In the Galicia Region, however, a classical field-based radiata breeding programme has continued, along with research on the field progeny trials (e.g. Codesido and Fernández-López 2009).

7.5 Demonstration and Marketing of Genetic Gain

7.5.1 Gain

As the breeding programme advanced it has become more important to quantify the genetic merit of improved material. This involves two levels:

- individual candidates, for a range of traits, first for selection and then for offering to clients who might want to make their own choices of seed-orchard parents or else of clones for mass-deployment in clonal forestry
- crop-level performance of improved seedlots or even clones, as a more rigorous measure of achieved genetic gain.

Competitive interactions, of the sort that can occur among progenies in progeny tests or between clones in clonal tests, mean that gains in crop-level growth are likely to be less than the apparent gains at the level of the individual. Research on this aspect, however, has been limited with radiata in recent years.

7.5.1.1 Individual-Tree Performance

For practical reasons, the performance of large numbers of progenies or clones needs to be tested using small plots. The resulting efficiency of ranking the candidates, however, comes at the cost of competitive interactions being likely to inflate growth-rate differences for growth variables, notably wood volume.

For testing and ranking selection candidates, field designs have evolved, helped by the availability of vastly improved computing power. Complete block replicates in forest-tree trials tend to be so large that they cannot partition off environmental variation at all efficiently. Small block units, as in incomplete block designs, which include alpha-designs (Williams and Matheson 1994), allow better partitioning off of environmental effects. Moreover, with the lower replication needed for testing clones, the loss of direct comparisons among subsets of candidates as in the sets-in-replicates design used in New Zealand becomes much more important.

Actually, it is possible to compensate for sub-optimal field designs, as often exist historically, by using spatial analysis (White et al. 2007, p. 430). Such analysis uses the data to estimate and correct for local environmental effects, in order to obtain the best available information on selection candidates. While that usually requires major computing power, it is now far less of a limiting factor, to the extent that spatial analysis is often a routine part of analysing field data.

7.5.1.2 Crop-Level Gains

Such gains are relatively difficult to demonstrate rigorously. Indeed, incautious interpretation of progeny- or clonal test results can lead to gross overestimates of crop-level gains, as has recently been demonstrated for eucalypt clones (Stanger et al. 2011). However, thanks to the large-plot genetic gain trials that were first planted in 1978, insights could be achieved into crop-level genetic gains in radiata. Carson et al. (1999) showed substantial genetic gains for stem volume, and that they did not trail off greatly as the trees got older and competition had longer to operate. For a conventional open-pollinated seed-orchard lot (GF improvement rating 14) the observed gain in stem volume per hectare over "climbing select" (GF7) was 17% at age 15, and projected to 13% at age 30 and 12% at age 40; for a pair-cross (GF22) the corresponding figures were 28%, 22% and 20%. Interestingly, the gains were expressed much more strongly in the sectional area per hectare than in tree height, whereas the expectation had been that gains in wood volume would be driven by gains in height growth. However, the influence of the trade-off between wood density and stem volume in these stem-volume gains was not studied. The gains in growth were of course accompanied by major gains in the standard of tree form.

Since then, advances in knowledge of crop-level gains in wood volume production have been hampered by limited planting of large-plot genetic gain trials.

Even so, Kimberley et al. (2015) have confirmed that genetic gains in wood volume production in radiata were well sustained as the trees got older, were closely related to Improvement Rating, and varied little percentagewise among sites of varying productivity.

7.5.2 Certification and Assurance

7.5.2.1 New Zealand

The New Zealand certification scheme for genetic quality of radiata was brought under Registered Trademark as *GFPlus* in 1998. The ratings for Growth and Form continued, based mainly on progeny-test performance of the parents of the seedlots. This, however, became complicated by shifts over time in the emphasis given to different traits. With some higher GF ratings, which reflected more intensive selection for tree form, additional gains in growth rate were not always evident, especially when there was an element of selection for increased wood density or even better stiffness.

Underpinning the improvement ratings, or to some degree supplementing them, came estimates of genetic merit of the individual parents, in the form of breeding values. Adapting practice in animal breeding, using BLUP (Best Linear Unbiased Prediction) software, this was designed to cope with parents—and offspring—belonging to different generations and either different lineages or varying relatedness, and with varying levels of representation in field tests. First adopted in 1998 (Jayawickrama and Carson 1990), such estimates could behave erratically, depending on the level of cross-reference between field trials and the availability of new assessment data. Such estimates were on a trait-by-trait basis. They were also made without addressing genotype-site interaction, although this is often a minor limitation in New Zealand (Carson 1991).

Despite limited new planting, and the option of vegetative multiplication, the requirements for restocking of felled areas sustained a strong demand for genetically superior seed. Even so, substantial seed exports were made to northern Spain, where retirement of marginal land was leading to much afforestation.

7.5.2.2 Australia

In Australia, the Improvement Ratings embedded in the New Zealand certification scheme were never formally adopted. Instead, TREEPLAN was used from the outset, from 2002, to give explicit estimates of breeding values, using multi-trait information. As in New Zealand, genotype-site interaction is not yet accommodated in breeding value estimates, which can be more of an issue across Australia and is recognised as posing a major challenge.

7.5.2.3 Economic Returns

The major effort on assessing economic returns has been in Australia. This has tended to be addressed indirectly, largely in connection with assigning economic weights to different traits. Rigorous quantification of economic returns is complicated by a number of factors:

- difficulties in tracking all costs associated with genetic improvement programmes; the total costs include ones incurred by parties other than the actual breeders, and the costs of approaches that eventually fail in a process of trial and error
- continuing uncertainties over relative economic weights of different traits
- market imperfections, whereby trading prices may not reflect true value, a situation indicated by different sets of economic weights for growers and processors respectively
- unquantified payoffs, such as cheaper and safer harvesting and retention of competitive viability in industry enterprises
- shortage of recent field trials that show genetic gains at whole-crop level; especially as comparisons among operational stands representing different levels of genetic improvement have given equivocal results
- continuing uncertainties as to how certain trade-offs, notably between wood density and stem volume at the whole-crop level, play out at harvest age
- changes in practices, such as setting harvest ages, which can change the relative economic weights of different traits
- wide variation among site categories and among industry members in relative economic weights of different traits.

Claims for economic rates of return have varied widely. After in-house tracking of the cumulative historical costs of New Zealand's radiata breeding programme, and studying results of early genetic gain trials, an estimated benefit-cost ratio for radiata breeding of 46: 1 was arrived at, but a later in-house study gave an estimate of only 14: 1. Various other figures for economic benefits have been produced (e.g. Wu et al. 2007a), albeit not to any common criterion.

There was no doubt about early imperatives to improve tree form and/or adaptation to local conditions, according to different regions where radiata is grown, although much could be achieved over time by a combination of semi-natural and silvicultural selection. Intensive breeding has produced material that is strongly and unreservedly preferred by forest growers, who are typically prepared to pay extra for it. Nowadays, the immediate imperative is to improve wood properties, to maintain commercial competitiveness of both growers and processors. For the longer term, there remains the imperative to prepare for changes in breeding goals, which may come abruptly, if there is a biotic crisis to address. Unfortunately, the attractions of growing genetically improved forest crops can create strong disincentives to prepare for unexpected future developments.

7.6 Advances in Knowledge of Genetic Architecture

7.6.1 Variation among Natural Populations

Main advances have come from catch-up work in Australia, where some quite old provenance trials from the 1978 seed collection have been assessed recently (Gapare et al. 2012a,b). Among the native provenances, Año Nuevo has tended to perform best on moist, relatively fertile sites. On drier, less fertile sites Monterey has tended to do slightly better than Año Nuevo, whereas Cambria has done relatively well on some dry sites. Cambria and the Island populations have tended to perform poorly on moist, summer rainfall sites where dothistroma needle blight is a hazard (Gapare et al. 2012a). Any advantage the Cedros provenance may have for hot, dry sites has tended to be negated by susceptibility to post-planting mortality. Wood properties have generally compared among the provenances much the same as they have in New Zealand. However, Cedros has not shown superior wood stiffness commensurate with its superiority for wood density, unlike Guadalupe (Gapare et al. 2012b).

7.6.2 Within-Population Variation and Inheritance

Updated information on the variability and inheritance of individual traits in radiata is summarised in Table 7.4.

7.6.2.1 Wood Properties

Concerns over wood properties of radiata have centred on several problems: poor stiffness (Modulus of Elasticity, or MoE) affecting structural grades; dimensional instability in service which affects both structural and appearance grades; and some features affecting appearance grades, namely resin pockets, bird's-eye grain and internal checking during drying. Density had been assumed to be the main determinant of stiffness, which it is in the outerwood. However, in the corewood, which represents a substantial proportion of stem volume unless trees are beyond any commercially accepted harvest age, there is another important determinant of stiffness, namely microfibril angle (MfA), the inclination of the smallest-scale cellulose strands from the fibre axis. MfA being difficult to determine cheaply, MoE needs to be addressed in its own right; even so, it is now usually addressed through a proxy in the form of acoustic properties of the wood. Among the other problems with wood properties, especially in corewood, dimensional stability is governed by both spirality of the wood grain and uneven longitudinal shrinkage (LS) upon drying, both of which are most marked in corewood. Indeed, poor MoE

Table 7.4 General importance and indicative information obtained since around 1990 on variability and heritability of selected individual traits in radiata pine (Mainly after Wu et al. 2008, Kumar et al. 2008 and Mead 2013 and sources)

Trait	Economic importance	Status[a]	Coefficient of variation (%)[b]	Heritability	Importance of	
					Non-additive gene effects	Genotype-site interaction
Wood stiffness	Very high	BG	~10	0.6	•	•
Acoustic velocity	Indirect	Sel	10	0.4	••	•
Microfibril angle	Indirect	Inc	10	~0.6	○?	○
Wood collapse	High −ve	Sp	~40	0.2	•••?	?•
Longitudinal shrinkage (−ve)	High −ve	Sp	~40	0.3	•••	?
Resin pockets	Locally −ve	Sp	–	–	?	?
External resin bleeding	Indirect	Sel	≥50	0.3–0.4	•	?
Resistance to						
Mg deficiency	Uncertain	No	–	0.7	?	?
Pitch canker	Locally high	Sp	–	~0.4	?	?
Essigella	Locally high	Sp	–	0.4	•	•

[a]BG denotes key breeding-goal trait: Sel - Key selection trait for indirect breeding-goal selection; Inc – often addressed incidentally by selection trait(s); - Sp – addressed in specific situations; No – not generally addressed to date in breeding
[b]Tree-to-tree, if there is a meaningful measure

has been associated with LS (Ivković et al. 2009), the former being appreciably heritable (Gapare et al. 2008). Internal checking, which occurs during drying, only becomes evident after wood is planed smooth, and affects appearance badly; while generally affecting corewood, it does not affect the wood after it becomes heartwood. Heartwood, however, can be objectionable because of its appearance and difficulty of preservative treatment. Resin pockets, while of no real consequence for structural timber can cause serious losses of potentially high-value appearance grades, while bird's-eye grain almost exclusively affects corewood and in radiata is viewed by the market as a defect rather than an attribute.

Obtaining good knowledge of inheritance of traits, just like being able to select intensively, depends on being able to evaluate large numbers of individuals. This in turn depends on cheap and reasonably accurate assay technologies, for either the actual traits of interest or reliable proxies. This posed special challenges with wood-property traits (Sect. 7.7.1), especially ones that affect wood quality more directly. Stiffness can now be assayed on standing trees at reasonable cost, using proprietary instruments such as TreeTap and FAKOPP. Susceptibility to internal checking, which results from fibre collapse during drying because of cohesiveness of capillary water columns, has been very challenging to assay. It is partly associated with low wood density, so can be reduced by selecting for high wood density, but high wood density is not generally needed for appearance grades. A rough assay, however, can be obtained by measuring tangential shrinkage when increment cores dry rapidly. Such a measure of susceptibility to collapse has shown strong broad-sense heritability but only low narrow-sense heritability (Kumar et al. 2008), suggesting considerable non-additive gene effects. Indeed, some heritability has been demonstrated for actual incidence of internal checking (Kumar et al. 2010). Incidence of resin pockets is indicated, albeit far from precisely, by external resin bleeding (Cown et al. 2011).

Stiffness has shown significant heritability in its own right (e.g. Dungey et al. 2006; Baltunis et al. 2007; Matheson et al. 2008; Kumar and Burdon 2010; Gapare et al. 2012b) but is less heritable and more subject to genotype-site interaction than wood density. The general picture appears to be one of performance-related wood properties to be less heritable and more subject to non-additive inheritance than wood density, especially in the corewood (Kumar et al. 2008).

Pilot work has been done to develop very early screening for wood density, stiffness (as acoustic velocity), and shrinkage behaviour (which governs dimensional stability), by forcing young seedlings out of the vertical to study the development and properties of the resulting wood (Apiolaza et al. 2011). Modest heritabilities were observed for these variables; however, a very strong and favourable genetic correlation was observed between velocity and shrinkage, but in the opposite wood rather than in the induced compression wood. The value of the results will depend on genetic correlations with later performance of the trees.

7.6.2.2 Uncomfortable Trade-Offs

With very few exceptions, wood properties have exhibited varyingly adverse genetic correlations with growth variables (Kumar et al. 2008; Wu et al. 2008; Gapare et al. 2009; Burdon 2010) (Table 7.5). Typically, these correlations have been more adverse with stem diameter than with height; moreover, because of the contribution of the square of diameter to stem volume, and the coefficients of variation (standard deviations divided by the means) for height and diameter, the genetic correlations of diameter and stem volume with other traits are often almost identical. Adverse genetic correlations between economically important traits make life particularly difficult for the breeder, for several reasons. They can constrain attainable breeding goals, in terms of gains simultaneously obtainable in the traits concerned (Sect. 4.3.2). They also mean that the breeder needs good information on comparative economic weights; otherwise, effective selection can actually reduce economic value. Moreover, with wood properties, especially density, it becomes a matter of having to weigh up the wood properties versus stem volume measurements in selecting quite young trees (Burdon 2010). That is fine for the

Table 7.5 Approximate pattern, in terms of sign and strength, of genetic correlations between traits. Pluses denote positive or favourable correlations, minuses negative or adverse correlations, 0 roughly neutral (Adapted from Burdon (1992) and Wu et al. (2008))

	HT	DIAM	BRF	IL	BDA	STR	DEN	STF	STA
Height (HT)	1								
Stem diameter (DIAM)	+++	1							
Branching frequency (BRF)	+*	++*	1						
Internode length (IL)	(−)	(−)	(− −)						
Branch diam./angle (BDA)	−	+	++	(− −)	1				
Stem straightness (STR)	++	+	+	(−)	++	1			
Wood density (DEN)	−	−	0	(0)	0	0	1		
Wood stiffness (STF)	(−)	−	0	(0)	0	0	++	1	
Dimensional stability (STA)	(−)	(−)	(0)	(0)	(−)	(0)	(0)	(++)	1

*Much less pronounced on slower-growth sites
Parentheses indicate evidence is indirect or sketchy (Sect. 7.6.2)

wood properties, which are most problematic when the trees are young, but the volume production that matters is at final harvest, and the projection of early, single-tree volumes to crop performance at final harvest is basically guesswork, the more so because of the genetic trade-offs.

7.6.2.3 Other Aspects

With new diseases having appeared, tree-to-tree variation and heritability of resistance is of much interest. For pitch canker, length of lesions in inoculated seedlings showed a narrow-sense heritability of ca. 0.4 (Matheson et al. 2006), although it is not known how good a proxy the response is for field resistance. Heritabilities of resistance to *P. pinifolia* and RNC are now being researched in Chile and New Zealand respectively.

Selecting trees while young can give more rapid genetic gain, provided the early-age trait shows both satisfactory heritability and adequate genetic correlation with the mature-age trait—in addition to being able to evaluate enough individuals. The two traits can be the same at different ages, but they do not have to be, provided the above conditions are met. If extremely early selection is effective, it might be applied to shortlist the candidates that are subjected to further evaluation, as in field tests.

A pattern of age-age genetic correlations had already been established for growth variables in radiata, whereby selection at about one-third the harvest age is efficient in terms of gain per unit time, even if markedly suboptimal in gain per generation. With wood properties, as with resistance to certain diseases, the expression of traits at quite early ages can actually be of direct economic importance.

Even so, interest exists in extremely early evaluation. This requires at least reasonable heritabilities at a very early age, combined with worthwhile age-age genetic correlations, conditions that have generally not been demonstrated. On the other hand, wood properties in quite young trees, around 6–7 years old, are of interest in their own right since they involve the troublesome corewood (e.g. Wu et al. 2007b), while trees that are younger cannot generally be used for intermating in any population.

7.6.3 Genotype-Environment Interaction

Genotype-environment interaction has often been prevalent for radiata in Australia, and yet it has largely defied efforts to discern any pattern (e.g. Matheson and Raymond 1984; Cullis et al. 2015). Historically, phosphorus deficiency has implicated as a driver of interactions (Fielding and Brown 1961; Burdon 1975; Burdon et al. 1997a,b, 1998), but correction of this deficiency by using fertiliser has now become routine. Occasional exceptions have arisen when specific maladaptations of individual clones or families have been exposed on particular sites. Recent work, however, has revealed a pattern of interactions in New South Wales (Raymond 2011; Gapare et al. 2012a), in that performance on moist, high-altitude sites correlated quite poorly with performance on dry, low-altitude sites. The detailed basis for such a pattern of interaction, however, is unclear. A general problem in both New Zealand and Australia was limited overlap between field trials in representation of progenies (Raymond 2011; Apiolaza 2012; Gapare et al. 2012b, Cullis et al. 2015).

As in past studies, diameter growth appeared to be the trait that showed the most interaction. In some cases, there were obvious causes, such as disease-prone parents being differentially affected among sites of widely varying disease hazard, but such obvious causes were not always evident. Tree-form traits, notably branching pattern, tended to be less subject to interaction, while certain wood properties, notably density, often showed minimal interaction, with rankings varying very little among sites (Apiolaza 2012). However, some wood properties of major practical interest, such as stiffness or shrinkage behaviour upon drying, could show quite marked interaction, although those interactions may involve different levels of expression of genetic variation rather than changes of genotypic rankings.

7.6.4 Performance of Species Hybrids

The one easily produced species hybrid of radiata, with the knobcone pine (*Pinus attenuata*) (Fig. 7.4), has been tested on a small scale in New Zealand over many years. On many sites it had too much of knobcone pine's susceptibility to dothistroma needle blight to warrant any consideration. However, it continued to show

Fig. 7.4 Hybrids between radiata and knobcone pines, combining almost the full growth potential of radiata with much superior resistance to snow damage; as such they promise an effective extension of the site tolerances of radiata, but are not an option where dothistroma blight can be a factor (Photo RDB)

promise in the semi-continental climates of central South Island, where dothistroma is not a factor, but frost, drought and snowfalls pose problems for radiata. Over the years, the hybrid appeared better adapted to these harsh sites, but a revelation came in 2006 (Dungey et al. 2011), following a heavy fall of wet snow in an 8-year-old hybrid trial. On two harsh sites, the hybrids, which had outgrown the knobcone parent and kept up well with radiata, showed to full advantage. The snow caused severe damage in the pure radiata, compared with almost none in knobcone and very little in the hybrids. In a separate trial, on a harsh, droughty site, hybrids of knobcone and Cedros Island radiata have been showing special promise (N.J. Ledgard unpubl.); intriguingly, this was despite the site being boron-deficient and the pure Cedros provenance being prone to this deficiency. Anecdotal observations (RDB and WJL 2001) of a planting of the radiata and knobcone hybrid combination in northern Sierra Nevada, California, have shown the same great superiority for snow resistance in the hybrids compared with radiata.

The hybrids' tolerance of snow and cold, and presumably drought, effectively mean a substantial extension of the areas where radiata can be grown safely. Actually, no current demand exists for planting hybrids on such semi-continental sites, even though such sites can be very productive on account of some good soils, good rainfall and abundant sunshine. Moreover, the hybrids are not expected to share the unwelcome invasiveness of various exotic conifers in central South Island, New Zealand.

Other interspecific hybrids are of interest (Dungey et al. 2003), especially ones with Mexican pines of the closely related subsection of *Pinus*, the Oocarpae. These species, coming from summer-rainfall areas, can be expected to have very different spectra of disease resistance from radiata which is already grown widely in areas of known disease hazard and may be highly vulnerable to some new disease. Such pines are known to be in varying degrees more resistant to pitch canker (Hodge and Dvorak 2000). Of greatest interest has been the hybrid with *P. tecunumannii* but it has been difficult to produce and commitment of resources has been limited, although CAMCORE are giving active support to South Africa in trying to cross radiata with canker-resistant species. Easier to produce but less attractive is the hybrid with *P. greggii.*

7.7 Evolution and Differentiation of Breeding Goals

7.7.1 The Blitz on Wood Properties

From the very early days, industry parties had balked at making calls for genetic improvement of wood properties in radiata, largely because they were unwilling to contemplate any consequent sacrifice in stem volume production (Burdon 2010). Eventually, after stockings and harvest ages had been lowered, the imperative for genetic improvement of certain wood properties became clear. Indeed, operational screening of logs for wood stiffness was adopted for pricing logs and allocating them to process lines (Fig. 7.5).

Thus, by 1997 there was a strong call for genetic improvement of radiata's wood properties, in Australia as well as in New Zealand (e.g. Cave and Walker 1994; see also Sects. 6.3.2.2 and 6.7.2.4). Answering this call, however, was impeded by several factors. For properties of immediate economic interest, notably stiffness, inheritance was uncertain. While the importance of microfibril angle

Fig. 7.5 Demonstration of using HitMan instrument for determining wood stiffness of a radiata log, as basis for log sorting

(MfA) as a determinant of stiffness especially in corewood had become increasingly clear (Butterfield et al. 1997), and there has been a technical advance in its determination (Evans 2006) the costs and difficulty of direct operational assays of selection candidates for MfA have remained effectively prohibitive. Whether the heritability of stiffness as such was high enough to justify seeking genetic improvement, except purely through selection for density, was not fully clear. Also uncertain was the magnitude of adverse genetic correlations with growth rate. The roles of basic properties in influencing product-performance properties, such as dimensional stability of sawn timber in service were not always clear.

Moreover, uncertainty as to the economic importance of wood properties relative to that of volume growth, along with uncertain genetic correlations, meant the two key classes of information were missing for being able to define realistic breeding goals and assure positive economic gain from selection. Furthermore, the direct assay technology for evaluating selection candidates was expensive or unreliable or both, while reliable and easily determined proxy traits often remained to be confirmed.

Since then, assay technology has improved radically, and various proxies for quality-related traits have been confirmed, which have allowed both greater selection intensities for wood quality and much better information on inheritance. As mentioned in Sect. 6.7.2.4, various anatomical properties and stiffness of wood can be evaluated electronically on increment cores, using the SilviScan instrument; however, cost per core and service availability are problems. Stiffness can now be measured in standing trees, using different acoustic tools (Sect. 7.6.2.1), which measure time of sound transmission between two points on the lower bole. To a lesser degree, progress has been made in assay technology for shrinkage properties (as a proxy for dimensional stability) and resin pockets. Grain spirality, which also affects dimensional stability, has remained a difficult trait to assay, despite promising early indications of high heritability (Burdon and Low 1992: Sorensson et al. 1997). A very recent—and not fully documented—development in New Zealand is "Discbot" hardware and software for evaluating physical, chemical and anatomical wood properties, on stem cross-sectional discs. Compared with SilviScan it evaluates the entire section as opposed to individual radii, and has the additional capability of evaluating grain spirality. The destructive nature of taking discs can be mitigated by taking the discs from the ends of cut logs or by having individuals clonally replicated so a tree can be sacrificed without destroying all members of a genotype.

While the range of wood properties for which selection can be made has broadened, there remains the issue of how to weigh them up against potential improvement in other traits, especially as there are typically adverse genetic correlations between wood properties and growth traits. The main efforts to derive economic weights for different traits have been in Australia, where the focus on producing light structural timber is especially strong. In particular, a joint study by CSIRO and STBA (Ivković et al. 2006) indicated that for growers stem volume production was the most important economic trait, while for sawmillers stiffness was. Apart from the fact that the study depended on strong assumptions concerning how volume and stiffness trade off at harvest age, this divergence points to a flawed log pricing structure. In New Zealand, where there is typically a broader mix of end-products, such a study is more complicated. In Chile, the greater relative importance of pulpwood would also cause complications.

Recently Apiolaza et al. (2011) and Sharma et al. (2015) have started exploring the possibility of evaluating wood properties on extremely young trees. Trees under two years old were placed on a lean, to study compression wood formation, stiffness, and longitudinal shrinkage which mainly governs dimensional stability. Heritable variation was evident in the trees, although the practical value for selection remains to be confirmed. Such selection would likely have its main application in starting the progressive culling of clones for deployment in clonal forestry, rather than in the actual breeding population.

Meanwhile much research has been undertaken, and will continue, on trying to develop the use of DNA sequences for selecting for wood properties (Sect. 7.11.1.4).

7.7.2 *Disease and Pest Resistance*

While some degree of selection had become routine in New Zealand and New South Wales for resistance to dothistroma needle blight and cyclaneusma needle cast (Burdon et al. 2008), disease resistance had not been a prime concern for genetic improvement. In the last few years, however, biotic events have raised new concerns over a need to mount, at short notice, selection programmes to achieve rapid gains in disease resistance. Not only that, experience has taught that the diseases might well be hitherto unknown. This has spurred interest in cross-resistance to a range of fungal pathogens, but awaits substantial research verification, Moreover, the potential for genetic engineering to introduce several genes of large effect that collectively assure durable resistance now appears very attractive (Burdon and Wilcox 2011), provided such genes can be identified. The identification will remain a challenge for some time to come, but if it is achieved the prospects are now very good for being able to insert such genes by transformation (the favoured form of genetic engineering).

A very proactive move was testing Australian and New Zealand select material in California for resistance to pitch canker (Sect. 7.3.5), but that was unsuccessful because almost no pitch canker infection was evident with all observed dieback being attributable to infection by the long-known pathogen *Diplodia pinea*.

A new avenue for studying disease resistance has opened up in recent years, involving endophytes, which are fungi that live harmlessly or even beneficially in the tissues of plants. Endophytes are actually remarkably prevalent in conifers (e.g. Ganley and Newcombe 2006), and manipulating endophyte populations, if achievable, may confer protection against a range of pathogens (Eyles et al. 2010).

7.7.3 *Specialisation or Not, and Deployment*

A perennial debate has continued over whether, in the presence of substantial genotype-environment interaction, to select parents or clones for specific conditions or for a broad range of conditions. Either approach, however, depends on good knowledge of the patterns of interaction, in terms of what environmental factors drive interaction. For specialised selection this means being able to define site categories within which interaction may be minor. For selecting broadly adapted genotypes, one wants to be able to define an environment, or a small set of environments, where good performance can guarantee broad adaptation (Li et al. 2017).

Actually, specialisation of selections for sites may be indicated in the absence of any interaction. For several traits, radiata shows massive environmental effects. According to site types, wood properties may or may not be severely limiting. On many cold but fertile sites, wood density and stiffness tend to be too low in radiata for structural timber grades. There, one could either select intensively for density and/or stiffness, at the expense of genetic gain in some other traits, or else one could focus on producing appearance grades which would require different tending regimes if not a separate "breed."

Actually, significant selection on a "horses-for-courses" basis can be done without having formally differentiated breeds. Even within a breed, genetic segregation will guarantee considerable variation among individual offspring. For field deployment, therefore, a customer can choose individually the parents for a seed orchard (and/or mass vegetative multiplication of seedlings) or clones for direct vegetative propagation. For choosing orchard parents on this basis, good estimates of breeding values are paramount; likewise with genotypic values for deployment in clonal forestry.

7.8 Strategy and Management of Total Genetic Resources

7.8.1 Native-Population Resources (CONSERVE)

Large plantings had been made from the 1978 seed collections in the natural stands, to establish both numerous provenance trials and gene-resource blocks. In Australia, these plantings were mainly provenance trials, with smaller numbers but larger areas of gene-resource plantings which, together with some earlier trials, brought the total to 67 provenance plantings (Eldridge 1998). In New Zealand, provenance trials from that collection numbered 23; these were complemented by larger, gene-resource plantings which covered a total of 135 ha between eight other sites. Together with earlier provenance trials these plantings from the 1978 collection brought the total up to 47 provenance plantings in New Zealand (Eldridge 1998).

Provenance trials had been assessed more or less systematically in New Zealand, the aim being to assess them at comparable tree heights, such that almost all were assessed by age 12 from planting. Only one trial rated as an outright failure, although several others were a limited success. Some were reassessed at age 15, mainly confirming earlier comparisons. In addition to being the subject of the formal provenance trials, native-provenance seedlots were included as ancillary material in various progeny trials involving breeding-population material. Overall, a clear picture emerged of the adaptive profiles of the various provenances in relation to New Zealand conditions (Burdon et al. 1997a, 1998 – also Sect. 6.7.1.1).

In Australia, some trials were assessed quite promptly, and an early picture emerged of tolerances of salinity and of the root pathogen *Phytophthora cinnamomi* (Sect. 6.7.1.2). Many other trials, however, were largely neglected, and some were lost to disasters, generally fire; the latter fate was shared by some of the gene-resource plantings. Nevertheless, a picture has eventually emerged of adaptive profiles of provenances relative to Australian conditions.

In both countries, the performance of the native-population collections for both growth rate and tree form tended to be substantially inferior to that of intensively select breeding-population seedlots. This was no surprise for breeders; indeed, it was a further vindication of their efforts. Industry parties, however, having less of a focus on the cryptic variation that might be needed in future circumstances, were put off by the indifferent performance of fully wild material, such that both the RPBC and the STBA were very reluctant to commit time and money to maintaining and renewing these gene resources. This work repeatedly appeared in the proposals, only to be pushed off the end of the list of priorities. Eventually, however, a meeting was called, in the form of the CONSERVE Workshop held in Canberra in 1998 (Matheson et al. 1999). The following principles were agreed on:

- reliance on open pollination to perpetuate the collections
- planting large blocks (≥20 ha, effective population ≥2000) per subpopulation, or else replicating between sites
- rotations extended as long as possible
- priority to care of Cambria among mainland populations
- Guadalupe population in hand, with recent new conservation plantings
- stored seed from Cedros to be sown for planting.

A few plus trees from mainland provenance material in New Zealand had been selected and archived in 1986, along with 47 from the Guadalupe provenance. Then in 1994, further plus trees were selected, almost all in gene-resource plantings from the Californian mainland. In New Zealand, 37, 57 and 34 trees were selected from within the Año Nuevo, Monterey and Cambria provenances respectively (Bian et al. 2011), and open-pollinated seed collected, while a few of the seed parents have since been archived as clones. Some "secondary" provenance trials have since established from these seed collections, but results to date have added little to the existing picture of provenance differences. In the last four years, in the course of felling, a further seed collection of open-pollinated seed has been made, generally from 40 or so of the better trees in each of the 13 subdivisions of the mainland populations. The seed is in storage, with a separate lot for each seed parent, with a view to eventually re-establishing the gene resources as proposed in Shelbourne et al. (1986).

Overall, commitment to maintaining large genetic resources of native-population material has been limited. The gap in genotypic merit between wild and domesticated material is increasing the opportunity costs of maintaining wild material, and thence the reluctance to grow it. Yet large stocks of wild material would be needed to give scope for the very intensive selection that would be needed to mitigate that gap. And there have been no serious efforts to generate the large, complex interpopulation hybrid swarms that would expose the full range of potential gene combinations to intensive natural and artificial selection. The institutional resistance, or inertia, has doubtless been accentuated by varying combinations of rapid changes in forest ownership and a widespread short-term commercial focus.

Meanwhile, the advent of pitch canker in California, and the consequent biosecurity concerns, have placed native stands off limits for further gene-resource imports into Australia and New Zealand.

7.8.2 *Structuring of Breeding Populations*

In both New Zealand and Australia the breeding plans, based upon the structuring of the breeding populations, have been significantly revised.

7.8.2.1 Australia

The STBA came up with a new breeding plan (White et al (1999), which superseded that adopted in 1993 (Boomsma 1997), and which has undergone some further evolution (Wu et al. 2007a). Developments include

- dropping the stratification, adopted in 1993, into the Nucleus and Main populations
- retention of two unrelated sublines to guarantee future outcrossing
- adoption of the "rolling front" feature (Borralho and Dutkowski 1998) in place of discrete generations, to smooth the workload and accelerate the capture of genetic gain
- interim maintenance of the three-breed structure of the breeding population, but eventual relegation of the breeds structure to the level of field-deployed crops
- reduction of population size to a census number of 340
- planned infusion of new genetic material in selections from the 1978 seed collections made in the natural populations.

Underpinning both the management of genetic material and the marketing of seed or planting stock has been the system of calculating estimated, or "predicted" breeding values of individual members of the population for various traits, using the TREEPLAN® software of PlantPlan Genetics Pty Ltd.

An ancillary feature has been a small side programme that embraced deliberate use of quite intense inbreeding (Wu et al. 2004 and preceding papers in the series). Two generations of inbreeding were practised, which ranged in intensity from matings between half-sibs to two generations of selfing. Inbreeding depression was evident, but with widely different responses among lineages to increased inbreeding. While optimism has been expressed over the prospects of inbreeding to "purge" inbreeding depression in radiata (Wu et al. 2004), this approach involves additional costs and wastage of genetic material, and is not being adopted operationally.

7.8.2.2 New Zealand

A strategy review of the radiata breeding programme had been undertaken in 1990 (Jayawickrama and Carson 1990). It reaffirmed the adoption of subdividing the breeding population into just two unrelated sublines, a feature that that was incorporated soon after the production of the 1986 Breeding Plan (Shelbourne et al. 1986). It adopted the stratification of the breeding population into the elite Nucleus and the Main components, but specialised breeds (Structural Timber and Clear Cuttings) were designated as components of the Nucleus, along with the top-ranking Growth and Form component. The lower-rated Main was not differentiated into breeds. Controlled pair-crossing was planned throughout. A separate pair-crossed breeding population was to be maintained for the Guadalupe provenance, based on 67 plus-tree selections made among New Zealand plantings. Numbers of parents envisaged per generation were 225 for the Main population, plus 24 from each of two specialised breeds.

Further revision of the strategy for the breeding population came from a review begun in 2004 and essentially completed in 2007 (Dungey et al. 2009). Features of this included:

- retaining the two-subline structure (Red and White "superlines")
- testing the two superlines on disjunct sites
- 250 parents per subline
- retaining the differentiation into Elite and Main strata
- 24 parents per subline within the elite
- dropping of the specialised breed designation (although the breed differentiation would not be extinguished immediately)
- Dropping of controlled crossing in main population, in favour of open pollination, in expectation that pedigree reconstruction will become feasible
- retention of controlled double-pair crossing in the Elite stratum
- testing of Elite progenies on 10 sites (five per superline)
- clonal replication within Elite offspring (20 clones × 2 cuttings × 5 sites)
- supplementing these clones with seedlings in the Elite
- testing of Main progenies on eight sites (four per superline)
- 20 seedlings per family per site in Main.

In addition, the separate pure Guadalupe breeding population, based on 67 parents, was left to stand. So, too, was a pure Cedros breeding population, based on 30 parents, which had been put in place since the 1990 review.

The proposal for clonal replication of individual offspring was controversial, hence its adoption for only part of the Elite stratum.

7.8.2.3 Chile

In keeping with the structure of the breeding Cooperativa, company breeding programmes have tended to remain largely self-contained, with the actual breeding continuing largely along the originally planned lines. Despite the widespread success of seed orchards, vegetative multiplication has been vigorously adopted in some quarters to speed up the operational capture of genetic gain (e.g. Balocchi 1997).

Genetic resistance to the pine shoot moth has been explored, even enlisting New Zealand researchers to confer it by genetic engineering (Grace et al. 2005), but the single resistant transgenic cell line was not put into operational use.

7.9 Advances and Problems in Propagation Technology

7.9.1 Vegetative Propagation

7.9.1.1 Advances

Producing cuttings of radiata for vegetative multiplication of scarce seedlots was by 1998 a reasonably "mature technology." So too was production of radiata plantlets using in-vitro culture for rapid initial multiplication. Neither technology, however, could actually beat the process of maturation and the problems it posed. For control of maturation, and convenient storage of live clones, cryopreservation at ultra-low temperatures was already being developed for radiata (Hargreaves et al. 1997, 1999). Quite soon, it became broadly feasible, in that the various types of in-vitro cultures, namely embryos, cotyledons and shoot-tip meristems, can all be revived after removal from cryogenic storage (Hargreaves and Menzies 2007). However, embryogenesis (Fig. 7.6), as a platform for obtaining very high multiplication rates starting from fully juvenile material, was harder to achieve satisfactorily with radiata. Pursued since the late 1980s, starting with embryos of immature seeds, it would work very well with spruces, but not with radiata. A big problem was poor "genotype capture." In many whole families it would fail. Even among the remaining families, many of the seeds could not be induced to produce embryos that could be successfully reared through into plantlets. However, a

Fig. 7.6 Embryos (white) appearing on callus tissue (brownish) cultured from embryos excised from partly developed radiata seed

major breakthrough was made in increasing genotype capture for embryo initiation (Hargreaves et al. 2009, 2011). Better timing of collection of immature cones, appropriate choice of dissection technique and, above all, a modified culture medium allowed a genotype capture rate that was no longer a significant limitation. While the finding was just for embryo initiation, this is typically the limiting step in producing plantlets.

7.9.1.2 Some Stumbles

Maturation led to a failure of the initial attempts at clonal forestry with radiata (Fig. 7.7). Although control of maturation had since become quite well understood technically, practical management of it in producing planting stock has not always succeeded. Even with hedging and serial propagation, maturation will proceed insidiously, and its impact can be underestimated. That, in combination with efforts to trim production costs, can lead to a decline in the quality of nursery stock, with weaker plants and unbalanced root systems. In turn, such deterioration in quality of planting stock exacerbates the effects of slipshod establishment practice in the field, especially on difficult sites. Such establishment practice can reflect

Fig. 7.7 Early New Zealand clonal test, with single-tree plots. Good resolution of clonal differences was obtained, but by the time the best-performing clones could be selected they could no longer be repropagated reliably and cheaply

a combination of over-zealous cost-cutting and declines in field-based skills, the latter resulting from continuing rapid changes of field personnel with casualization of the labour force. Indeed, there have been significant areas of crop failure with radiata in New Zealand, requiring replanting after several years, with other, less-affected crops still significantly compromised.

Thus, despite great advances in developing software for devising prescriptions and planning field operations, success in growing crops can be seriously compromised by injudicious cost-cutting and associated losses of on-the-ground skills.

7.9.2 Seed Production and Hybridisation

With the improvements in seed-orchard siting, the capability for accelerated seed ripening by "curing" cones, and the availability of mass vegetative multiplication of scarce seed or seedling stocks, research on routine seed production became less of a priority. However, characterising stages of seed development became an issue for vegetative multiplication by embryogenesis (Hargreaves et al. 2009), to capture reliably almost all genotypes of interest.

Another aspect of seed production is production of hybrids with other species. With radiata this is potentially of great practical importance. Very few species hybridise at all readily with it, yet there are great attractions in being able to combine its advantages with making good its biotic vulnerabilities by incorporating resistances from other species. Among other species with such resistances are a number that appear closely enough related to make the prospect of hybrids tantalising. Yet, among the mainstream institutions for breeding radiata any strong commitment to overcoming crossability barriers has tended to be given low priority.

7.10 Where to for Clonal Forestry?

Despite the advances made in vegetative propagation (Menzies and Aimers-Halliday 2005), control of maturation remained a problem. Clones had often been maintained as *in-vitro* cultures or as hedges, so as to allow continued propagation for a number of years, delaying maturation in varying degrees but never halting it. Thus, continued production of high-quality cuttings or plantlets at reasonable cost has remained more difficult, and was often failing. And, where planting stock are of indifferent quality, the margins for error in site preparation, standard of planting and follow-up weed control are greatly reduced.

The development of a reliable cryopreservation system has provided the one guarantee of clonal storage in a juvenile state, which can be used in parallel with clonal evaluation (Fig. 7.8). Such use, however, depends on setting up clonal storage prior to establishing field tests, and maintaining the storage facility, requiring careful logistical preparation. With embryogenesis, the greatly improved techniques

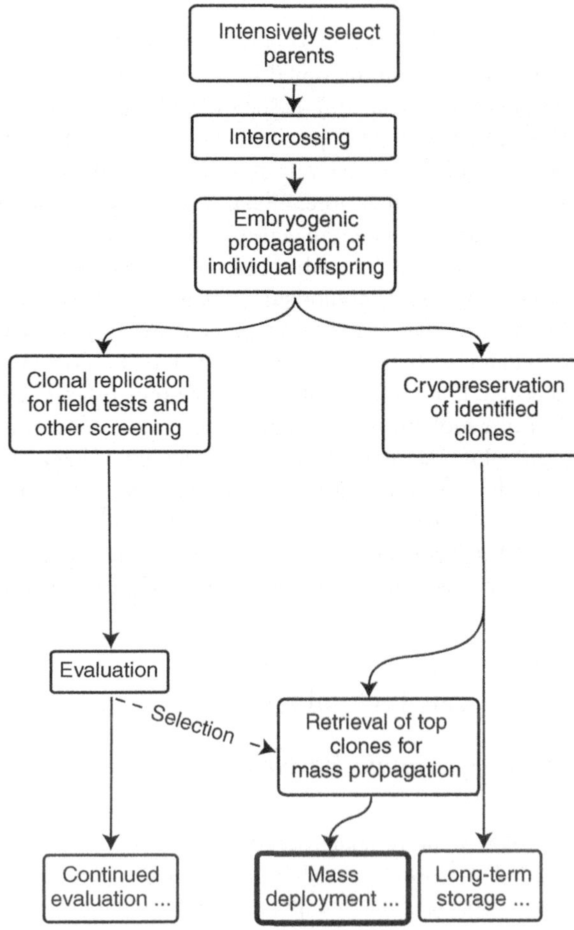

Fig. 7.8 Flow chart of operations including use of cryopreservation for storage of clones so as to halt maturation during clonal evaluation, with approximate time frame denoted in the vertical axis

(Hargreaves et al. 2009) now mean genotype capture is greatly improved and is no longer an important limitation. This avoids loss of genetic gain in the breeding objective through having to add propagability as a selection criterion, and/or a risky narrowing of the genetic base in the commercial crops.

Despite the big progress in cryogenics and embryogenesis, an ideal solution has not been achieved. That would be an ability to completely rejuvenate vegetative material of any tree at any age. This goal remains elusive, but it has never been addressed for radiata with major institutional commitment.

Even with this limitation, and some special problems of managing biotic risks (Aimers-Halliday & Burdon 2003; Burdon and Aimers-Halliday 2003), clonal forestry retains some major attractions for radiata. Outstanding clones are indeed outstanding, consistent with the expected scope for clones to capture non-additive as well as just additive genetic gain. Having logs of known wood properties, despite all the within-log variation and the technology for individual-log assay for stiffness, would still be extremely welcome. Clonal forestry offers the best prospect of precise control of branching pattern for allowing optimal recovery of clear-cuttings from unpruned trees. Moreover, monoclonal blocks offer more uniform log sizes, and they reduce the focus on competitive ability which may not totally align with crop productivity. It is thus fully understandable that vigorous pursuit of clonal forestry is proceeding in New Zealand (Sect. 7.4.2) and Chile (Sect. 7.4.3). Furthermore, clonal forestry is the context for any foreseeable operational use of genetic engineering with radiata (Burdon and Lstibůrek 2010).

While some of the advantages of clonal forestry apply to using a mosaic of monoclonal blocks, comprising enough clones for effective risk spread, the case for blocks rather than intimate mixtures of clones is not overwhelming. There may well be situations where failed clones cannot be conveniently abandoned or salvage-harvested. For deployment, clonal mixtures promise better protection against downsides of genotype-site interaction, yet, with really good knowledge, clonal forestry would allow the most effective exploitation of interaction. Certainly, clonal material offers the best information on the magnitude and patterns of interaction, because clonal trials give the highest expected "signal-to-noise" ratio.

Further evidence of similar behaviour and genetic parameters between juvenile clones and seedlings (Baltunis et al. 2009) is reassuring.

7.11 Gene Technology

7.11.1 Genomic Research

By 1998, a preliminary genomic map for radiata was available (Devey et al. 1996). There were also genomic tools for locating specific DNA sequences from another species that are present and functional in both it and another species; indeed, two or more species can be used simultaneously to attract complementary funding and provide complementary information.

However, unravelling a complete genomic sequence of a pine is a forbidding challenge. The genome is huge (at ca 2.4×10^{10} base pairs, around 10 times the size of the human genome), and the pattern of massively repeated sequences greatly exacerbates the problems of trying to arrive at a complete sequence. Hence the attractions of being able to apply genomic information from other plant species, not just pines or even other conifers, are major.

Actually, some of the very early genomic research with radiata focussed on the organelle genomes, with emphasis on chloroplasts, these genomes being much smaller than the nuclear genome and therefore more convenient to study. Advances in genomic technology and bioinformatics have since facilitated study of the nuclear genome, which is much more informative, so we do not review the studies of organelle genomes.

Among plants, the cress, *Arabidopsis thaliana*, has become the favourite model species for genomic research, because of its very small size and short lifespan, together with a very small genome. Not only can its genomic architecture be studied very easily, but effects of variant genes can be tracked through the developmental processes whereby they are expressed. Remarkably, despite very different developmental organisation, certain genes appear to be present and have similar roles in both *Arabidopsis* and pines (Krutovsky et al. 2006).

7.11.1.1 Comparative Genomics

The genomic stability of most conifers, and of pines in particular, means that much genomic information on one pine species can easily be cross-referenced (Burdon and Wilcox 2011 and references therein). By 1999, it was evident that the genomic structure of radiata and loblolly pines was very similar (Devey et al. 1999b), in terms of organisation into chromosomes, and the general arrangement of genes within chromosomes. It evidently holds for a range of pine species, if not across a wide range of conifer genera. This offers the prospect of leveraging genomic information acquired in one species for application in one or more other species. Genomic studies, however, are subject to countervailing institutional pressures. Some parties like to keep such intellectual property as proprietary information, in the face of the fact that radical advances will depend enormously on collaboration among many parties, even big ones. Against the attempts to appropriate genomic information, there are parties, notably US government agencies, that are committed to the public domain. There is in USA the Loblolly Pine Genome Project that falls within the wider Conifer Genome Project, while some other pine species, notably maritime pine (*Pinus pinaster*) and Scots pine (*Pinus sylvestris*) have also been the subject of intensive genomic studies. The scope for linking genomic research on radiata to that on other pines, especially loblolly pine, is thus considerable.

Given the pairing behaviour of DNA strands, and modern capabilities for DNA sequencing, there is now a range of techniques for locating the same or roughly homologous sequences of DNA in different species, which offers much promise of circumventing the obstacle posed by conifer genomes being too large to sequence *in toto*.

7.11.1.2 DNA Fingerprinting and Clonal Verification

For the tree breeder, it is important to be sure that identity of clonal material is correct and that pollen parentage off offspring is verified. Using the wrong clones as parents in breeding-population crosses, or in seed orchards, could lose much potential gain, the more so as a breeding programme becomes more advanced. For such fingerprinting, microsatellites, or simple sequence repeats, are DNA sequences that can be used like bar codes, as in DNA forensic work. Such sequences are non-coding DNA, characterised by variable numbers of repeats of particular short DNA base sequences. These sequences effectively show numerous alleles at a locus, such that as few as six well-chosen loci suffice to identify almost any clone (Bell et al. 2004). In fact, such fingerprinting has consistently identified misidentifications among clones in a number of tree breeding programmes, including ones involving radiata (Burdon et al. 2008). As a breeding programme progresses, eliminating such misidentifications becomes more and more important, as the genetic gains at issue increase.

Just as clonal identity needs verifying, so does both seed- and pollen-parent identity, which can be done in the same way for family seedlots.

7.11.1.3 Pedigree Reconstruction

Going one step further than verification of pollen parentage, it is in principle possible to reconstruct pedigrees. With the seed parent, this is relatively straightforward, being usually a matter of verification, with the bonus that seed megagametoyphytes (or conifer endosperm) can be used very easily for this purpose. With pollen parentage, it can be much more challenging than verifying a single pollen parent, because with open-pollinated families the potential pool of pollen parents is huge—more so with wind-pollinated species like radiata than with insect-pollinated species including eucalypts. Even so, a finite number of pollen parents may account for a high proportion of the pollinations, in which case individuals of unknowable pollen parentage might be disregarded without serious loss of genetic gain.

Indeed, the prospect of being able to reconstruct pedigrees has contributed to the acceptance in the RPBC (Dungey et al. 2009) of labour-saving open-pollination for the Main breeding population. However, much development work remains to be done before the pedigree reconstruction could be done with acceptable reliability.

7.11.1.4 Search for DNA Markers for Selection

By 1998 high hopes were held (Sect. 6.13.2) for **marker-assisted selection**^{G} (MAS) or even **marker-based selection**^{G} (MBS) for earlier and/or greater capture of genetic gain. While MAS uses genomic data in conjunction with phenotypic data, MBS uses genomic data alone and can therefore be done on younger offspring. Neither form of selection depends on identifying actual genes governing variation. Rather, they both depend on chromosomal regions containing one or a number of genes of net effect being consistently linked with one identifiable DNA marker; by ideal if improbable good luck, the marker could characterise the functional variant, being a quantitative trait nucleotide (QTN) if the marker is a single nucleotide polymorphism (SNP). Whatever the functional DNA variant(s), a chromosomal region governing expressed variation is termed a quantitative trait locus (QTL). Early theoretical work (Lande and Thompson 1990) had indicated that MAS/MBS would have the greatest advantage in situations with QTL of large effect in a context of low heritability, the problem being that proof of concept can be hardest when heritability is low. Anyway, theoretical work for a radiata breeding scenario (Wilcox et al. 2001) showed promising economic benefits from adopting MBS.

Practical development of MAS/MBS, however, encountered major problems (Burdon and Wilcox 2011). Such selection is dependent on linkage disequilibrium (LD) between the markers and the QTL, with consistent associations between particular QTL variants and particular marker alleles. But with an outbreeding species like radiata LD, while present within families, extends across populations over only minute segments of chromosomes (Kumar et al. 2003; Dillon et al. 2010). This is in sharp contrast with the situation in various crop plants which are managed as inbreeders, and to a lesser extent, the situation in dairy and poultry breeding. Even within families of radiata it can be hard to find QTL that can be reliably detected and quantified. Nominal family members can prove to be misidentified, as has happened. There was one pair-cross family that was expected to show a large QTL for wood density, and while a chromosome region was identified as a likely QTL the evidence was not conclusive (Kumar et al. 2000). Indeed, large-effect QTL appear to be very uncommon in conifers (Burdon and Wilcox 2011). Moreover, QTL of small effect are extremely difficult to detect reliably; false positives can in theory arise very easily, and reported cases of promising QTL can easily prove illusory in the light of additional data; this has all too often happened in the search for human "disease" genes.

Problems of detecting QTL in conifers have shifted research efforts in the direction of association genetics, trying to identify associations across whole outbreeding populations. In principle, this is still addressing chromosomal linkage, which may work within very small subpopulation units. In practice, however, it may largely amount to a search for SNPs that represent QTNs. That, on the face of it, is like looking for the proverbial needle in a haystack, but researchers are now looking to narrow the search by using candidate genes. These are genes that have been identified as being belonging in the developmental pathways for traits of interest, typically in plants that have already been studied in more detail. Such a study has been undertaken by Dillon et al. (2010), who tested in radiata associations between multiple variants of around 40 such genes involved in wood-forming pathways and 13 wood-core traits. A number of statistically significant associations were found, but none accounted for more than 6.5% of trait variation, although two associations were repeated in a verification population. However, scope remains for further verification, e.g. in a range of environments.

Recently, however, emphasis has shifted towards the pursuit of genomic selection (Sect. 7.2.2). This depends on cross-referencing very large numbers of genomic markers with phenotypic data (Sects. 9.2.1 and 9.3.4.4), to give global estimates of breeding values or genotypic values of candidate offspring without waiting for phenotypic information (Isik 2014). No actual reliance is placed on verifying phenotypic effects associated with specific markers. The aim is that inaccuracy of these genomic estimates can be more than compensated for by the time gained in achieving genetic advance. However, genomic markers need to be cross-referenced with phenotypic data that are an adequate guide to harvest-age performance.

7.11.1.5 Gene Expression Studies and "Gene Discovery"

Often candidate genes may be identifiable from their known roles in other plant species, leaving their roles in a species like radiata to be confirmed. Alternatively, or complementarily, "gene discovery" can be pursued directly within the subject species (Burdon and Wilcox 2011). This was done with radiata by Cato et al. (2006), studying differences in expression of genes associated with variation in wood properties among different parts of the tree, using some prior knowledge of the genes' roles. Study of expression pathways, be it to confirm roles of candidate genes or for fresh gene discovery, can involve a suite of "-'omics", i.e., transcriptomics, proteomics and metabolomics, involving the various stages of gene expression. This can be particularly important, and yet challenging, where there are alternative biochemical pathways that create redundancies, as occur in wood formation (Boudet 2000). Full confirmation of the roles of certain genes, however, may require incorporating them by genetic engineering (Sect. 7.11.2).

7.11.1.6 Some Convergence With Other Plant Breeding

Historically, forest tree breeding has aligned more closely with large animal breeding programmes, rather than with most plant breeding. The emphasis has been on progressive population improvement, rather than on producing discrete and uniform cultivar varieties. Partly because of this, and the fact that forest trees are typically outbreeders, as farm animals necessarily are, the quantitative breeding methodology used has been largely similar. A major difference among forest tree breeders and animal breeders has been that forest tree breeders readily accept that differences between sites in heritabilities can be genuine, whereas animal breeders tend to see variation in heritability estimates as reflecting sampling error about essentially constant values.

In recent years, however, the emphasis on genomic studies has in one respect narrowed the historical gap between tree breeding and most other plant breeding. On the other hand, the inbreeding systems of many domesticated crops, and the very finite population sizes of certain animal breeds, have both generated levels of linkage disequilibrium (LD) that are crucial to most marker-assisted (or marker-based) selection. By comparison, the extremely limited LD in most conifers leaves obstacles for both genomic selection and gene discovery.

7.11.2 Genetic Engineering

Genetic engineering (GE) has some big attractions with forest trees. It offers the prospect of introducing very specific new attributes that are not available within a species, and in quick time. Even if the attributes can be introduced by hybridisation, the process can be very slow, and crossing can bring in from the "donor" species huge numbers of unwanted genes, which can only be purged by generations of further crossing and selection. Risks, however, are involved, in possible side-effects of the process of genetic transformation and of the introduced "transgenes." The latter are potentially catastrophic, if manifestation is delayed and risk management is not highly rigorous (Burdon and Walter 2004). Such risk management, however, can largely negate the theoretical time savings. Perceptions of the nature of the risks associated with GE vary widely among different groups of people. In fact, the major perceived risks often involve the environment (Fladung et al. 2012) rather than crop security. Research on environmental risks with GE radiata has been actively pursued (Sect. 7.11.2.2). Between risk-management considerations and ideological objections to genetic engineering—and even to any domestication of forest trees—the rule is for genetic engineering to be subject to tight regulatory protocols.

The technology for genetic transformation with radiata has advanced in that *Agrobacterium* can now be used as a vehicle for the transgenes. Selection of successfully transformed cells for regeneration of transgenic plants is achieved by incorporating resistance to a non-clinical antibiotic (traditionally kanamycin) in the transgene "construct" and using a culture medium containing that antibiotic. Advances in in-vitro culture technology have made the whole process easier. Moreover, advances in genomic technology has also made it much easier to confirm stable incorporation and actual function of particular transgenes. A recent advance, in the CRSPR technology to achieve "gene editing" (Dance 2015), is making it possible to implement genetic engineering without involving various features that were seen as either incurring certain risks or prompting ethical objections.

The basic amenability of radiata to some form of genetic engineering had been confirmed by 1998 (Walter et al. 1998). Practical demonstration of genetic engineering to confer resistance to a herbicide has been achieved in New Zealand in a tightly regulated field trial in 2013 (C. Walter et al. in prep.), despite sabotage by environmental activists who managed to breach security. Stark contrasts have been demonstrated in herbicide resistance between transformed and untransformed members of the same clones.

7.11.2.1 Target Traits

Target traits for genetic engineering of radiata have evolved. Originally, they were simply "reporter" genes that could be shown to have integrated stably into the recipient culture. Since then, the traits addressed by the New Zealand Forest Research Institute for radiata have included: herbicide resistance (Fig. 7.9); insect resistance in the *Bacterium thuringensis* (*Bt*) toxin; variants for cell-wall polymers (especially lignin); and reproductive development involving pollen production and, to a lesser extent, conelet production. Incorporation of herbicide resistance was achieved quite early (Bishop-Hurley et al. 2001), and insect resistance was also been incorporated successfully (Grace et al. 2005) albeit on only a single cell line. Achieving satisfactory reproductive sterility is more challenging, but if

Fig. 7.9 Two young plants of radiata after treatment with an effective but rapidly biodegradable herbicide: on left, killed control: on right, genetically engineered for resistance to the herbicide and unaffected (Photo Christian Walter)

achieved, it has a two-fold attraction: avoiding diversion of resources into reproduction and providing a containment mechanism for transgenes.

7.11.2.2 Associated Research

Testing for success in incorporating herbicide resistance is relatively quick and straightforward. For other, traits, however, it is more complicated. There are various ways in which genetic transformation can be used, which include inserting completely new genes or up-regulating or down-regulating existing genes. Given the complex and often redundant pathways of gene expression, up- or down-regulating genes can serve as a research tool to study the true role of particular genes. Transformation for genes governing cell wall polymers has been accompanied by culturing tissue *in vitro* to produce tracheary (wood-fibre) elements, verifying the anticipated changes in cell-wall chemistry (Wagner et al. 2007). Still to be tested, however, are the effects of such transformation of field performance and economic value of the wood (cf Wagner et al. 2009; Voelker et al. 2011). Insect resistance has been confirmed with feeding studies involving defoliating insects. The success of transformation to control reproduction will remain the subject of a relatively long-term study, done in conjunction with observing a regulatory protocol of removing all reproductive structures.

Field trials of transgenic plants have been subject to detailed study of possible environmental side-effects which have been postulated (often vehemently) by opponents of genetic engineering. Studies to date (e.g. Schnitzler 2010; Lottmann et al. 2011; Shi et al. 2012) have shown no sign of unintended, let alone adverse, environmental side-effects.

7.11.2.3 Regulatory Context

New Zealand has had the main announced commitment to genetic engineering with radiata. However, the work is done under very tight regulatory controls, in terms of the Hazardous Substances and New Organisms (HSNO) Act, which is administered by the Environmental Protection Agency (EPA), which until 2011 was called the Environmental Risk Management Authority (ERMA). The present situation mirrors the report of a Royal Commission on Genetic Modification, which sat during 2000 and reported in mid-2001. While supporting the development of the new technology, it adopted a strongly precautionary position. Approvals for genetic engineering research, when sought, are subject to public notification and submissions, with a proportion of the public implacably opposed to genetic engineering. Laboratory containment protocols are tight and field containment protocols tighter still—and almost prohibitively expensive to implement. In the interests of public transparency field trial sites are publicly identified as well as being elaborately fenced. Public notification, however, has twice led to very expensive field trials studying effectiveness and possible environmental side-effects being destroyed by environmental activists.

In Australia the political climate for genetic engineering is if anything more hostile, with states superimposing their own legislative controls over federal ones, and it has not been seriously addressed for radiata. Chile has on one occasion commissioned New Zealand researchers to attempt to introduce resistance to the European pine shoot moth, which yielded just the one individual that appeared to have been successfully transformed (Grace et al. 2005). We are not aware of any other reports of genetic engineering of radiata in Chile.

A recent court decision in New Zealand (NZHC 2014) has exposed an inadequacy in the regulatory legislation for facilitating approval of genetic modification that employs new technology to circumvent certain traditional objections.

7.12 Summary of Domestication Progress

Afforestation with radiata has lagged, and in some cases has been reversed, through a combination of factors, of which some have operated more indirectly than others. The factors include: institutional instability, international financial crises, an international boom in dairying, booms (some abortive) in afforestation with eucalypts, and neglect of environmental benefits of forestry in government policies.

Radiata gained attention for its potential to sequester carbon for combating global warming. Availability of carbon credits would reduce growing costs, especially with regimes designed to maintain wood quality, and could favour growing much bigger areas of radiata. The ways in which carbon trading and carbon liabilities have been managed in New Zealand, however, have led to deforestation rather than the expected net encouragement of the carbon sequestration.

Knowledge of inheritance patterns has continued to advance, albeit patchily. Adaptive profiles of natural populations in relation to Australian environments have become clearer. Some broad patterns of genotype-environment interactions have emerged there, but knowledge remains generally limited concerning the exact environmental factors that generate the interactions.

Emphasis has increased on genetic improvement of wood properties, with much research into targeting appropriate properties and refining assay technologies. Adverse genetic correlations between growth performance and various wood-quality traits have become better understood, and, while not always strong, tend to be pervasive. These pose major challenges for tree breeders. Ideally, breeders should become able to assign correct economic weights to adversely correlated traits. Major uncertainty concerning comparative economic weights, however, may require some alternative defensive approach(es) to guarantee gains in net economic worth. One such approach would be the pursuit of differentiated breeds, or at least differentiated deployment populations. This can accommodate different economic weights according to sites, wood-processes or end-uses, all of which can be important issues for radiata. For deployed material it also recognises that optimal values for individual traits can be conditional upon values for other traits, although formal optimisation would be very challenging. Anyway, a differentiated "portfolio" should guarantee some protection against worst-case outcomes.

Genetic fingerprinting has been developed for quality control to avoid loss of genetic gain through misidentification, and is a prospective tool for averting the need for laborious controlled pollination without sacrificing potential genetic gain.

Progress has continued in development and adoption of clonal forestry, but its fully satisfactory implementation still poses challenges. However, major technical progress has been made in cryopreservation to control maturation state, with much better "genotype capture" to avoid amenability to cryopreservation becoming an important selection criterion. It allows clones to be conserved for future revival while they are screened for the breeding-goal traits. Nevertheless, the goal of being able to rejuvenate any clone, whatever the maturation state, remains elusive.

Radiata remains the context for developing new software-based technologies associated with modern plantation forestry, for modelling growth and product outturns as decision- and planning aids, for using remote sensing for crop inventory, and for precision control of machinery in the field and in wood processing.

Progress has also continued on developing genetic transformation as a genetic-engineering (GE) technology for genetic improvement. A pilot success has been achieved in genetic transformation of radiata to produce syringyl lignin instead of guaiacyl lignin, which would make chemical pulping easier and cheaper. However, compatibility of this change with field fitness is quite unproven. Moreover, any quest to apply GE faces very elaborate and expensive regulatory protocols in New Zealand and Australia.

Also, DNA markers are being vigorously explored as a tool for earlier and more cost-effective genetic selection. This poses major challenges because of the essentially wild state of the genetic system, a huge genome, an apparent paucity of individual genes of large effect, and the delays and difficulties in obtaining good field-performance information for detecting and verifying informative genomic regions. Disease resistance, however, appears a promising avenue for such genomic selection.

Fresh biotic alarms have occurred in recent years, mainly involving diseases caused by *Phytopthora* spp. They continued a pattern involving pathogens with no prior record of being significant. These alarms have prompted fresh research on breeding for disease resistance, including a search for cross resistance that would confer simultaneous protection against multiple pathogens.

Since the mid-1990s tensions have continued between those charged with advancing breeding programmes along classical lines and those researching new DNA technology for genetic improvement. Allocating resources between the two camps remains a fraught issue. The latter have made big advances, but "killer applications" remain elusive.

While the domestication of radiata has come a long way, it has much further to go, as we explain in the final chapter.

References

Adams T, Brack C, Farrier T et al. (2013) So you want to use LiDAR? – A guide to how to use LiDAR in forestry. N Z J For 55(4):19–23

Aimers-Halliday J, Burdon RD (2003) Risk management for clonal forestry with *Pinus radiata*: Analysis and review. 2: Technical and logistical problems and countermeasures. N Z J For Sci 33:181–204

Anon. (1998) Privatisation for Victorian state plantations. N Z J For 43(2):6

Anon. (2013) Updated Chile forest incentive scheme winners and losers. South Hemisph For Ind J 18(3):4

Apiolaza LA (2012) Basic density of radiata pine in New Zealand: growth and environmental factors. Tree Genet Genomes 8:87–96

Apiolaza LA, Chauhan SS, Walker JCF (2011) Genetic control of very early compression and opposite wood in *Pinus radiata* and its implications for selection. Tree Genet Genomes 7:563–571

Arauco 2012. Resumen Público Plan Maestro de Ordenación, September 2012. www.arauco.cl.

Balocchi CE (1997) Realised versus operational gain: the role of propagation strategies. In: Burdon & Moore, pp 253–255.

Baltunis BS, Wu HX, Powell RM (2007) Inheritance of density, microfibril angle and modulus of elasticity of juvenile wood of *Pinus radiata*. Can J For Res 37:2164–2174

Baltunis BS, Wu HX, Dungey HS et al. (2009) Comparisons of genetic parameters and clonal value predictions from clonal trials and seedling base population trials of radiata pine. Tree Genet Genomes 5:269–278

Beèche M, Lanfranco D, Zapata M et al. (2012) Surveillance and control of the sirex woodwasp: the Chilean experience. In: Slippers B, de Groot P, Wingfield MJ (eds) The sirex woodwasp and its fungal symbiont: research and management of a worldwide invertebrate pest. Springer, Dordrecht, The Netherlands, pp 229–245

Bell JC, Powell M, Devey ME et al. (2004) DNA profiling, pedigree lineage analysis and monitoring in the Australian breeding program of radiata pine. Silvae Genet 53:130–134

Bian L, Gapare WJ, Ivkovič M et al. (2011) Genetic variation between and within *ex-situ* native-provenance collections of *Pinus radiata* D.Don planted in Australia and New Zealand. Silvae Genet 60:276–285

Bishop-Hurley SL, Zabkiewicz JA, Grace LJ et al. (2001) Conifer genetic engineering: transgenic *Pinus radiata* (D.Don) and *Picea abies* (Karst) plants are resistant to the herbicide Buster. Plant Cell Rep 20:235–243

Boomsma DB (1997) Development of the Southern Tree Breeding Association breeding population. In: Burdon & Moore, pp 211–216
Borralho NMG, Dutkowski GW (1998) Comparison of rolling front and discrete generation breeding strategies for trees. Can J For Res 28:987–993
Boudet A-M (2000) Lignins and lignification: selected issues. Plant Physiol Biochem 38:81–96
Burdon RD (1975) Foliar macronutrient concentrations and foliage retention on radiata pine clones on four sites. N Z J For Sci 5:148–152
Burdon RD (1992) Genetic survey of *Pinus radiata*. 9. General discussion and implications for genetic management. N Z J For Sci 22:274–298
Burdon RD (2010) Wood properties and genetic improvement of radiata pine. N Z J For 55(2):22–27
Burdon RD, Aimers-Halliday J (2003) Risk management for clonal forestry with *Pinus radiata*: analysis and review. 1: Strategic issues and risk spread. N Z J For Sci 33:156–180
Burdon RD, Carson MJ (1999) Conservation and management of genetic resources of commercial forests in New Zealand: challenges of institutional changes and new technology. In: Mátyás C (ed) Forest genetics and sustainability. Kluwer, Dordrecht, The Netherlands, pp 235–246
Burdon RD, Low CB (1992) Genetic survey of *Pinus radiata*. 6. Wood properties: variation, heritabilities and interrelationships with other traits. N Z J For Sci 22:228–245
Burdon RD, Lstibůrek M (2010) Integrating genetically modified traits into tree improvement programs. In: El-Kassaby YA (ed) IUFRO biotechnology task force state-of-the-art report. FAO, Rome, pp 123–134
Burdon RD, Moore JM (eds) (1997) IUFRO '97 Genetics of radiata pine. In: Proceedings of NZ FRI-IUFRO conference 1–4 December and workshop 5 December, Rotorua, New Zealand. FRI Bulletin No. 203. p 354
Burdon RD, Walter C (2004) Exotic pines and eucalypts: perspectives on risks of transgenic plantations. In: Strauss SH, Bradshaw HD (eds) The bioengineered forest. RFF Press, Washington, DC, pp 52–75. (Chapter 5)
Burdon RD, Wilcox PL (2011) Integration of molecular markers in breeding. In: Plomion C, Bousquet J, Kole C (eds) Genetics, genomics and breeding of conifers. Science Publishers, Jersey, British Channel Islands and Enfield, NH, USA. pp 276–322
Burdon RD, Firth A, Low CB et al. (1997a) Native provenances of *Pinus radiata* in New Zealand: performance and potential. N Z [J] For 41(4):32–36
Burdon RD, Broekhuizen P, Zabkiewicz JA (1997b) Comparison of native-population and New Zealand land-race samples of *Pinus radiata* using cortical oleoresin monoterpenes. In: Burdon & Moore, pp 50–56
Burdon RD, Firth A, Low CB et al. (1998) Multi-site provenance trials of *Pinus radiata* in New Zealand. In: *Forest Genetic Resources*. No. 26 FAO, Rome, p 3–8
Burdon RD, Carson MJ, Shelbourne CJA (2008) Achievements in forest tree genetic improvement in Australia and New Zealand 10: *Pinus radiata* in New Zealand. Aust For 71:263–279
Butterfield BG (ed.) (1997) Microfibril angle in wood. Proceedings IAWA/IUFRO Workshop on the significance of microfibril angle to wood quality, Westport, New Zealand, November 1997. Christchurch, New Zealand. p 401
Carson SD (1991) Genotype × environment interaction and optimal number of progeny test sites for improving *Pinus radiata* in New Zealand. N Z J For Sci 21:32–49
Carson SD, García OP, Hayes JD (1999) Realized gain and prediction of yield with genetically improved *Pinus radiata* in New Zealand. For Sci 45:186–200
Castedo-Dorado F, Lago-Parra G, Lombardero MJ et al. (2016) European gypsy moth (*Lymantria dispar dispar* L.) completes development and defoliates exotic radiata pine plantations in Spain. N Z J For Sci 46:18. doi:10.1186/s40490-016-0074-y
Cato S, McMillan L, Donaldson L et al. (2006) Wood formation from the base to the crown in *Pinus radiata*: gradients of tracheid wall thickness, wood density, radial growth rate and gene expression. Plant Mol Biol 60:565–581
Cave ID, Walker JCF (1994) Stiffness of wood in fast-grown plantation softwoods: the influence of microfibril angle. For Prod J 44(5):43–48

Coutinho TA, Steenkamp ET, Mongwaketsi K et al. (2007) First outbreak of pitch canker in a South African pine plantation. Austral Plant Pathol 36:256–261

Cown DJ, Donaldson LA, Downes GM (2011) A review of resin features in radiata pine. N Z J For Sci 41:41–60

Crane PE, Hopkins AJM, Dick MA et al. (2009) Behaviour of *Nectria fuckeliana* causing a pine canker disease in New Zealand. Can J For Res 39:2119–2128

CSIROpedia (2013) http://www.csiropedia.csiro.au/display/CSIROpedia/Forestry+and+Forest +Products 11 July

Codesido V, Fenández-López J (2009) Implications of genotype × site interaction on *Pinus radiata* breeding in Galicia. New For 17:17–34

Cullis BR, Jefferson PA, Thompson R, Smith AB (2014/2015) Factor analytic and reduced animal models for the investigation of additive-by-environment interaction in outcrossing plant species, with application to a *Pinus radiata* breeding programme. Theor Appl Genet 127:2193–2210

DAFF (Department of Agriculture, Forestry and Fisheries – Forestry Branch) (2010) Report on commercial timber resources and primary roundwood processing in South Africa: 2008/2009. DAFF (Department of Agriculture, Forestry and Fisheries – Forestry Branch), Pretoria

Dance A (2015) Core concept: CRISPR gene editing. Proc Natl Acad Sci USA 112(20):6245–6246

De Diego N, Sampedro MC, Barrio RJ et al. (2013) Solute accumulation and elastic modulus changes in six radiata pine breeds exposed to drought. Tree Physiol 33:69–80

Devey ME, Bell JC, Smith DN et al. (1996) A genetic linkage map for *Pinus radiata* based on RAPD, RFLP and microsatellite markers. Theor Appl Genet 92:673–679

Devey M, Matheson C, Gordon T (eds) (1999a) Current and potential impacts of pitch canker in radiata pine. In: Proceedings of the IMPACT Monterey workshop Monterey, California, 30 November to 3 December 1988. CSIRO Forestry and Forest Products, Australia, Technical Report No. 111. p 120

Devey ME, Sewell MM, Uren TL et al. (1999b) Comparative mapping in loblolly and radiata pine using RFLP and microsatellite markers. Theor Appl Genet 99:656–662

Dick MA, Williams NM, Bader MK et al. (2014) Pathogenicity of *Phytophthora pluvialis* to *Pinus radiata* and its relation with red needle cast disease in New Zealand. N Z J For Sci 44 (6): doi:10.1186/s40490-014-0006-7

Dillon SK, Nolan M, Li W et al. (2010) Allelic variation in cell wall candidate genes affecting solid wood properties in natural populations and land races of *Pinus radiata*. Genetics 185:1477–1487

Dungey HS, Carson MJ, Low CB et al. (2003) Potential niches for inter-specific hybrids with *Pinus radiata* in New Zealand. N Z J For Sci 33:295–318

Dungey HS, Matheson AC, Kain D et al. (2006) Genetics of wood stiffness and its component traits in *Pinus radiata*. Can J For Res 39:1165–1178

Dungey HS, Brawner JT, Burger F et al. (2009) A new breeding strategy for *Pinus radiata* in New Zealand. Silvae Genet 58:28–38

Dungey HS, Low CB, Ledgard NJ et al. (2011) Alternatives to *Pinus radiata* in New Zealand high country: early growth and survival of *Pinus radiata*, *P. attenuata* and their F_1 hybrid. N Z J For Sci 41:61–69

Durán A, Gryzenhout M, Slippers B et al. (2008) *Phytophthora pinifolia* sp. nov. associated with a serious needle disease of *Pinus radiata* in Chile. Plant Pathol 57:715–727

Durán A, Gryzenhout M, Drenth A et al. (2010) AFLP analysis reveals a clonal population of *Phytophthora pinifolia* in Chile. Fungal Biol 114:746–752

Dyck B, Thomson P (1999) Carter Holt Harvey's millennium forestry. N Z J For 43(4):2–4

Eldridge KG (1998) Australian radiata provenance trials: an annotated catalogue of Australian provenance trials and genetic conservation plantings of *Pinus radiata* of Californian origin, up to May 1977. Client Report No. 416. CSIRO FFP, Canberra. p 25

Espinoza SE, Martínez VA, Magni CR et al. (2014) Genetic control of growth, biomass allocation, and survival under drought stress in *Pinus radiata* D.Don seedlings. Tree Genet Genomes 10:1045–1054

Espinoza SE, Magni CR, Santilices RE et al. (2016) Changes in drought tolerance of *Pinus radiata* in Chile associated with provenance and breeding generation. Ann For Sci 73:267–275

Evans R (2006) Wood stiffness by X-ray diffractometry. In: Stokke DD, Groom LH (eds) Characterization of the cellulosic cell wall, Wiley, Hoboken, NJ, USA. pp 138–148

Eyles A, Bonello P, Ganley R et al. (2010) Induced resistance to pests and pathogens in trees. New Phytol 185:893–908

Fallow B (2013) Pace of forest to dairy conversion gathers. New Zealand Herald 15 April 2013

Ferguson I (2014) Australian plantation inventory: ownership changes, availability and policy. Aust For 77(1):1–15

Ferguson I (2015) Forestry R&D in Australia: a response to Kile et al. (2014), Aust For 78 (2):111–113

Fielding JM, Brown AG (1961) Tree-to-tree variations in the health and some effect of superphosphate on the growth and development of Monterey pine on a low quality site. Forestry and Timber Bureau, Australia, Leaflet No. 79. p 19

Ganley RJ, Newcombe G (2006) Fungal endophytes in seeds and needles of *Pinus monticola*. Mycol Res 110:318–327

Ganley RJ, Watt MS, Manning L et al. (2009) A global climatic risk assessment of pitch canker disease. Can J For Res 39:2246–2256

Gapare WJ, Baltunis BS, Ivković M et al. (2009) Genetic correlations among juvenile wood quality and growth traits and implications for selection strategy in *Pinus radiata* D.Don. Ann For Sci 66:606. doi:10.1051/Forest/209044

Gapare WJ, Ivkovic M, Powell MB, McRae TA (2008) Genetics of shrinkage in juvenile trees of *Pinus radiata* D.Don from two test sites in Australia. Silvae Genet 57:145–151

Gapare WJ, Ivković M, Dutkowski GW et al. (2012a) Genetic parameters and provenance variation of *Pinus radiata* D.Don 'Eldridge collection' in Australia 1: growth and form traits. Tree Genet Genomes 8:391–407

Gapare WJ, Ivković M, Dillon S et al. (2012b) Genetic parameters and provenance variation of *Pinus radiata* D.Don 'Eldridge collection' in Australia 2: wood properties. Tree Genet Genomes 8:895–910

Grace LJ, Charity JA, Gresham B et al. (2005) Insect-resistant transgenic *Pinus radiata*. Plant Cell Rep 24:103–111

Hall SS (2016) Editing the mushroom. Sci Am 314(3):48–55

Hargreaves C, Menzies M (2007) Organogenesis and cryopreservation of juvenile radiata pine. Chapter 6. In: Jain SM, Häggman H (eds) Protocols for micropropagation of woody trees and fruits. Springer. pp 51–65

Hargreaves CL, Smith DR, Foggo MN et al. (1997) Cryopreservation of zygotic embryos of *Pinus radiata* and subsequent plant regeneration. In: Burdon & Moore, pp 281–284

Hargreaves CL, Smith DR, Foggo MN et al. (1999) Development of protocols for cryopreservation of zygotic embryos of *Pinus radiata* and subsequent plant regeneration. N Z J For Sci 29:54–63

Hargreaves CL, Reeves CB, Find JI et al. (2009) Improving initiation, genotype capture and family representation in somatic embryogenesis of *Pinus radiata* D.Don by a combination of zygotic embryo maturity, media and explants preparation. Can J For Res 39:1566–1574

Hargreaves CL, Reeves CB, Find JI et al. (2011) Overcoming the challenges of family and genotype representation and early cell line proliferation in somatic embryogenesis from control-pollinated seed of *Pinus radiata*. N Z J For Sci 41:97–114

Hayes J, Andersen C (2007) The scion permanent sample plot (PSP) database system. N Z J For 52(1):31–33

Hodge GR, Dvorak WS (2000) Differential responses of central American and Mexican pine species and *Pinus radiata* to infection by the pitch canker fungus. New For 19:241–258

Hopkins AJM, Dick MA, Carlson CA et al. (2012) Early investigations into the infection courts used by *Nectria fuckeliana* to enter *Pinus radiata* stems. Eur J Plant Pathol 132:537–548

Horgan G (2004) Central North Island forest ownership puzzle unravelled. N Z J For 49(1):33–34

Ivković M, Wu HX, McRae TA et al. (2006) Developing breeding objectives for radiata pine structural wood production. 1. Bioeconomic model and economic weights. Can J For Res 36:2920–2931

Ivković M, Gapare WJ, Albarquez A et al. (2009) Prediction of wood stiffness, strength, and shrinkage in juvenile wood of radiata pine. Wood Sci Technol 43:237–257

Isik F (2014) Genomic selection in forest tree breeding: the concept and an outlook to the future. New For 45:379–401

Jayawickrama KJS, Carson MJ (1990) A breeding strategy for the New Zealand Radiata Pine Breeding Cooperative. Silvae Genet 49:82–90

Johnson IG, Cotterill IM, Raymond CA et al. (2008) Half a century of *Pinus radiata* tree improvement in New South Wales. N Z J For 52(4):7–13

Katz A (2013) A strategic view of New Zealand log exports. N Z J For 58(3):14–20

Kile GA, Nambiar EKS, Brown AG (2014) The rise and fall of research and development for the forest industry in Australia. Aust For 77:142–152

Kimberley M, West G, Dean M et al. (2005) The 300 index – a volume productivity index for radiata pine. N Z J For 50(2):13–18

Kimberley MO, Moore JR, Dungey HS (2015) Quantification of realised genetic gain in radiata pine and its incorporation into growth and yield modelling systems. Can J For Res 45:1676–1687

Krutovsky KV, Troggio M, Brown GR et al. (2006) Conserved ortholog sets in forest trees. Tree Genet Genomes 3:61–70

Kumar S, Burdon RD (2010) Genetic improvement of stiffness in radiata pine: a synthesis of results. N Z J For Sci 40:185–197

Kumar S, Spelman RJ, Garrick DJ et al. (2000) Multiple-marker mapping of wood density loci in an outbred pedigree of radiata pine. Theor Appl Genet 100:926–933

Kumar S, Echt CS, Wilcox PL et al. (2003) Testing for linkage disequilibrium in the New Zealand radiata pine breeding population. Theor Appl Genet 108:292–298

Kumar S, Burdon RD, Stovold GT (2008) Wood properties and stem diameter of *Pinus radiata* in New Zealand: genetic parameter estimates of clonal and seedling material. N Z J For Sci 38:88–101

Kumar S, Cown DJ, Ivković M et al. (2010) Susceptibility to intra-ring checking in *Pinus radiata*: potential for genetic improvement. N Z J For Sci 40:61–70

Lande R, Thompson R (1990) Efficiency of marker-assisted selection in the improvement of quantitative traits. Genetics 121:185–199

Lanfranco DM, Bűchner JF, Aguilar AM et al. (1991) Parasitoidos nativos en el control de la polilla del brote del pino (*Rhyacionia buoliana*): avances en la identifcacíon del complejo y de sus estrategias de desarollo. Bosque 12(1):69–74

Li Y, Suontama M, Burdon RD, Dungey HS (2017) Genotype by environment interactions in forest tree breeding: review of methodology and perspectives on research and application. Tree Genet Genomes 13:60. p 18

Liebenberg F, Bientema NM, Kirsten JF (2004) South Africa. Agricultural science and technology indicators, ASTI country brief no. 14, p 10 <http://www.asti.cgiar.org>

Lottmann J, O'Callaghan M, Baird M, Walter C (2011) Bacterial and fungal communities in the rhizosphere of field-grown genetically modified pine tree (*Pinus radiata* D.Don). Environ Biosafety Res 9:25–40.

Manley B, Maclaren P (2009) Modelling the impact of carbon trading legislation on New Zealand's plantation estate. N Z J For 54(1):39–44

Matheson AC, Raymond CA (1984) The impact of genotype × environment interactions on Australian *Pinus radiata* breeding programs. Aust For Res 14:11–35

Matheson C, Spencer D, Eldridge K (1999) A workshop on issues and strategies to conserve the genetic resources of *Pinus radiata ex situ*. For Genet Res 27:75–78. FAO, Rome

Matheson AC, Devey ME, Gordon TL et al. (2006) Heritability of response to pine pitch canker of seedlings of radiata pine. Aust For 69:101–106

Matheson AC, Gapare WJ, Ilic J et al. (2008) Inheritance and genetic gain in wood stiffness in radiata pine assessed acoustically in young standing trees. Silvae Genet 57:56–64

Mead DJ (2013) Sustainable management of *Pinus radiata* plantations. FAO Forestry Paper No. 170. FAO, Rome, p xv + 246

Menzies MI, Aimers-Halliday J (2005) Propagation options for clonal forestry with conifers. In: Walter C, Carson MJ (eds) Plantation forest biotechnology for the 21st century. Research Signpost, Trivandrum, Kerala, India, pp 255–274

Mitchell RG, Steenkamp ET, Coutinho TA et al. (2011) The pitch canker fungus, *Fusarium circinatum*: implications for South African forestry. South For 73(1):1–13

Monge JJ, Parker WJ, Richardson JW (2016) Integrating forest ecosystem services into the farming landscape: a stochastic economic assessment. J Environ Manage 147:87–99. doi:10.1016/j.jenvman.2016.01.030

Montalbán IA, De Diego N, Moncaleán P (2012) Enhancing initiation and proliferation in radiata pine (*Pinus radiata* D.Don) somatic embryogenesis through seed family screening, zygotic embryo staging and media adjustments. Acta Physiol Plant 34:451–460

Morales R (2009) Detection of *Nectria fuckeliana* in Chile associated to stem cankers and malformations in *Pinus radiata* plantations. Bosque 30:106–110

Morris A (2010) A review of pitch canker fungus (*Fusarium circinatum*) as it relates to plantation forestry in South Africa. SAPPI Forests Research Shaw Research Centre, Howick, South Africa Research Document8/2010. p 35

NEFD (2013) National exotic forest description as at 1 April 2012. Ministry of Primary Industries, Wellington, New Zealand

Neilson DA, Evans J (2009) Policy, institutional and ownership issues. In: Evans J (ed) Planted forests: uses, impacts and sustainability. FAO, Rome and CABI, Wallingford, UK, pp 91–111

NZHC (New Zealand High Court) (2014) Case CIV2013-485-877, decision NZHC 1067, Wellington, para 74

NZIER (New Zealand Institute of Economic Research (Inc.)) (2000) Developing forest ownership through privatisation. Working Paper 2000/3, ISSN 1170 2583, Wellington. p 19

NZIER (New Zealand Institute of Economic Research) (2017). Plantation forestry statistics: contribution of forestry to New Zealand. Report prepared by C. Nixon, D. Gamperle, D. Pambudi and P. Clough. Wellington. p 73.

Productivity Commission (2008) Financial performance of government trading enterprises, 2004–05 to 2006–07. Commission Research Paper, Canberra, July

Ramos SM, Lanfranco Leverton D (2010) El complejo de parasitoides de la polilla del brote del pino en Chile: pasado, presente y posible escenario futuro (The parasitoid complex of the pine shoot moth: past, present and possible future scenario). Bosque 31:1000–1008

Raymond CA (2011) Genotype by environment interaction for *Pinus radiata* in New South Wales, Australia. Tree Genet Genomes 7:819–833

Rodríguez R, Sánchez F, Gorgoso J et al. (2002) Evaluating standard treatment options for *Pinus radiata* plantations in Galicia (north-western Spain). Forestry 75:273–284

Schnitzler FR, Burgess EPJ, Kean AM, Philip BA et al. (2010) No unintended impacts of transgenic pine (*Pinus radiata*) treesin above ground invertebrate communities. Environ Entomol 39:1359–1368

Sharma M, Apiolaza LA, Chauhan S et al. (2015) Ranking very young *Pinus radiata* families for acoustic stiffness and validation by microfibril angle. Ann For Sci 73:393. doi:10.1007/s13595-015-0529-y

Shelbourne CJA, Burdon RD, Carson SD, et al. (1986) Development plan for radiata pine breeding. New Zealand Forest Service, Forest Research Institute, Rotorua. p 142

Shi S, O'Callaghan M, Jones E et al. (2012) Investigation of organic anions in tree root exudates and rhizosphere microbial interactions using in situ and destructive sampling techniques. Plant Soil 359:149–163

Smith M (2012) Campbell group completes purchase of South Australia state forest assets. South Hemisph For Ind J 18(2):5

Sorensson CT, Burdon RD, Cown DJ, et al. (1997) Incorporating spiral grain into the New Zealand radiata pine breeding programme. In: Burdon & Moore, pp 180–191

Stanger TK, Galloway GM, Retief ECL (2011) Final results from a trial to test the effect of plot size on *Eucalyptus* hybrid clonal ranking in coastal Zululand, South Africa. South For 73(3–4):131–135

Sutton WRJ (1999) Wink Sutton reviews CHH's millennium forestry. N Z J For 43(4):4–5

Tsehaye A, Buchanan AH, Walker JCF (1997) Sorting of logs using acoustics. Wood Sci Technol 34:337–344

Voelker SL, Lachenbruch B, Meinzer FC et al. (2011) Reduced wood stiffness and strength, and altered stem form, in young antisense 4CL poplars with reduced lignin contents. New Phytol 189:1096–1109

Wagner A, Ralph J, Akiyama T et al. (2007) Exploring lignifications in conifers by silencing hydroxycinnamomoyl-CoA: shikimate hydroxycinnamomoyltransferase in *Pinus radiata*. Proc Natl Acad Sci USA 104:11856–11861

Wagner A, Donaldson L, Kim H et al. (2009) Suppression of 4-coumarate-CoA ligase in the coniferous genome *Pinus radiata*. Plant Physiol 149:370–383

Wallace B (2001) Battle of the titans: Sir Ronald Trotter, Hugh Fletcher and the rise and fall of Fletcher Challenge. Penguin Books, Auckland, New Zealand, p 256

Walter C, Grace LJ, Wagner A et al. (1998) Stable transformation and regeneration of transgenic plants of *Pinus radiata* D.Don. Plant Cell Rep 17:460–469

Watson M, Withers TM (2012) Buddleia takes a beating. What's New in Biological Control of Weeds 62: 7, Scion, Rotorua, New Zealand

Watson MC, Kriticos DJ, Drayton GM et al. (2008) Assessing the effect of *Essigella californica* on *Pinus radiata* at two sites in New Zealand. N Z Plant Protect 61:179–184

Watt P, Watt M (2011) Applying satellite imagery for forest planning. N Z J For 56(1): 23–25

Watt MS, Adams T, Marshall H et al. (2013) Modelling variation in *Pinus radiata* stem volume and outerwood stress-wave velocity from LiDAR metrics. N Z J For Sci 43:1–7. doi:10.1186/1179-5395-43-1

Wharton TN, Kriticos DJ (2004) The fundamental and realized niche of the Monterey pine aphid, *Essigella californica* (Essig) (Hemiptera: Aphididae): implications for managing softwood plantations in Australia. Divers Distribut 10(4):233–262

White TL, Matheson AC, Cotterill PP et al. (1999) A nucleus breeding plan for radiata pine in Australia. Silvae Genet 48:122–133

White TL, Adams WT, Neale DB (2007) Forest genetics. CABI Publishing, Wallingford and Cambridge, MA, p x + 682

Wilcox PL, Carson SD, Richardson TE et al. (2001) Benefit-cost analysis of DNA marker-based selection in progenies of *Pinus radiata* seed orchard parents. Can J For Res 31:2213–2224

Williams ER, Matheson AC (1994) Experimental design and analysis for use in tree improvement. CSIRO, Victoria, Australia, p 174

Wingfield BD, Coutinho TA, Viljoen A (1998) Pitch canker: a South African perspective. In: Devey et al, pp 62–69

Wu HX, Owen JV, Albarquez A et al. (2004) Inbreeding in *Pinus radiata* – V. The effects of inbreeding on fecundity. Silvae Genet 53:80–87

Wu HX, Eldridge KG, Matheson AC et al. (2007a) Achievements on forest tree improvement in Australia and New Zealand 8: Successful introduction and breeding of radiata pine in Australia. Aust For 70:215–225

Wu HX, Powell MB, Yang JL et al. (2007b) Efficiency of early selection for rotation-aged wood quality traits in radiata pine. Ann For Sci 64:1–9

Wu HX, Ivković M, Gapare WJ et al. (2008) Breeding for wood quality and profit in *Pinus radiata*: a review of genetic parameter estimates and implications for breeding and deployment. N Z J For Sci 38:56–87

Yao RT, Harrison DR, Harnett M (2017) The broader benefits provided by New Zealand's forests. N Z J For 61(4):7–15

Zas R, Serrada R (2003) Foliar nutrient status and nutritional relationships of young *Pinus radiata* D.Don plantations in northwest Spain. For Ecol Manag 174:167–176

Chapter 8
In Retrospect

8.1 Amenability of Radiata to Domestication

For domestication, a forest-tree species will almost certainly need to grow rapidly, to give returns on outlays soon enough to represent an acceptable rate of return on capital. The species must also be able to produce wood of acceptable quality when grown rapidly and harvested relatively young, and to produce stems of form that allows easy harvesting and processing and the production of satisfactory wood products. Ease of propagation and tolerance of transplanting are important, although those features may be much enhanced by improved nursery- and field-establishment practice. Site tolerances generally need to be broad, largely because of the pervasive demands on high-quality land for other, more exacting land uses. Ability to provide environmental services, notably erosion control and maintenance of water quality, can represent an important bonus. In general, radiata meets these requirements to at least a large degree, despite the limitations in its site tolerances and in the purposes for which its wood can used satisfactorily. Actually, while radiata is basically a utility timber, the uses to which it is suited account for a very high proportion of total wood consumption. Moreover, it can provide appearance-grade products of quite high value.

Of note has been the commercial acceptance of the species' wood in the absence of significant commercialisation of wood from its native stands, which doubtless made commercial acceptance more gradual than it would otherwise have been.

The very restricted natural range, which reflects an extremely localised habitat niche, contrasts sharply with the scale on which radiata is now grown internationally. To a large extent that reflects an accident of plant geography whereby *Pinus* remained a Northern Hemisphere genus despite it being superbly adapted to large areas in the Southern Hemisphere.

That radiata is grown almost entirely as an exotic species has caused concerns. In fact, its preferred status over large areas has survived a number of alarms over

R. Burdon et al., *Domestication of Radiata Pine*, Forestry Sciences,
DOI 10.1007/978-3-319-65018-0_8

fungal diseases and insect pest outbreaks, and in some cases has even been enhanced. However, some of these biotic events have led to contractions of the regions where radiata is planted, and remain a risk factor to be weighed up and accommodated in strategic planning.

8.2 Different Drivers of Planting in Different Countries

While radiata has been grown successfully in several different countries, the drivers of its planting, and the trajectories of domestication, have differed markedly among the countries, although individual enthusiasts were influential in all the main grower countries.

In Australia a regional timber shortage in South Australia drove early domestication in the late 19th century, but much later a major and more widespread surge in planting from the end of the 1960s was prompted by a national perception of a need for a much larger domestic supply of softwood timber.

In New Zealand, the domestication began largely in planting for shelter and firewood. Early concerns, in the late 19th century, over future timber supplies generated sporadic political responses; however, it was only from 1913 onwards that a need to replace rapidly dwindling supplies of native timber dominated policy and action. A crash programme of planting by the state from the mid-1920s was largely matched by a private investment boom which was even more focussed on radiata. Then, towards the end of the 1960s, another boom began, driven partly by government encouragement to broaden the national export base beyond agricultural products. That boom ended in the mid-1980s, when government incentives were abruptly terminated. After a comparative lull came a third, intense boom driven by high log prices, but which was short-lived, peaking in the late 1990s.

In South Africa the paucity of native timber supplies was a strong driver of exotic plantation establishment which, however, only began on a large scale after 1910. Planting of radiata was tempered, by around 1930, by increasingly evident issues of species adaptation. In recent years planting has been further constrained by concerns over invasiveness of pines and other exotic tree species in iconic native vegetation, and by the threat posed by pitch canker.

In Chile the exposure of the economy to the vagaries of the world copper market was a prime driver, along with a strong perception of a comparative (and absolute) advantage for growing radiata, a need for retirement and rehabilitation of degraded farmland, and some regional unemployment problems. Large-scale planting began towards 1940.

In northern Spain there was a combination of widespread shortages of sawlogs, fragmented land holdings favouring part-time farming, the outstanding growth rate compared with native tree species, and the site tolerances and generally good tree form. Planting increased steadily from around 1915.

8.3 Commercial Forestry as a Business Model

Commercial forestry represents a vexing business model, at least in temperate climates and on the sort of land that is typically not pre-empted for more demanding land uses, a scenario that is precisely the context for growing radiata. At least in the early days, the properties of the plantation-grown timber crops were highly suspect. Added to that, the time lags between investment and returns have often meant that market signals have provided little or no reliable guide, especially when timber markets were still dominated by the availability of the remaining virgin timber from natural forests. Moreover, the technology for growing plantation crops was often far from fully refined, as was the optimisation of financial returns even if good market information was available.

In that context, it has typically been seen as a role of government to engage directly in afforestation or to incentivise private planting by tax breaks and/or direct grants. Other drivers of state involvement have been desires to create employment and/or to engage in soil conservation and land rehabilitation. A notable exception was the private planting boom around 1925–1935 in the pumicelands of New Zealand, where an unsolved animal micronutrient problem made good-quality land essentially valueless for pastoral farming.

Forestry ventures also pose their own problems for drawing up taxation rules. In such ventures, the usual convenient accounting distinction between operating costs and capital investments becomes blurred. The resulting complications can come to a head in trading forest assets, which has generated elaborate taxation rules in New Zealand but which have defied efforts to avoid all manifest anomalies (Bilek 1999; Manley 2002; Moore 2016). An associated complication is that growing crops of trees represent an important element of capital, which over time assumes the character of saleable stock as the annual value gain of a crop represents a declining rate of return on its existing asset value.

Incentivising forestry by governments is now technically favoured by the desirability of carbon sequestration in standing forest and favouring the use of wood as a building material, the latter to both burn less fossil fuel in producing building materials and achieve sequestration through wood in service. Political and technical management of greenhouse gas emissions and sequestration, however, has often been questionable, to put it mildly.

Incentives created by governments can produce various unintended consequences, notably investment booms that are driven much more by subsidies and tax breaks than by the technical and economic realities, for example, the largely ill-fated boom in planting Tasmanian blue gum in Australia (Ferguson 2013).

Some of the state support for afforestation and tending has been in the nature of "picking winners." This has certainly been the case in Chile, where planting of radiata has figured very prominently, and was done in a context of extreme exposure to the world copper market, along with other factors. To a degree "picking a winner" figured in New Zealand's second forestry boom, but that was alongside a concern over the country's exposure to the world markets for agricultural products.

In recent years, picking winners has become anathematised among neoliberal economic circles that look to market signals for almost all guidance. Unlike New Zealand in recent years, Chile never faltered in state support for forestry, and that must surely stand as a big success in picking winners. There, the government incentives have worked well in the long term.

8.4 Role of Radiata in Development of Plantation Forestry

With the widespread exhaustion of virgin forests as timber supplies, plantation forestry has become an imperative. While domestication of forest trees in the form of plantations with clearcut harvesting systems faces much public opposition, such systems often mimic quite closely the natural regeneration ecology of the species concerned. Being eminently suited to plantation forestry, radiata has naturally figured very prominently in the development and refinement of modern plantation management systems. Likewise, radiata has figured prominently in developing strategy, methodology and technology for tree breeding. Coming much later than the early large-scale plantings, tree breeding is the other main plank of domestication alongside the development of plantation management inputs.

Despite its inherent suitability for plantation forestry, radiata has posed considerable challenges in economic optimisation of both growing and wood processing and utilisation, with traditional practices developed in Europe being exposed as financially inefficient for its growing. With future pressures to intensify production, improve economic returns and yet achieve environmental sustainability, stern challenges will remain for both forest managers and tree breeders.

8.4.1 Forest Management Systems

8.4.1.1 Establishment Practices and Nursery Systems

Natural regeneration, while often occurring spontaneously, has fallen from favour, it giving erratic results and limited control over the genetic make-up of crops. That gave an impetus for renewed efforts to refine nursery practices, which was the subject of intensive research in New Zealand in the 1970s. Alongside that came parallel efforts in site preparation, and post-planting tending. There, the benefits of improved nursery systems and field establishment practices have included significant expansion of the areas where the species can be grown successfully, often by mitigating harsh ground-level microclimates that created establishment problems. In Australia, site preparation was often greatly modified, notably by ceasing the burning of logging residues, to avert productivity declines after harvesting the preceding rotation of pines.

Radiata is very easily transplanted as bare-rooted stock, which was for long the focus of improving nursery and establishment practices. In recent years, however, the use of container-grown stock has found some favour for maximising the benefits of very intensive genetic-improvement measures.

8.4.1.2 Use of Fertilisers

In traditional forestry, use of soil fertilisers was largely eschewed. For radiata, however, fertiliser use has caused some dramatic expansion of the land areas where it can be grown successfully and has since become the preferred species. This has certainly occurred in Australia from the 1930s, beginning with the recognition and correction of zinc deficiency and later with widespread correction of phosphorus deficiency; in New Zealand from the 1950s with phosphorus deficiency; later in South Africa with phosphorus deficiency; and in Chile from the 1980s with large but less decisive benefits. Adapting technology from agricultural aviation in New Zealand was very important. Indeed, the success with fertilisers in growing radiata has doubtless contributed to the wider acceptance of using fertilisers in established plantation forests.

8.4.1.3 The Tending Regime Conundrum

The traditional approach of thinning to produce intermediate yields of wood has posed problems for forest managers. Waiting till thinnings represented trees that were big enough to be harvested at an immediate profit was long the accepted practice. Often, however, immediate profits were not to be made from such an operation. And, even if such profits could be made, waiting till the thinning could give a net current return could incur cryptic opportunity costs that would severely compromise the value of the final crop and thence the total economics of the plantations. It became very clear that delays in thinnings of radiata were to be avoided—a lesson learnt in South Africa in the 1930s and strongly reinforced in New Zealand during the 1960s—but forest owners and managers often held on in futile hopes of profitable first thinnings. Thus, from the early 1970s, the practice of heavy pre-commercial thinnings became very widespread in New Zealand where markets for thinnings were poor, terrain often difficult and extraction costs high. Indeed, there was widespread adoption of the Direct Sawlog silvicultural regime which included no commercial thinnings. In the other grower countries, where conditions were often very different and attitudes conservative, this radical departure from tradition tended not to be followed. In Australia, where policy often mandated maximum wood yields, commercial thinnings tended to be retained and final harvests delayed.

Pruning of lower branches to produce knot-free timber had major attractions in radiata, with its rapid growth allowing quick occlusion of pruning scars, and its wood being very suitable for appearance-grade uses. Often, however, pruning operations were ill-timed, but in the late 1950s and the 1960s in New Zealand pruning schedules were developed that integrated tightly with the Direct Sawlog thinning regime, and well-timed pruning became very widespread practice in New Zealand plantations from then on. However, the amount of pruning has declined substantially since around 2000 in the face of cost pressures and market uncertainties. In Australia and Chile, where there was greater emphasis on producing structural timber and/or pulpwood, pruning never became nearly as widespread as in New Zealand.

Aggressive silvicultural regimes, with much lower than traditional stockings, often proved to be overdone in New Zealand, with disappointing total wood yields, excessively large branches and some poor wood quality. Causes were various, but in recent years there has often been a shift towards more conservative regimes. There were pervasive pressures to shorten rotations in order to increase rates of financial return. The pressures were especially strong in New Zealand after privatisation of forest assets from the 1990s. The research on silvicultural regimes in New Zealand during the 1960s and 1970s, however, was already strongly oriented towards financial rates of return. Later on some forest owners and managers overreached themselves, failing to achieve makeable returns through pursuing rates that proved unrealistic.

From the 1970s large areas of pasture land in New Zealand and Australia were converted into radiata plantation, but the elevated soil fertility from pastoral farming led to problems with tree form—especially with genetically unimproved stock—and wood properties.

8.4.1.4 Modelling of Growth and Outturns

The research that led to the development in New Zealand of "radical" silvicultural regimes for radiata, although extremely influential, was only a beginning. It was followed up in the development of decision-aid tools for forest planning. Yield modelling had been proceeding, in the case of New Zealand, on the basis of data from a national network of regularly measured sample plots. From 1979, however, this work was intensified with modelling to predict not only yields but more detailed crop outturns under alternative management regimes. Such models have been subject to progressive—and still-continuing—refinement over the years to give more accurate and user-friendly software packages for forest planners to make management decisions and predict yields and properties of forest outturns. The models need to accommodate regional differences in the behaviour of radiata and progressive changes over time in the genetic make-up of the crops. They are informed by the results of a very wide range of research, not only on measured sample plots, but on many aspects of the species' biology.

8.4.1.5 Scope for Ongoing Refinements

The very existence of large areas of intensively managed radiata means that they have become the context for refinements based on new and ever more sophisticated management technology ("precision forestry"). Remote-sensing technologies continue to become increasingly valuable for forest management, in monitoring tree growth and health, and for predicting outturns as a basis for management planning. Developments include global positioning systems, other navigational aids, video systems and robotics to make harvesting more efficient and safer, while recording of wood outturns is being increasingly automated.

8.4.2 Genetic Improvement

8.4.2.1 The Needs and Opportunities

That forest trees are still essentially wild organisms means that natural genetic variation is available for breeders to exploit. At the same time, practices of convenience for seed collection—or propagation by cuttings—are often conducive to dysgenic selection which tends to progressively lower the genetic quality of the growing stock. Thus, even to prevent actual genetic deterioration, active eugenic measures are typically needed. Moreover, plantation forestry provides special opportunities to manipulate the genetic quality of the growing stock, for better or for worse. Furthermore, genetic variation among individual trees can be more strongly expressed in early-harvested plantations than in virgin forest.

In past history there were examples of local or sporadic interest in capitalising on genetic variation in forest trees. Early in the 20th century genetic improvement of poplars was actively pursued, based on the combination of rooting cuttings being the method of choice for propagation and the fact that interspecies hybrids tended to show superior vigour. Monoclonal plantations of poplars were easy to produce and manage, but were prone to disease outbreaks in which every tree would get badly affected.

The development of genetics as a science in its own right, with the neo-Darwinian synthesis of the 1920s reconciling the discrete nature of genes found by Mendel with the typically continuous variation among individuals, provided a platform for rigorous breeding methodology. The resulting quantitative genetic model set a framework within which the results of genetic experiments could be analysed informatively and the results of alternative breeding measures could be predicted and compared. This model is a process model, and while only a crude approximation to reality, it has proved remarkably powerful and robust in its predictions. It was also eminently suited to tree breeding, in which individuals could be compared through progeny performance only by collection and analysis of voluminous data.

Large-scale, systematic tree breeding programmes began in the early 1950s, following the "Scandinavian model." It began by very intensive selection of "plus" trees in natural stands or plantations, and these were propagated by grafting into seed orchards for producing genetically superior seed, while progeny tests could allow the parent clones to be culled in a measure called roguing. For future progress various options were available and needed to be explored by theory and practical experience.

8.4.2.2 Breeding of Radiata

Radiata has been an outstanding subject for genetic improvement. Despite its virtues, it often showed a glaring need for genetic improvement, in respect of tree form and health. Individual trees varied widely in a number of respects, and there were strong visual indications that much of this variation was genetic. These indications have been well confirmed in field genetic trials. Intensive breeding of radiata began in the 1950s in Australia, New Zealand and South Africa, but much later in Chile (late 1970s, after an abortive start a few years earlier).

A special circumstance was the existence of only five, discrete natural populations of radiata. By good fortune, the Californian origins of the exotic plantation stocks (land races) were two generally well-adapted native populations, and so represented a good starting point for intensive breeding.

Despite setbacks, and learning from mistakes, the seed orchards did produce genetically improved seed, meeting New Zealand's planting requirements by around 1986. Australia was approaching that situation by 1983, when several seed orchards were lost in catastrophic fires in South Australia. Chile benefited greatly from New Zealand experience in the siting and management of seed orchards, so the companies were largely meeting their planting requirements from seed orchards by the early 1990s. South Africa had only modest planting programmes with radiata, making seed-orchard production less of an issue. Spain had problems finding satisfactory seed orchard sites, and radiata breeding languished in the Basque Autonomous Region. An upsurge of plantings of radiata in Galicia around 2000 was done mainly with surplus orchard seed from New Zealand.

Selection traits for radiata were health, good vigour and lack of malformation as prerequisites. Beyond that, varying emphasis was placed on growth rate and branching behaviour, but for many years little real call generally came from industry for selecting for wood properties. Later, resistance to foliage diseases has been selected for strongly in order to serve some regions.

From operational progeny testing, and more fundamental population studies, much had been learned by 1980 about the heritabilities of various traits and how those traits are intercorrelated. That confirmed how best to go about improving various traits and indicated what multi-trait breeding goals are realistic.

Intensive selection generally gave strong responses, with big improvements in growth rates and/or tree form according to the emphasis in selection. Systematic genetic gain trials began in New Zealand in 1978, to monitor and quantify genetic improvement, and later to provide a basis for predicting crop outturns from genetic improvement.

While synergies between management inputs and genetic improvement have not yet matched those with various agricultural crops and domesticated animals, the improved tree form meant that fewer trees needed to be planted per hectare, with flow-on savings. Also, the scarcity of seed of the very best genetic quality provided an additional incentive to obtain better yields of planting stock from improved nursery practice.

Capturing immediate genetic gain without eroding the genetic variability that would be needed for continuing genetic gain in changed breeding goals poses a big strategic challenge for tree breeding. The time frames involved mean that strategic errors cannot be corrected quickly or easily, a consideration that led tree breeding strategy to become a discipline in its own right from the late 1960s, with radiata breeding figuring prominently in this process. By the early 1970s a scheme had been devised for organising genetic material into a hierarchy of populations designed to meet both short- and long-term needs for genetic improvement.

As a complement to operational breeding based on local, land-race stocks, material from the entire natural range of radiata was tested together to compare the five native populations for growth, adaptation and quality traits, and major seed collections were made from native stands in 1968. Native-population material tends to be inferior in growth rate and bole form to the products of breeding programmes, but its conservation and exploitation would likely confer long-term benefits in adaptation and technical properties. Managing the trade-off between level of genetic improvement and retention of genetic diversity was always going to be a challenge. However, after widespread privatisation of radiata forest, forest owners have generally been reluctant to bear the costs of seriously addressing this challenge.

Recourse to wild genetic material by classical breeding, as has been practised with various domesticated plant crops, tends to be very slow with forest trees, especially if there are threats of disease transfer, placing a greater premium on forward preparation. In principle, there is now the gene technology to transfer specific genes from wild into domesticated material, but there remain the challenges of identifying the requisite genes, especially if they are needed not just individually but in specific combinations.

Breeding goals have evolved over time, in response to biotic events, changed markets, impacts of new management practices, and improved techniques for assaying certain traits, notably several wood properties. From the mid-1990s the emphasis on genetic improvement of wood properties has much increased in the three main grower countries.

8.4.2.3 The Evolving Role of Radiata Breeding

With the evolution of breeding goals has come an evolution of the role of genetic improvement in radiata. To a degree, breeding of radiata has become a means of making good the adverse effects on tree form and wood properties that result from crop-management practices designed to boost productivity and reduce optimal harvest age. Such practices include the use of fertilisers to boost site productivity and adoption of reduced stockings to accelerate stem diameter growth and achieve log-size specifications at earlier ages. Indeed, while selective breeding can certainly boost productivity, the main improvements in productivity often arise from management inputs.

Self-sufficiency for genetically improved planting stock from seed orchards, which was achieved in New Zealand in 1986, meant that the huge secondary gene resources in genetic unimproved populations were subject to progressive attrition. So at about this time a major selection operation was mounted in the remaining "land-race" stands with 1000 plus trees selected afresh. Similar operations were also mounted in Australia.

8.4.2.4 Towards Advanced Delivery Systems for Genetic Gain

Before seed orchards came into seed production, collection of seed from the best trees in plantations had offered preliminary genetic improvement. Then the original seed orchards came into full seed production and superseded the select-tree seed collections. However, while the seed orchards eventually delivered, they were a cumbersome system. They also gave very limited control over which parents mated with which. From the late 1970s alternative orchard systems were explored in New Zealand and Western Australia which entailed better control of pollination. In the mid-1980s, using a much better seed-producing site, alternative seed orchard systems were developed in New Zealand which produced seed sooner and with controlled pollination. As a back-up option, progeny tests were generally available as seedling seed orchards.

Vegetative propagation, using grafts and then sometimes cuttings, was used from the outset for creating clonal seed orchards. Long-neglected, however, was the potential of easy propagation of cuttings from young seedlings. First practised with radiata in the late 1830s, its feasibility was rediscovered in Australia and New Zealand in the 1930s. But only from around 1980 did its potential start to be seriously exploited. Using it to multiply up seedlings it could be used to "extend" pilot quantities of seed of the very best genetic quality, thereby capturing genetic gain with less lead time. Its use burgeoned in Australia from the mid-1980s, a bit later in New Zealand, and from the mid-1990s in Chile.

Continued developments in the technologies of vegetative propagation, along with new breeding-orchard technologies, had far-reaching effects. Multi-trait capture of genetic gains in planting stocks thereby came faster. More precise targeting of both traits and parents, through more-informed choice of which parents to use for large-scale propagation, and more appropriate deployment of breeds and clones to sites and tending regimes, enhanced both speed and amounts of captured genetic gain. Much better control of maturation of vegetative tissue was offered by successful cryopreservation, but ideal control still remains elusive, even to 2017. Even so, cryopreservation has brought advances towards allowing full clonal forestry, which would allow the option of finding, cloning and deploying a few fortuitous offspring among progeny that combined traits it would take many generations to incorporate reliably into breeds. More radically, the ability to rely entirely on controlled crossing for producing planting stocks not only allowed better-targeted pair-crossing of parents, but it could also revolutionise breeding-population management through making possible two-subline systems.

8.4.2.5 Roles of Modern Biotechnology

Among conifers, radiata has proved comparatively amenable to *in-vitro* propagation technology, albeit less so than spruces (*Picea* spp.). Development of *in-vitro* propagation was achieved for radiata by around 1980, and its operational use was pursued by a company in New Zealand from 1983. While tissue-culture plantlets could be raised for field planting, they could also be used as a preliminary multiplication phase before mass-propagation as nursery cuttings.

Genomic selection, using DNA sequences instead of or in conjunction with field-performance data, potentially allows very early selection for various attributes, which is highly attractive for a long-lived species with delayed expression of certain traits. It has been explored from the mid-1990s for radiata. So far, a combination of factors has prevented operational use, but it is now being explored with renewed vigour. Continued advances in gene technology should eventually allow practical application, with disease resistance being an area of notable promise. However, DNA markers are now important for quality control, verifying identity of clones and seed parentage, and should eventually allow pedigree reconstruction which could save the labour and costs of controlled pollination. Inability to induce super-early flowering, as part of "flowering on command and command only" poses a major barrier to taking full advantage of genomic selection.

Genetic engineering can confer specific attributes that are missing in a species, without introducing large numbers of unwanted and unidentified genes. From the late 1990s its feasibility with radiata has been demonstrated in New Zealand, with successful incorporation of herbicide resistance. Further research and eventual application, however, have remained subject to very stringent regulatory restrictions. These restrictions have greatly slowed progress towards operational development, and have indirectly led to expensive research to verify biological safety.

8.5 Pursuit of Clonal Forestry

Clonal forestry, based on mass commercial propagation of proven clones, offers benefits in potential genetic gains and avoidance of unwanted genetic variability for more precisely targeted deployment and management of crops. Radiata could be propagated easily enough as rooted cuttings to offer promise of achieving clonal forestry. This was first pursued seriously in Australia in the late 1950s and in New Zealand in 1966. Its effective implementation, however, has long been frustrated by the phenomenon of maturation, whereby after a few years clones lose the potential for cheap and easy propagation, although maturation does benefit tree form. While maturation can be slowed by hedging seedlings close to ground level it cannot be halted or reversed thereby. And it can proceed even in *in-vitro* culture, sometimes quite rapidly. Thus, while some degree of clonal forestry could be practised from the early 1980s, it was very imperfect. Cryopreservation technology, which was developed for radiata in the last 10 years, has allowed maturation to be halted, which means that clones can continue to be used indefinitely. Since 2000 there has been renewed activity on commercial development of clonal forestry, accompanied by some shifts to using container-grown rather than bare-rooted planting stock. Reversal of maturation, without having to involve seed production which incurs unwanted genetic variation, would give much more flexibility, but that goal has remained elusive.

8.6 The Wood-Quality Issue

The properties of plantation-grown radiata wood have created ongoing problems, with poor stiffness and dimensional stability in solid-wood products. These problems involve interplay between management regimes and genetic improvement. They have been much exacerbated by forest growers and managers reducing harvest ages in striving for higher economic returns. For most purposes, young trees produce much inferior wood compared with older ones, yet this has tended not to be reflected in timely price signals and therefore has often been inadequately factored in during attempts to optimise tending schedules and harvest ages. The tree breeders, while keenly aware of coming problems with wood properties, were handicapped by lack of guidance from industry; moreover, they did not foresee the extent of the decreases in harvest ages and the full effects on wood properties of the associated tending regimes. Nor were the breeders encouraged by primitive and expensive technology that was available in the past for assaying wood properties. These problems were most acute in New Zealand, where soil fertility tended to be higher and silvicultural regimes more aggressive. A recent focus on genetic improvement of wood properties, in Australia, New Zealand and Chile, has been helped by much better assay technology, although a close watch for unintended consequences of genetic shifts in wood properties will be needed. Achieving good wood properties while greatly increasing wood production through boosting soil fertility will pose ongoing challenges. Actually, the problems with wood properties of radiata have been closely paralleled in USA in plantations of loblolly pine which are much more extensive than those of radiata although producing wood less intensively.

Despite the problems encountered, radiata generally compares well for wood quality with other site-tolerant tree species grown to similar ages.

8.7 Influence of Political Structures

The story in radiata's main grower countries has inevitably been influenced by political structures. Australia was a set of essentially independent colonies until it formally became a Commonwealth at the beginning of 1901, but it retained an important legacy in the states with their own governments and separate forest services, and thence a longstanding fragmentation of tree breeding work. It also created an often-fraught divide between federal research agencies and the individual state agencies. New Zealand abandoned a federal structure early on, in 1876, and Chile never had a federal structure. In South Africa, radiata is now grown just in one province, and after privatisation of plantations in 1993 the government structure has had little relevance. In northern Spain, the regional government structure would have had only a very secondary influence on the location of radiata plantations, but an important role in the R&D.

8.8 Forest Ownership and Institutional Structures

Afforestation initiatives in Australia, New Zealand and South Africa were taken largely by governments establishing the forest plantations, but were supplemented by some booms in private investment, especially one in New Zealand around 1925–1935. In Australia, private afforestation, mostly with radiata, was undertaken by wood processors who wanted to secure wood supplies for the future. In Chile the initial planting boom, beginning in the 1940s, was implemented essentially by private interests who responded to generous government incentives.

In New Zealand, and to a lesser extent Australia, governments were heavily involved in establishing industrial plant for processing radiata wood, a matter of urgency after the effective moratorium imposed by World War II.

In Chile, one private company pioneered the large-scale processing of wood from radiata plantations, but apart from that, Chile's development of industrial processing plant depended greatly on the combined efforts of its own government and FAO. After an interlude with an avowedly socialist government in the early 1970s, with much nationalisation of forest and processing plant, privatisation began in the late 1970s. Although neoliberal economics took hold there in the late 1970s, successive governments remained steadfast in active support of the forestry sector. Added to that, two Chilean family business empires became increasingly dominant, effectively insulating the forestry sector from foreign control. Investment has continued strongly in industrial plant, with capital mobilised to build new and expanded pulping and sawmilling plant, and the forestry sector has remained firmly committed to vertical integration of forest growing and wood processing and marketing. While the state largely exited forest research in the late 1970s, the strength of the main forestry corporates has allowed a resurgence of research in the universities and university/industry consortia. Oddly, domestic use of wood as a building material has never burgeoned as might be appropriate for such an earthquake-prone country. Moreover, the focus on pulping has contributed to a big expansion of the eucalypt plantation estate.

In New Zealand and Australia the fashion for neoliberal economics took hold from the mid-1980s, and it has had much greater impact on the forestry sectors there. After initial corporatisation, the eventual privatisation of state-owned forest has eventually led to globalisation, with overseas parties exercising controlling interests over much of the forest estate. Privatisation was also followed by volatility in forest ownership, with numerous sales of large forest assets, often with disruptive changes and reorganisation of management personnel. Investment in new or renovated processing plant has mostly languished, especially in New Zealand where a dairying boom has led to erosion of the export competitiveness of timber products—not helped by overseas tariffs on sawn timber—leaving logs as the main export line. The resulting shortage of sawmill residues is now stifling the revitalisation and expansion of chemical pulp production which is potentially profitable. In Australia, the waning of industry support for forestry and forest products research (Kile et al. 2014) has even led the government to close its own research establishment for those areas. Vertical integration of the sector has been largely abandoned in both countries. The volatility of forest ownership, different types of forest owner, and abandonment of vertical integration all parallel some overseas developments in commercial forestry, especially in North America.

In South Africa privatisation of state forests, including radiata plantations, occurred in 1993 and, much as in New Zealand, has been followed by ownership and institutional changes. Increasing trouble with pitch canker since around 2000 has reduced the commitment to radiata.

In Spain, pine plantation ownership has remained mostly fragmented, with fluctuating support from regional government, and market structures for logs have remained largely haphazard.

Institutions carrying out radiata breeding have changed with time, not in close parallel with changes in forest ownership. Almost all operational breeding of radiata actually began within government research agencies, in an unusual mix of research and operations, to be later separated off from research "providers." A separation of the breeding and research came early in Chile with the breeding Convenio in 1976, although that was based at a university. Such separation began in Australia with the creation in 1983 of the STBA, which was immediately given strong impetus by a seed-supply crisis after catastrophic fires. In New Zealand the separation began in 1987 with the disestablishment of the Forest Service and the institution of partial "user pays" for tree breeding, but its completion was slow and sometimes tortuous. The separation in South Africa came relatively late. Changes did not all work smoothly. With existing systems it is often easy to recognise their disadvantages. In considering alternative systems, it is so often much easier to see advantages than too foresee disadvantages that often emerge upon implementation. Also, new systems can easily be devised with ulterior purposes in mind that may not be served without compromising the objectives of the systems that they replace.

8.9 Significance for Economic and Environmental Sustainability

Plantations of radiata represent a readily renewable resource. They can be easily regenerated, and will grow well with modest management inputs, especially as their nutrient requirements are frugal compared with those of most crops. As such they have a high degree of economic sustainability.

Plantation forests are pre-eminently commercial, and although they are often decried as ecological deserts they have often performed very positive environmental roles. On a wider scale, their presence has relieved certain pressures for destructive exploitation of natural forests. This happened notably within New Zealand from the 1960s, but it also operates globally, according to what the pressures are on natural forests in what countries. Plantation forests can have their own amenity values, being used for a range of recreational activities, and are known to provide surprisingly good habitat for native plant and animal species. Their benefits for soil conservation, mitigating flood peaks and maintaining water quality are generally substantial. By producing wood, plantations can play a two-fold role in countering global warming: substituting for building materials that require large fossil fuel inputs to produce, and sequestering carbon in both standing crops and wood in service. In all these respects radiata has played a significant part towards environmental sustainability.

Many individual enthusiasts, scientists and foresters have contributed to the enormous advances in plantation forestry over the last half-century or so that we have described. Knowledge gained through research and experience with radiata is now being globally applied, notably in temperate and sub-tropical regions with eucalypts. Radiata, repeatedly and accurately described by Jack Fielding as a "marvellous tree", has had a conspicuous role, directly and indirectly, in efforts to cater for the needs of a rapidly burgeoning global population facing what have been described as "humanity's ten greatest challenges" (Cribb 2017).

References

Bilek EM (1999) Forest tax policy: the cost of bush and forest valuation in New Zealand, or it ain't perfect yet. N Z J For 44(2):19–21

Cribb J (2017) Surviving the 21st century: humanity's ten greatest challenges and how we can overcome them. Springer, Switzerland, p. 253 http://www.springer.com/us/book/9783319412696

Ferguson I (2013) Australian plantation inventory: ownership changes, availability and policy. Aust For 76(3–4):183–193

Kile GA, Nambiar EKS, Brown AG (2014) The rise and fall of research and development for the forest industry in Australia. Aust For 77:142–152

Manley B (2002) Relationship between discount rates to be applied to before-tax and after-tax cashflows. N Z J For 46(1):28–32

Moore H (2016) Options for changing the cost of standing timber provisions. N Z Tree Grow 30 (2):18–28

Chapter 9
The Future

9.1 Introduction

Much has been achieved in the domestication of radiata, it being among the most domesticated of all forest trees. A broad array of technologies has contributed in both growing radiata as an intensively managed crop and achieving genetic improvement, the two main planks of domestication. Refinements in the technology for processing and utilisation of the wood, however, are also part of the broader issues of domestication. Such refinements may obviate the need for improving certain wood properties by genetics or management inputs. On the other hand, the same refinements may help to identify further genetic improvements that need to be made. In this context, much of the further progress in domestication is likely to involve genetic improvement.

Radiata plantations have also provided the context for many technological developments in plantation forestry. Those developments have been in fields that include management and decision-aid systems, harvesting technology and ergonomic research. Radiata will doubtless feature in further developments. That said, we do not address the full gamut of ramifications.

So far, breeding has served in part to mitigate the ill-effects of intensive management practices designed to both boost yields and otherwise improve profitability by reducing harvest ages. In the longer term, the significance of breeding for this role and for mitigating the impacts of other factors that detract from the yield or value of the crop will presumably increase. The identity of these factors will vary with location and site. They may include new pests and diseases, climate change, either intensive fertilizer use or nutrient depletion, and adverse physical changes in soils. That much forest plantation will doubtless occupy inherently poor soils is relevant; lapses in management can quickly squander the results of decades of careful husbandry (e.g. O'Hehir and Nambiar 2010).

Despite the future importance of continued genetic improvement, replenishment of nutrient resources will remain an issue. Fertilizer will often be needed to boost

R. Burdon et al., *Domestication of Radiata Pine*, Forestry Sciences,
DOI 10.1007/978-3-319-65018-0_9

or maintain yields on finite land areas. Depletion of some nutrients can easily be made good by a combination of minimizing losses, geological weathering, and replacement with fertilizers. Nitrogen and phosphorus, however, remain problematic. Nitrogen fertilizer will tend to be expensive because of its high energy requirements. Significant N fixation (6–24 kg ha^{-1} yr^{-1}) can occur in radiata plantations in the absence of leguminous ground vegetation (Turner and Lambert 2011), but this will depend on general soil fertility and adequate humus. Depending on leguminous vegetation to fix nitrogen will require light which in turn means less than full site occupation by the pines. Recent, unpublished work indicates that applying nitrogen to mid-rotation stands can be more cost-efficient than was believed, possibly with little downside for wood properties, while leaching of the nitrogen into waterways should not be important in such stands. Phosphorus fertiliser may become scarce in future (Vaccari 2009; Ashley et al. 2011), which is a worry because radiata is relatively phosphorus-demanding among pines. Yet the phosphorus demands for radiata crops are likely to be very minor compared with those for food crops and pasture. Moreover, genetic variation within radiata in phosphorus demands could presumably be exploited if need be. Furthermore, judicious use of phosphorus fertilizer for radiata should not lead to troublesome eutrophication in streams and lakes.

However, for both setting the agenda for genetic improvement and managing forest estates, prediction of crop performance is going to be very important. Such predictions will in turn depend on modeling growth and wood properties through to harvest age. Such modeling is admittedly well advanced, but there remains much scope for refinement, especially as the genetic makeup of plantations will be subject to continuing change.

Before considering what future developments in the domestication might be, it is appropriate to take more detailed stock of the present state of domestication. In important respects it is still very incomplete, which poses plenty of remaining challenges and opportunities. Let us, then, look at the key respects in which radiata is still essentially a wild organism.

9.2 Domestication Gaps and their Implications

Forest trees in general are genetically in essentially a wild state. With limited exceptions, their natural variability is essentially intact, and most species are resistant to management under the inbreeding regimes that are needed to produce genetic uniformity in various major crop plants. However, where clonal replication is an option, it can avoid unwanted variation arising from the genetic segregation that occurs during seed production. Provided an infrastructure of a breeding population is maintained and properly managed, it is possible to both retain genetic diversity and yet enjoy the benefits of crop uniformity. Moreover, the vast majority of forest trees are not even subject to plantation forestry, and few of them even look like promising candidates for the future.

Radiata, despite the considerable domestication that has been achieved, remains largely a wild organism in three general respects: the state of its genetic system, its reproductive biology and its physical architecture. We now review these three respects and their implications.

9.2.1 The State of the Genetic System

Radiata is still very much an outbreeder, dependent on cross pollination among essentially unrelated individuals, with much genetic variation among individuals, and high heterozygosity (effectively genetic variation between duplicate chromosomes within individuals). All this means that plenty of continued genetic gain can be expected from classical selective breeding. By the same token, the species is not amenable to close and prolonged inbreeding in order to ensure crop uniformity and/or to capture dramatic heterosis (hybrid vigour) as is widely practised in maize breeding. Repeated self-fertilisation can give widely varying outcomes among inbred lines of radiata (cf Wu et al. 2004 and preceding papers in series). In some cases it is thereby possible to "purge" highly deleterious genes ("hard genetic load"), after which the lines may regain self-fertility and vigour. In other cases, inbred lines may slide over time to extinction, a situation that is thought to reflect epistatic genetic load, through which continued inbreeding causes insidious loss of function among various alternative developmental pathways. The potential of intensive inbreeding as a practical breeding tool in radiata is thus very problematic. In any case, use of inbreeding is not needed for obtaining desired crop uniformity if it is possible to practise mass-propagation of clones.

Associated with the wild-state outbreeding is a virtual lack of linkage disequilibrium (LD) between genetic loci (Burdon and Wilcox 2011), except within families. This means that between sites (or loci) on individual chromosomes the co-occurrence of different alleles (alternative forms of genes) is essentially random. As a result, the use of selectively neutral genetic markers, which have no direct bearing on fitness or economic traits, for selecting superior individuals is not generally possible. Use of association genetics, relating particular DNA variants at particular loci to trait expression, is possible in principle. For that, purely empirical studies look unpromising, but use of candidate genes based on known roles in other plant species is more promising (Burdon and Wilcox 2011). The increasing practicality of genome-wide scanning, using very large numbers of DNA markers, is prompting the pursuit of **genomic selection** whereby very early selection can be done just on DNA information with a reliability that is acceptable in relation to the time saved. This is being done in centralised modern dairy-herd breeding (e.g. Hayes et al. 2009a; De Mello et al. 2014), where LD is much greater. For forest trees, with very limited LD, the population sizes for which genomic selection will work satisfactorily are still problematic, and may be smaller than what are strategically appropriate. For successful application of such genomic selection, the dramatic reduction in cost identifying DNA markers is crucial, but the huge size of the pine genome will surely be a complication.

9.2.2 Reproductive Biology

Both sexual reproduction (production of seed) and vegetative reproduction (classically, grafting or producing cuttings) are involved.

Seed production is essential for selective breeding, since it involves the genetic recombination that allows the breeder to produce ever better individuals to choose from and thus provide cumulative genetic gain. And for various breeding purposes, the earlier in the tree's life seed can be produced, the better. On the other hand, producing pollen, conelets and thence seed typically consumes considerable resources. While hard to demonstrate rigorously, this surely represents a diversion from producing wood—to a degree, "making love, not wood." Moreover, sexual reproduction, especially in the phase of pollen shedding and dispersal, is likely to breach containment of transgenes if genetic engineering is practised.

An ideal for long-term domestication is to have flowering (and thence pollen and seed production) "on command and command only." Whether this is a realistic goal is controversial. Extremely precocious flowering has not been induced by using gibberellin on pines in the way that it can be induced in Cupressaceae. However, genes that can cause such flowering can evidently exist in *Pinus* (Smith and Konar 1969), although their amenability to management is quite unproven. Complete suppression of sexual reproductive structures in field-fit pines remains only a goal. Also, even if it is achieved, some adjustment of crown architecture may be needed (Burdon and Walter 2004), because the production of pollen in pines does influence crown architecture by creating zones on twigs that are bare of foliage. With elimination of pollen-producing structures these bare zones would disappear, which might mean that tree crowns become denser than is optimal.

Sexual propagation, using hybridisation, is the traditional means of combining desired attributes of different populations or species. We would certainly like to overcome the crossability barriers that currently allow hybridisation of radiata with very few other species, but that will almost certainly entail strong research commitment.

Vegetative propagation technology of radiata needs further refinement. With grafting, delayed incompatibility can be troublesome, and is not fully overcome, but it is no longer a major limiting factor. However, other forms of vegetative propagation, namely rooting of cuttings and in-vitro propagation technologies, are still beset with very important limitations. Young seedlings are easily multiplied as nursery cuttings, and this can be continued over a number of years with technologies that slow the process of maturation and allow continued propagation. Practising embryogenesis from seed tissues, in conjunction with cryopreservation using liquid nitrogen, allows repropagation of clones over long periods of time (Sect. 7.9.1). Embryogenesis also provides a platform for genetic engineering (Sect. 7.11.2). However, a restriction remains on the practice of clonal forestry in which radiata clones of choice can be mass-propagated on an unlimited scale over an indefinite time frame, because achieving this still depends on using fully-juvenile embryo material for cryopreservation. The ideal remains the ability to achieve reliable control over maturation whereby any individual tree, whatever the age, can be mass-propagated at will from whatever maturation state. In practice, this would entail being able to achieve full rejuvenation of any individual without incurring the genetic recombination associated with seed production.

9.2.3 Physical Architecture

In the wild, a tree needs to be able to get its crown where it can intercept enough light to compete and eventually reproduce itself. In the case of a light-demanding, pioneer species, like radiata, that means being able to compete vigorously with both other members of its own species and with other species. To achieve that, with a safety margin to avoid climatic damage (e.g. windthrow, or wind- or snow-breakage), highly efficient physical architecture is required. That architecture embraces not just the general form of the tree, but also patterns of within-tree variation in wood properties.

There are shade-tolerant tree species, whose competitive ability does not depend on rapid growth from an early age. Among conifers, these include many true firs (*Abies* spp.). Such species can, as stands get older, be extremely productive, but they are generally not amenable to the intensive domestication that has proved possible with radiata. When grown as plantations, their typical slow early growth delays effective site occupation so as to both erode per-annum wood production and delay harvest, both effects having serious implications for the economics of commercial forestry.

As competitors, forest trees typically have a highly efficient mechanical architecture, although the architectural strategies are very diverse among species. In pines, the radial and vertical gradients in wood properties, involving density, stiffness and fibre dimensions, can readily be explained as mechanical adaptations. These effectively mean that trees are achieving their competitive status by producing the minimum necessary amount of wood substance to present their crowns to the light while ensuring mechanical robustness. And the within-tree variations in wood properties, while they make for mechanical efficiency of intact stems, can be highly undesirable for wood processing and end-product performance; quality control can be badly affected, and solid-wood products can be dimensionally unstable during drying and in service.

It is instructive to digress into a development in crop plant breeding. With many crop plants, gains in yield have come mainly in harvest index (that is, the proportion of biomass that is of commercial value) rather than in primary biomass production (Evans 1993). A classic case is the dwarf wheats of the Green Revolution. The short stalks mean less diversion of resources into essentially unusable straw. They also mean that growth responses to fertiliser do not get vitiated by lodging which tends to happen with traditional long-strawed wheats, much to the detriment of grain yield. Herein is a particularly strong synergism between genetic improvement and management inputs. Yet the very features that make those dwarf wheats efficient grain producers means that they are poor competitors, acutely dependent on weed control.

So far, tree breeding practice has, in pursuit of increased growth, entailed selection for competitive ability. The high percentage of primary biomass that the bole represents in trees like radiata much reduces the scope for productivity gains through increasing harvest index. Nevertheless, there are almost certainly some productivity gains to be made through exploiting some divergences between competitive ability and productivity. Paradoxically, such gains could come from achieving mechanical inefficiency (or "overdesigning") in trees. In such a scenario, wood can be produced in excess of mechanical requirements, and without the within-tree variations. The "crop ideotype" tree is suggested as having the following characteristics: a fat, cylindrical bole, with highly uniform wood, and far more strength and stiffness than is needed for mechanical stability; modest height after early stages of growth (cf Libby 1987; Tuskan 2007); a crown with light branches, possibly drooping, with renewal of crown not depending on continued height growth; and no diversion of resources into reproduction. Fast early growth, however, will still be needed, to achieve the rapid site occupation that is a prerequisite for high productivity on an economic time scale. And the righting mechanisms needed to recover from minor mechanical damage will still need to function. But even with these features the crop ideotype will be a poor competitor. That, however, should be no disadvantage if the trees are grown in monoclonal blocks, within which their competitive limitations would not matter. Pursuit of the theoretical crop ideotype, however, rests uncomfortably with the observation that, in largely undomesticated radiata, stem diameter is less favourably correlated with quality traits than is height (Burdon 1992, table 3), suggesting a stiff challenge in achieving truly intensive domestication.

The issue of mechanical architecture also intersects with manipulation of soil fertility. Experience with growing radiata on highly fertile ex-pasture sites is that productivity can be strongly boosted, with an increase in wood volume of 15–25% at harvest (Maclaren and Knowles 1999; Maclaren and West 2005). Unfortunately, tree form can be badly affected, with gross branching and much malformation, and wood properties can be very badly affected (Burdon 2010a; Hawke 2011). The gross branching makes the trees subject to top breakage. Moreover, while some flexibility of the stem and crown is desirable for reducing wind resistance, excessive flexure resulting from wood of lower stiffness is likely to make trees more vulnerable to the gustiness caused by the wind turbulence that occurs within the stand canopy. In short, the developmental responses that are adaptive under low to moderate soil fertility appear to become counterproductive under high soil fertility. Achieving adaptation to the higher fertility that can boost productivity, without adverse effects on physical architecture, will be a challenge for the future.

Radiata belongs with almost all other forest tree species in that physical architecture has not been addressed in connection with genetic improvement, except at a very crude level. Yet if the issue is to be addressed in depth, radiata must be very high among the list of forest tree species involved.

9.2.4 Overview

Making substantial advances on these fronts poses formidable technical challenges, and will surely take a long time. If achieved, however, such advances will automatically bring a much higher level of domestication. In addition, some advances, especially in *in-vitro* propagation, will enable or facilitate the use of new technologies that allow greater and faster genetic improvement than ever before. That, however, is likely to involve genetic improvement of properties that were hitherto intractable for tree breeding, now that some of the most easily achievable gains have already been secured. Such developments will further raise the level of domestication. Some will save labour and costs. Others, however, may entail additional costs, but domestication is all about being prepared to make greater inputs in order to reap better returns. Pursuit of higher returns is liable to incur increased risks. The risks are probably manageable, technically. Concerns remain, however, over how well institutions will manage such risks (Burdon 2010b; Dungey et al. 2015).

9.3 Main Issues and Drivers of the Future

Many drivers of change exist. Human population pressures and the associated need to increase the production or utilization of renewable resources will be paramount. Within this context, many drivers of current or future change can be listed and outlined, under the following major issues:

- What will the species be grown for?
- Where will it be grown?
- How it will be grown?
- What will be the course of continuing genetic improvement?
- What will be the impacts of institutional and political factors?

Some of the drivers, while listed under particular subheadings, will have implications for one or more others. Having listed the drivers, we will then look at the likely changes, under the same subheadings. Cutting across the technical and purely economic issues are political and institutional issues.

9.3.1 What Will the Species be Grown for?

Likely drivers are:

- flux in future markets: uncertain emphasis on different product classes
- advances in processing and utilisation technology, including new bioproducts
- alternative sources of wood and other biomass, potentially affecting comparative advantage of radiata
- energy demands and bioenergy options
- carbon forestry revenues
- other environmental services.

Much debate has centred on the future importance of **solid-wood products** (traditionally sawtimber, plywood, poles) relative to **pulp and reconstituted-wood products**. Many have predicted that solid-wood products will decline in importance. Yet, for softwoods, the markets for solid-wood products are what generally continue to make plantation forestry profitable. This profitability is likely to be enhanced by adopting new categories of solid-wood products such as laminated veneer lumber (LVL) and cross laminated timber (CLT). When compared with certain hardwood species, radiata, because of its wood properties, the lack of super-fast site occupation and thus the relatively slow culmination of productivity, and the size to which logs can be grown quite quickly, looks to have a future comparative advantage for solid-wood products.

For the most part, pulp- and reconstituted-wood products are seen as future by-products of growing radiata for solid-wood products. This is especially so now that short-fibred hardwood pulps are becoming increasingly valued for providing high-quality printing papers. Among solid-wood products, there are ones for structural purposes and others for appearance purposes. While structural products often dominate current production, it can be difficult on some sites to produce high-quality structural timber from radiata. Moreover, appearance products can command the real price premiums.

Use of wood for **bioplastic feedstocks** has also been mooted. While this may become significant, it seems doubtful if much of the future radiata crops would be grown specifically for this purpose, instead of being a means of using arisings from solid-wood crops.

Growing radiata as a dedicated, renewable **energy crop** looks similarly unattractive. The species lacks the extremely rapid early growth that enables many eucalypts (and some other hardwoods) to produce high yields of biomass on very short rotations. Arisings from producing more traditional wood products, however, are seen as good sources of biomass for energy, although economic production of liquid fuels from wood cellulose has yet to be achieved. Even now, the lignin that is dissolved in chemical pulping is burnt to produce very significant energy for pulp mills, and much available biomass is not traditionally harvested. However, intensive harvesting of such biomass, which can include twigs and even foliage, would almost always deplete soil organic matter as well as mineral nutrients, with potentially serious consequences for future site productivity.

Values of ecosystem services provided by planted forests have been reviewed in a studies by Monge et al. (2016) and Yao et al. (2017, 2017), largely in a context of radiata plantations as a prospective land use. Considering realistic (but non-traded) values of such services, the current economic disadvantages of forest plantations as a land use largely disappeared. We now consider some specific services.

Radiata has advantages for **carbon sequestration**, which is widely accepted as an important tool for combatting global warming caused by the current rapid build-up of atmospheric carbon dioxide (Whitehead 2011), even though there will be some countervailing effect of afforestation increasing the absorption of sunlight. Advantages of radiata for carbon sequestration are its attractiveness for growing for solid-wood products on longish rotations, its ability to sustain high wood production in relatively old crops, and its ability to thrive on relatively poor soils. This general attraction, however, is a bonus rather than a self-sufficient justification for growing the species, but it has important implications for how the species is likely to be grown (Sect. 9.3.3). In much of New Zealand, growing radiata on rotations of 60 years or longer (Fig. 9.1) for increased carbon sequestration looks technically feasible (Woollons and Manley 2012).

As with carbon sequestration, the value of radiata for **soil stabilisation** (Fig. 9.2) and moderating flood-peak streamflows, is an important bonus. The advantages for

Fig. 9.1 View of 65-year-old radiata stand at Rotorua, New Zealand during harvesting, showing feasibility of long rotations. However, such rotations incur high proportions of unwanted heartwood, and delay sequestration of carbon as wood in service (Photo John R. Moore)

Fig. 9.2 Aerial photo showing negligible soil erosion in a basin planted with radiata compared with abundant slipping in an adjoining basin in pasture on fertile but erodible hill country in the North Island of New Zealand (Courtesy PF Olsen Ltd (Back cover New Zealand Journal of Forestry 59(2) 2014))

both carbon sequestration and soil stabilisation may actually be more significant in determining locally where the species is grown than in governing how it is then grown for wood products. Situations will presumably continue to exist where planting radiata can achieve other forms of soil rehabilitation, where topsoil has been lost and even some salination is threatening. There will, however, be cases where radiata can cause unacceptable reductions in water yields from catchments. However, log harvesting operations do entail risks of significant mass flows of sediment and logging debris which are environmentally harmful and, even if very uncommon, incur very bad publicity.

Large areas of land remain in New Zealand where planting radiata would achieve soil stabilization, although government policy has not been addressing this issue effectively. In Chile, large areas of land remain where afforestation with radiata could be used to rehabilitate degraded soils, which is now being more specifically encouraged by government. In Australia, where salination of farmland is a widespread problem, remediation by planting radiata is unproven and, even if it works, would need to be weighed up against reduced water yields.

Radiata will doubtless retain a role for farm shelter in New Zealand, and to a lesser extent in Australia. Shelterbelts are typically grown to provide substantial volumes of timber as a bonus, effectively occupying for wood production areas well beyond the span of the tree crowns. Sacrifices in potential timber production, however, will be needed in order to ensure continuity of shelter over time.

9.3.2 Where Will the Species Be Grown?

Several main influences are seen:

- pressure on availability of land: demands for agriculture, horticulture and conservation estates
- climate change: warming, storm intensities, increases or decreases in rainfall
- biotic developments: arrival of previously absent pathogens and pests
- availability of other tree species that are commercially competitive with radiata.

Of these, the second and third can be strongly interrelated because of the influence the relationship between hosts and their pathogens and pests, so they are reviewed together. Finally, the expected changes in global and regional distribution of radiata plantations are reviewed.

9.3.2.1 Classes of Land—Availability and Suitability

By and large, the better the land and the easier the terrain, the better forest plantations will perform, although there are exceptions. More importantly, however, plantation forestry must compete for land, economically and politically, with alternative land uses.

Very high quality land may have no special advantage for plantation forestry; indeed, it can have significant disadvantages, apart from being too expensive. Many tree species, especially pines, need only modest soil fertility to grow well, higher fertility being conducive to problems with both tree form and basic wood quality.

Many very fertile horticultural soils may also lack the good drainage that radiata needs. For other crops, both agricultural and horticultural, the advantages of high soil fertility and/or easy terrain are often much more crucial than for growing forest plantations. And there is land where climates are well suited to horticulture, such as growing grapes and other fruits, but unsuited to intensive wood production. In the long term, land of generally modest fertility and often of rolling to steep terrain will remain available for growing radiata, despite the greater difficulties and higher costs of harvesting. There may be areas where forest cover reduces catchment water yields, through increased interception and transpiration losses, so much as to lead to prohibition of growing radiata (cf Mead, p. 48).

In New Zealand, complex land-use issues are involved. There some recent changes in land use have occurred, with thousands of hectares of radiata forest being converted into dairy farms. This happened on easy terrain, with a boom in dairy prices, and on soils where such a conversion was relatively easy. Also, the regulatory machinery was ill-equipped to review and restrict such land-use changes in order to head off likely environmental ill-effects. On the other hand, large areas of marginally economic pastoral farmland may be converted to radiata, especially with incentives for carbon sequestration (Levack 2016). The demands on land for commercial production in some areas, and for conservation values in others, will tend to favour intensification of growing practices for commercial crops, although the type of land that will be available for forest plantations will tend to militate against that. The net effects of this trade-off remain to be seen, but are likely to vary from case to case. A complicating factor in New Zealand is the price regime for marginal land for sheep and beef farming. Despite the marginal status, the land prices are effectively governed by those for pastoral farming, yet those prices bear little relationship to current profitability, settling to values that mean returns of 2% or less on the farming operation, well below normal commercial discount rates. The prices that get paid presumably reflect a combination of two interdependent factors: preparedness to pay a premium for the right to run one's own farming operation and an expectation that this will ensure capital gains upon selling the properties. Since such land prices greatly affect forest profitability (Evison 2008, 2013; Richards 2013), afforestation of such land may need some active incentives in the absence of a functional carbon-credit market. Recent volatility of dairy commodity markets may help with market corrections, but not in the longer time frames needed for orderly commercial forestry. Moreover, any government failure to impose "polluter pays" will militate against both timely carbon sequestration to mitigate climate change and maintaining water quality in streams and lakes.

Political antipathy can exist within rural communities to both overseas interests and local urban-based investors owning large land holdings. Maori ownership of large areas of land currently or potentially carrying plantation forest might be expected to favour plantation forestry. Yet some of the owners want such land to revert to native forest, essentially for its cultural value. Probably more significant, though, is an upsurge of interest in manuka, a native scrub species, for producing honey or essential oil (Orme 2017). Long prized for its flavor, that honey is now even more prized for pharmaceutical properties, leading to a boom in research on its genetics and culture, and to various speculative ventures. Should such ventures succeed well, much land could be diverted from commercial forestry. Otherwise, much time could be lost in rationalising land use.

9.3.2.2 Climate Change and Biotic Developments

Where radiata can now be grown is well understood, because its attributes have led to it being tried so widely. Mild temperate climates, away from tropical and subtropical summer-rainfall conditions, are seen as where radiata will be grown in the future. Radical changes of its global distribution, even with foreseen climate change, seem unlikely. Some changes, however, may become possible if radiata's biotic tolerances can be further improved by being able to hybridise it with additional species, or being able to use genetic engineering, thereby extending or at least safeguarding its current range.

In many areas where radiata is grown on a large scale, some known and often serious fungal pathogens are not yet present. Western gall rust (WGR, caused by *Endocronartium harknessii* J.P.Moore) is a case in point. The likelihood of it being introduced and becoming established in Southern Hemisphere countries is greatly reduced by the strong asynchrony of infection seasons between hemispheres. If, however, WGR did cross the Equator there would almost certainly be large areas where it would be extremely dangerous, being likely to cause much deformation of the lower boles or even tree mortality. WGR can be devastating on radiata north of radiata's natural range in California, where conditions are moister, and within the natural stands of radiata WGR is especially prevalent after unusually wet springs, which roughly match humidity regimes that prevail over large areas where radiata does well as an exotic. Since WGR is associated with the juvenile or semi-juvenile growth phase, its impact should be mitigated by growing material of provenance origin that shows more rapid onset of maturation (cf Old et al. 1986) or vegetative propagules (cuttings or plantlets) with appreciable maturation (cf Zagory and Libby 1995).

Globalisation of trade, and even of tourism, will continue to increase the risk of new pathogens or pests becoming established in grower countries, despite the best efforts in border biosecurity. Yet despite the known existence of hazardous pathogens and pests, the biotic alarms with radiata have mostly arisen with pathogens and even insect pests with no prior history of being serious.

Overall, climate change is seen more as accentuating certain biotic hazards, and mitigating some others, rather than of itself creating completely new ones.

9.3.2.3 Competing Species for Plantation-Forestry Roles

Even where radiata will continue to thrive, other species will be competing for parts of its niche as a plantation-forestry species. A number of eucalypt species have been gaining increasing popularity for plantation forestry, even displacing radiata in parts as the species of choice. Factors include: the very rapid early growth of various eucalypts, the associated amenability to growing as short-rotation pulpwood crops, the economic attraction of very short rotations, the favoured status that their short-fibred pulp has acquired, and the continuing "biotic honeymoons" of freedom from certain insect pests enjoyed by eucalypts on land masses remote from Australia. Countervailing factors for eucalypts include the site sensitivity often shown by them; the associated preference for growing them on easy country, often land with a horticultural potential, compared with growing them on steep sites with often low soil fertility; the limited long-term carbon sequestration under short-rotation pulpwood regimes; the difficulties for satisfactory conversion to solid-wood products that are posed largely by growth stresses; and the prospect of biotic honeymoons ending. Indeed, the biotic honeymoon for some eucalypts in New Zealand ended around a century ago with insect pests becoming established from Australia across the Tasman Sea ("The Ditch"), a process that continues now (Kay 2005) and doubtless into the future. Establishment of insect pests, however, can be slower than that of the fungal pathogens that have often affected radiata. Moreover, research is proceeding to improve eucalypts genetically for producing solid-wood products. Problematic, however, is the future impact on eucalypts of myrtle rust which is caused by a South American fungus that is spreading rapidly in some other parts of the world.

More locally, coast redwood could gain increasing favour, at least in New Zealand, as an alternative to radiata on some site categories. Its long-term productivity can be very high, it has very few pests and diseases, its timber can be of high value, and its coppicing behaviour can both avert replanting costs and ensure continued soil stabilisation after harvest. However, it tends to be much more site-demanding, is more difficult and expensive to establish initially, and requires longer rotations to produce quality timber. In New Zealand, Douglas-fir has a major advantage for some areas in much superior resistance to snow damage.

9.3.2.4 Prospective Shifts in Distribution

Some significant shifts in the distribution of radiata plantations are expected both within and among grower countries. Climate change could bite at both levels. Within Australia, likely shifts have been addressed by Booth (1990), Booth et al. (2002) and Pinkard et al. (2010), and within New Zealand by Kirschbaum et al. (2012) and Watt et al. (2012), through either direct climatic effects or effects of pathogens and pests.

The likely climate-related changes are various, with both local contractions and local extensions of plantation areas. Some areas will become too dry, and better for, say, olive growing. Other areas, however, may become too warm and humid, and/or subject to violent, damaging storms. On the plus side, there are expected to be areas where rainfall increases to acceptable levels without undue heat and humidity, or where the present-day risks of frost or snow damage become acceptably low.

Among countries, changes could be additionally driven by other factors, notably: uncontained incursions of pests and diseases, whether pathogens have effective vectors; and political and institutional factors.

9.3.3 *How Will the Species Be Grown?*

A number of likely drivers are involved here:

- interplay between genetic improvement and nursery production systems
- environmental issues: sustainability of site productivity, biodiversity, public reaction
- advances in plant protection: new-generation herbicides, new biocontrol options
- biotic developments: unpredictable spread of diseases and pests, but new control options
- precision field technology
- carbon forestry: own economics, impacts on stocking and rotation ages
- cost structures: manual operations more expensive, favouring mechanization
- advances in crop modeling techniques.

Plantation forestry, growing even-aged crops that are clearfelled, is expected to remain the norm for radiata. While clearfelling draws much public criticism, routine implementation of uneven-aged, continuous-canopy silviculture is generally highly impractical with a species like radiata. Indeed, clearfelling and replacement by even-aged stands mimics the natural regeneration ecology better. The norm in nature is evidently for fires to kill significant areas of radiata forest which get replaced after fire causes release of the cumulative seed bank in the closed cones. Replacement stands are typically much healthier than ones resulting from regeneration under an existing canopy. In the bigger picture, successful domestication of forest trees has been predominantly with species that are at least in part of pioneer status ecologically.

Artificial regeneration is generally a central feature of modern plantation forestry. In radiata, use of bare-rooted nursery stock has been greatly refined, yet there have been recent moves towards use of container-grown stock to help achieve quicker and more efficient capture of genetic gain from clonal selection, with a bonus of extending the safe planting season. This is now being implemented in some major nurseries in Chile and New Zealand, at least. Its adoption creates a strong incentive to develop mechanization in what is largely a new area.

Clearfelling can certainly create problems with soil erosion after harvest, but these are typically transient, and so affect only very small portions of a forest estate at any one time. Likewise, elevated but transient leakage of soil nutrients following clearfelling typically affects only small sub-catchments at any given time. Surprisingly, clearcut areas can perform an important role in maintaining landscape biodiversity. They produce favoured habitat for certain bird species, a specific case involving radiata being the endangered but iconic New Zealand falcon (Seaton et al. 2009). Moreover, radiata plantations even support populations of the rare New Zealand long-tailed bat (Borkin and Parsons 2010), and some very dense populations of the kiwi (Kleinpaste 1990), although averting ill-effects of harvesting operations poses challenges.

Control of competing vegetation, namely weeds, is likely to become more of an issue in the future, involving plantation forests in general and not just radiata. Typically, weeds are easily dispersed and far more easily established than eliminated, creating the prospect of an ever-increasing suite of weed species that depress tree growth by competition and/or inflate plantation-growing costs. Mechanised clearing of vegetation will remain an option, but it must be used carefully on steep terrain, and even on easy terrain it can entail scalping of topsoil. A more strategic approach, timely control of incipient weed infestations, often brings costs and criticism from anti-herbicide activists but no immediate profits for forest owners nor plaudits for the managers. More weed species will tend to create a need for broader-spectrum herbicides, and yet various traditional herbicides have been outlawed or likely to become so. Genetic engineering, to incorporate resistance to broad-spectrum, environmentally-benign herbicides remains a possibility, and its feasibility has already been demonstrated (Bishop-Hurley et al. 2001). Although evolution of herbicide tolerance in weeds is emerging as a problem with agricultural cropping, this seems much less of a danger in forest plantations where the need for weed control is typically for short periods at the start of each rotation. Future use of higher stockings may be justified in part by how it can help control weeds for current access and to ease crop re-establishment.

Use of fire to remove logging residues and help control weed growth has been widely discontinued, while fire can actually favour some troublesome weeds. Such burning provokes public objections, to both the smoke and perceptions of ecological ill-effects. Apart from the smoke pollution, burning entails immediate losses of nutrients and release of carbon dioxide. Yet fire, despite the immediate loss of site nutrients that it entails, may be conducive to sustained soil health and fertility, through generating soil charcoal or biochar. Indeed, fire is part of the natural ecology of pines in general, and the trade-offs in its exclusion or use are unclear.

Control of pests and diseases will surely remain a very important issue, given the exotic status of radiata plantations, how modern transport facilitates incursions, and the attractions of growing radiata where it will perform at all acceptably. Genetic resistance would be preferred, if available within a reasonable time and if proof against genetic shifts in pests or pathogens. Chemical control, as with spraying, may well retain a role. Biological control, however, has big technical and ethical attractions. Insect pests are often amenable to biological control, and this has already been implemented in radiata (e.g. for sirex—Kaya et al. 1993; Burdon 2001). In addition, there are exciting prospects for biological control of fungal pathogens. In horticulture, applying spore suspensions of certain fungi can already control others. It is now known that conifer tissues can contain rich suites of endophyte fungi, of which at least some may confer resistance to pests and diseases. Manipulating these populations may afford future biological control (Ganley 2010; Burdon 2011).

There will almost certainly remain some conflicts arising from the fact that on the more productive sites, with relatively high rainfall, risks of fungal diseases will tend to be elevated (Burdon 2010b), in a classic risk-reward relationship. But on drier sites, where disease is less of a hazard but inherent productivity lower, insect pests will tend to be more of a hazard.

Use of fertilizers is likely to be more significant in future, especially to replace nutrients that will be lost in the course of more complete harvesting and use of biomass. Precision systems should make possible better-targeted applications. Raising general fertility, with elevated nitrogen availability, can enhance wood volume production much more than height growth. This accords with classical experimental findings that optimum nitrogen levels for diameter growth exceed those for height growth in pines (e.g. Roberds et al. 1976). Actually, height growth and stem volume production can be significantly uncoupled among regions in radiata, with lower optimum temperatures for volume production (Burdon 2001; Palmer et al. 2010), and in loblolly pine (DeBell et al. 1989). Also, using fertilizers to improve growth without unduly compromising wood properties appears to be a challenge on at least some sites. Indeed, the availability and desirability of using fertiliser for growing radiata can pose some major conundrums. As a true pine, radiata can tolerate relatively low soil fertility, which will commend it to relatively low-quality land. Nevertheless, productivity of radiata can surely be boosted substantially over large areas by raising soil fertility, yet requiring much less fertility than most food and pasture crops, meaning that water quality in streams and lakes would not necessarily be jeopardised by using fertiliser in plantations. Indeed, the finite area of available land will argue strongly for raising soil fertility for growing radiata if it is at all economic. On the other hand, pines appear to be generally ill-adapted to high soil fertility in respect of wood quality, so the fertility levels needed for optimal productivity of radiata are very inimical to wood quality. Thus, for substantial areas of land a major challenge may lie in genetic manipulation of wood properties such that wood quality remains good even under elevated fertility.

The issues of wood quality pre-eminently involve the corewood zone, which in radiata will pose ongoing challenges (Moore and Cown 2017), not only for the forest managers abut also for the tree breeders.

The emerging (but sputtering) market for carbon sequestration favours both the adoption of higher stockings and backing off from very short rotations. This, however, is largely reinforcing trends that have been emerging from the realization that low stockings and shorter rotations have been compromising both yields and wood properties to a greater degree than had been assumed. Another reinforcing factor is the increased role of USA-based TIMOs (timber investment management organisations) for whom economically optimal rotations are longer than for conventional corporates.

Other aspects of management regimes will depend strongly on anticipated markets and end-uses. Very problematic is the future of pruning in order to produce clear timber. As a highly manual operation it is becoming relatively more and more expensive. Many forest growers, who see the costs, but not the final returns, are flinching from this operation, especially in the light of recent market behaviour (a weak market in USA for high-quality clearwood, but a strong market in China for knotty logs). Moreover, the sacrifice in wood volume with pruning accentuates its economic dependence on strong premiums for clear timber. Modifying the branching pattern by genetic selection is an option, but has little promise for the bottom 3 m of the stem, where branching behaviour tends to rule out high-quality clearwood in unpruned stems but wood properties militate against quality structural timber (Burdon et al. 2004). Pruning to just 3 m would fit well with the proposal (e.g. Xu and Walker 2004) to segregate as a separate log the bottom 3 m of the bole, but this would pose problems for harvesting, logmaking and log transport. In Chile, the continuing level of attack by the pine shoot moth (Mead 2013) will now militate against growing trees of the "long-internode" branching habit which recover less readily from leader damage.

Underpinning many of the future field operations will be adaptations of "precision agriculture" technology. Obvious examples—and only examples—will include navigational aids for aircraft that are used for spraying and applying fertilizers. Use of remote-sensing technologies is being much aided by the availability of small Unmanned Aerial Vehicles (UAVs or "drones"); this allows detailed ground profiling of forests which is extremely valuable for forest management and potentially very useful for tree breeding.

Growing clonal material has important attractions, provided risk-spread multi-clonal portfolios are deployed and grown judiciously. Clonal forestry, as it is called, offers greater potential genetic gains. It can also be used to avoid unwanted genetic variation. Avoiding such variation can substantially simplify processing and marketing. In addition, appropriate use of genetic uniformity should allow far better control of branching pattern, with benefits for clear timber production (Burdon 2008). More radically, it may serve the production of crops that are even better producers of wood (Sect. 9.2.3), especially if clones are deployed as mosaics of monoclonal blocks.

Straddling the issues of how the species will be grown and what genetic improvement can be sought and achieved is modelling of crop performance, for predicting outturns under alternative growing regimes and processing options. Such modelling has advanced dramatically in the last few years, with increased computing power and advances in remote-sensing technologies (Sect. 7.3.3). It will doubtless be subject to further refinement, leading to better-informed decisions by forest managers. Challenges will remain in adapting the models to accommodate continuing changes in genetic populations. More fundamentally, better models, based largely on better genetic information, should serve to project genetic gain to harvest age and quantify, at the level of the crop, the trade-offs involving adversely correlated traits. Indeed, quantifying such trade-offs satisfactorily can often be crucial for defining appropriate breeding goals as well as selecting the appropriate individuals.

9.3.4 *Genetic Improvement Issues*

Likely drivers are as follows:

- evolving breeding goals
- advances in assay technologies, for example wood properties and disease or pest resistance
- advances in data analysis, largely incremental
- advances in gene technology: pedigree reconstruction, selection tools, hybridization techniques, genetic engineering
- the bioinformatic revolution
- advances in propagation technology: mass-propagation systems, *in-vitro* technology, control of maturation state, hybridization, platform for genetic engineering, control of flowering.

9.3.4.1 Evolution of Breeding Objectives

Future breeding objectives must include pursuit of growth rate, good tree form, desirable wood properties and resistance to known diseases or pests that affect productivity and/or raw-material quality. Beyond that, unless the trees are grown on extremely short rotations, the tree breeders and silviculturalists are facing major uncertainties in both the biology and the markets.

Biological uncertainties are various. They include:

- imperfect knowledge of the genetic variation in traits of potential interest
- often very imperfect knowledge of genetic correlations between different traits, which can be particularly important with adverse genetic correlations
- limits to what are realistic breeding goals, including those imposed by adverse genetic correlations
- detailed patterns and drivers of genotype-environment interaction
- biotic events, such as the arrival of new pathogens, which may impose new breeding goals, noting that major biotic events affecting radiata have repeatedly involved pathogens with no prior record of being significant
- potential side-effects on field fitness (e.g. mechanical stability of standing trees) of certain improvements in wood properties
- impacts of impending climatic changes.

Market uncertainties are largely self-evident. Consumer fashions may change, while the economic fortunes of client countries that have their own consumer preferences may fluctuate. The development of alternative or substitute products may depress demand, or else declining supplies of alternatives can boost demand. New processing technologies can produce market competitors, or they may boost the value of one's own wood outturn.

Despite the uncertainties, certain breeding objectives are undisputed. Faster growth is always welcome, provided it is not unduly at the expense of wood quality. Straight, healthy trees, free of malformation are always wanted, too. Improvement of wood properties, especially stiffness and dimensional stability, is generally very welcome, most of all in the corewood zone. Fortunately, low microfibril angles are conducive to both good stiffness and dimensional stability, which are especially desirable for structural and appearance products alike. That said, carbon forestry, by lengthening optimal rotations, will improve average wood properties substantially, an exception being increased heartwood content which is not wanted by processors of radiata wood (Kennedy et al. 2013).

More debatable is the optimum branching pattern (Burdon 2008). Future price premiums for radiata clearwood are uncertain despite some good current premiums for appearance-grade timber, pruning costs are becoming problematic, and yet production of clearwood without pruning faces biological constraints. While the short-internode branching pattern confers superior adaptation to many New Zealand and Chilean sites, it is not conducive to obtaining good yields of clear timber as clearcuttings without pruning. Yet the other extreme of branching pattern, namely a very long-internode habit, tends to incur penalties in growth rate and tree form (Burdon 1992). If there is an "intermediate optimum" for branching habit, which gives good yields of clearcuttings without much sacrifice of potential growth rate and tree form, it should be achievable through clonal forestry.

Gains in productivity, while always welcome in themselves, are even more so in a context of carbon forestry.

Between the biological uncertainties, market uncertainties, often complex production systems for processing and their ramifications, choice of appropriate strategies and algorithms for genetic selection poses very complex challenges (Ivković et al. 2010a, b), especially for radiata. On these challenges, the last word will not be said anytime soon.

9.3.4.2 Assay Technologies

Improvements sought in wood properties often relate to performance in service. Stiffness can now be assayed on standing trees (Grabianowski et al. 2006), bypassing problems of obtaining cheap, large-scale assays for microfibril angle, and there have been advances in achieving cheap, non-destructive assays for other key wood properties. Similar improvements can be expected in assays for growth performance, climatic tolerances, and disease or pest resistance. Even if DNA sequences eventually become the basis for selecting for these various attributes, this will almost certainly depend on enhanced assays of the actual trees as intermediate steps in developing applications of the gene technology. Reliable cross-reference between field performance and individual genetic variants (represented in DNA polymorphisms) will be essential for such selection, but a more immediate task will be developing assays that are good proxies for field performance.

9.3.4.3 Classical Data Analysis

Data analysis for field genetic experiments, and other empirical trials, has undergone major advances in recent years. Even so, further, incremental advances can be expected, especially as computing facilities become still more powerful. Integrating data on physical performance of trees with DNA data will be the target of much future effort.

9.3.4.4 Use of Gene Technologies

Gene technologies have several potential roles in genetic improvement. These include:

1. Verifying or discovering genetic identity
2. Providing a basis for selection, either in lieu of or in conjunction with assessing field performance and all the delays this can entail
3. Helping identify feasible breeding goals
4. New hybridization technology
5. Use of genetic engineering (GE), either operationally or to inform conventional breeding.

Of these, (2) and (5) represent particularly fast-evolving fields of research and technology, making any pronouncements especially tentative.

Verifying genetic identity is already a well-developed and widely-proven technology (Burdon et al. 2008). It can avoid much loss of genetic gain, by avoiding use of wrong clones in seed orchards or of crosses coming from the wrong parents. For the future, there is the prospect of identifying the pollen parents of individuals resulting from natural pollination, thereby maintaining the advantages of full pedigree without the labour and costs of traditional controlled crossing. That allows better assessment of the genetic merit of selection candidates for the next generation, control over unwanted inbreeding, and much better maintenance of genetic diversity for the future. Actually, there is a potential to improve upon classical pedigree information through being able to pick up random departures from the expectation that an individual receives one-quarter of its genes from each grandparent.

Being able to **select individuals on the basis of genes** has the attractions of saving on costs of expensive field trials and saving time spent waiting for reliable field expression of traits. That attraction is become greater with steadily decreasing costs of obtaining DNA sequence information. However, this depends on relationships being established between DNA information and field performance, which itself can take a long time and involve very large numbers of individuals which may be very costly to produce (Burdon and Wilcox 2011). Hopefully, this process can be shortened by narrowing down the list of possible genes of interest, using candidate genes that have been shown, very often in other types of plant, to exert effects of interest.

However, recent experience with animal breeding, and the dramatic reduction in costs of DNA sequencing, have shifted the emphasis towards **genomic selection** (Hayes et al. 2009a; Grattapaglia and Resende 2011; De Mello et al. 2014). Rather than searching for and trying to verify quantitative trait loci of large effects, the emphasis is now on genome-wide screening to use for selection the sequences in chromosomal segments associated with phenotypic effects. Also of value is the marker information provided on the genomic contributions from various ancestors (**Realised Relationship Matrices**, e.g. Hayes et al. 2009b). Already used for large-scale dairy-herd breeding (Sect. 9.2.1), genomic selection has potentially worthwhile reliability for a species like radiata as soon as embryos or seedlings can be genomically assayed non-destructively. Even if the tree breeder remains dependent on some field evaluation of selection candidates, genomic selection may help greatly in shortlisting the candidates. Taking full advantage of such a capability, however, would only become feasible with greatly accelerated sexual reproduction, on a basis of "flowering on command and command only."

Major challenges remain, though, for the tree breeder pursuing genomic selection (cf Nakaya and Isobe 2016). For large, modern dairy-breeding programmes, commercial herds are the breeding populations, and abundant good phenotypic data on individuals come available from the monitoring now entailed in routine herd management. For tree breeding, in which breeding populations are separate from commercial crops, comparable data are certainly not obtainable in the course of routine forest management, and for some traits good phenotypic data can take considerable time to become available. Improved remote-sensing technologies are likely to help, though. Limited linkage disequilibrium in largely undomesticated trees, together with the large conifer genomes (Sect. 9.2.1), pose additional complications, which may restrict satisfactory application of genomic selection to only small subsets of forest-tree breeding populations.

What breeding goals are feasible will depend very much on pathways of gene action, which can now be traced in much greater detail. Of special importance are cases involving traits that are adversely correlated. If the adverse genetic correlations between traits involve fundamentally antagonistic effects of genes exerting desired effects on the different traits, the scope for simultaneous gain in the traits may be severely restricted, meaning that solutions other than genetic improvement may need to be sought. A specific question, affecting radiata along with other conifers, is whether producing the syringyl rather than guaiacyl lignin, which makes for the easy pulping of broadleaved trees, is compatible with the rest of conifer metabolism. This objective, however, is not accessible through conventional breeding, but rather through genetic engineering (Sect. 9.3.4.6).

New hybridization technology may make available a new range of genetic diversity for future genetic improvement of radiata. At present, it is crossable with very few other species. Among those species is knobcone pine, *Pinus attenuata*. Hybrids with knobcone pine, while very susceptible to dothistroma needle blight, are far more resistant than radiata to snow damage (Dungey et al. 2011). Such resistance promises a major extension of where radiata can be safely grown. Among the suite of Mexican pines, a few can be crossed with radiata with some difficulty. Several others, which are also closely related, could confer valuable attributes if hybridization could be achieved. In particular, those Mexican pines, coming from summer-rainfall areas, will have very different spectra of disease resistance compared with radiata, which could be crucial in the event of a biotic crisis affecting radiata. However, recourse to using such species in areas where radiata is now grown will require much more research commitment than is generally being made at present.

Genetic engineering (GE) has several attractions. It can be used to confer attributes that are not available from the existing variation within a species. It also brings in very short and highly specific sequences of DNA, instead of the unknown and largely unwanted genetic "baggage" that comes when hybridization is practiced in order to bring in some new attribute. Indeed, it could be an alternative to actual hybridization as a means of incorporating attributes from other species. However, achieving the appropriate specificity depends on knowing the exact sequences that are needed. The basic feasibility of GE with radiata has been well demonstrated (Sect. 7.11.2). As an operational tool, however, it is still unproven, and the lack of unintended adverse effects on the trees and their products remains to be confirmed (Ahuja 2014), although such risks would appear very low for herbicide resistance. Regulatory restrictions on field release of genetically engineered forest-tree material remain a huge constraint (Strauss et al. 2015), adding enormously to development costs. These restrictions are partly driven by political resistance that will not disappear at all readily. Less contentious, perhaps, is the potential of GE as a research tool to inform more conventional breeding. The scope of using GE, and the issues associated with is use, are addressed in more detail in Sect. 9.3.4.6.

Both selection on the basis of DNA sequences and targeting of GE will likely be helped greatly by much research on **gene expression**, whereby relationships can be established between DNA sequences and properties of the actual trees. This, however, could be hampered by the running down of a substantial workforce engaged on the physiology of radiata during the 1980s.

Use of gene technologies offers the prospect of avoiding many laborious activities in genetic improvement, and greatly foreshortening certain time frames. Yet developing such technologies will, at least in the short to medium term, entail a big increase in commitment to genetic improvement, unless it is done much at the expense of classical breeding activity. Using such gene technologies further intensifies domestication, and domestication is all about increasing inputs in order to achieve better returns.

With advances in gene technology, and with accumulated experience, the scope for GE to achieve environmental as well as economic benefits is becoming increasingly clear (Sect. 9.3.4.6).

9.3.4.5 DNA Sequencing and Bioinformatic Advances

The costs of DNA sequencing are falling dramatically, which opens up huge possibilities for using DNA technology in genetic improvement, in identifying feasible breeding goals, genetic selection, and targeting genetic engineering. Optimism, however, must be tempered by the challenges of genomic sequencing in a species like radiata and of managing and analyzing the prodigious amounts of DNA sequence data that can be increasingly generated.

Radiata, like other pines and conifers in general, has a huge genome (Sect. 7.11.1). Why this should be so is a mystery. Anyway, attempting a complete genomic sequencing is currently prohibitive, and interpreting a complete sequence is a massive challenge. There is, however, something in our favour (Ritland et al. 2011). The genomes of pines, indeed Pinaceae in general, are remarkably similar. The chromosomes are much the same, and the various genes do essentially the same things, and are arranged in essentially the same chromosomes and in very similar sites within the chromosomes. So genomic information obtained in one pine species can be very helpful for studying genomes of other pines. Indeed, progress has been made in cross-referencing genomic information between radiata, loblolly pine (*Pinus taeda*) from south-eastern USA, and maritime pine (*P. pinaster*) in France, the work on each species drawing in its own research funding. While loblolly pine can attract the biggest funding, radiata may otherwise have greater overall advantages as a "model species" for research. In fact, there is no need to choose a single pine as a model species for research, because several can be used in conjunction to exploit all their individual advantages.

Dealing with genomic sequence data is termed bioinformatics. Managing and analyzing the sequence data that will come, and interpreting the findings, are already becoming far bigger tasks than just acquiring the data. The increasing availability and dwindling costs of computing power will help greatly, but devising efficient systems for managing and processing the data will still be a great challenge.

9.3.4.6 Role of Genetic Engineering

Operational uses of GE in radiata will presumably come in time, but the nature of the uses seems uncertain. A classic use, conferring herbicide resistance, is problematic, because radiata is comparatively tolerant of weed competition at establishment, and it also tolerates some herbicides which can therefore operate selectively. Conferring resistance to insect pests is a popular target for GE in some plants, notably cotton. In radiata it has been sought as a solution to a problem in Chile, but without achieving practical application; moreover, insect pests are often amenable to biological control. Conferring reproductive sterility, avoiding all diversion of resources into reproduction, is very attractive in principle (Burdon and Libby 2006), and promises the huge benefit of containing the transgenes, which addresses an important regulatory barrier to use of GE. Whether sterility can be conferred completely and reliably is contentious. Also uncertain would be the ecological impacts of preventing the dispersal of mineral nutrients that occurs through the wind dispersal of pollen grain. The ability to induce super-precocious flowering (cf Kean 2010) in material that is otherwise sterile would be extremely welcome for the breeder, but achieving and controlling it presents an additional challenge.

Another research target for GE in radiata is producing syringyl lignin in place of guaiacil lignin, which would make chemical pulping easier and cheaper. While it has been achieved in the laboratory (Wagner et al. 2015) much remains to prove that it can be done without appreciable ill-effects on field fitness and the suitability for solid-wood products.

Incorporating resistance to certain fungal diseases may be yet another target for use of GE. Ideally, one wants to incorporate resistance genes of large effect. One also wants a diversity of resistance mechanisms, "pyramiding" resistance genes (Burdon and Wilcox 2011), in order to ensure durability of resistance against emergence of new and more dangerous strains of the pathogen(s). Achieving this, however, will depend on big advances in knowledge of resistance mechanisms.

Operational use of GE is foreseen in at least the medium term in the context of clonal forestry (Burdon and Lstibůrek 2010), using for commercial plantations finite numbers of well-characterised clones containing the desired transgenes. This contrasts with the situation with annual crops like soybean, cotton and canola, in which the transgenes are stably integrated such that the crops "breed true" for the genetic transformations. In forest trees, reaching the point of breeding true can take a long time, and the need for it can be obviated by clonal propagation. However, advances in DNA technology now promise to remove this constraint.

Genetic engineering currently faces vehement political opposition from "green" organisations, largely on environmental grounds in the case of forest trees. We are concerned, though, that such opposition is misplaced, to the point of being downright counterproductive to central causes of those groups. Some key goals of GE are surely environmentally beneficial, in direct and indirect ways. Among the more direct benefits are the use of GE-conferred insect resistance to avert the need for objectionable insecticides or of herbicide resistance to favour the use of low-toxicity, non-persistent herbicides. Indirect environmental benefits should accrue from greater and more reliable production on the land area devoted to plantation forestry. While much has been made of the postulated risks of GE, those risks have generally not materialized in closely monitored situations (e.g. Häggman et al. 2014). Moreover, recent advances in gene technology have made it possible to achieve GE on a much more targeted basis, and with interventions that now avoid most of the perceived risk factors. Indeed, it is now becoming possible to release GE crop or fruit cultivars in which extraneous DNA sequences have been excised (Kean 2010). Still more recently, "gene editing" with CRISPR technology can be used to completely bypass any use of extraneous DNA sequences (e.g. Dance 2015; Hall 2016). There will remain some risks to be managed with GE, but they must be balanced against the potential for GE to serve as a risk-management tool. While the precautionary principle has been applied to place the onus of proof on safety of GE, proper application of the principle is becoming less one-sided, and is likely to eventually reverse. Thus the use or at least the availability of GE is potentially part of an environmental imperative.

9.3.4.7 Delivery Systems for Genetic Gain

Clonal seed orchards, as originally conceived, are largely obsolete for radiata. Given the ease with which mass vegetative multiplication can be practised in the species, it can be used to deliver the leading-edge genetic gain that can be achieved in small-scale production of seed in crossing archives. New forms of *in-vitro* propagation not only extend the range of options for mass multiplication of material without repeated recourse to seed production. For instance, when combined with cryopreservation, vegetative propagation can give greater control over maturation which is effectively halted at cryopreservation (liquid-nitrogen) temperatures. In this connection, a favoured form of *in-vitro* propagation is embryogenesis—dissecting out seed embryos and inducing the resulting culture to produce large numbers of fresh embryos which may be reared through into plantlets. Other in-vitro technologies can be used as alternatives to embryogenesis or in conjunction with it. Pines, however, are not yet as amenable to embryogenesis as spruces are, but that may change.

Mass-propagation of proven clones, namely clonal forestry, can capture greater genetic gains than are possible with propagation by seed, and avoids unwanted genetic variation that will always exist among seedlings. Fully successful clonal forestry ideally requires total control of maturation, which ideally includes being able to rejuvenate material from old trees to the state of a freshly germinated seedling. While that occurs every time a seed is produced, the challenge is to achieve it without the genetic recombination that occurs in the course of seed production. In practice, one would want to rejuvenate propagable material of any tree at any age, which remains to be achieved with radiata along with all other pines. Given the special significance that clonal forestry is likely to assume, we will soon revisit it to address some broader issues that it involves.

In-vitro propagation systems can serve another purpose, in providing material on which GE can be performed. This fits nicely with the fact that operational use of GE with forest trees currently seems best implemented in a context of clonal forestry (Burdon and Lstibůrek 2010).

9.3.4.8 Optimising Field Deployment

Given the market uncertainties, the large impact of site on important traits such as certain wood properties and tree form and disease incidence, and generally adverse genetic correlations between growth rate and various wood properties, it is unrealistic to deploy the same set of genetic material everywhere. Instead, it is appropriate to deploy genetic portfolios, sets of genetic material that are tailored to different sites and/or different markets. Hopefully, a market-based portfolio may allow the grower to cater efficiently for niche markets that are highly profitable but susceptible to gluts. Any set of deployed material, however, should contain its own risk spread: a minimum of seven unrelated clones has been proposed (Libby 1982), but a figure of 16 or so is widely regarded as significantly safer. There seems no justification, though, for the hundreds of clones that are prescribed by law or regulation in some countries.

9.3.4.9 Strategic Genetic Management

Whatever the potential for use of novel gene technology, it is still foreseen that its use will be superimposed upon a platform of classical breeding programmes. Hence the hierarchy of populations (Sect. 5.5.3), accommodating the trade-off between level of genetic improvement and genetic diversity, will likely remain in some form. To recapitulate, the production populations represent the top of the hierarchy, with the greatest genetic gain but least diversity. They are underpinned by breeding populations, representing somewhat less gain but greater diversity, and which are the "engine room" for achieving continued genetic gain. In turn, breeding populations are underpinned by genetic resources, representing the least gain but the greatest diversity. Maintaining such a hierarchy, as a "metapopulation," makes it possible to enjoy genetic gains without substantial loss of genetic diversity, thereby protecting very long-term genetic gain, and providing for imposed changes in breeding goals. However, it has its costs, both direct costs and opportunity costs, placing demands on commitment to managing various risks. Even with strong commitment, detailed structures of population hierarchies are likely to change with new technologies. For instance, a high degree of pedigree reconstruction should avert the need for laborious and costly controlled crossing, while some traditional restrictions on moving material between levels in the hierarchy (Sect. 5.5.3) may be lifted.

The very success of genetic improvement is likely to jeopardize conscientious genetic management (Burdon 2010b). The availability of greatly improved stock for commercial planting increases the opportunity costs of maintaining less-improved or largely unimproved material that would be the reservoirs of genetic diversity. Given the degree to which radiata has become domesticated, the genetic stewardship of population management is of special importance. These considerations create a special need for strong political and institutional commitment to forward-looking population management. Yet, as we shall see (Sect. 9.3.5), various political and institutional developments are militating against such commitment.

9.3.5 Institutional and Political Changes and Challenges

The future for growing radiata and its further domestication will be in changing political and institutional contexts, which will be intertwined with changing economic circumstances. It is possible to review such changes and the directions that they are currently taking. While that may give pointers to the future domestication, detailed predictions must be speculative.

Forest ownership and corporate structures, along with research and development (R&D) organizations, are key institutional factors. Changes in them are often partly driven by politics, although development of new technologies can also exert strong influences on the R&D organizations. Non-government organizations (NGOs) can be intensely political.

For safeguarding and improving the productivity, quality and profitability of radiata, the institutional and political picture is a concern. Taking many of the opportunities, and meeting needs mentioned earlier, will require staff of high calibre, with an appropriate mix of skills, operating in well-run organizations that have adequate funding and ongoing commitment. Meeting these requirements will require much lead time, yet some key parts of the skills base are being eroded on a global scale.

9.3.5.1 Forest Ownership

Forest ownership has been undergoing major changes in recent years, and will doubtless continue to do so. In both New Zealand and Australia there has been massive privatization of state-owned forest. This has been driven largely by political doctrine that market forces can solve most problems, and by global pressures to reduce tax takes. Essentially complete in New Zealand, privatisation has some way to go in Australia, where it is likely to proceed further than at present. Since privatization, forest ownership has tended to be volatile, forests often changing hands and sometimes repeatedly. Vertical integration, with forest growing and wood processing vested in the same ownership, has become the exception. Globalisation of business has become a prominent factor, in that forest ownership has often become dominated by large, offshore interests. For such interests, radiata wood can be seen more as a commodity, rather than as a source of high-value products. Moreover, for those interests, managing the risks of depending massively on the one species can often be accommodated by global spread of risk. Even among smaller forest owners, who own a very significant part of the New Zealand forest estate, forestry interests often form parts of risk-spread investment portfolios. The risk is basically one of a biotic crisis, probably some new disease, displacing radiata from much of the range where it now thrives. Such a risk is important, because of potentially dire consequences despite low probability of eventuating. The spread of such risk that is enjoyed by individual forest owners contrasts with the risk exposure of forestry sectors and even the national economy in the case of New Zealand (Burdon 2010b).

In Chile, forest ownership has largely stabilized into the hands of Chilean nationals, while vertical integration is considerable, albeit more within groups of commonly-owned companies rather than within individual members of company groups.

In Spain, the forest ownership is mostly fragmented into small ownerships which, however, enjoy considerable support at various levels of government. Choice of tree species, between radiata and eucalypts, remains volatile.

In South Africa much of the radiata plantation area has gone into conservation estate, where restoration of native vegetation is often being pursued. Between this factor, and problems with pitch canker, this is a country where the future role of radiata may well be sharply reduced.

Economic forces have already influenced the pattern of ownership, and market perceptions will affect how the species is grown. The American market has dominated the demand for high-value appearance-grade products, but in recent years has been weak. Products from lower-grade logs have dominated the booming Chinese market and still dominate a burgeoning Indian market. But it is becoming increasingly unlikely that these Asian markets will continue to be dominated by demand for lower-grade log products. Indeed, growing now for these markets in their present form could prove costly.

9.3.5.2 R&D Institutions

The widespread privatisation of forest estate has been accompanied by institutional changes involving R&D organizations. Such changes have often had much the same political drivers as privatisation of forests, and the subsequent changes in forest ownership have had their own impacts. In addition, new technology has brought its own changes within R&D organizations. Outright and rapid exiting of governments from the R&D has not occurred, but funding has commonly diminished and funding mechanisms have tended to change.

Genetic improvement of radiata has increasingly devolved onto tree breeding associations or companies, such as the Southern Tree Breeding Association in Australia and the Radiata Pine Breeding Company in New Zealand, such that breeding operations are no longer embedded within R&D organisations. Even where government support for genetic improvement and other R&D continues, it may be very much conditional upon industry support. In Australia, such industry support essentially ceased for the now disbanded CSIRO Forest and Forest Products Division. Research arms of state agencies have greatly contracted, and forestry research via Cooperative Research Centres is now discontinued.

Within Chile, R&D capabilities relating the growing, genetic improvement and processing of radiata have become minimally vested in government agencies as such. There is major corporate involvement, in Bioforest within the Arauco group of companies, and within the Mininco group. A high proportion of the actual research is done within various universities, which can pose problems of co-ordination. Uptake of overseas research has been significant. The breeding cooperative is still in existence, but the dominance of its role is now reduced. A return to high level of local ownership of forest resources and processing plant is probably a significant advantage.

In Spain, with the fragmentation of forest ownership, government agencies, especially at the provincial level, but with support from the European Union, are playing prominent roles in R&D. While various avenues of biotechnology are being pursued, the platform of physical breeding-programme infrastructure is limited.

In South Africa breeding work with radiata is now limited, and is likely to remain so, except that efforts are being made to hybridise it with pine species that are resistant to pitch canker. Successful hybridization, or achieving resistance through genetic engineering, could bring a major resurgence of its use.

9.3.5.3 Impacts of New Technology

Developments in DNA technologies, along with budgetary restrictions, have led to a cultural divide between traditional tree breeders and the "molecular" camp. That divide has been exacerbated by a perception among the breeders that molecular work has drained off their funding, with consequent opportunity costs in respect of achieving genetic gain. Managing the problems of communication and associated issues of resource allocation will remain a challenge for some time to come.

The traditional model of cooperative organizations undertaking tree breeding has come under severe stresses. Some of the stresses arise from institutional changes, but advances in technology have played a major part. The availability of new technology has widened the gaps in level of technological engagement among cooperative members, which accentuates temptations for some individual members to appropriate new technology for themselves. Specialised biotechnology companies have arisen, sometimes spun off from forest growers' own biotechnology arms, and with various fresh alliances formed. Competition among such companies, while typically a spur, can be a barrier to the cooperation on which technological advances can depend heavily. Moreover, proprietary appropriation of biotechnology can contribute to adverse perceptions in a public that is inherently wary of new biotechnologies.

9.3.5.4 Non-Governmental Organisations (NGOs)

NGOs can be very influential politically. They can influence politicians directly, or else indirectly through manipulating measures of public opinion. They can also influence consumer purchases. Various NGOs are heavily involved with environmental causes, and in forestry they have set up schemes of environmental certification of timber, only endorsing timber or wood products that have been assessed as having been produced on an environmentally sustainable basis. But, however laudable their ultimate goals may be, at least some of the NGOs have problems with the phenomenon of plantation forestry. Even acceptance of plantation forestry came reluctantly, and various practices aimed at safeguarding and intensifying wood production are not endorsed. Insofar as such non-endorsement may compromise the role of plantation forestry in relieving pressures of exploitation on natural forest, such stances on the part of NGOs can be highly counterproductive (Sect. 9.3.4.6). Needless to say, the growing of radiata, being in the forefront of plantation forestry, has been heavily embroiled in these issues. Hopefully, the stances of certain NGOs will continue to evolve in directions that serve their goals better.

9.4 The Clonal Forestry Goal

Clonal forestry, namely the commercial deployment of finite numbers of proven clones, has already featured several times in this chapter. However, it warrants special coverage, with some recapitulation, because its implementation is a key part of the highest possible level of domestication. It offers a means of capturing genetic gains that are not otherwise available. These come from favourable non-additive gene effects which are almost all lost in the course of seed production. Some of the most intriguing non-additive gene effects involve epistasis, resulting from interactions between the effects of genes at different loci. Such effects, however, are very hard to quantify and characterise, unless they involve massive effects of particular genes.

The genetic uniformity of clonal material promises major advantages for harvesting and processing. Because wood properties are typically so heritable, genetic uniformity allows big reductions in tree-to-tree variation. Even for growth variables, far better uniformity may be achievable. For stem diameter growth, and thence volume production, the fact that all the non-additive gene effects contribute to heritability with clonal systems means that genetic variation is no longer part of the "noise" variation. In addition, avoiding certain physical interactions that occur between different genotypes can further reduce noise variation. Using clones to minimize variation in stem size, however, will demand excellent practice in the nursery and in field establishment, but the latter may be difficult to achieve on many of the planting sites that will be available. Above all, though, the more predictable the wood properties of individual trees, the more effectively can processing be optimized.

Large-scale clonal forestry with a species like radiata would require risk-spread "portfolios" of clones, as a matter of basic prudence. Deployment of such portfolios will need to be addressed at both local and landscape scales, the latter scale involving greater specialisation of clones for definable site categories. Such matching of clones to sites at once promises greater genetic gains. At the same time, the genetic uniformity does not provide a "cushion" against the presence of some ill-adapted genotypes segregants whose failure can be easily tolerated in natural or artificial thinning of growing stands. This makes especially important good matching of clones to sites, or else identification of clones that are very broadly adapted.

More subtle benefits may come from clonal forestry. Some of the key prospective benefits of clonal forestry involve traits that involve interactions among individual trees. Competitive interactions strongly affect growth of individual stems. Where different clones have different growth curves, much unwanted variation in log size may be difficult or impossible to avoid if different clones are grown in intimate mixture. Moreover, it is in monoclonal blocks that it should be possible to exploit divergences between productivity as a crop and competitive ability. With resistance to certain types of climatic damage, such as wind damage, superiority may not be clearly expressed when clones are grown in mixture, because the fate of individual trees can be so strongly influenced by that of their neighbours.

Empirical evaluation of the performance of individual clones as crops, rather than essentially as competitors, is quite impractical for the potential large numbers of candidate clones. A challenge remains, then, to characterise the "crop ideotype" so that far greater numbers of candidate clones can be evaluated with sufficient reliability on the basis of how they conform to that ideotype.

Several of the theoretical advantages of clonal forestry accrue in monoclonal blocks. Even so, it is not fully clear whether or when mosaics or intimate mixtures of clones will eventually be preferred in radiata. The appropriate choice is likely to depend on specific circumstances, and may change over time.

The technical platforms for fully satisfactory clonal forestry with radiata are not yet totally assured. If any tree, whatever its age, can be rejuvenated at will such that it is immediately amenable to mass vegetative multiplication, we would have a full answer. But such rejuvenation with pines remains to be achieved. Short of that, the capability of using cryopreservation to store individuals in a fully juvenile state and allow indefinite mass-propagation would make clonal forestry much more practicable. Recent progress in these directions (Hargreaves et al. 2009) is very promising.

As mentioned earlier, clonal forestry is seen as the likely context for any commercial implementation of genetic engineering.

In a final note of caution, clonal forestry, by depending on use of proven clones, is inherently behind the leading edge of advancing genetic gain for additive gene effects. Its net benefits must therefore outweigh the significance of this lag. That lag, however, may be avoided by committing every candidate clone to cryopreservation from before or very soon after germination. While this is logistically demanding, storage of individual clones can be discontinued immediately after culling.

9.5 Concluding

The growing of radiata has seen fluctuations in rates of planting and even in total areas of plantation. These fluctuations have resulted from various factors: biotic, political, economic and institutional. In parts of the world, especially southern and eastern Africa, there have been substantial retrenchments of the areas where it is grown, first through diseases catching up with plantations in inherently unsuitable climates, and latterly through introduction of a new disease in the presence of vectors. On the other hand, one major biotic alarm, with *Dothistroma* in New Zealand, actually strengthened the position of radiata as the preferred species. Also, use of fertilisers reversed some initial plantation failures and substantially extended the range of sites where radiata can thrive. Further biotic alarms will doubtless occur and contribute to fluctuations in planting. Political decisions and wars have contributed more or less directly to booms and lulls in new planting. Less directly, political decisions have led in recent years to changes and volatility in forest ownership, leading to some major lulls in planting. The global financial crisis has also contributed to lulls, partly through a consequent collapse in prices for carbon credits. A boom in short-fibred pulp, along with the financial attractions of growing short-rotation eucalypt crops, and the "biotic honeymoon" of exotic eucalypts, have also eroded or stalled the status of radiata in recent years. Nevertheless, the areas of infertile land remaining, their catchment values, the likely need for solid-wood products, and the advantages of solid-wood crops for carbon sequestration would all point in favour of future expansion of the role of radiata. The outcome of the 2015 UNFCCC conference in Paris, at which the acceptance of climate change as a problem gained international agreement, should lead to a boost in afforestation. Climatic change will surely cause some shifts in the areas where radiata can be grown successfully, rather than necessarily affecting the total area (cf Mead 2013), but genetic improvement should be able to extend the range of suitable climates. The anticipated pluses and minuses for the future role of radiata are summarized in Box 9.1.

Box 9.1 Strengths, Limitations, Threats and Opportunities for Future Role of Radiata

Basic strengths

Outstanding growth potential for conifer, with sustained productivity
Significant range of climatic tolerances
A pine's tolerance of limited soil fertility
Tolerance of significant weed competition
Amenability to selective herbicides
Ease of handling in nursery and transplanting
General ease of wood processing and utilisation, especially for solid-wood products
Genetic variability well sampled and now available
Domestication achieved to date
Existing scientific knowledge base on species
Carbon sequestration potential

Current limitations

Climatic tolerances
Biotic vulnerabilities
Early relative growth rate still modest
Wood-quality problems
Diversion of resources into reproduction
Competitor, not crop ideotype
Imperfect control of maturation
Very limited crossability
Problems posed by huge genome

Threat factors and hazards

Pest/pathogen incursions: with potential for biotic crises
Climatic change: contractions in geographic range; increased risks of climatic damage and biotic crises; extending geographic ranges of pests, pathogens and weeds
Potential phosphorus shortage: affecting part of species' range, but problems subordinate to those of food production
Weeds: extended climatic ranges; continuing incursions
Biosecurity barriers: restricting access to germplasm
Competing land uses: farming (on easy terrain), eucalypt plantations
Increased mechanisation: degrading economics of pruning

(continued)

Box 9.1 Strengths, Limitations, Threats and Opportunities for Future Role of Radiata (continued)

Political pressures: opposition to herbicides and wood preservatives, subsidies to alternative land uses, mis-perceptions of environmental hazards

Institutional factors: compromised risk management

Genetic improvement: militating against risk management in protecting diversity

Economic pressure for earlier harvests: compromised wood properties; advantages accruing to eucalypts; reduced carbon sequestration

Opportunities: factors and influences

Genetic improvement: gains in productivity; improved wood quality; enhanced pest/disease resistance; extended site tolerances (direct and biotic)

Propagation technology: allowing accelerated breeding; faster and better capture of genetic gain; more efficient production of planting stock

Climatic change: some extensions of geographic range; premium on carbon sequestration, both direct and favouring wood over competing materials; accentuated call for soil and catchment protection

Radical upgrade of domestication: flowering on command and command only, redesign of tree architecture; genetic engineering

Environmental services: existing calls for soil and catchment protection and rehabilitation; wildlife habitats

Enhanced management systems and technology: remote sensing, software systems, nutrient budgeting

Precision technology: better application of protective sprays and fertiliser, and other operations, assisted by new remote-sensing systems

Novel DNA technology: both informing and supplementing classical breeding

Overcoming crossability barriers: radical gains in biotic resistances

Biocontrol: of pests, weeds and pathogens

Improved biosecurity diagnostics: direct biosecurity gains, better germplasm availability.

Radiata provides a case history of domesticating a forest tree species, in both the achievements so far and the challenges ahead. Both the history and the future embrace a wide range of issues which are varyingly intertwined: scientific and technological; economic, political and institutional; and environmental and even humanitarian. We will not, however, trace these various issues afresh; rather, we just re-emphasise two relating to domestication *vis-à-vis* the wild state. Domestication of forest trees offends the sensibilities of many who see forests as essentially wild ecosystems. Intensive cultivation is likely to reduce biodiversity where it is practised, although mitigation of losses will often be possible. Of far greater importance is that intensive use of very large areas of land will be an imperative for the world's human population. The more intensive such use can be, without disastrous environmental consequences, the more land should remain available for recreation and biological conservation. As Brockerhoff et al. (2008) have argued, plantation forest can in various ways play a very positive role for biodiversity. There are direct ways in which the presence of the cultivated tree species can favour organisms whose perpetuation is desired; less directly, minor modifications in plantation management can serve to favour biodiversity and general amenity values; and quite indirectly the availability of plantation-grown wood can relieve at least some exploitation pressures on natural forests. Within radiata, there are the challenges of achieving the benefits of domestication, on the one hand, and maintaining genetic diversity both for its own sake and for providing for the distant future and unpredictable developments, on the other. Reconciling this apparent conflict is quite possible technically. Achieving that, however, faces stern political and institutional challenges.

References

Ahuja MR (2014) Regulation of transgene expression and containment in forest trees. Pp 335–365 in: Ramawat KG, Merillon J-M, Ahuja MR (eds) Tree biotechnology. CRC Press, Boca Raton

Ashley K, Cordell D, Mavinic D (2011) A brief history of phosphorus: from the philosopher's stone to nutrient recovery and reuse. Chemosphere 84:737–746

Bishop-Hurley SL, Zabkiewicz JA, Grace LJ et al. (2001) Conifer genetic engineering: transgenic *Pinus radiata* (D.Don) and *Picea abies* (Karst) plants are resistant to the herbicide Buster. Plant Cell Rep 20:235–243

Booth TH (1990) Mapping regions climatically suitable for particular tree species at the global scale. For Ecol Manag 36:47–60

Booth TH, Jovanovic T, New M (2002) A new world climatic mapping programme to assist species selection. For Ecol Manag 163:111–117

Borkin KM, Parsons S (2010) The importance of exotic plantation forest for the New Zealand long-tailed bat (*Chalinolobus tuberculatus*). N Z J Zool 37(1):35–51

Brockerhoff EC, Jactel H, Parotta JA et al. (2008) Plantation forests and biodiversity: oxymoron or opportunity? Biodivers Conserv 17:925–951

Burdon RD (1992) Genetic survey of *Pinus radiata*. 9. General discussion and implications for genetic management. N Z J For Sci 22:274–298

Burdon RD (2001) *Pinus radiata*. In: Last FT (ed.) Ecosystems of the World, Vol. 19, Tree Crop Ecosystems. Elsevier, Amsterdam. pp 99–161

Burdon RD (2008) Branching habit in radiata pine – breeding goals revisited. N Z J For 53 (1):20–23

Burdon RD (2010a) Wood properties and genetic improvement of radiata pine. N Z J For 55 (2):22–27

Burdon RD (2010b) Conservation and management of potentially resistant germplasm: a key but easily neglected part of a robust biosecurity strategy. N Z J For Sci 40:115–122

Burdon RD (2011) Shoot dieback epidemics of *Pinus radiata* in New Zealand: is there an endophyte story? N Z J For Sci 41:131–139

Burdon RD, Carson MJ, Shelbourne CJA (2008) Achievements in forest tree genetic improvement in Australia and New Zealand 10: *Pinus radiata* in New Zealand. Aust For 71:263–279

Burdon RD, Libby WJ (2006) Genetically modified forests: from Stone Age to modern biotechnology. Forest History Society, Durham, NC, USA. p xii + 79

Burdon RD, Lstibůrek M (2010) Integrating genetically modified traits into tree improvement programs. Pp 123–134 (Chapter 7) In: El-Kassaby YA (ed) IUFRO biotechnology task force state-of-the-art report. FAO, Rome

Burdon RD, Walker JCF, Megraw RA et al. (2004) Juvenile wood (*sensu novo*) in pine: conflicts and possible opportunities for growing, processing and utilization. N Z J For 49(3):24–31

Burdon RD, Walter C (2004) Exotic pines and eucalypts: perspectives on risks of transgenic plantations. Pp 52–75 (Chapter 5) In: Strauss SH, Bradshaw HD (eds) The bioengineered forest. RFF Press, Washington

Burdon RD, Wilcox PL (2011) Integration of molecular markers in breeding. Pp 276–322 (Chapter 7) In: Plomion C, Bousquet J, Kole C (eds) Genetics, genomics and breeding of conifers. Science Publishers, St Helier, Jersey, British Channel Islands and Enfield

Dance A (2015) Core concept: CRISPR gene editing. Proc Natl Acad Sci USA 112(20):6245–6246

DeBell DS, Harms WR, Whitesell CD (1989) Stockability: a major factor in productivity differences between *Pinus taeda* plantations in Hawaii and the southeastern United States. For Sci 35:708–719

De Mello F, Kern FL, Bertoli CD (2014) Progress in dairy cattle selection. Adv Dairy Res 2:110. doi:10.4172/2329-888X.10001110

Dungey HS, Low CB, Ledgard NJ et al. (2011) Alternatives to *Pinus radiata* in New Zealand high country: early growth and survival of *Pinus radiata*, *P. attenuata* and their F_1 hybrid. N Z J For Sci 41:61–69

Dungey HS, Yanchuk AD, Burdon RD (2015) A 'reality check' in the management of tree breeding programmes. Pp 461–479 In: Fenning T (ed) Challenges and opportunities for the world's forests in the 21st century. Springer Science+Business Media, Dordrecht

Evans LT (1993) Crop evolution, adaptation and yield. Cambridge: New York, Cambridge University Press. p 500

Evison D (2008) A method for comparing investment returns from major rural land uses including forestry. N Z J For 53(3):27–32

Evison D (2013) Some observations on the economics of commercial forestry in New Zealand. N Z J For 58(3):3–13

Ganley R (2010) Controlling diseases without pesticides. Biosecurity 99 (August 2010: 12–13). MAF Biosecurity of New Zealand, Wellington

Grabianowski M, Manley B, Walker JCF (2006) Acoustic measurements on standing trees. Wood Sci Technol 40:205–216

Grattapaglia D, Resende MDV (2011) Genomic selection in forest tree breeding. Tree Genet Genomes 7:241–255

Häggman H, Sutela S, Walter C et al. (2014) Biosafety considerations in the context of the development of GE trees. Pp 491–524 In: Fenning T (ed) Challenges and opportunity for the world's forests in the 21st century. Forestry sciences 81. Springer, Dordrecht

Hall SS (2016) Editing the mushroom. Sci Am 314(3):48–55

Hargreaves CL, Reeves CB, Find JI et al. (2009) Improving initiation, genotype capture and family representation in somatic embryogenesis of *Pinus radiata* D.Don by a combination of zygotic embryo maturity, media and explants preparation. Can J For Res 39:1566–1574

Hawke M (2011) Agroforestry – the Tikitere experience. N Z Tree Grow 32(2):23–26

Hayes BJ, Bowman PJ, Chamberlain AJ et al. (2009a) Invited review: genomic selection in dairy cattle: progress and challenges. J Dairy Sci 92(2):433–443

Hayes BJ, Visscher PM, Goddard ME (2009b) Increased accuracy of artificial selection by using the realised relationship matrix. Genet Res (Camb) 91(1):47–60

Ivković M, Gapare WJ, Wharton T et al. (2010b) Risks affecting breeding objectives for radiata pine in Australia. Aust For 73:265–278

Ivković M, Wu HX, Kumar S (2010a) Bioeconomic modelling as a method for determining economic weights for optimal multiple-trait selection. Silvae Genet 59:77–90

Kay M (2005) The risk of growing eucalypts in New Zealand. Pp 229–236 In: Lieutier F, Ghaioule D (eds) Entomological research in mediterranean forest ecosystems. INRA, France, Editions Quae

Kaya HK, Bedding RA, Akhurst RJ (1993) An overview of insect-parasitic and entomopathogenic nematodes. Chapter 1 In: Bedding RA, Akhurst RJ, Kaya HK (eds) Biological control of insect pests. CSIRO, East Melbourne

Kean S (2010) Besting Johnny Appleseed. Science 328:301–303

Kennedy SG, Yanchuk AD, Jefferson PA (2013) Relationship between heartwood traits with diameter growth, implications for genetic selection in *Pinus radiata*. Tree Genet Genomes 9:1313–1319

Kirschbaum MUF, Watt M, Tait A, Ausseil A-G (2012) Future wood productivity of *Pinus radiata* in New Zealand under expected climatic changes. Global Change Biol 18:1342–1356

Kleinpaste R (1990) Kiwis in a pine forest. Pp 97–103 In: Fuller E (ed) Kiwis: a monograph of the Apterygidae. Raymond Harris-Ching SoTo Publishing, Auckland

Levack H (2016) Will forestry be a winner from the Emissions Trading Scheme review? *N Z Tree* Grow 37(2):28–31

Libby WJ (1982) What is a safe number of clones per plantation? Pp 342–360 in: Heybroek HM, Stephan BR, von Veissenberg K (eds) Resistance to diseases and pests in forest trees, Proceedings third International Workshop on the Genetics of Host-Parasite Interactions in Forestry, Wageningen, the Netherlands, 14–21 Sept 1980. Pudoc, Wageningen

Libby WJ (1987) Do we really want taller trees? Adaptation and allocation as tree-improvement strategies. The HR MacMillan Lecture in Forestry. University of British Columbia, Vancouver, p 15

Maclaren JP, Knowles RL (1999) Economics of final crop stocking at the Tikitere field trial. Part 1: Volume and quality comparisons. N Z J For Sci 29:165–174

Maclaren JP, West GG (2005) Farm site effect. Pp 76–77 In: Colley M (ed) NZIF forestry handbook, 4th ed. New Zealand Institute of Forestry, Christchurch

Mead DJ (2013) Sustainable management of *Pinus radiata* plantations. FAO forestry paper No. 170. FAO, Rome. p xv + 246

Monge JJ, Parker WJ, Richardson JW (2016) Integrating forest ecosystem services into the farming landscape: a stochastic economic assessment. J Environ Manag 147:87–99. doi:10.1016/j.jenvman.2016.01.030

Moore JR, Cown DJ (2017) Corewood (juvenile wood) and its impact on wood utilisation. Curr For Rep 3:107–118

Nakaya M, Isobe SN (2016) Will genomic selection be a practical method for plant breeding? Ann Bot (Lond) 110:1303–1316

O'Hehir J, Nambiar S (2010) Productivity of successive rotations of *Pinus radiata* plantations in South Australia over a century. For Ecol Manag 259:1857–1869

Old KM, Libby WJ, Russell JH, et al. (1986) Genetic variability in susceptibility of *Pinus radiata* to western gall rust. Silvae Genet 35:145–149

Orme S (2017) Manuka – a rapidly growing industry. N Z Tree Grow 38(1):10–13

Palmer DJ, Watt MS, Kimberley MO et al. (2010) Mapping the productivity of radiata pine. N Z Tree Grow 31(2):18–19

Pinkard L, Battaglia M, Howden M et al. (2010) Adaptation to climate change in Australia's plantation industry. CSIRO, East Melbourne, p 78

Richards K (2013) Using science to inform the place of forestry in the landscape. N Z J For 58 (2):7–15

Ritland K, Krutovsky KV, Tsumura Y et al. (2011) Genetic mapping in conifers. Pp 196–238 In: Plomion C, Bousquet J, Kole C (eds) Genetics, genomics and breeding in conifers. Science Publishers Inc, Enfield

Roberds JH, Namkoong G, Davey CB (1976) Family variation of growth response of loblolly pine to fertilising with urea. For Sci 22:291–299

Seaton R, Holland JD, Minot EO, Springett BP (2009) Breeding success of New Zealand falcons (*Falco novaeseelandiae*) in a pine plantation. N Z J Ecol 33(1):32–39

Smith WH, Konar RN (1969) Initiation of ovulate strobili in cotyledon-stage seedlings of *Pinus elliottii*. Can J Bot 47:624–626

Strauss SH, Costanza A, Séguin A (2015) Genetically engineered trees: paralysis from good intentions. Science 349:794–795

Turner J, Lambert MJ (2011) Analysis of nutrient depletion in a radiata pine plantation. For Ecol Manag 262:1327–1336

Tuskan GA (2007) Bioenergy, genomics, and accelerated domestication: a US example. In the role of agricultural biotechnologies for production of bioenergy in developing countries. Proceedings, FAO Seminar, Rome, 12 October 2007. http://www.fao.org/biotech/seminar-oct2007.htm

Vaccari DA (2009) Phosphorus famine: the threat to our food supply. Sci Am 300(6):42–47. (June 2009)

Wagner A, Tobimatsu Y, Phillips L, et al. (2015) Syringyl lignin production in conifers: proof of concept in a pine tracheary element system. Proceedings of the National Academy of Sciences of the USA. doi:10.1073/pnas.1411926112

Watt MS, Palmer DJ, Bulman LS (2012) Predicting the severity of Dothistroma needle blight on *Pinus radiata* under future climate in New Zealand. N Z J For Sci 41:207–215

Whitehead D (2011) Forests as carbon sinks—benefits and consequences. Tree Physiol 31: 893–902

Woollons RC, Manley BR (2012) Examining growth dynamics of *Pinus radiata* at old ages in New Zealand. Forestry 85(1):79–86

Wu HX, Owen JV, Albarquez A et al. (2004) Inbreeding in *Pinus radiata* – V. The effects of inbreeding on fecundity. Silvae Genet 53:80–87

Xu P, Walker JCF (2004) Stiffness gradients in radiata pine trees. Wood Sci Technol 38(1):1–9

Yao RT, Harrison DR, Harnett M (2017) The broader benefits provided by New Zealand's forests. N Z J For 61(4):7–15

Zagory D, Libby WJ (1995) Maturation-related resistance of radiata pine to western gall rust. Phytopathol 75:1443–1447

Species Common Names with Latin Names

List cross-referencing species common names with Latin names (excluding some species that are mentioned once only).

Category	Latin name	Synonyms	Common name(s)
Fungi & oomycetes	*Allantophomopsis pseudotsugae* (M.Wilson) Nag Raj	*Phomopsis strobi* Syd.	phomopsis
	Cyclaneusma minus (Butin) di Cosmo et al.	*Naemacyclus minor* Butin	cyclaneusma; naemacyclus
	Diplodia pinea Kickx	*Sphaeropsis sapinea* (Fries) Dyko et. Sutton	diplodia
	Dothistroma pini Hulbary	*D, septosporum* (Doroguine) Morelet *Mycosphaerella pini* Postrup ex Monk *Schirrhia pini* Funk et Parker	dothistroma; dothi; red band needle blight
	Nectria fuckeliana (C.Booth) Castl. & Bossman	*Neonectria fuckeliana* (C.Booth) Dick & Crane	nectria
	Austropuccinia psidii (Winter) Beenken	*Puccinia psidii* Winter *Uredo rangelii* JA.Simpson et al.	myrtle rust guava rust eucalypt rust
	Fusarium circinatum Nirenberg & O'Donnell	*Fusarium subglutinans* f. sp. *pini*	pitch canker
	Phytophthora pinifolia Durán et al.		'daño foliar del pino'
	Phytophthora pluvialis Reeser et al.		red needle cast
Trees or shrubs	*Buddleja davidii* Franch.		buddleia
	Pinus attenuata Lemm.	-	attenuata; knobcone pine (US)
	P. elliottii Engelm.	-	slash pine

(continued)

R. Burdon et al., *Domestication of Radiata Pine*, Forestry Sciences,
DOI 10.1007/978-3-319-65018-0

(continued)

Category	Latin name	Synonyms	Common name(s)
	P. muricata D.Don	-	muricata pine; muricata; bishop pine (US)
	P. nigra ssp. laricio (Poiret) Maire	*P. laricio* Poiret	Corsican pine; Corsican
	P. patula	-	patula pine; patula
	P. pinaster Aiton	-	maritime pine; pinaster
	P. ponderosa Laws.	-	ponderosa pine; ponderosa; western yellow pine (USA)
	P. radiata D.Don	*Pinus insignis* Dougl. [Various other names]	radiata pine; radiata; insignis; Monterey pine (USA)
	P. sylvestris L.		Scots pine
	P. taeda L.	-	loblolly pine
	Pseudotsuga menziesii (Mirbel) Franco	*Pseudotsuga taxifolia* Britton	Douglas-fir; Oregon pine
	Leptospermum scoparium		manuka; tea tree
Insects	*Essigella californica* Essig	-	Monterey pine aphid
	Rhyacionia buoliana Schiff.	-	European pine shoot moth
	Sirex noctilio Fabricius	-	sirex
	Lymantria dispar dispar L.	-	European gypsy moth

Note: Lists of synonyms and common names are not exhaustive; see also CAB International (2002) and/or consult Internet. For radiata, see also Forde (1964)

References

CAB International (Compil.) (2002) Pines of silvicultural importance. CAB International, Wallingford, UK. p 531

Forde MB (1964) Variation in natural populations of *Pinus radiata* in California. Part 4. Discussion. N Z J Bot 2:486–501

Glossary[1]

Additive gene effects Those gene effects whereby offspring tend to be intermediate between their two parents, which can be captured through reproduction by seed, and form the basis of cumulative genetic gain through selection over repeated generations.

AFLP A type of DNA marker that can serve as a 'genetic fingerprint', but often not representing 'functional' diversity in itself.

Agroforestry* Combination of producing both timber and pasture, embarked upon in radiata to produce pasture as an intermediate yield in place of commercial thinning.

Alleles Alternative forms of a gene at a specific chromosomal locus.

Backwards selection Re-selection of parents based on the performance of their offspring in **progeny tests**.

Bole The entire main **stem** of the tree.

Branch cluster A group of branches, which can include seed cones, occurring together at a particular position up the **bole** or along a larger branch. While appearing to be a whorl in pines, the branches actually occur close together in several parallel spiral arrangements.

Breed* In radiata, a group of individuals characterised by having been genetically selected for a specific breeding goal. This has superseded an earlier concept based on groups of parent clones selected at specific dates.

[1] **Notes**

1. Some of the definitions, marked with asterisks, are in some degree specific to the context of this book rather than being universally applicable. Not included are some terms that are used and explained just within a single passage.
2. There are certain terms in the text which are denoted in boldface and suffixed by a superscripted G. These instances direct the reader to refer the Glossary. They appear normally at the first substantive mention of each term concerned; however, where there are long gaps between mentions, this flag for a term in the text may be repeated.

R. Burdon et al., *Domestication of Radiata Pine*, Forestry Sciences,
DOI 10.1007/978-3-319-65018-0

Breeding goal A set of attributes, characterized by genetic value(s) of a certain trait or traits, that is/are pursued by the breeder. Note the distinction that can exist between breeding-goal traits and **selection traits**.

Breeding population Population of select parents subject to recurrent cycles of selection, mating, progeny trials, selection and so on, to achieve cumulative genetic gain (from additive gene effects) over successive generations. Provides selections for the **production population** and backed up by gene resources.

Breeding value A measure of the expected contribution of the parent's offspring to changing the population mean for a given trait or index of traits. Governed by **additive gene effects**, it is commonly expressed in units of additive genetic standard deviation about the mean for the parental candidate population. The term may be alternatively to the value of the parent, or to one-half of that, the latter being the expected value of its offspring if the parent is mated to a random sample of its population

C-effects (see also **epigenetic effects**) Effects common to ramets of a clone that are not due to genetic covariation. These may include true maternal effects, a non-random common-environment in the nursery, or a common physiological condition (notably **maturation state** or 'physiological age') of donor tissue used for cloning. Such effects that are specific to individual ramets of a clone may be denoted **c-effects**.

Chromosome Strand of DNA including genes and non-coding sequences, along with protein, that forms a discrete part of the **genome**. Radiata has two sets of 12 chromosomes.

Clearcutting (i) Piece of clear (defect-free) sawn timber cut from between defects (typically knots in radiata)
(ii) The practice of clearfelling areas of stand.

Climbing select Term applied to seed collected by climbing the best trees in commercial stands.

Clonal forestry Mass-propagation and deployment of known (and preferably well-proven) **clones**. Can be grown in risk-spread 'portfolios' as mixes or mosaics of single clones. Clones can be misleadingly termed "varieties".

Clone A set of individuals that are genetically identical, comprising the **ortet** and the **ramets** that have been propagated from it.

Coefficient of variation Standard deviation of the population in question divided by its mean. The standard deviation is not necessarily phenotypic; e.g. a genetic coefficient of variation is the square root of the genetic variance divided by the mean.

Commercial plantation Plantation established with the prime aim of producing saleable timber.

Commercial thinning Thinning involving extraction of wood with aim of achieving immediate profit.

Common-garden Term applied to plantings that allow proper comparison of different groups of plant material growing in the same environment.

Conifers Literally, cone-bearing trees. In practice, trees belonging to gymnosperms in which unfertilised ovules are open to the air instead of being enclosed in ovaries. Also, in connection with their timber, termed **softwoods**. While varying widely, none produce truly hard wood. Generally easier to process and use as lumber than **hardwoods**.

Conservation As this term has evolved, it now implies management of the forest to maintain, restore or enhance functioning forest ecosystems and the services they provide (cf **preservation**).

Control-pollinated Pertaining to a family of a seed parent produced using pollen of a single known tree (pair-cross) or a known mix of pollens.

Corewood The first few rings from the pith, which in radiata are inferior for most wood uses. The first five rings are often termed inner corewood, while rings 11-15 are often termed transitional to **outerwood.** Historically, corewood has been widely—but incorrectly—termed "juvenile wood".

Correlation breaker Wide deviant from a genetic correlation, typically applied to favourable deviants from adverse correlations.

Cotyledons The first leaves to appear in a seedling, differing in anatomy and arrangement from all subsequent leaves.

Creaming See **Highgrading.**

Cryopreservation Preserving living material at ultra-low temperature. Keeping alive and 'reviving' pose challenges, but effectively halts **maturation**.

Cultivar (i) Applied in agriculture or horticulture to stabilised plant variety if propagated as seed, or a clone if routinely propagated vegetatively. Widely subject to formal registration safeguarding Plant Breeders' Rights.

(ii)* Applied in forestry to a **breed** or **clone**, traditionally characterised by distinctiveness, uniformity and stability, features that assure at least some predictability of performance.

Cutting (i) A portion of shoot, pre-treated or not, taken to produce roots and start growing as an independent plant in a rooting medium.

(ii) A plant that has been successfully rooted in this way (syn. **steckling**).

[Wood] Density Specific gravity, at a defined moisture content, **basic density** being oven-dry weight divided by green (completely undried) volume.

Diallel cross A crossing scheme that, in its complete form, entails crossing a set of parents in all possible pairwise combinations. It may or may not include self-pollinations or crossing each pair of parents in reciprocal combinations. Incomplete variants include partial diallels and disconnected diallel subsets.

Dieback (die-back) Death of the tip of a shoot. Has various causes, but if extensive or repeated and involving the leader will generally cause bad malformation in the form of forks or multiple leaders.

Dimensional stability Freedom of wood from twisting and warping upon drying and general changes in moisture content. Main immediate causes of instability are grain spirality and longitudinal shrinkage on drying.

Diploid The state of tissue containing two sets of the basic number (12 in radiata) of chromosomes, which is the state of the entire visible tree.

DNA Deoxyribonucleic acid, the 'secure repository' of genetic information, some of which is transcribed via **RNA** into proteins, some just into RNA, and much having no currently known function but does contain 'bar code' information.

Dominance A measure of the departure of the effect of a **heterozgote** at a **locus** from the average effect of the two **homozygotes**. In the cases reported by Mendel, dominance (or recessiveness) was 100%.

Duff The partly-decomposed dead leaves and other organic detritus in the lower litter layer on top of the soil in a forest, typically invaded by soil fungi and other decomposers.

Dysgenic (i) Pertaining to practices or influences that are conducive to selection of genetically undesirable parents (opposite: **eugenic**).

(ii) State of population(s) influenced by dysgenic selection (e.g. highgrading or creaming).

Economic (i) Pertaining generally to the science and/or phenomena of economics.

(ii) **Profitable**, according to a specified, broad criterion. e.g. positive Net Present Value, or positive return on capital.

(iii) Pertaining to analysis of profitability (syn. **financial** analysis).

Effective population size (i) **Status number**, based on proportions of ancestry contributed by the various founder individuals (assumed unrelated)

(ii) **Inbreeding effective size** A number of individuals, which assuming random variation in their contribution to next generation would lead to a specific accumulation of inbreeding.

Elite Relating to trees with superiority proven by progeny testing. (ii) Relating to populations of known high genetic superiority.

Embling A plant produced by embryogenic culture.

Embryo The part of a seed from which the seedling develops. Embryos can be induced to form artificially in **embryogenesis**

Embryogenesis Inducing formation of fresh embryos. Platform for both **genetic engineering** and **cryopreservation**.

Environmental variation Variation among individuals arising from effects of environment in which they grow. Some environmental effects can be quantified and allowed for in evaluating individuals, but residual, microenvironmental effects represent 'noise' variation.

Epigenetic effects Effects that reflect the activity state to which actual genes are switched, often manifested as **maturation** state.

Epistasis A form of non-additive genetic variation involving interactions between gene effects at different loci, such that gene effects at one loci are conditional upon the **allele**(s) present at some other **locus**/loci.

Eugenic (i) Pertaining to practices or influences that are conducive to selection of genetically desirable parents.

(ii) State of population(s) influenced by eugenic selection. (opposite: **dysgenic**)

Extirpation Local extinction of a species or population.

Family forestry Strictly, growing blocks of plantation composed of single families, of which either the seed parent only or both parents are known, with a view to achieving some greater uniformity and allowing more precise deployment and tailoring of tending regimes. Term sometimes applied to known mixes of families.

Fascicles The bundles in which pine needles (leaves) are arranged, characterized by a sheath at the base. Very unusually, different populations of radiata have different numbers of needles in fascicles (three in mainland populations and two in island populations).

Felling select Term applied to seed collected from the best trees, after they have been felled, in commercial stands. Was adopted where trees were too big and tall for safe and economic climbing.

Financial (i) Pertaining to availability and/or provision of finance.

(ii) Pertaining to analysis of profitability (see **Economic** (iii)).

Flower Structures producing pollen and containing the ovules that are fertilized by pollen to produce seed. In conifers these are not flowers in the strict sense, but strobili. Male strobili, or catkins, produce pollen, and female strobili develop after fertilisation into the characteristic [seed] cones.

Form (i) Tree form, relating to combination of bole straightness, level of malformation and branching characteristics.

(ii) Bole form, relating to straightness and level of malformation.

(iii) A taper function relating bole volume to combination of tree height and diameter at breast height.

Forwards selection Selection among the offspring of the hitherto latest generation of select parents, to produce the next generation of select parents.

Founder effect A random genetic change in a derived population compared to its origin population as a result of having started from a small sample of the origin population. When founding numbers of the new population exceed about 20 unrelated breeding individuals, founder effects usually become unimportant.

Gene (i) Basic unit of heredity, containing coding (specifying a protein) and regulatory regions of DNA.

(ii) **Allele**.

Gene resource Population managed or preserved to conserve maximum genetic diversity (cf **Genetic resources**).

General combining ability The average departure of a parent's offspring from the population mean when the parent is crossed with random members of the population. It corresponds the whole or the half of the parent' **breeding value**, depending on how breeding value is defined.

Genetic connection The presence of common genetic parentage that can provide the basis for rigorous comparison between different groups of genetic material.

Genetic correlation The measure of the degree to which genetic variation in one trait is tied to genetic variation in another. Selection purely for one trait is expected to lead to correlated responses in other traits in relation to their genetic correlations with the selection trait.

Genetic diversity The genetic variability among individuals or populations. Within an unstructured population diversity will be a function of population size. Divergences can exist between levels of functional diversity, involving performance in the field, and of marker diversity revealed in the laboratory.

Genetic engineering Achieving **genetic modification** through processes other than normal sexual reproduction involving pollination. In practice, has meant genetic transformation by inserting short, very specific DNA sequences into recipient genotypes. There is now increasing emphasis on pursuing **gene editing**, targeting changes in individual base pairs and avoiding insertion of any DNA extraneous to the objective.

Genetic gain Desirable change in population mean resulting from selection and/ or genetic engineering.

Genetic modification (i) Any genetic changes in population means or variances resulting from human influences.

(ii) **Genetic engineering.**

Genetic parameters Genetic variances and **heritabilities** for individual traits, genetic covariances and **genetic correlations** between traits, genetic variances being linked to environmental variances by heritabilities, and genetic covariances being linked to environmental covariances by genetic correlations.

Genetic resources (i) **Gene resources**

(ii) Totality of populations managed by the breeder, alternatively 'metapopulation'.

Genome The genetic material, comprising genes and non-coding DNA, arranged in chains within chromosomes. The main genome is located in the cell nucleus, but two small, subsidiary genomes which are not subject to regular gene recombination are located in cell organelles.

Genomic selection Selection based on DNA variants (markers) that are located throughout the genome, using some combination of inferred DNA contributions from various ancestors and relationships observed between markers and phenotypic values.

Genomics Study of **genomes**.

Genotype The genetic makeup of an individual, representing its particular set of genes and making its particular contribution to **phenotype**.

Genotype–environment (or genotype x environment, or genotype by environment) interaction ("G × E") The phenomenon whereby comparative performance of different genotypes varies according to environments.

Graft A **ramet** of clone representing the scion (a bud or shoot of the clone) and the rootstock (usually a seedling),

Half-sibs (cf **Full-sibs**) Relatives with a single parent in common, a true half-sib family having a random sample of the population as the remaining parents (cf **open-pollinated** family).

Haploid The state of tissue having only a single set of **chromosomes** (12 in all true pines), unlike the **diploid** complement (double set) of free-growing material. **Gametes** are haploid.

Hardboard A dense, hard panel board produced at high pressure from freshly produced mechanical pulp, relying on spontaneous adhesion of the fibres.

Hardwoods Trees belonging to the dicotyledonous group within the angiosperms, which bear true flowers, generally have broad leaves, and have a more complex wood structure than **softwoods** (conifers), but among which there is an extreme range of wood density and hardness. Often more amenable to chemical pulping than softwoods, but generally more difficult to process and use as lumber.

Harvest index The tree biomass that is recoverable and usable in the process of harvesting as a proportion of (a) the current total biomass, or (b) the total primary biomass, including turnover, that is produced in the lifetime of the crop.

Heritability The proportion of variation (variance) that is genetic, **narrow-sense heritability** pertaining to additive gene effects and being applicable to most systems of propagation by seed, **broad-sense heritability** pertaining to total genetic effects and being applicable to clonal propagation systems.

Heterozygote An individual that, at the **locus** in question, has different **alleles** present in the duplicate copies of the gene.

Highgrading The selective harvesting of the better trees in a population, which may leave the poorer trees as parents of the next generation. Alternative term is **creaming**.

Homozygote An individual that, at the **locus** in question, has the same allele in both of the duplicate copies of the gene.

Indirect selection Selection on values for one or more traits to achieve genetic gain in some target trait(s). It can be superior to direct selection for target traits if selection traits are both highly heritable and closely correlated genetically with the target traits, especially if the selection traits can be assessed on more candidates and/or at earlier ages.

Internode* More properly termed "shoot cycle" but for our purposes, the zone of a stem between two successive **branch clusters** (which are not actually nodes in the strict botanical sense). Of special interest in radiata because they can often be long enough to allow cutting of significant lengths of clear timber, but long internodes tend to be associated with large, steep-angled branches

Introgression Gene flow from one species or population into another by the repeated crossing of hybrids with one of the parental species or populations.

Juvenile wood Traditional but technically incorrect term for **corewood**. More appropriately applied in radiata seedlings to wood in the bottom 3 m of the bole.

Land race (or **Landrace**) 1.1 Populations that are non-trivially different from their origin populations and from each other as a result of human activity. Usually applied to material growing outside area of origin.

Linkage disequilibrium (LD) The departure from independent assortment of alleles at different loci as a result of the loci being on the same chromosome. The disequilibrium is a function of the distance apart on the chromosome and the number of generations since the initiation of the closed breeding line or population. In radiata, LD exists mainly within groups of relatives, and hardly at all within populations at large.

Locus, plural loci A specific site on a specific chromosome, where alternative forms of the gene represent different **alleles**.

[Forest] Management (i) Coordination of establishment, protecting, tending and harvesting of stands comprising the forest estate, based on predicting and planning of expenditure, wood yields and cash flows.

(ii) Operations involved in growing crops (see **silviculture**) and harvesting.

Marker-Assisted Selection (MAS) Selection using DNA markers that are linked to chromosome regions controlling significant amounts of trait variation, in conjunction with phenotypic information. When used in absence of phenotypic information it is termed **Marker-Based Selection (MBS)**.

Maternal effects Environmentally induced covariation among offspring of the same maternal parent. In radiata, this commonly results from a good or bad start for most or all offspring as a result of seed weight, which in turn is a function of both genetic makeup of the parent tree and its environment when the seeds were produced.

Mating design Scheme for crossing parents. The simplest to implement is [natural] open-pollinated. Other options include polycrosses, single-pair crossing, hierarchical crossing, factorial crossing, and diallel crossing, the most complex being the complete diallel.

Maturation Changes in the properties of tree as it gets bigger and older. In radiata these changes include different appearance and anatomy of leaves and appearance of buds, less amenability to rooting as cuttings, onset of reproduction, and some different wood properties. In radiata maturation level is mainly related to distance along the shoot from seedling root collar.

Meristems Tissues where growth occurs, e.g. 'growing points' at shoot- or root tips for extension growth, cambium for radial growth,

Microfibril Strands of cellulose, which represent multiple chains of sugar molecules, within the tracheid (wood fibre) walls. Embedded in a matrix of lignin which contributes stiffness and hardness to the wood.

Microfibril angle (MfA) The angle between the long axis of the wood fibre (tracheid) and the direction of the cellulose microfibrils. Generally, the microfibril angle refers to that occurring in the dominant S2 layer of the cell wall. High microfibril angle is conducive to both poor stiffness and distortion on drying.

Microsatellites Repeats of short, simple, sequences of DNA bases (**SSR**s) that are arranged end to end, characterised by numbers of such tandem repeats. Until recently, favoured DNA 'fingerprints' for verifying clonal identity and parentage of progeny.

Modulus of elasticity (MoE) Measure of wood stiffness.

Multiclonal variety A set of characterized clones assembled to meet the same breeding objective and field-deployment niche, their number being designed to achieve risk spread.

Mycorrhizae Root-tip structures formed by a mutually beneficial (symbiotic) association between the plant and certain fungi. Of special importance to the plants for acquiring scarce phosphorus from the soil, with the fungi depending on the plant for food in the form of sugars.

Natural selection Preferential survival of individuals with particular expressions of traits that are in some degree heritable, outside contexts of deliberate human intervention. Together with the genetic variation arising from **mutation** natural selection is the key driver of evolution.

Non-additive gene effects Gene effects whereby offspring tend to depart from intermediacy between their parents, and which can be captured only in part with specific crosses but in full through selection and commercial deployment of clones.

Nurse tree A species that is planted together with a preferred species to provide shelter or soil conditions to facilitate establishment and good early growth of the latter. May be cut out or allowed to become suppressed when this purpose has been served.

Ontogenetic Pertaining to developmental stage (or **maturation** state) in the life cycle of the tree.

Open-pollinated ("OP") Applying to families (or progenies) from seed parents that have been naturally pollinated. With wind pollination, such families in radiata typically differ by roughly half their [additive genetic] **breeding values** and can be used to estimate **narrow-sense heritabilities** and **genetic correlations** between traits. See also **half-sibs**.

Origin [of genetic material] The native geographic location from which ancestral material was collected (cf **source**) (See also **provenance**).

Ortet The initial individual (seedling) that may be vegetatively propagated to produce a **clone**. See **ramet.**

Outerwood Wood outside the **corewood,** namely beyond the first few rings out from the pith. Typically deemed to comprise around ring 16 outwards in radiata, it has generally good technical properties. Historically, outerwood has been widely but somewhat inappropriately termed "mature wood".

Phenotype The tree as it is, being the resultant of the effects of its **genotype** and its environment.

Physiological ageing See **maturation**.

Plantlet (syn. Plantling) Vegetative propagule produced from *in-vitro* culture. (cf **cutting**)

Pleiotropy The phenomenon of a gene or genes affecting more than one trait, being a prime cause of genetic correlations between traits.

Plus trees Trees selected, generally in unimproved stands, as highly superior phenotypes (on growth, form and health, and sometimes on wood properties), but whose genetic superiority (or **breeding value**) has not been proven by progeny testing.

Polygenic Pertaining to inheritance of trait(s) governed by large numbers of genes (polygenes) that individually exert very small effects. Appears to be the typical situation in radiata.

Polymix A physical mix of pollens from different parents, a polymix cross resulting from pollinating a seed parent with such a mix.

Polymorphism Presence of more than one non-rare allele of a gene.

Preservation Isolating from all human influences with the objective of maintaining a truly natural state. Often, however, impracticable. (cf **conservation**)

Production population Highly select material, usually as **seed orchards** or stool-beds for cuttings, drawn from the **breeding population**, to provide commercial planting stock.

Profitable (i) Pertaining to monetary returns exceeding immediate associated costs. (ii) **Economic** (iii), according to broader criterion.

Progeny test A planting designed to evaluate parent candidates by comparing the performance of their offspring. A parent-ranking test.

Progeny trial A planting that employs pedigreed progeny with the intent of finding and selecting the parents of the next generation. More broadly, any designed planting using known progeny for some purpose(s), which will often include studying **genetic parameters**.

Propagule Collective term applied to any viable plant, including seedlings, grafts, rooted cuttings, or plantlets of various categories produced by *in-vitro* techniques.

Provenance (i) The native geographic location of a natural population.
(ii) Genetic material obtained in current or some past generation from such a natural population.

QTL (Quantitative trait locus) A zone within the genome in which DNA polymorphism is associated with some detectable phenotypic effect on one or more traits.

Ramet A vegetative propagule, traditionally graft or cutting, of an **ortet**. A **clone** is composed of the ortet and its ramets.

Ramicorn A large, steep angled branch, the more severe cases merging into the category of unequal forks.

RAPD A type of DNA marker.

Reflexed Pointing backwards; with the prickles (mucros) on a pine cone this means that the cone is not prickly to handle; or with the whole cone pointing backwards from the shoot tip.

Response to selection The shift or gain in the population mean resulting from selection, which can be predicted from the product of the **selection differential** (difference between the mean s of the select parents and the whole candidate population) and the heritability.

RFLP A type of DNA marker.

RNA (Ribonucleic acid) Immediate transcription product from **DNA** which is either further transcripted to protein or performs several other roles, some poorly understood.

Scion The part of a graft, normally a twig or bud, that is grafted onto that **rootstock**. The scion contains the genetic material that produces the desired seeds in a seed orchard, or the desired properties of the compound tree such as peaches in an orchard or gunstock wood in a walnut orchard.

Seed orchard An area that has been planted specifically for producing seed of highly superior genetic quality, or a field trial of select genetic material that has been converted to production of genetically superior seed.

Seed-production area Part of an operational plantation or natural forest that is converted to efficient production of seeds. It is normally sited also for easy access. Management includes thinning to remove inferior trees and other treatments designed to boost seed production and facilitate seed collection. Normally an interim measure until seed orchards produce sufficient needs for program needs.

Selection differential Difference between mean of select individuals and the mean of the candidate population from which they were chosen.

Selection intensity Selection differential in standard deviations for the candidate population, but sometimes applied to the proportion of the candidate population chosen.

Selection traits The traits used as the basis for selection, which are often proxies for the **breeding-goal** traits.

Sexual propagation/reproduction In plants, the process pollination and subsequent seed production, with the key feature of genetic recombination.

Simple sequence repeats (SSRs). See **Microsatellites**.

Silviculture (i) The operations of raising nursery stock, field establishment, protection against pest and diseases, and thinning and pruning of forest stands, representing applied ecology.

(ii) Just field establishment, thinning and pruning. (cf **Forest management**)

Slash Residues left on site from thinning and logging operations, mainly comprising all parts of branches, the cones and the unharvested remnants of stems.

Softwoods See **conifers**.

Somaclonal variation Variation within clones that generally arises from propagation history and persists among propagation lines, as either **epigenetic** variation or genuine DNA mutations in vegetative material.

Source Where seed or clonal material has come from, e.g. location of an exotic stand, a seed orchard, or a nursery stool bed, but may include [native] **origin**.

Specific combining ability The departure of the average of a pair-cross from the average of the parents **general combining abilities**, which typically reflects around one-quarter of the non-additive genetic variance of a population.

[**Standard deviation** Square root of variance. With a perfect normal distribution 68.27% percent of the population falls between one standard deviation above and below the mean.

Steckling German term for rooted cutting

Stem More widely used synonym for **bole**.

Stiffness Resistance to bending of the material, namely to deformation caused by a given force adjusted for the dimensions of the specimen. Technically expressed as **Modulus of Elasticity** (MoE). Low stiffness often restricts the value of radiata timber for demanding structural uses.

Strength The capacity of wood to withstand a force without breaking, namely the force that it can withstand in relation to dimensions of the specimen. Technically expressed as Modulus of Rupture.

Strobilus (plural strobili) Reproductive structure of conifer. Male strobili (or catkins) produce pollen, and female strobili (or conelets) develop into seed cones after fertilization by pollen.

Sublines Subsets of a breeding population that are totally unrelated, but within which inbreeding can be tolerated. Maintained as a means of ensuring complete outcrossing at any point in the future.

Tracheids Wood-fibre cells of a conifer, serving both as units in the mechanical structure of wood and to translocate water and mineral nutrients from the root towards the crown tips. Such translocation ceases when wood becomes heartwood.

Transgenes Sequences of DNA that are artificially inserted into a recipient genotype, not necessarily whole genes.

Tree form (i*) Combination of absence/presence of forking, straightness/crookedness, pattern of taper, and branching habit

(ii) Taper pattern of **bole** or **stem.**

Variety (i) Taxonomic entity representing a population that is distinctive but less so than a subspecies.

(ii) A distinctive **cultivar** (e.g. in the case of a fruit that is grown as a crop in its own right). The term is often applied to clones, but in radiata individual clones are more appropriately grown commercially as members of risk-spread clonal portfolios and not stand-alone crops.

Vegetative multiplication ("VM") Mass vegetative propagation of genetically mixed lots (families or mixes of families) without regard to genetic identity of individuals. Used in radiata to extend highly improved seedlots with additional, scarcity value.

Vegetative propagation Propagation not involving the sexual reproduction that entails the genetic recombination involved in producing seeds. Traditionally done by rooting cuttings or grafting, but now widely possible using *in-vitro* culture.

Index

R. Burdon et al., *Domestication of Radiata Pine*, Forestry Sciences,
DOI 10.1007/978-3-319-65018-0

B

C

G

H

I

J

K

L

M

N

The manufacturer's authorised representative in the EU is Springer Nature Customer Service Centre GmbH, Europaplatz 3, 69115 Heidelberg, Germany. If you have any concerns regarding our products, please contact ProductSafety@springernature.com

Printed and bound by CPI Group (UK) Ltd, Croydon, CR0 4YY
22/09/2026
02221048-0012